Ecological Forestry Management

Ecological Forestry Management

Edited by **Malcolm Fisher**

Callisto Reference
New York

Published by Callisto Reference,
106 Park Avenue, Suite 200,
New York, NY 10016, USA
www.callistoreference.com

Ecological Forestry Management
Edited by Malcolm Fisher

International Standard Book Number: 978-1-63239-708-9 (Hardback)

Printed in China.

Contents

Preface

The world is advancing at a fast pace like never before. Therefore, the need is to keep up with the latest developments. This book was an idea that came to fruition when the specialists in the area realized the need to coordinate together and document essential themes in the subject. That's when I was requested to be the editor. Editing this book has been an honour as it brings together diverse authors researching on different streams of the field. The book collates essential materials contributed by veterans in the area which can be utilized by students and researchers alike.

Forestry is the practice of managing, planning, using, conserving and preserving the forests. Forestry management deals with administrating all the legal, social, scientific and technological aspects of forest regulation. It also includes managing urban values, wilderness, wood products, forest genetic resources, etc. This book traces the progress of this field and highlights some of its key concepts and applications. It unravels the recent studies conducted in this area. The topics covered in this book will provide the readers new insights in the subject. It is an essential guide for both academicians and those who wish to pursue this discipline further.

Each chapter is a sole-standing publication that reflects each author's interpretation. Thus, the book displays a multi-facetted picture of our current understanding of applications and diverse aspects of the field. I would like to thank the contributors of this book and my family for their endless support.

Editor

1

An air-drying model for piled logs of *Eucalyptus globulus* and *Eucalyptus nitens* in Chile

Horacio E. Bown[1*] and Jean-Pierre Lasserre[2]

Abstract

Background: Water accounts for around half of the total mass in living trees, and therefore large savings can be achieved if logs are dried before transporting or burning.

Methods: An air-drying model for piled logs of *Eucalyptus globulus* Labill. and *Eucalyptus nitens* H.Deane & Maiden was developed. Daily moisture content (MC) loss was modelled based on meteorological variables and pile characteristics. The trial had a factorial design with two species, two debarking treatments (with or without bark) and two log lengths (244 and 350 cm). Independent trials started in July 2007 (winter), October 2007 (spring) and January 2008 (summer). There were five replicate piles per season and treatment. Wood pile masses were weighted weekly or twice weekly using a crane, a 10,000 kg balance and chains to hold the piles.

Results: The main and interactive effects of seasons and treatments on daily MC loss were highly significant. However, the effect of season (climate) was far greater than the main effects of treatments or the season × treatment interaction. Overall, *E. globulus* dried 20 % faster than *E. nitens*, debarked logs dried 8 % faster than barked logs and 244 cm logs dried 3 % faster than 350 cm logs. Daily MC loss for the current day was better explained by a power function of MC at the start of the day, daily air relative humidity, daily air temperature and the number of logs per square metre of pile cross-section (or an equivalent average log diameter).

Conclusions: The air-drying model for piled logs can be used to predict drying times (days) to achieve a given target moisture content, providing a new tool for decision-making in forest transport and industrial planning.

Keywords: Log air-drying model; Environmental variables; Eucalyptus; Management decisions

Background

Chile produces around 41 million cubic metres of timber each year from around 2.4 million ha of planted forests. *Pinus radiata* D.Don (1.470 million ha), *Eucalyptus globulus* Labill. (541,000 ha) and *Eucalyptus nitens* H.Deane & Maiden (232,000 ha) are the most widely planted species (Instituto Forestal 2014). *Pinus radiata* is the main source for sawn timber and long fibre for the pulp industry, while *E. globulus* and *E. nitens* are the main providers of short fibre for pulp production, and marginal providers for veneer and other uses (Lignum 2005).

Chilean forest companies commonly dry large quantities of pulp logs at the roadside before being transported to pulp and paper plants, and recently to industrial boilers for the generation of steam, heat and electricity. The rationale for this strategy is that water accounts for around half of the total mass in living trees, and therefore large savings in transport costs are achieved when logs are dried at the roadside. Also, industrial boilers require logs under a 50 % moisture content on a dry basis (MC) to operate efficiently as the calorific power of wood increases as MC decreases. Moreover, yard handling would become unfeasible when considering drying and storing large log quantities for long periods of time.

Air-drying times vary widely from a couple of weeks to several months depending on species, dimensions, local weather conditions and time of year when the material is stacked (Simpson and Wang 2004). Rainfall, temperature, radiation and relative humidity all contribute to the wetting/drying process of wood piles. Also species exhibit a fivefold range in drying times, and water loss slows down as wood water content decreases

* Correspondence: hbown@uchile.cl
[1]Faculty of Forestry & Conservation, University of Chile, P.O. Box 9206, Santiago, Chile
Full list of author information is available at the end of the article

(Simpson and Wang 2004). Additionally, different log diameters and lengths, debarking and piling specifications may strongly affect air-drying rates. Therefore, the aim of this study was to predict air-drying times for piled logs of *E. globulus* and *E. nitens*, specifically to (i) assess the influence of species, log diameter and length, and debarking on daily MC loss; (ii) correlate daily MC loss with meteorological variables; and (iii) discuss the possibility of applying such a model under different environmental conditions. The development of an air-drying model for piled logs may assist decision-makers in predicting drying times (days) to achieve a given target moisture content, and provide a new tool for decision-making in forest transport and industrial planning in Chile.

Methods

Trial specification

An experiment was set to determine MC loss of air-drying eucalypt wood piles in continental Concepción (Rucamanqui, latitude 37° 15′ S longitude 71° 55′ W) in Chile. The climate is Mediterranean with cold wet winters and hot dry summers. Rainfall for the year 2007 was particularly low (696 mm) compared with the previous year (1407 mm). Annual air temperature for 2007 was on average 10.8 °C, with a monthly minimum of 4.3 °C (June) and a monthly maximum of 19 °C (January). Rainfall in January and February 2008 was 10.1 and 7.3 mm, respectively.

Piles were about 1.4 m wide, 1 m tall and either 2.44 or 3.50 m long. Treatments were made out of the combination of three factors: species (*E. globulus* or *E. nitens*), debarking (with bark or without bark) and length (244 or 350 cm). Wood piles were stacked with freshly harvested logs (Fig. 1) at three dates: 14-July-2007 (winter), 6-October-2007 (spring) and 17-January-2008 (summer) and were measured for 167, 146 and 43 days, respectively. The total sample size was 120 piles (3 seasons × 2 species × 2 debarking treatments × 2 lengths × 5 replicates).

For convenience, piles were grouped by length. Within each length, different combinations of species and debarking treatments were alternated. Piles were separated by poles anchored in the soil (Fig. 1a). Wood pile masses were weighed weekly or twice weekly using an electronic dynamometer with a capacity of 10 tonnes, hanging from a crane and towing ropes to hold the piles (Fig. 1b). Meteorological information was obtained from an automatic weather station located at a close distance from the piles.

Data analysis

All analyses were undertaken using The R System for Statistical Computing (R Development Core Team, 2010). Variables were tested for normality and homogeneity of variance, and transformations were made as necessary to meet the underlying statistical assumptions of the models used. Analysis of variance (Proc GLM) was used to examine the main and interactive effects of species, lengths and debarking treatment on log moisture content loss. Analyses were conducted at the pile level. Tukey's least significant difference test was used to distinguish among individual means where applicable with a confidence level of $P \leq 0.05$. The effects of species, debarking treatments and lengths on slopes and intercepts of the linear relationship between average daily MC loss and number of logs per square metre

Fig. 1 a Layout of log piles of *E. globulus* and *E. nitens* set for the air-drying experiment. Piles were about 1.4 m wide, 1 m tall and 244 cm or 350 cm in length. Trials were started in winter, spring and summer with three factors: species (*E. globulus*, *E. nitens*), debarking (with bark, without bark) and length (2.44 m, 3.50 m). **b** Detail of pile-weighing procedure and the dynamometer cross-section were tested for significance by analysis of covariance.

Daily MC loss modelling

Following Simpson and Wang (2004), daily moisture content (MC) loss (ΔM) per day (% green basis) was modelled as

$$\Delta M = aM^b\ T^c\ H^d\ N^e \qquad (1)$$

where M is the MC at the beginning of the day (%, green basis), T is the daily average air temperature (°C), H is the daily average air relative humidity (%), N is the number of logs per square metre cross-section and a, b, c, d and e are parameters to be determined by non-linear regression. Dummy variables were included in parameter "a" of model (1) in order to account for the effect of species, debarking and log lengths, as: $a = \alpha_0 + \alpha_1 D_1 + \alpha_2 D_2 + \alpha_3 D_3$, where D_1, D_2, D_3 are dummy variables representing species (0 = *E. globulus*, 1 = *E. nitens*), debarking (0 = without bark, 1 = with bark) and lengths (0 = 244 cm, 1 = 350 cm), respectively; and α_0, α_1, α_2 and α_3 are parameters to be determined by non-linear regression.

Results

Treatment effects

There were inherent treatment differences in pile volume and mass when first stacked. When freshly stacked, log piles ranged from 1748 to 4299 kg, 2.7 to 6.2 m^3 cord (stacked) volume, 1.6 to 3.9 m^3 solid volume and 9 to 35 logs m^{-2} in the cross-section (or equivalently 14 to 33 cm in average log diameter) (Table 1). The initial wood moisture content ranged from 47 to 60 % (average 54.3 %) on a green basis (89 to 149 % on a dry basis, average 119.6 %). Basic wood density (±1 standard error, $n = 30$) was significantly greater for *E. globulus* (514 ± 6 kg m^{-3}) than *E. nitens* (491 ± 5 kg m^{-3}, $P = 0.003$).

Log moisture content generally decreased over time in all piles across seasons, species, lengths and debarking treatments. Moisture content loss over time for piles of either *E. globulus* or *E. nitens* (350 cm log lengths/with bark) stacked in summer, spring and winter are shown in Fig. 2. Other log lengths and debarking treatments followed the same pattern as the one illustrated in Fig. 2. Log MC decreased for both species at higher rates for logs stacked in summer than spring than winter (summer > spring > winter). Also, log MC decreased at decreasing rates as logs became drier for piles stacked in summer and spring (faster at the beginning and slower as time progressed). In contrast, moisture content of logs stacked in winter decreased slowly while weather was cold and damp to decrease at increasing rates as weather conditions improved (Table 1, Fig. 2). Overall, *E. globulus* dried faster than *E. nitens.*

The main effects of season ($F_{2,95} = 456–1769$) on daily log MC loss were far greater than the main effects of treatment ($F_{7,95} = 15–62$) or their interaction ($F_{14,95} = 7–29$), as shown by their F values and P range (Table 1). Daily log MC loss was on average threefold greater for piles stacked in summer (0.53 % per day, green) than in winter (0.17 % per day, green), while two times greater in summer than in spring (0.25 % per day, green). In contrast, the treatment effects were relatively small compared to season: *E. globulus* dried 3–47 % (average 20 %) faster than *E. nitens*, debarking increased the drying rate by 3–12 % (average 8 %) and logs 244 cm long dried 1–4 % (average 3 %) faster than logs 350 cm long (Table 1).

Because log diameter (or the number of logs per square metre in the pile cross-section) had such a strong influence on the wood moisture content loss per day, covariance analysis was carried out to separate the effects of the covariate, log diameter or number, ($F_{1,100} = 60$, $P < 0.001$) from the effect of the factors (species, lengths, debarking, seasons). Slopes ($F_{2,100} = 3.8$, $P = 0.02$) but not intercepts ($F_{2,100} = 2.9$, $P =$ 0.06) of the linear relationship between daily MC loss and the number of logs per square metre in the pile cross-section were significantly influenced by season (Fig. 3). Slopes and intercepts of these linear relationships were not influenced by species, lengths or debarking treatments when tested within each season ($F_{7,100} = 0.9–1.3$, $P > 0.28$). However, it is worth noting that for summer, *E. nitens* showed an insignificantly yet lower slope than *E. globulus* (Fig. 3).

Air-drying model

Model (1) was fitted to the experimental data of daily MC loss on a green (ΔM_{green}) and dry basis (ΔM_{dry}). The general model for ΔM_{green} was:

$$\Delta M_{green} = a_{green} M_{green}^{2.3979}\ T^{0.2268}\ H^{-5.6734}\ N^{0.7987},$$
$$r^2 = 0.67,\ P < 0.001$$

with

$$a_{green} = 119020 – 26671\ D_1 - 40969\ D_2 - 6814\ D_3$$

where M is the MC at the beginning of the day (%, green basis), T is the daily average air temperature (°C), H is the daily average air relative humidity (%) and N is the number of logs per square metre cross-section. Dummy variables were included in parameter "a" of model (1) in order to account for the effect of species, debarking and log lengths, as: $a = \alpha_0 + \alpha_1 D_1 + \alpha_2 D_2 + \alpha_3 D_3$, where D_1, D_2, D_3 are dummy variables representing species (0 = *E. globulus*, 1 = *E. nitens*), debarking

Table 1 Initial and final condition of air-drying log piles ($n = 120$) of *E. nitens* and *E. globulus* in two lengths (244 cm, 350 cm) with and without bark in Rucamanqui, Los Angeles, Chile

Season	Species	Treatment	*n*	Days	Green mass start (kg)	Water loss (kg)	Water loss (kg day^{-1})	Total MC loss (%, green)	Daily MC loss (% day^{-1}, green)	Total MC loss (%, dry)	Daily MC loss (% day^{-1}, dry)
Winter	*E. nitens*	244 bark	5	167	1950 ± 71e	529 ± 16d	3.2 ± 0.1d	27.2 ± 0.4bc	0.16 ± 0.003bc	62.9 ± 0.7a	0.38 ± 0.004a
		244 debark	5	167	2172 ± 115de	608 ± 39cd	3.6 ± 0.2cd	27.9 ± 0.6bc	0.17 ± 0.004bc	59.8 ± 1.2a	0.36 ± 0.007a
		350 bark	5	167	2848 ± 103bc	658 ± 20cd	3.9 ± 0.1cd	23.1 ± 0.5d	0.14 ± 0.003d	50.8 ± 1.4b	0.30 ± 0.008b
		350 debark	5	167	2534 ± 171cd	693 ± 49c	4.1 ± 0.3c	27.4 ± 0.6bc	0.16 ± 0.003bc	57.8 ± 1.1ab	0.35 ± 0.006ab
	E. globulus	244 bark	5	167	2387 ± 68cde	641 ± 15cd	3.8 ± 0.1cd	26.9 ± 0.6c	0.16 ± 0.003c	63.0 ± 1.6a	0.38 ± 0.009a
		244 debark	5	167	2751 ± 42c	877 ± 19b	5.3 ± 0.1b	31.9 ± 1.1a	0.19 ± 0.006a	69.0 ± 2.9a	0.41 ± 0.018a
		350 bark	5	167	3225 ± 123ab	899 ± 31b	5.4 ± 0.2b	28.0 ± 1.1bc	0.17 ± 0.006bc	64.5 ± 2.8a	0.39 ± 0.017a
		350 debark	5	167	3435 ± 70a	1051 ± 41a	6.3 ± 0.2a	30.6 ± 0.9ab	0.18 ± 0.005ab	65.6 ± 2.3a	0.39 ± 0.014a
		Mean	40		2663 ± 83	744 ± 28	4.5 ± 0.2	27.9 ± 0.5	0.17 ± 0.003	61.7 ± 1.0	0.37 ± 0.006
Spring	*E. nitens*	244 bark	5	146	2526 ± 81d	993 ± 30cde	6.8 ± 0.2cde	39.4 ± 0.8ab	0.27 ± 0.005ab	93.7 ± 1.8a	0.64 ± 0.012a
		244 debark	5	146	2942 ± 100cd	958 ± 36de	6.6 ± 0.2de	32.6 ± 0.5cd	0.22 ± 0.004cd	73.1 ± 1.7c	0.50 ± 0.012c
		350 bark	5	146	3796 ± 152a	1178 ± 65bc	8.1 ± 0.4bc	31.1 ± 1.3d	0.21 ± 0.009d	73.0 ± 2.8c	0.50 ± 0.019c
		350 debark	5	146	3761 ± 112a	1506 ± 57a	10.3 ± 0.4a	40.1 ± 1.2a	0.27 ± 0.008a	85.1 ± 2.4ab	0.58 ± 0.016ab
	E. globulus	244 bark	5	146	2487 ± 102d	912 ± 22e	6.2 ± 0.2e	36.8 ± 0.9ab	0.25 ± 0.006ab	86.2 ± 1.7bc	0.59 ± 0.012bc
		244 debark	5	146	2668 ± 92cd	992 ± 25cde	6.8 ± 0.2cde	37.2 ± 0.6ab	0.26 ± 0.004ab	80.5 ± 1.4bc	0.55 ± 0.010bc
		350 bark	5	146	3133 ± 67bc	1119 ± 30cd	7.7 ± 0.2cd	35.7 ± 0.4bc	0.24 ± 0.002bc	84.4 ± 1.7bc	0.58 ± 0.011bc
		350 debark	5	146	3577 ± 192ab	1330 ± 65ab	9.1 ± 0.4ab	37.2 ± 0.5ab	0.25 ± 0.003ab	77.8 ± 2.7c	0.53 ± 0.018c
		Mean	40		3111 ± 90	1124 ± 34	7.7 ± 0.2	36.3 ± 0.5	0.25 ± 0.004	81.7 ± 1.2	0.56 ± 0.009
Summer	*E. nitens*	244 bark	5	43	2537 ± 124b	466 ± 31d	10.8 ± 0.7d	18.3 ± 0.5d	0.43 ± 0.012d	41.1 ± 1.4c	0.96 ± 0.032c
		244 debark	5	43	2468 ± 30b	463 ± 17d	10.8 ± 0.4d	18.8 ± 0.6d	0.44 ± 0.015d	39.1 ± 1.5c	0.91 ± 0.036c
		350 bark	5	43	3296 ± 85a	593 ± 26c	13.8 ± 0.6c	18.0 ± 0.4d	0.42 ± 0.008d	39.9 ± 0.6c	0.93 ± 0.014c
		350 debark	5	43	3373 ± 174a	635 ± 24bc	14.8 ± 0.6bc	18.9 ± 0.5d	0.44 ± 0.013d	38.9 ± 1.4c	0.91 ± 0.033c
	E. globulus	244 bark	5	43	2261 ± 56b	593 ± 5c	13.8 ± 0.1c	26.3 ± 0.5bc	0.61 ± 0.011bc	59.8 ± 1.3a	1.39 ± 0.030a
		244 debark	5	43	2445 ± 62b	727 ± 19b	16.9 ± 0.5b	29.7 ± 0.4a	0.69 ± 0.009a	58.7 ± 1.0ab	1.37 ± 0.023ab
		350 bark	5	43	3035 ± 104a	753 ± 28b	17.5 ± 0.6b	24.8 ± 0.6c	0.58 ± 0.013c	53.0 ± 1.5b	1.23 ± 0.035b
		350 debark	5	43	3233 ± 148a	900 ± 43a	20.9 ± 1.0a	27.8 ± 0.3ab	0.65 ± 0.008ab	53.8 ± 1.0b	1.25 ± 0.023b
		Mean	40		2831 ± 76	641 ± 24	14.9 ± 0.6	22.8 ± 0.7	0.53 ± 0.017	48.1 ± 1.4	1.12 ± 0.033
ANOVA/	Covariate				–	**	**	***	***	***	***
ANCOVA	S				***	***	***	***	***	***	***
	T				***	***	***	***	***	***	***
	S×T				***	***	***	***	***	***	***

Main and interactive effects of season (S) and treatment (T) were assessed by analysis of variance and covariance. The covariate was the number of logs per square metre in the pile cross-section. Significant differences are presented as *ns* non-significant, * significant at $P < 0.05$, and *** significant at $P < 0.001$. Values are presented as means (± 1 SE) for seasons and treatments. Separation of means was determined by a Tukey test when applicable. Different letters indicate significant differences at $P < 0.05$

(0 = without bark, 1 = with bark) and lengths (0 = 244 cm, 1 = 350 cm), respectively.

The relationship between average log diameter, d, and number of logs per square metre of pile cross-section, N, ($d = 14.4182 + 45.1968\ e^{-0.12\ N}$, $r^2 = 0.91$, $P < 0.001$) can be replaced into (1) if d rather than N is preferred to be used in the calculations.

Model (1) was also fitted to predict daily MC loss on a dry basis (ΔM_{dry}) as

$$\Delta M_{dry} = a_{dry} M_{dry}^{2.3570}\ T^{0.2460}\ H^{-5.7448}\ N^{0.8422},\ r^2 = 0.63,\ P < 0.001$$

with

$$a_{dry} = 55319 - 10470\ D_1 - 22015\ D_2 - 3641\ D_3$$

There was an observed upper bound of moisture content loss ($\Delta M_{\mathrm{green}}$) of 1.97 % per day on a green basis

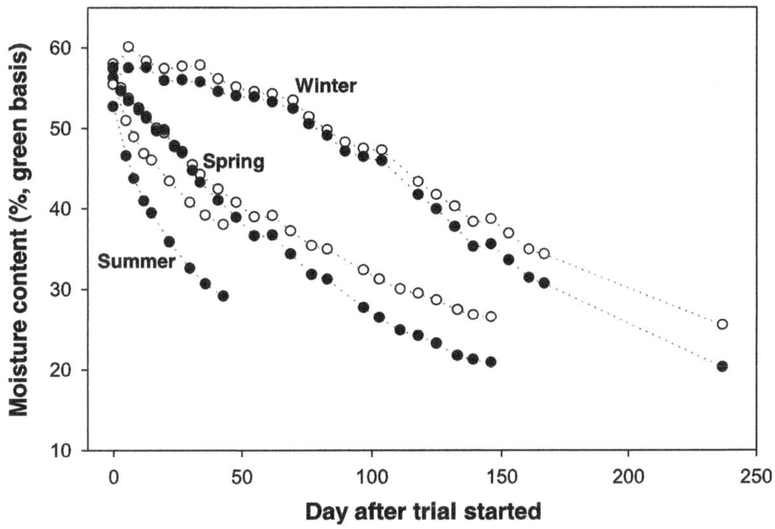

Fig. 2 Changes in wood moisture content (%, green basis) for six randomly selected 350 cm logs piles with bark. *Open symbols* represent *E. nitens* while *closed symbols* represent *E. globulus*

(3.96 % per day on a dry basis) given by the experimental data that should be acknowledged within the model. In practice, this means than any estimate over this value should be set to 1.97 % (3.96 % dry basis), and this upper bound would be required to ensure that realistic values of daily MC loss are estimated.

The model was apparently unbiased for a traditional residual analysis (Fig. 4 a) and also when predicting moisture content for each log pile for the duration of the experiment (Fig. 4 b). Rainfall was uncorrelated with the log moisture content loss.

Air-drying times for *E. globulus* and *E. nitens* with and without bark are shown in Fig. 5. The graph shows estimated times (days) to reduce the moisture content on a green basis from 60 to 50, 60 to 40, 60 to 30 and 60 to 20 %, for logs stacked in any day of the year. One important observation is that piles stacked in late summer and early autumn had greatly extended drying times,

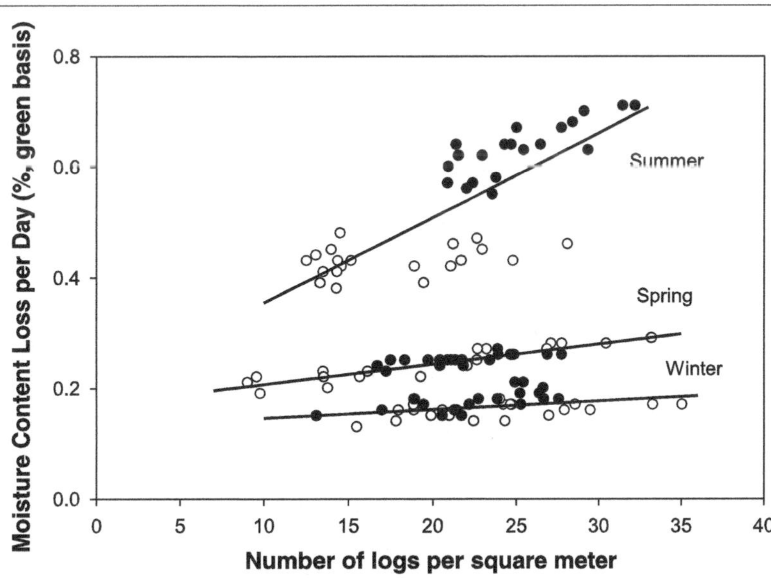

Fig. 3 The linear relationship between average log moisture content loss per day (%, green basis) and the number of logs per square metre in the pile cross-section. Analysis of covariance showed that slopes, but not intercepts, of these linear relationships were influenced by the season in which the piles were stacked. Various treatments (species, lengths, debarking) did not have any significant influence on these linear relationships. Logs stacked in summer, $y = 0.20140 + 0.01534\,x$, $r^2 = 0.61$, $P < 0.001$; spring, $y = 0.17089 + 0.00363\,x$, $r^2 = 0.85$, $P < 0.001$; winter, $y = 0.13097 + 0.00155\,x$, $r^2 = 0.16$, $P < 0.01$. *Open symbols* represent *E. nitens* while *closed symbols* represent *E. globulus*

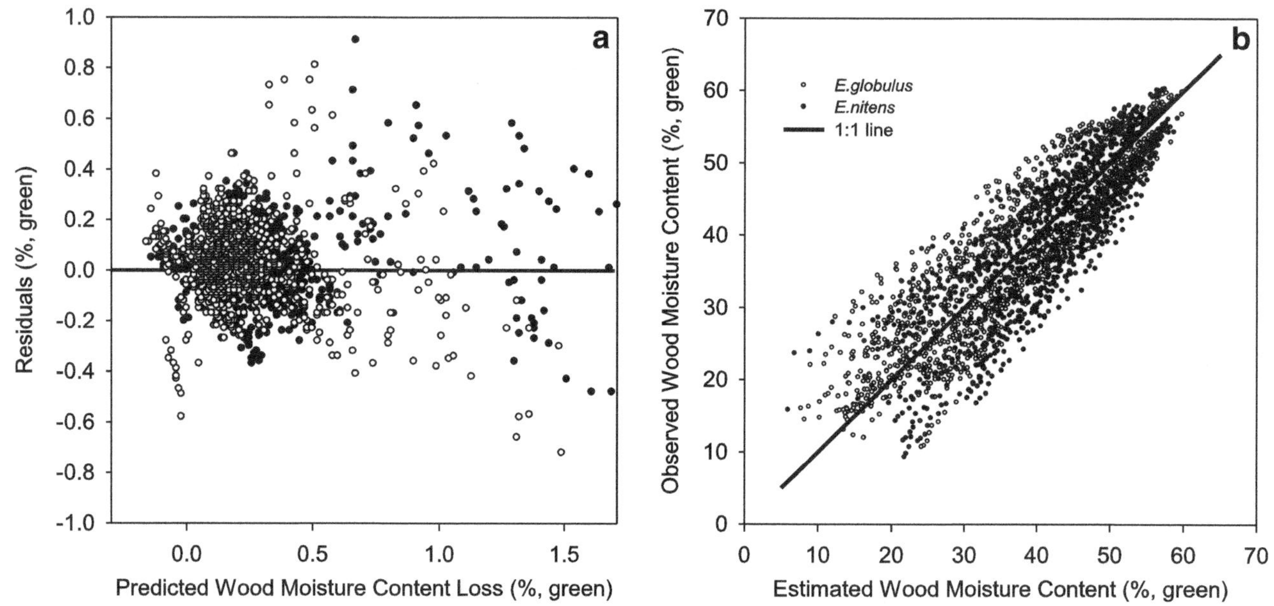

Fig. 4 a Residuals of model used to estimate daily MC loss: $\Delta M_{green} = (119{,}020—26{,}671\ D_1 - 40{,}969\ D_2 - 6814\ D_3)\ M^{2.3979}\ T^{0.2268}\ H^{-5.6734}\ N^{0.7987}$, $r^2 = 0.67$, $P < 0.001$; where D_1, D_2 and D_3 are dummy variables representing species (*E. globulus* = 0, *E. nitens* = 1), debarking (without bark = 0, with bark = 1) and lengths (244 cm = 0, 350 = 1); **b** the linear relationship between observed and estimated longer-term log moisture content (%, green basis) for *E. globulus* and *E. nitens* for the whole experimental data

particularly for low target moisture contents (20–30 %, green basis). In contrast, piles stacked in late spring to early summer required the least drying times compared with other stacking dates.

The influence of log diameter (or the number of logs per square metre of cross-section) on the air-drying time of *E. globulus* logs (244 cm) with bark is shown in Fig. 6a. For instance, for logs stacked in May, a 30 % reduction in green moisture content would be achieved in 206 days for an average log diameter of 18 cm (22 logs m^{-2}), and this period would increase by 70 days for average log diameters of 30 cm (9 logs m^{-2}). Air-drying time was also strongly influenced by initial log moisture content for *E. globulus* (Fig. 6b). For instance, for logs stacked in April, a 30 % reduction in green moisture content would be achieved in 218, 248 and 290 days starting with initial wood moisture contents of 60, 55 and 50 % (green basis), respectively. A similar pattern was observed for the air-drying time of *E. nitens* logs (244 cm) with bark in relation to either log diameter (or the number of logs per square metre of cross-section) (Fig. 6c) or initial log moisture content (Fig. 6d), although with extended drying times compared to *E. globulus*.

Discussion

Drying times of *Eucalyptus* logs can vary up to eightfold depending on weather conditions. Such large time variation poses tremendous uncertainty in flow planning and quality control (Simpson and Wang 2004). Additionally, moisture content has an impact on transport costs and bioenergy yields, and therefore accurately determining air-drying times becomes critical for minimising costs and maximising returns. As expected, there is a clear seasonal effect with logs drying faster when stacked in late spring to early summer than in late summer to early autumn, which can be explained by observed values of air relative humidity and temperature being the main environmental drivers of log moisture content loss. Although climate (season) was the main drive of log moisture content loss, there were also important effects of species and debarking treatments on this variable. *E. globulus* dried on average 20 % faster than *E. nitens*, and debarking enhanced moisture content loss by 8 %. For instance, in order to reduce 20 %, the moisture content on a green basis, *E. globulus* logs, with bark, harvested/stacked in January would take 22 days compared to 29 for *E. nitens*. Similarly, debarking would reduce drying times from 22 to 17 days in *E. globulus* and from 29 to 20 days in *E. nitens* to achieve a 20 % reduction in moisture content for piles freshly harvested/stacked in January. Visser et al. (2014) also found faster drying of debarked logs of *P. radiata*, suggesting that radial water movements are a significant contributor to the log drying process. Log lengths did not have a significant effect on moisture content loss in our study, although 244-cm logs dried on average 3 % faster than 350-cm logs.

Generally, species with low-density wood dry faster than higher density ones (Simpson and Hart 2001). However, the opposite result was found in the current study. *E. globulus* (which has a basic wood density of 514 kg m^{-3}) dried on average 20 % faster than *E. nitens*,

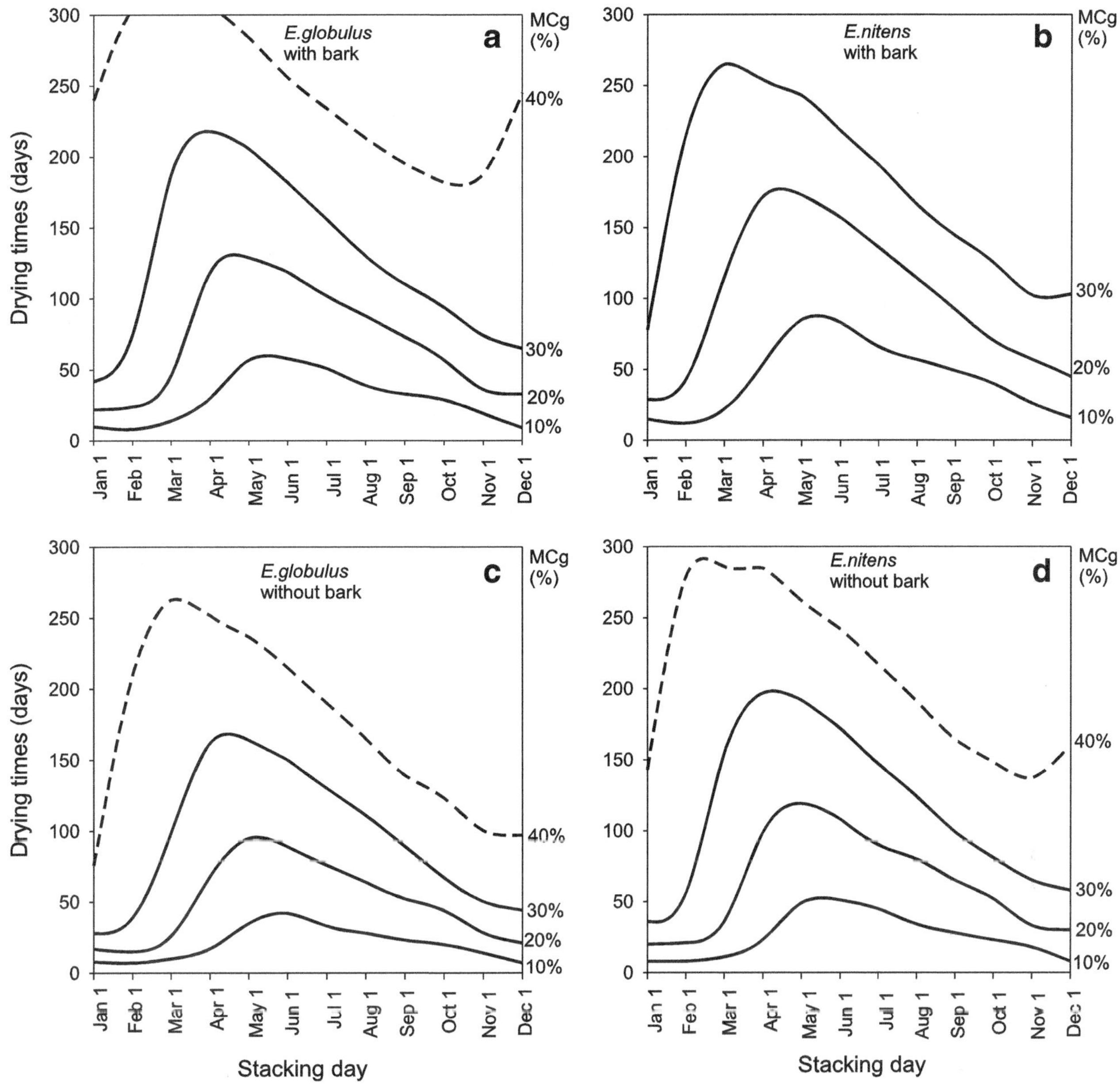

Fig. 5 Estimated air-drying times to achieve reductions of 10, 20, 30 or 40 % in log moisture content (green basis, percent MCg, right axis) stacked on any day of the year in Rucamanqui, Los Angeles, Chile. Calculations were made for an initial moisture content of 60 % on a green basis, log lengths of 244 cm and 22 logs per square metre (~18 cm diameter logs). *Dashed lines* represent extrapolated values. Panels (**a**) and (**c**) correspond to drying times of *E. globulus* with and without bark; while (**b**) and (**d**) to *E. nitens* with and without bark, respectively

(which has a 5 % lower basic density (491 kg m^{-3})). This result may indicate that water movement is more related to fibre conductivity than wood density. Muneri and Raymond (2001) reported that *E. globulus* exhibits greater fibre length and coarseness than *E. nitens* which may point in such direction.

The daily MC loss decreased at decreasing rates as the log MC dropped. For instance, logs of *E. globulus* with bark would take 22, 26 and 35 days in order to reduce 20 % moisture content starting with and initial wood moisture content of 60, 55 and 50 % on a green basis, respectively (provided they are stacked in January). Log diameter also showed to have a strong influence on drying times, e.g. 20 % reductions in moisture content in 18 cm logs would take 22 days if stacked in January, and that time would almost double for 30 cm logs. A similar pattern was observed for *E. nitens*. This was also observed by Visser et al. (2014), for *P. radiata* logs, arguing that smaller diameter logs and split logs exhibit a larger surface area to mass ratio and a smaller distance water

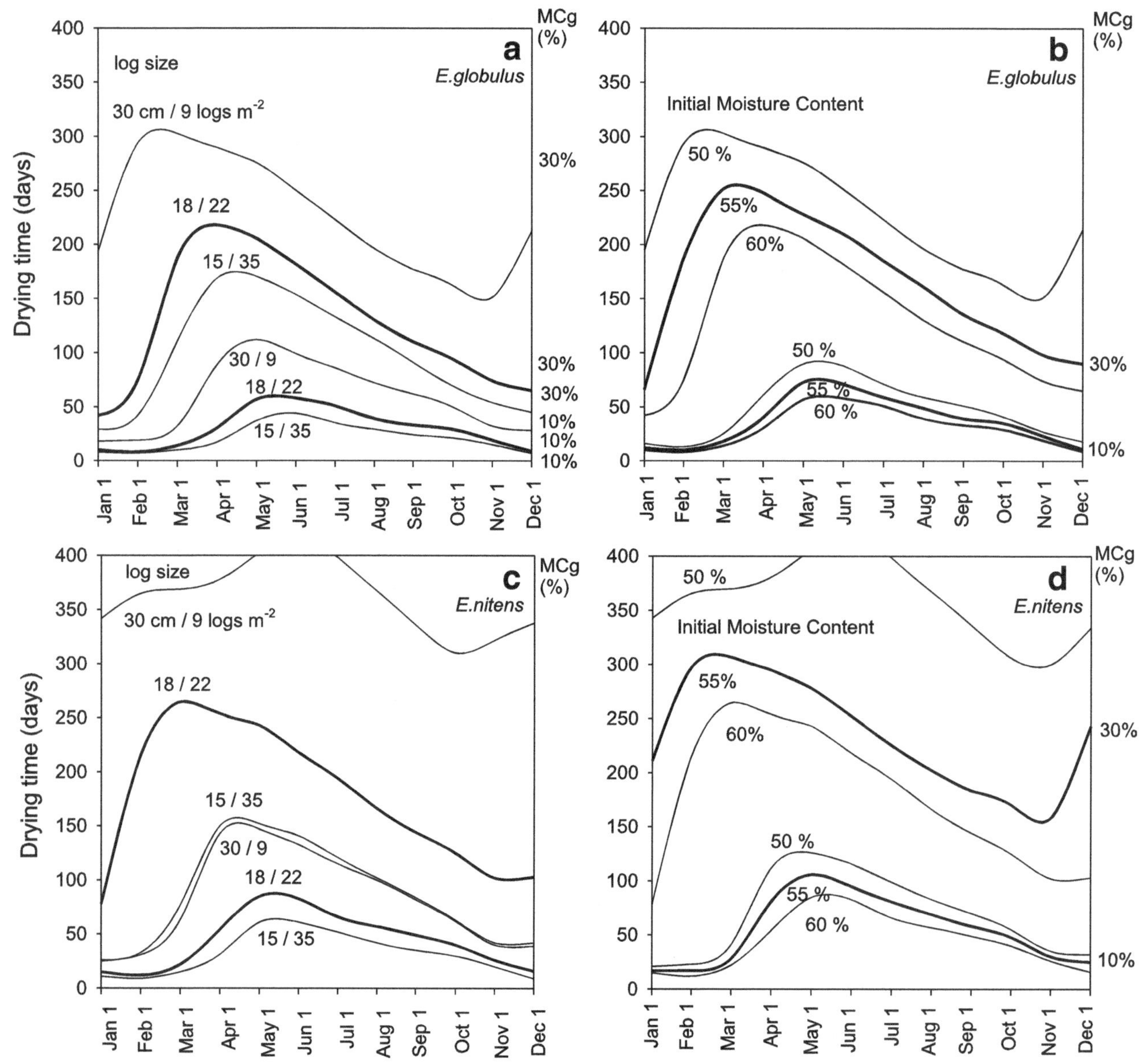

Fig. 6 **a** Influence of log diameter (or the number of logs per square metre of pile cross-section) on predicted air-drying times to achieve a reduction of either 10 or 30 % in log moisture content (percentage, green basis) of *E. globulus* stacked on any day of the year in Rucamanqui, Los Angeles, Chile. Calculations were made for an initial moisture content of 60 %. **b** Influence of initial log moisture content on predicted air-drying times to achieve a reduction of either 10 or 30 % in wood moisture content (percentage, green basis) stacked on any day of the year in Rucamanqui, Los Angeles, Chile. *Panels* **c**, **d** are the equivalents of (**a**) and (**b**) for *E. nitens*. Predictions were made for 244 cm logs with bark

must move from the inner log to the surface in order to evaporate. Since wood-free water flows primarily along the stem evaporating form the log ends (Walker 2006), there must be a secondary movement across the stem if small diameter logs, large split logs and debarked logs dry faster than large whole logs (Abbot et al. 1997, Visser et al. 2014).

Daily MC loss (ΔM) scaled positively following a power function with the MC at the beginning of the day (M), average air temperature (T) and the number of logs per square metre in the cross-section (N), and scaled negatively with air relative humidity (H). Simpson and Wang (2004) found the same variables to significantly explain daily MC loss, but found that a power model fitted best Ponderosa pine (*Pinus ponderosa* Douglas ex C.Lawson) while a linear model was most suitable for Douglas-fir (*Pseudotsuga menziesii* (Mirb.) Franco). Rietz (1970) considered also wind speed and rainfall to explain air-drying rates of lumber. Surprisingly, in our study, rainfall was uncorrelated to air-drying rates, which we assumed was explained because most water intercepted by the piles was evaporated, while only marginal amounts of water infiltrated the logs.

Intensively tended forest plantations of *E. globulus* and *E. nitens* in Chile provide biomass for the short- fibre pulp industry and more recently for industrial boilers for the generation of electricity and heat. The companies are looking to reduce the moisture content of the wood in order to diminish the transport cost of pulp logs and to increase the energy power of the biomass. Kiln-drying would be largely uneconomical because of large log diameters (long residence times) and low log prices not justifying high kiln-drying costs. Hence, air-drying becomes the natural alternative to kiln-drying for fuelwood (Filbakk et al. 2011). However, air-drying times are known to vary largely from several weeks to several months depending on species, dimensions, local weather conditions and time of year when the material is stacked (Simpson and Wang 2004). Most studies on air-drying times are based on stacked lumber, and only a few, notably Simpson and Wang (2004), have determined air-drying times for logs, and to the best of our knowledge none has worked with logs of *Eucalyptus* species.

The model fitted for *E. globulus* and *E. nitens* in the current study may be potentially applied to other sites provided that some validation can be obtained. This approach was the same originally developed by Simpson and Hart (2001), who applied a model fitted for only a few locations to many climates across the USA. Once underlying processes describing air-drying rates are explained by environmental variables, then we may correctly assume that the same processes would explain changes for other climates. Reinforcing this idea, Filbakk et al. (2011) found that air-drying rates of whole trees of broadleaved species at three sites in Norway were mainly explained by air temperature and precipitation independent of site. The simplicity of the model developed in this study, based on four variables only, may provide guidance to decision-makers in the forestry industry to minimise transport costs based on currently available meteorological data. This is something that managers may consider since moisture content has an impact on transport costs and energy yields. The model is now routinely used by Forestal Mininco S.A. to predict log air-drying times at the roadside across different environmental conditions.

Conclusions

In summary, the effect of season (climate) was far greater than the effect of species, debarking or log length on log moisture content loss. Overall, *E. globulus* dried 20 % faster than *E. nitens*, debarked logs dried 8 % faster than barked logs and 244 cm logs dried 3 % faster than 350 cm logs. Daily MC loss was explained by a power model of MC at the beginning of the day, average daily air temperature, average daily air relative humidity and the number of logs per square metre cross-section of the piles (or equivalently average log diameter). The model needs an upper bound of 1.97 % daily MC loss on a green basis (3.96 % on a dry basis). The model can be used to estimate air-drying times of piles stacked any day of the year, to any final MC, for a wide range of log diameters (in the approximate 14- to 33-cm range), and for different environmental conditions; providing a new tool for decision-making in forest transport and industrial planning.

Competing interests

The authors declare that they have no competing interests.

Authors' contributions

Both JPL and HEB designed the experiment, monitored data collection and did intial data analysis. JPL wrote material and methods and substantially contributed in other sections of the manuscript. HEB made figures and tables and wrote the introduction, results and discussion. All authors have read and approved the final manuscript.

Acknowledgements

This research was sponsored by Forestal Mininco S.A. The experiments and measurements undertaken for this paper comply with the current laws of Chile. None of the authors have any competing interests in the manuscript.

Author details

[1]Faculty of Forestry & Conservation, University of Chile, P.O. Box 9206, Santiago, Chile. [2]Forestal Mininco S.A., Los Canelos 71, San Pedro, Concepción, Chile.

References

Abbot, P, Lowore, J, Khofi, C, & Werren, M. (1997). Defining firewood quality: a comparison of quantitative and rapid appraisal techniques to evaluate firewood species from a Southern African savanna. *Biomass & Bioenergy, 12*, 429–437.

Filbakk, T, Høibø, O, & Nurmi, J. (2011). Modelling natural drying efficiency in covered and uncovered piles of whole broadleaf trees for energy use. *Biomass and Bioenergy, 35*, 454–463.

Instituto Forestal. (2014). *Anuario Forestal 2014. [Technical Report]. Boletín Estadístico No 144* (p. 159). Concepción: Instituto Forestal. http://wef.infor.cl/publicaciones/publicaciones.php. Access date: August 23, 2015.

Lignum. (2005, April). *Eucalyptus nitens, Nueva Especie Latifoliada Industrial. Revista Lignum.*

Muneri, A, & Raymond, CA. (2001). Nondestructive sampling of *Eucalyptus globulus* and *E. nitens* for wood properties; II. Fibre length and coarseness. *Wood Science and Technology, 35*, 41–56.

Rietz, RC. (1970). *Air drying lumber in a forklift yard. [USDA forest service research note Fpl-0209]*. Madison, WI, USA: US Department of Agriculture, Forest Service Forest Products Laboratory.

Simpson, WT, & Hart, CA. (2001). Method for estimating air-drying times of lumber. *Forest Products Journal, 51*(11/12), 56–63.

Simpson, WT, & Wang, X. (2004). Estimating air-drying times of small-diameter ponderosa pine and Douglas-fir logs. *Forest Products Journal, 54*(12), 24–28.

Visser, R, Berkett, H, & Spinelli, R. (2014). Determining the effect of storage conditions on the natural drying of radiata pine logs for energy. *New Zealand Journal of Forestry Science, 44:* 3.

Walker, J. (2006). *Primary wood processing* (2nd ed.). Dordrecht, Netherlands: Springer Verlag.

2

Modelling spiral grain angle variation in New Zealand-grown radiata pine

John R. Moore*, Dave J. Cown and Russell B. McKinley

Abstract

Background: Spiral grain angle (SGA) is a wood property that has a strong influence on end-product quality, particularly for solid timber, and most commercial log and timber grading rules restrict the amount of visible surface sloping grain. The aim of this study was to develop parametric models that can be used to predict the intra- and inter-stem variation in SGA in radiata pine (*Pinus radiata* D.Don) trees growing in New Zealand.

Methods: Empirical models were developed using a dataset that contained records from over 1,100 trees that had SGA measured in both the radial direction and at different heights up the stem. Linear and nonlinear model forms were evaluated for their ability to predict the radial variation in SGA.

Results: Most values of spiral grain angle were between −5° and +10°, with a few extreme values (up to 20°) observed. A simple linear model based on cambial age was able to account for the radial variation in SGA outside of the innermost growth rings, but only explained 26 % of the overall variation in SGA. Including a relative height term in this linear model increased the proportion of variation in SGA explained to 30 %. A variance components analysis showed that 78 % of the variation in SGA occurred within individual stems, with only 7 % of the variation due to differences between sites.

Conclusions: Our results confirmed that SGA decreased from the inner growth rings to the bark and increased with height up the stem, with low values of SGA only found in the lower peripheral part of the stem. Given that the data came from a wide range of sites, the relatively small amount of inter-site variation was unexpected. The models developed here can be incorporated into a growth and yield simulation system to enable forest managers to compare the potential impacts of different factors on the size of the corewood zone in a tree containing wood with high SGA.

Keywords: Spiral grain; Grain angle; Radiata pine; Wood properties; Radial variation

Background

Spiral grain is a regular feature of many tree species, and remains one of the most important factors influencing wood utilisation (Denzler et al. 2015). However, the factors controlling its formation are poorly understood (Harris 1989; Kubler 1991; Skatter and Kučera 1997). Spiral grain angle (SGA) is defined as the orientation of wood fibres or tracheids with respect to the longitudinal axis of the tree stem (Kozlowski et al. 1967). Changes in the grain angle within stems are a well-known feature of old trees and of the corewood of conifers, and are normally attributed to pseudotransverse and anticlinal cell division in the vascular cambium (Bannan 1966; Larson 1994; Schulgasser and Witztum 2007). Spiral grain is an important defect is sawn timber and is a primary source of twist distortion during drying (Ormarsson and Cown 2005).

Given both the lack of understanding of the nature of spiral grain in trees and its impact on end-product performance, there is a need to understand how it varies within and among trees. In radiata pine (*Pinus radiata* D.Don), as in other species (Brazier 1965; Northcott 1957; Pape 1999; Sall 2002), there is a degree of systematic variation in spiral grain angle within stems (Cown et al. 1991a; Cown et al. 1991b). Most stems have near zero values of SGA adjacent to the pith followed by increasing positive angles (upwards to the left as viewed from the bark) over the next few rings to a maximum at

* Correspondence: john.moore@scionresearch.com
Scion, Private Bag 3020, Rotorua, New Zealand

around rings six to ten, before declining again to about the 15th ring and often changing to negative angles as the trees mature. The characteristic within-tree pattern varies between species and individuals (Harris 1989; Pape 1999; Sall 2002). Longitudinal trends in radiata pine SGA are less well characterised. Some authors report minimal variation with tree height (Tsehaye and Walker 1995). However, most studies in radiata pine, including Cown et al. (1991b) and Cown et al. (2010) noted that higher values of SGA were generally observed in the upper portion of the stem. Similar results were reported in Sitka spruce (*Picea sitchensis* (Bong.) Carr.) by Pape (1999). Although circumferential variation in grain angle has been noted on log exteriors and inside radiata pine stems (Cown et al. 2010; Watt et al. 2013), the degree of variability in within-stem and between-stem patterns and its potential importance in radiata pine is largely unknown. Despite all of the SGA measurements that have been made, good models to predict the within- and between-tree variation in radiata pine do not exist.

Empirical models describing the radial trends in SGA have been developed for several species, including Sitka spruce (Fonweban et al. 2013) and Norway spruce (*Picea abies* (L.) Karst.) (Gjerdrum and Bernabei 2009; Gjerdrum et al. 2002). Despite the large amount of data that have been collected on SGA in radiata pine in numerous published and unpublished studies, there has not been a systematic analysis of these data in order to quantify the within and between-tree variation in SGA and to develop empirical models of the radial and longitudinal variation in SGA. The only empirical models that have been developed were fitted to data from 90 trees sampled in Kaingaroa Forest (Tian et al. 1995). Therefore, the objectives of this study were to: (1) collate historic data on spiral grain variation in radiata pine in New Zealand; (2) develop models that can explain the intra-stem variation (radial and longitudinal) in SGA for different sized trees; and (3) investigate the extent to which these intra-stem patterns are affected by site and silvicultural factors and the need for including future adjustments in the models.

Methods

Data sources

A database was assembled containing spiral grain angle measurements collected during numerous published (Cown and McConchie 1981; Cown et al. 1991b; Lausberg et al. 1995; Young et al. 1991) and unpublished studies. The latter have often been written up as internal or client reports, but as they are not officially published cannot be cited in the references. However, such reports may contain valuable information so details of relevant ones are provided (see Additional file 1 for more details). The data included in this study were collected from 57 stands that ranged in age from 11 to 35 years and were located across the full range of latitudes from the extreme north to the extreme south of New Zealand (Fig. 1, Table 1). Data were obtained from 1535 individual radiata pine trees. In most cases, at least ten trees were sampled per stand and measurements were made at four heights up the stem of each tree (normally the base, 1.4 m, 5 m and then at 5 m intervals to a top diameter of 100 mm). Spiral grain angle data were obtained from the original studies. In all cases, the SGA measurements had been made using the axis of the disc as the reference, according to the method described by Brazier (1965). Each selected growth ring had been assessed by exposing the latewood with a chisel and scribing along the grain with a "grain detector" (Harris 1989). Grain angle had been measured with respect to the upper surface of the disc. Values for equivalent rings on either side of the pith had then been averaged to remove the effect of disc skew. Spiral grain angle was never measured in the ring closest to the pith as it was always assumed to be zero degrees. Values were measured on every second ring starting from ring two in some studies but on every fifth ring starting from ring five in others.

Total tree height was not measured in almost all of the studies. Because of the different tree ages in the

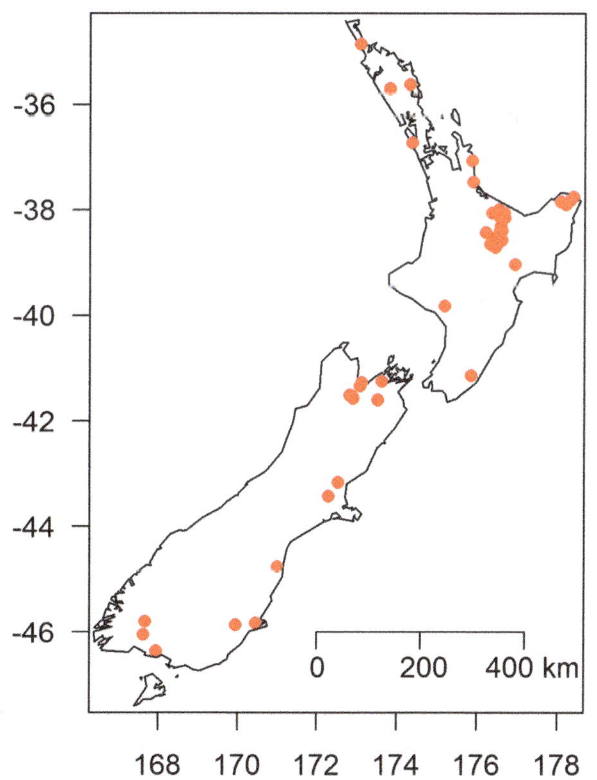

Fig. 1 Location of the 57 sites where data had previously been collected on spiral grain angle

Table 1 Summary of the data used to develop models of spiral grain angle

Region	Number of stands	Age range (yrs)	No. of trees	Diameter at breast height (cm)	Estimated total height (m)
North Island					
Northland	7	11-33	261	24.0-68.0	13.5-44.7
Central North Island	28	11-35	918	22.9-82.0	11.2-51.0
East Coast/Hawke's Bay	5	17-26	161	22.5-75.5	33.8-47.5
Southern North Island	2	25	16	31.5-66.2	31.2-45.0
South Island					
Nelson	7	25-33	114	22.7-70.5	30.9-51.1
Canterbury	3	25	30	28.8-60.5	24.7-36.0
Otago/Southland	5	25-26	35	35.3-66.8	23.3-39.9
Overall	57	11-35	1535	22.5-82.0	11.2-51.1

dataset and the need to be able to predict the intra-stem patterns in SGA for a wide range of tree sizes, it is useful to know the relative height that each sample was taken from as well as the absolute height. To overcome this problem, total height was predicted for each tree using the same approach that was taken for the dataset used to develop a model of radiata pine microfibril angle (Moore et al. 2014). This involved using a compatible volume and taper function (Katz et al. 1984) along with data on the diameter and heights of each disc sampled from the tree. For each tree, the estimated value of total height was increased incrementally and the difference between the actual and predicted diameters of the discs determined. The predicted height of each tree was selected, such that the difference between actual and predicted disc diameters was minimised. Total height could not be predicted for those trees where SGA had only been measured on a disc taken from a single height. This reduced the number of trees in the final dataset to 1133.

Data analysis

Because the data had a hierarchical structure (*i.e.* growth rings within discs, within trees, within stands), a mixed modelling approach was adopted similar to that used in other studies that have developed ring level wood properties models (*e.g.* Auty et al. 2013; Jordan et al. 2005; Moore et al. 2014). A variance components analysis was undertaken to determine how much of the variation in spiral grain angle was attributable to each stratum (*i.e.* site, tree, disc or ring) in the dataset. A number of different model forms were then evaluated for their ability to explain the radial trends in SGA. Once the most suitable model form was identified, the second step was to determine whether the model parameters varied as a function of height within the tree. Model selection was based on visual appraisal of plots of the normalised residuals versus fitted and explanatory variables (Pinheiro and Bates 2000) and Akaike's information criterion (AIC; Akaike 1974). The AIC is used when comparing models with different numbers of parameters fitted to the same dataset and the model with the lower AIC is generally preferred. Parameter estimates were obtained using the maximum likelihood method, and only those parameters that were significant ($p < 0.05$) were retained in the final models. In order to estimate how much of the variation in SGA was explained by the models, two sets of fit indices (R^2) were calculated using the equations given in Parresol (1999). In the first set, the predicted values were estimated from only the fixed effects terms of each model, and in the second, they were calculated from both the fixed and random effects. All statistical analyses were carried out using functions contained in the nlme library (Pinheiro et al. 2012) of the R statistical programming environment (R Development Core Team 2013).

Three different model forms were used to explain the radial variation in SGA. The first was a simple linear function that is based on the approach taken by Gjerdrum et al. (2002) to model SGA in Norway spruce. The difference in the current study is that radial distance from pith was replaced with ring number from the pith, so that the model has the following form:

$$y_{ijkl} = \gamma_0 + \gamma_1 CA_{ijkl} + c_{0,i} + c_{0,ij} + c_{0,ijk} + \varepsilon_{ijkl} \qquad (1)$$

where, y_{ijkl} and CA_{ijkl} are the mean SGA (degree) and cambial age (years), respectively, of the lth annual growth ring in the kth disc of the jth tree from the ith site, γ_0 and γ_1 are the fixed effects parameters and $c_{0,i}$, $c_{0,ij}$ and $c_{0,ijk}$ represent the random effect of γ_0 at the ith site, j^{th} tree in the ith site and k^{th} disc in the jth tree in the ith site, respectively; ε_{ijkl} is the random error due to the lth annual ring of the kth disc from the jth tree at the ith site. The second model form was based on the approach taken by Tian et al. (1995), who modelled SGA as a function of the natural logarithm of cambial age:

$$y_{ijkl} = \delta_0 + \delta_1 logCA_{ijkl} + d_{0,i} + d_{0,ij} + d_{0,ijk} + \varepsilon_{ijkl} \quad (2)$$

where, y_{ijkl} and CA_{ijkl} are the mean SGA (degree) and cambial age (years), respectively, as defined in Eq. 1, δ_0 and δ_1 are the fixed effects parameters and $d_{0,i}$, $d_{0,ij}$ and $d_{0,ijk}$ represent the random effects of δ_0 at the site, tree and disc levels, respectively.

The third model form was the nonlinear function used by Fonweban et al. (2013) to model the radial variation in Sitka spruce spiral grain angle. This model has the following form:

$$y_{ijkl} = \left(\left(\omega_0 + o_{0,i} + o_{0,ij} + o_{0,ijk}\right) + \omega_1 CA_{ijkl}\right)e^{\left(-\omega_2 CA_{ijkl}\right)} + \varepsilon_{ijkl} \quad (3)$$

where, y_{ijkl} and CA_{ijkl} are the mean SGA (degree) and cambial age (years), respectively, as defined in Eq. 1, ω_0,ω_1 and ω_2 are the fixed effects parameters and $o_{0,i}$, $o_{0,ij}$ and $o_{0,ijk}$ represent the random effects of the ith site, the jth tree in the ith site and the kth disc in the jth tree in the ith site, respectively.

The fixed effects parameters in the three models given by Eqs. (1, 2 and 3) were also modified to include the influence of height. Linear, log-linear and quadratic height terms were added to the model following the approach described by Jordan et al. (2005). Because ring width data were not available for most of the sites, this variable could not be included in the models. In the two linear models (Eqs. (1 and 2)), terms representing both the relative height and the interaction between relative height and cambial age were added to the model, while in the nonlinear model given by Eq. (3) relative height terms (HT_{rel}) were added to each of the parameters, *i.e.*

$$y_{ijkl} = \gamma_0 + \gamma_1 CA_{ijkl} + \gamma_2 HT_{rel,ijk} + \gamma_3 HT_{rel,ijk} CA_{ijkl} + c_{0,i} + c_{0,ij} + c_{0,ijk} + \varepsilon_{ijkl} \quad (4)$$

$$y_{ijkl} = \gamma_0 + \gamma_1 CA_{ijkl} + \gamma_2 \log HT_{rel,ijk} + \gamma_3 \log HT_{rel,ijk} CA_{ijkl} + c_{0,i} + c_{0,ij} + c_{0,ijk} + \varepsilon_{ijkl} \quad (5)$$

$$y_{ijkl} = \delta_0 + \delta_1 \log CA_{ijkl} + \delta_2 HT_{rel,ijk} + \delta_3 HT_{rel,ijk} \log CA_{ijkl} + d_{0,i} + d_{0,ij} + d_{0,ijk} + \varepsilon_{ijkl} \quad (6)$$

$$y_{ijkl} = \delta_0 + \delta_1 \log CA_{ijkl} + \delta_2 \log HT_{rel,ijk} + \delta_3 \log HT_{rel,ijk} \log CA_{ijkl} + d_{0,i} + d_{0,ij} + d_{0,ijk} + \varepsilon_{ijkl} \quad (7)$$

$$y_{ijkl} = \left(\left(\omega_0 + o_{0,i} + o_{0,ij} + o_{0,ijk}\right) + \omega_3 HT_{rel,ijk} + \left(\omega_1 + \omega_4 HT_{rel,ijk}\right) CA_{ijkl}\right)e^{\left(-\left(\omega_2 + \omega_5 HT_{rel,ijk}\right) CA_{ijkl}\right)} + \varepsilon_{ijkl} \quad (8)$$

$$y_{ijkl} = \left(\left(\omega_0 + o_{0,i} + o_{0,ij} + o_{0,ijk}\right) + \omega_3 \log HT_{rel,ijk} + \left(\omega_1 + \omega_4 \log HT_{rel,ijk}\right) CA_{ijkl}\right)e^{\left(-\left(\omega_2 + \omega_5 \log HT_{rel,ijk}\right) CA_{ijkl}\right)} + \varepsilon_{ijkl} \quad (9)$$

In order to visualise the intra-stem pattern in SGA, the models were then applied to a single 29-year-old tree (DBH = 32.8 cm, height = 36.9 m) that had spatial information on ring number from the pith generated at 5 mm resolution in the radial direction and 100 mm resolution in the longitudinal direction. This information was generated using the Forecaster growth and yield modelling system (West et al. 2013).

Results

Most values of spiral grain angle were between −5° and +10°, with a few more extreme values (up to 20°) observed (Fig. 2). A linear, or slightly curvilinear, trend in SGA was observed at all heights in the stem, with values of 5-6° observed in the first few growth rings from the pith decreasing to zero or becoming slightly negative

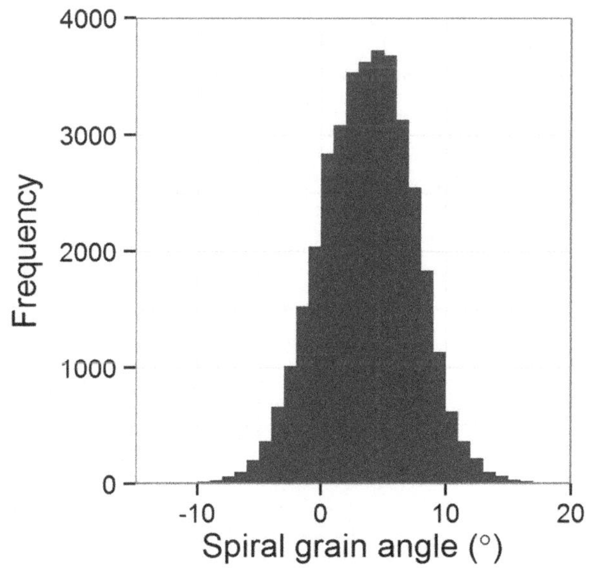

Fig. 2 Histogram showing the distribution of spiral grain angles for all trees in the radiata pine dataset

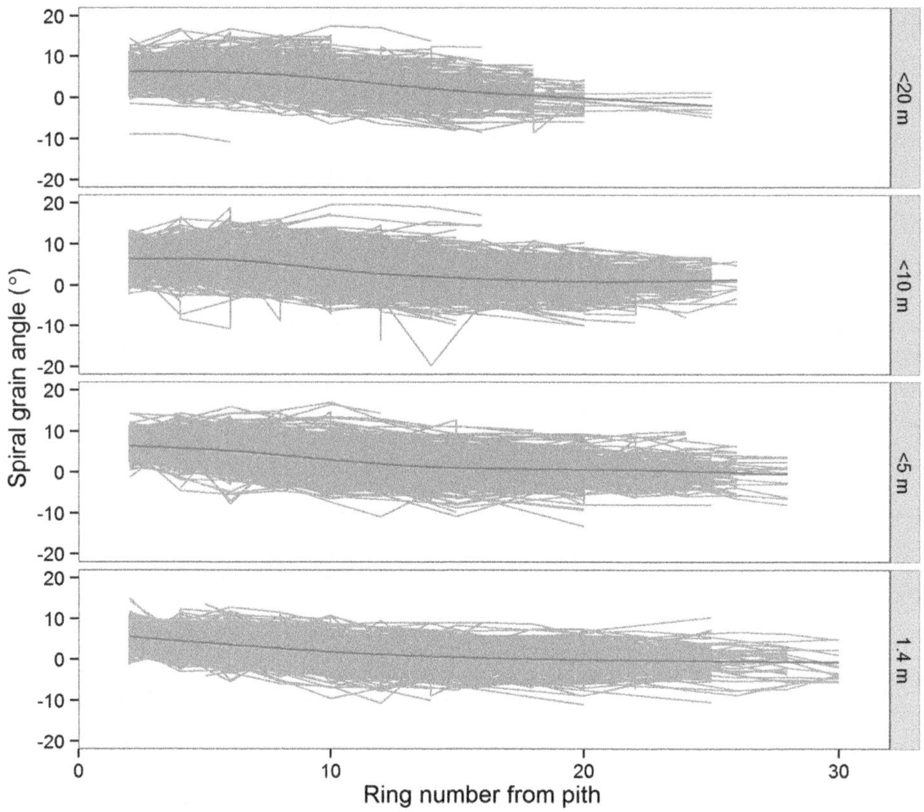

Fig. 3 Radial profiles in SGA for discs taken at different heights within the stem. Each grey line represents the actual observations from an individual tree and the black line is a locally-weighted smoothing function (lowess) fitted to all observations

with increasing ring number from the pith (Fig. 3). Most of the variation in SGA was due to radial variation within the stem (55 %), with a further 23 % due to vertical variation within the stem. Approximately 14 % of the variation in SGA was attributed to differences between trees within a stand, with only 7 % attributed to differences between sites.

Modelling radial variation

The simple linear model (Eq. (1)) appeared to perform the best of the three models evaluated, having the lowest AIC, and was able to predict approximately 26 % of the variation in SGA based solely on the fixed effect of cambial age (61 % including the random effects of site, tree and disc) (Table 2). The simple linear model indicates that, on average, the value of SGA is approximately 6.1° at ring two from the pith and on average decreases by 0.3° with each additional growth ring out from the pith (Table 3).

Table 2 AIC, log-likelihood and fit indices for the models to predict SGA given by Eqs. 1, 2 and 3

Model	AIC	Log-likelihood	Fit indices (R^2)			
			Fixed	Site	Tree	Disc
Eq. (1)	147338	−73661	0.26	0.31	0.51	0.61
Eq. (2)	149165	−74575	0.22	0.28	0.48	0.61
Eq. (3)	148444	−74213	0.24	0.28	0.43	0.43

Modelling variation with height

Adding relative height terms to the linear model (Eqs. (4 and 5)) increased the proportion of variation

Table 3 Parameter estimates and associated standard errors (s.e.) and tests of significance for the models given by Eqs. 1, 2 and 3

Parameter	Estimate	s.e.	*t*-value	*p*-value
Eq. (1)				
γ_0	6.708	0.156	42.883	<0.001
γ_1	−0.310	0.003	−121.942	<0.001
Eq. (2)				
δ_0	8.506	0.161	52.775	<0.001
δ_1	−2.316	0.020	−117.628	<0.001
Eq. (3)				
ω_0	7.299	0.095	77.009	<0.001
ω_1	−2.357	0.153	−154.468	<0.001
ω_2	−0.059	0.001	-62.263	<0.001

in SGA that was explained by these models to approximately 30 % (Table 4). This was greater than the amount of variation explained by the addition of height terms to the nonlinear models given by Eqs. (8 and 9), but slightly less than the amount of variation explained by the log-linear models given by Eqs. (6 and 7). However, based on the AIC criterion, the "best" model for explaining the intra-stem variation in SGA is given by Eq. (5). The model given by Eq. (4) had a very similar AIC and also produced similar predictions of SGA, while the models given by Eqs. (6 and 7) only produced positive estimates of SGA, and those given by Eqs. (8 and 9) produced some very large negative estimates of spiral grain (up to −10°). Therefore, the linear models given by Eqs. (4 and 5) were preferred over the more complex nonlinear models, even though they do not account for the observed variation within the first few rings from the pith. The parameter estimates for these models are given in Table 5.

Visualisation of model predictions

The intra-stem variation in SGA predicted by the model described by Eq. (5) is shown in Fig. 4. This shows that the inner part of the tree has values of SGA of approximately 7-8° with SGA increasing with height up the stem for a given cambial age. SGA decreases with increasing cambial age with the lowest values found near the base of the tree. The predicted pattern of intra-stem variation in SGA was very similar regardless of whether the longitudinal variation along the stem is modelled as a function of HT_{rel} or $\log(HT_{rel})$ (results not shown).

Discussion

Spiral grain is apparently more variable and more difficult to model than other important radiata pine wood properties such as density (Kimberley et al. in review) or microfibril angle (Moore et al. 2014). However, the amount of variation in radiata pine SGA that was explained by the models developed here was slightly higher than by the models developed by Gjerdrum et al. (2002) and Fonweban et al. (2013) for Norway spruce and Sitka spruce SGA, respectively.

Table 4 AIC, log-likelihood and fit indices for the models to predict SGA given by Eqs. 4, 5, 6, 7, 8 and 9

Model	AIC	Log-likelihood	Fit indices (R^2)			
			Fixed	Site	Tree	Disc
Eq. (4)	49898	−24939	0.29	0.35	0.56	0.58
Eq. (5)	49829	−24904	0.30	0.36	0.57	0.59
Eq. (6)	50003	−24991	0.29	0.35	0.56	0.59
Eq. (7)	50012	−24996	0.29	0.36	0.57	0.59
Eq. (8)	51143	−25561	0.27	0.34	0.50	0.50
Eq. (9)	51087	−25532	0.29	0.36	0.36	0.36

Table 5 Parameter estimates and associated standard errors (s.e.) and tests of significance for the models given by Eqs. 4 and 5

Parameter	Estimate	s.e.	*t*-value	*p*-value
Eq. (4)				
γ_0	5.831	0.278	20.955	<0.001
γ_1	−0.367	0.006	−57.453	<0.001
γ_2	1.556	0.218	7.126	<0.001
γ_3	0.341	0.027	12.779	<0.001
Eq. (5)				
γ_0	7.319	0.282	25.920	<0.001
γ_1	−0.199	0.013	−15.505	<0.001
γ_2	0.835	0.070	11.914	<0.001
γ_3	0.073	0.008	9.483	<0.001

The patterns of SGA predicted by the models generally confirms what has been described for radiata pine in the literature (*e.g.* Cown et al. 1991b; Harris 1989; Lausberg et al. 1995; Tian et al. 1995; Watt et al. 2013). In the radial direction, the models developed here showed that there is a predominantly linear decline in SGA with increasing ring number from the pith and the rate of decline decreases with increasing height up the stem. By setting SGA equal to zero in Eq. (4), the age at which SGA transitions from positive (*i.e.* left handed) to negative (*i.e.* right handed) was estimated to be 16 years at the base of the tree (*i.e.* at a relative height of zero). However, previous studies (*e.g.* Cown et al. 1991b; Watt et al. 2013) have reported near zero values of SGA near the pith followed by an increase to a maximum at ring 6–10 followed by a decline with increasing growth ring from the pith. This pattern was not predicted by the models, but this may be an artefact of the resolution of the data used to develop the models. Many of the trees in the dataset only had SGA measured on every fifth growth ring starting at ring five from the pith, while others had SGA measured on every second ring starting at ring two from the pith. The increase in SGA from the pith to ring 6–10 would not be apparent in the coarser resolution data. Therefore, caution should be exercised when making predictions of SGA in the innermost two to five growth rings using a linear model. Recent advances in measurement techniques (*e.g.* Riddell et al. 2012) now mean that it is possible to collect higher resolution data on the variation in SGA within a stem, particularly in the circumferential direction, which will enable a better understanding of the patterns of variation as well as providing better data for modelling this variation.

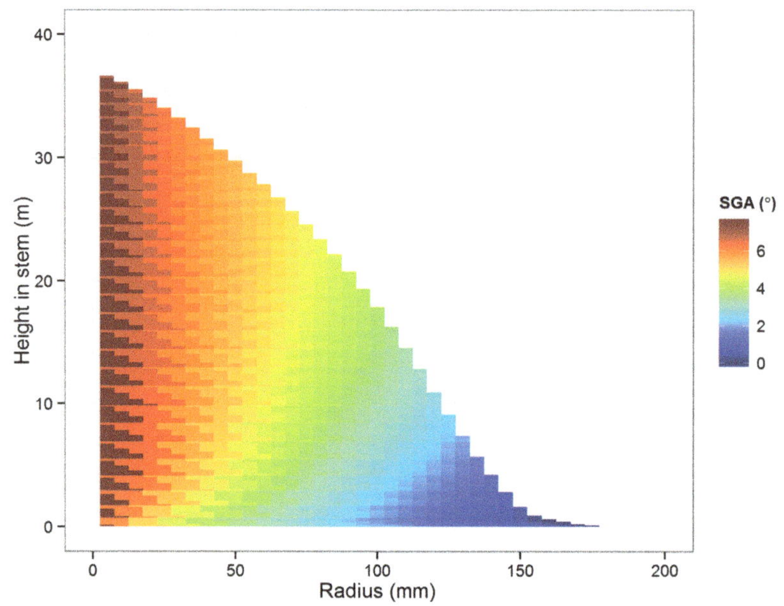

Fig. 4 Intra-stem pattern in SGA predicted by a modified logistic function with a logarithmic relative height term. Parameter estimates for this model (Eq. 5) are given in Table 5 in the text

Previous studies in radiata pine have noted that there is a large amount of tree-to-tree variation in the radial patterns of SGA (Cown et al. 1991b; Watt et al. 2013). The magnitude of this tree-to-tree variation is apparent in this study because when the random effects of site and tree were considered, the total amount of variation in SGA explained by the models almost doubled. The models developed in this study did not include ring width as an explanatory variable as ring width was not measured in most studies. Including a ring width term in future models may improve the amount radial variation in SGA that can be explained as several studies in other species have reported weak positive correlations between ring width and spiral grain angle (Fonweban et al. 2013; Pape 1999).

The models developed here showed that there was a linear increase in SGA with increasing height up the stem, which is consistent with has been reported in previous studies. In the longitudinal direction a general trend of increasing SGA has been noted in the previous studies (Cown et al., 1991b). More quantitative information was provided in a recent study by Watt et al. (2013) who found that SGA in 7-year-old radiata pine tree increased with height up the stem reaching a maximum at a height of 5 m. The additional amount of variation explained by the models when relative height terms were included is based on predicted values of relative height rather than measured values. These values have an error associated with them and it is recommended that tree height is measured in future studies investigating SGA, or other wood properties for that matter.

While site-level variation was relatively small for SGA (*i.e.* approximately 7.3 % of the total variation), it is still considered worthwhile to investigate this further as previous studies have shown that SGA may be affected by exposure to wind (Eklund and Säll 2000; Fonweban et al. 2013). Other site factors, such as soil type, temperature and rainfall distribution may also contribute and these vary widely across New Zealand, are strong determinants of stand productivity (Watt et al. 2010) and have also been shown to affect other wood properties, most notably density (Palmer et al. 2013). While there are known genetic differences in radiata pine SGA (Burdon and Low 1992; Gapare et al. 2007; Wu et al. 2008), the genetic origin of the trees in the dataset used to develop these models is very diverse it is unlikely for such differences to be observed over and above the other sources of variation in the dataset.

The results from this analysis highlight the issue of corewood in radiata pine which is the wood contained in the first ten or so growth rings from the pith (Cown 1992). In the corewood zone spiral grain angles are consistently high, often exceeding the nominal desired maximum of five degrees (Cown et al. 1991b). Previous utilisation studies in radiata pine have shown that the incidence of twist in sawn timber cut from the corewood zone is much higher than in timber cut from the outerwood zone (Haslett et al. 1991; Ormarsson and Cown 2005) due to the high SGA values, ring curvature, and high longitudinal shrinkage. This presents a challenge since silviculture in New Zealand has traditionally favoured rapid diameter growth, so at harvest age (*i.e.*

around 30 years) a substantial volume of each tree is comprised of corewood (30 - 80 % depending on regime and log height class (Cown 1992)). Over the range of stand densities at which radiata pine is typically grown, it is very difficult to restrict the size of the corewood zone.

Further work is required to better understand the patterns of variation in SGA within and between trees, and the drivers of this variation. The data used to develop the models here have low spatial resolution and, therefore, do not capture the full extent of variation that is known to exist within a stem (Cown et al. 2010). If the tree-to-tree variation that is currently included in the models as random variation could be at least partially explained by attributes that could be readily measured on trees or logs, then the predictive power of models could be greatly improved. Future work is planned to refine these models using better techniques to acquire data on SGA (Riddell et al. 2012) and purposely designed experiments that will enable the effects of environmental factors and radial growth rates on SGA to be better understood.

Conclusions

Based on the analysis of a large dataset of SGA measurements in radiata pine we found that approximately 78 % of the total variation in SGA was due to differences within the stem of a tree, with approximately 7 % due to differences among sites. This intra-stem variation could be modelled as a function of cambial age and relative height within the stem, with the "best" models explaining about 30 % of the variation in SGA. The modelled trends in SGA conformed to those reported by previous studies in radiata pine. High positive values of SGA were found in the first few rings from the pith and SGA declined with increasing ring number from the pith. Spiral grain angle was also found to increase with increasing height up the stem. Because of the silvicultural regimes that radiata pine is grown under (*i.e.* relatively wide initial spacing and short rotation lengths) the volume occupied by this high SGA wood is relatively high. The models developed in this study are able to be coupled to growth models enabling the impacts of site productivity and silvicultural regimes on the SGA distribution with a tree to be investigated.

Competing interests

The authors declare that they have no competing interests.

Authors' contributions

JRM conceived the idea for the overall analysis and undertook the data analysis and writing of the manuscript. DJC and RBMcK planned most of the wood quality assessment and contributed to the interpretation of the results and writing of the manuscript. All authors read and approved the final version of the manuscript.

Acknowledgments

The authors wish to acknowledge Wood Quality Initiative Ltd. who funded several of the studies where SGA data were collected and for allowing publication of the results. Funding for publication of the results was also provided by Future Forests Research Ltd. Dr Charles Sabatia provided assistance with the statistical analysis of the data. Three anonymous reviewers provided helpful comments on an earlier version of this manuscript.

References

Akaike, H. (1974). A new look at the statistical model identification. *IEEE Transactions on Automatic Control, 19*, 716–723.

Auty, D, Gardiner, BA, Achim, A, Moore, JR, & Cameron, AD. (2013). Models for predicting microfibril angle variation in Scots pine. *Annals of Forest Science, 70*, 209–218. doi:10.1007/s13595-012-0248-6.

Bannan, MM. (1966). Spiral grain and anticlinal division in the cambium of conifers. *Canadian Journal of Botany, 44*, 1515–1538.

Brazier, JD. (1965). An assessment of the incidence and significance of spiral grain in young conifer stems. *Forest Products Journal, 15*, 308–315.

Burdon, RD, & Low, CB. (1992). Genetic survey of *Pinus radiata*. 6. Wood properties: variation, heritabilities, and interrelationships with other traits. *New Zealand Journal of Forestry Science, 22*, 228–245.

Cown, DJ. (1992). Corewood (juvenile wood) in *Pinus radiata* - should we be concerned? *New Zealand Journal of Forestry Science, 22*(1), 87–95.

Cown, DJ, & McConchie, DL. (1981). Effects of thinning and fertiliser application on wood properties of *Pinus radiata. New Zealand Journal of Forestry Science, 11*, 79–91.

Cown, DJ, McConchie, DL, McConchie, MS, Young, GD (1991). Radiata pine wood properies survey. (FRI Bulletin No. 50 (Revised Edition)) Rotorua New Zealand Forest Research Institute.

Cown, DJ, Young, GD, & Kimberley, MO. (1991b). Spiral grain patterns in plantation-grown *Pinus radiata. New Zealand Journal of Forestry Science, 21*(2/3), 206–216.

Cown, DJ, Harrington, J, Bourreau, D, Haug, J, & Lee, J. (2010). Spatial variation in spiral grain: a single stem of *Pinus radiata* D.Don. *New Zealand Journal of Forestry Science, 40*, 211–224.

Denzler, JK, Weidenhiller, A, & Golser, M. (2015). Property relationships between spruce logs and structural timber. *Scandinavian Journal of Forest Research, 30*(7), 617–623. doi:10.1080/02827581.2015.1046479.

R Development Core Team (2013). R: a language and environment for statistical computing. In R Foundation for Statistical Computing (Ed.). Vienna.

Eklund, L, & Säll, H. (2000). The influence of wind on spiral grain formation in conifer trees. *Trees, 14*(6), 324–328. doi:10.1007/s004680050225.

Fonweban, J, Mavrou, I, Gardiner, B, & Macdonald, E. (2013). Modelling the effect of spacing and site exposure on spiral grain angle on Sitka spruce (*Picea sitchensis* (Bong.) Carr.) in Northern Britain. *Forestry, 86*(3), 331–342. doi:10.1093/forestry/cpt002.

Gapare, W, Hathorn, A, Kain, D, Matheson, C, & Wu, H. (2007). Inheritance of spiral grain in the juvenile core of Pinus radiata. *Canadian Journal of Forest Research, 37*(1), 116–127.

Gjerdrum, P, & Bernabei, M. (2009). Three-dimensional spiral grain pattern in five large Norway spruce stems. *Silva Fennica, 43*(3), 457–464.

Gjerdrum, P, Säll, H, & Storø, HM. (2002). Spiral grain in Norway spruce: constant change rate in grain angle in Scandinavian sawlogs. *Forestry, 75*(2), 163–170.

Harris, JM. (1989). *Spiral grain and wave phenomena in wood formation.* Berlin: Springer.

Haslett, AN, Simpson, IG, & Kimberley, MO. (1991). Utilisation of 25-year-old *Pinus radiata* Part 2: warp of structural timber on drying. *New Zealand Journal of Forestry Science, 21*(2/3), 228–234.

Jordan, L, Daniels, RF, Clark, A, & He, R. (2005). Multilevel nonlinear mixed-effects models for the modeling of earlywood and latewood microfibril angle. *Forest Science, 51*(4), 357–371.

Katz, A, Dunningham, A, & Gordon, A. (1984). *A compatible volume and taper equation for New Zealand* Pinus radiata *D. Don grown under the direct sawlog*

regime. (FRI Bulletin 67). Rotorua: New Zealand Forest Service, Forest Research Institute.
Kimberley, MO, Cown, DJ, McKinley, RB, Moore, JR (in review). Modelling variation in wood density within and between trees in stands of New Zealand-grown radiata pine. *New Zealand Journal of Forestry Science.*
Kozlowski, TT, Hughes, FJ, & Leyton, L. (1967). Movement of injected dyes in gymnosperm stems in relation to tracheid alingnment. *Forestry, 40*(2), 207–219. doi:10.1093/forestry/40.2.207.
Kubler, H. (1991). Function of spiral grain in trees. *Trees, 5*, 125–135.
Larson, PR. (1994). *The vascular cambium - development and structure* (p. 725). New York: Springer-Verlag, Berlin. ISBN 3-540-57165-5.
Lausberg, MJF, Cown, DJ, Gilchrist, K, Skipwith, J, & Treloar, CR. (1995). Physiological ageing and site effects on the wood properties of radiata pine. *New Zealand Journal of Forestry Science, 25*(2), 189–199.
Moore, JR, Cown, DJ, & McKinley, RB. (2014). Modelling microfibril angle variation in New Zealand-grown radiata pine. *New Zealand Journal of Forestry Science, 44*(1), 25. doi:10.1186/s40490-014-0025-4.
Northcott, PL. (1957). Is spiral grain the normal growth pattern? *Forestry Chronicle, 33*, 335–352.
Ormarsson, S, & Cown, D. (2005). Moisture-related distortion of timber boards of radiata pine: comparison with Norway spruce. *Wood and Fiber Science, 37*(3), 424–436.
Palmer, DJ, Kimberley, MO, Cown, DJ, & McKinley, RB. (2013). Assessing prediction accuracy in a regression kriging surface of *Pinus radiata* outerwood density across New Zealand. *Forest Ecology and Management, 308*, 9–16.
Pape, R. (1999). Influence of thinning on spiral grain in Norway spruce grown on highly productive sites in southern Sweden. *Silva Fennica, 33*(1), 3–12.
Parresol, BR. (1999). Assessing tree and stand biomass: a review with examples and critical comparisons. *Forest Science, 45*, 573–593.
Pinheiro, JC, & Bates, DM. (2000). *Mixed-effects models in S and S-PLUS*. New York: Springer.
Pinheiro, J, Bates, D, DebRoy, S, Sarkar, D, R Core Team (2012). *nlme: linear and nonlinear mixed effects models*. R package v.3.1-105.
Riddell, M, Cown, D, Harrington, J, Lee, J, & Moore, J. (2012). Assessing spiral grain angle by light transmission- Proof of concept. *IAWA Journal, 33*(1), 1–14.
Säll, H (2002). *Spiral grain in Norway spruce*. Acta Wexionensia No. 22/2002. Sweden: Växjö University Press.
Schulgasser, K, & Witztum, A. (2007). The mechanism of spiral grain formation in trees. *Wood Science and Technology, 41*(2), 133–156.
Skatter, S, & Kučera, B. (1997). Spiral grain - an adaptation of trees to withstand stem breakage caused by wind-induced torsion. *Holz als Roh - und Werkstoff, 55*(4), 207–213.
Tian, X, Cown, DJ, & Lausberg, MJF. (1995). Modelling of radiata pine wood properties. Part 1: spiral grain. *New Zealand Journal of Forestry Science, 25*(2), 200–213.
Tsehaye, A, & Walker, JCF. (1995). Spiral grain in Canterbury *Pinus radiata*: within- and between-tree variations and effect on mechanical properties. *New Zealand Journal of Forestry Science, 25*, 358–366.
Watt, MS, Palmer, DJ, Kimberley, MO, Hock, K, Payn, TW, & Lowe, DJ. (2010). Development of models to predict *Pinus radiata* productivity throughout New Zealand. *Canadian Journal of Forest Research, 40*(3), 488–499.
Watt, MS, Kimberley, MO, Harrington, JJ, Riddell, MJC, Cown, DJ, & Moore, JR. (2013). Differences in intra-tree variation in spiral grain angle for radiata pine. *New Zealand Journal of Forestry Science, 43*(1), 12. doi:10.1186/1179-5395-43-12.
West, GG, Moore, JR, Shula, RG, Harrington, JJ, Snook, J, Gordon, JA, & Riordan, MP. (2013). Forest management DSS development in New Zealand. In J Tucek, R Smrecek, A Majlingova, & J Garcia-Gonzalo (Eds.), *Implementation of DSS tools into the forestry practice* (pp. 153–163). Slovakia: Technical University of Zvolen.
Wu, HX, Ivković, M, Gapare, WJ, Matheson, AC, Baltunis, BS, Powell, MB, & McRae, TA. (2008). Breeding for wood quality and profit in *Pinus radiata*: a review of genetic parameter estimates and implications for breeding and deployment. *New Zealand Journal of Forestry Science, 38*(1), 56–87.
Young, GD, McConchie, DL, & McKinley, RB. (1991). Utilisation of 25-year-old *Pinus radiata*. Part 1: wood properties. *New Zealand Journal of Forestry Science, 21*(2/3), 217–227.

3

The influence of initial and post-thinning stand density on Douglas-fir branch diameter at two sites in New Zealand

J. C. Grace[1*], R. K. Brownlie[1] and S. G. Kennedy[2]

Abstract

Background: Branches support the foliage needed for tree growth, but if the branch diameter is too large it may constitute a defect when the tree is assessed for timber quality and when the tree is processed into boards. Consequently branch diameter is an important consideration when developing a silvicultural regime.
The objectives of this study were: (a) to develop site-specific models to predict branch diameter variation with position on the stem; and (b) to use the models to evaluate the influence of initial and post-thinning stand density on branch diameter in unpruned stands of Douglas-fir (*Pseudotsuga menziesii* (Mirb.) Franco) at two sites in New Zealand.

Methods: Branch diameters were measured using a photogrammetric technique, TreeD, on pre-selected sample trees from unpruned treatments in two silviculture experiments, one in the North Island and one in the South Island of New Zealand. The data were used to develop site-specific models to predict branch diameter along the stem. The models were then used to interpret response to a thinning at a stand age of approximately 10 years when the base of the green crown was still close to ground-level, and to make comparisons with a branch diameter limit of 40 mm.

Results: The models developed indicated that previously formed branches were influenced by the increased growing space created by thinning. At all three post-thinning stand densities (250, 500 and 750 stems ha^{-1}), the diameter of some branches on the mean tree within a stand were likely to exceed 40 mm. In unthinned treatments, the model indicated that an initial stand density of at least 1333 stems ha^{-1} would be needed to keep branch diameters on the mean tree below 40 mm along the whole stem at age 29 years.

Conclusions: This study indicates the importance of considering initial stand density, post-thinning stand density and timing of thinning when designing a silviculture regime that aims to control branch diameter.

Keywords: Douglas-fir; Branch diameter; Stand density; Thinning; Silviculture

Introduction

Douglas-fir (*Pseudotsuga menziesii* (Mirb.) Franco) is native to western North America where it is one of the most valuable and commercially important timber species (Miller and Knowles 1994). Outside its natural range it is an important commercial timber species in France, Germany and New Zealand (Maclaren 2009). Douglas-fir was first recorded in New Zealand in 1859, with plantations being established at various locations at the end of the 19th century/beginning of the 20th century (Miller and Knowles 1994). Currently it is the most important exotic timber species in New Zealand after radiata pine (*Pinus radiata* D.Don.). In April 2013, planted Douglas-fir forests covered 106,500 ha compared with 1,543,600 ha of planted radiata pine (Ministry for Primary Industries (MPI) 2013).

For any tree species, the size and arrangement of branches are important for both tree growth and timber quality (e.g., Larson 1962, 1969; Megraw 1985). The distribution of branches and foliage influences the amount of light intercepted by the foliage (Whitehead et al. 1990), and consequently tree growth. However, branches result in knots within the stem that constitute a defect

* Correspondence: jenny.grace@scionresearch.com
[1]Scion, Private Bag, 3020 Rotorua, New Zealand
Full list of author information is available at the end of the article

in both appearance and structural timber. There is reduced strength and stiffness in the vicinity of a knot, which is considered to be due to the arrangement of fibres around the knot (Phillips et al. 1981). Tustin and Wilcox (1978) measured a 'branch index' for the 2nd log in Douglas-fir. This index was the average diameter of 16 branches selected as follows. The largest branch was selected from each quartile in each 4 ft. (1.22 m) section of the 16 ft. (4.88 m) log. The position of the log was not specified but is presumed to be between 16 ft. (4.88 m) and 32 ft. (9.75 m). They concluded that branch diameter effects were 2.5 times greater than wood density effects in their influence on the variability of stiffness of timber framing. Ledgard et al. (2005) suggested that a desirable maximum branch diameter for unpruned Douglas-fir stands in New Zealand is 40 mm.

The terminology used to describe the structure of a Douglas-fir annual shoot varies. Jensen and Long (1983) described an annual shoot as follows: 'each shoot carries a terminal bud and a set of lateral buds from which next year's growth will emerge. Several (0–5) of these lateral buds occur in a tight whorl directly below the terminal bud, these buds will be referred to as "nodals". The remainder of the lateral buds are scattered along the length of the shoot, generally more than 15 mm from the terminal bud'. Jensen and Long (1983) referred to these as "non-nodals". Maguire et al. (1994) use the terms 'whorl (nodal) and interwhorl (internodal) primary branches', and noted that whorl branches grew more rapidly than interwhorl branches on young plantation-grown Douglas-fir located west of the Cascade crest in Oregon and Washington, USA. In the current study, the word 'cluster' has been used in preference to 'whorl'.

As the largest branch on a log has the greatest influence on log grade (Weiskittel et al. 2006), a modelling approach that predicts how branch diameter varies with height in the stem is considered more appropriate than a model that predicts branch index for a log. Models have been developed to predict how Douglas-fir branch diameter varies with depth into the tree crown (Maguire et al. 1994, 1999; Roeh and Maguire 1997; Ishii et al. 2000; Weiskittel et al. 2007a; Hein et al. 2008a, b) for some environments and silvicultural regimes in other countries, but it was considered important to develop models that are specific to New Zealand conditions (i.e., site and silvicultural regimes).

Decisions made during model development will affect the model's 'domain of applicability'. Three such decisions are: (1) which trees to sample; (2) which branches to model; and (3) what section of the tree crown or stem to consider. Tree selection criteria have included selection of trees across the DBH distribution (Maguire et al. 1994); selection of dominant and co-dominant trees (Hein et al. 2008a; Newton et al. 2012); and avoidance of trees with visible damage (Roeh and Maguire 1997; Hein et al. 2008a; Newton et al. 2012). In terms of branches modelled, nodal branches have been identified and modelled (Maguire et al. 1994, 1999; Roeh and Maguire 1997; Weiskittel et al. 2007a; Hein et al. 2008a, b), whilst Ishii et al. (2000) selected a set of maximum branches from all live and dead branches. The definition of 'crown' varies. For example, Ishii et al. (2000) defined crown depth as tree height minus lowest branch height. Weiskittel et al. (2007a) defined crown base as the tree height to the lowest live branch; whereas Maguire et al. (1999) did not give a precise definition for crown base. Regardless of sampling decisions, most of the models predict that branch diameter increases from the tip of the tree to a maximum value and then decreases again.

The objective of this study was to develop site-specific models to predict how branch diameter varies with distance from the tip of the tree. The models developed were then used to evaluate the influence of initial and post-thinning stand density on branch diameter in unpruned Douglas-fir at two sites in New Zealand, one in the North Island and one in the South Island. Site-specific models were developed as there was no reason to expect that branch diameters would be similar for two different sites. Furthermore, an individual tree-level model was developed rather than a stand-level model because branch diameters on an individual log are important for determining log grade.

Methods

Site location

The two sites selected for this study were the earliest designed Douglas-fir experiments established to examine the effects of different silviculture treatments on growth and yield. The experiments are located in Kaingaroa forest, North Island and Ribbonwood Station, South Island. Hereafter they are referred to as Kaingaroa and Ribbonwood. Environmental conditions for the two sites are given in Table 1. The Kaingaroa experiment was 10.92 ha within a 116 ha block planted in 1982 at 1666 stems ha^{-1}. The Ribbonwood experiment, planted in 1983, consisted of four blocks, each of about 2.5 ha and at different initial stand densities, namely: 833, 1250, 1333 and 2000 stems ha^{-1}.

Overall experimental design

The primary aim of these experiments was to evaluate the effects of final prune height, pruning severity at each lift, and the number of followers[1] on the growth of Douglas-fir. The silvicultural treatments applied reflected the intensive level of silviculture that was considered appropriate at that time for producing high quality sawlogs (James 1990). Unpruned treatments were also included.

Thinning to different stand densities occurred in 1993 when the Kaingaroa stand was 10.9 years old[2], the

Table 1 Site and silvicultural details for the unpruned treatments at Kaingaroa Forest and Ribbonwood Station

	Kaingaroa forest	Ribbonwood station
Latitude	38.5 °S	44.4 °S
Longitude	176.6 °E	169.8 °E
Elevation (m)	450	650
Aspect	NW	Between NE and SE
Slope	3°	5° to 17° average 10°
Mean daily temperature (°C)	11.2	6.8
Annual rainfall (mm)	1255	824
Daily mean windspeed (m s^{-1})	2.2	4.3
Year trial planted	1982	1983
Seedlot	NN 79/6	CY/C/79/11
Stand age (years) when the only thinning occurred	10.9	10.1
Size of PSPs	0.2142 ha in total with measurements confined to inner 0.08 ha	0.147 ha in total with measurements confined to inner 0.08 ha
Number of PSPs from which sample trees were selected	6	15
Site index (mean top height at age 40 years) (m) for selected PSPs	39.0 Range: 38.1 – 40.7	30.8 Range: 30.0 – 40.7

Mean daily temperature, annual rainfall, and daily mean windspeed data are from National Institute of Water and Atmospheric Sciences Ltd. (NIWA) (2012)

Ribbonwood stand was 10.1 years old, and when the base of the green crown was still close to ground level at both sites.

At the same time, Permanent Sample Plots (PSPs) (Hayes and Anderson 2007) were established in each treatment to monitor tree growth in response to the different silvicultural treatments. Their size is given in Table 1.

All PSPs have been measured 14 times between 1993 and 2012 inclusive. At each measurement, tree diameter at 1.4 m (DBH) was measured on all trees. Total height and green crown height were measured for selected trees. Green crown height is defined as the point midway between the lowest green branch and the lowest whorl where the majority of branches are green and is used to represent the base of the green crown. The data collected at each measurement were entered in the PSP database (Hayes and Andersen 2007). Within the database system, the data were analysed to estimate mean top height (the average height of the 100 largest DBH stems ha^{-1}); mean height (the height of the tree with mean DBH) and the height of individual trees. Mean crown height, which is the average green crown height for the trees measured, was also calculated. However, branch diameter is not a routine PSP measurement so a sampling scheme for branch diameter measurements needed to be developed as described later.

Selection of sample PSPs

This study examined branch diameter only in unpruned PSPs since pruning of Douglas-fir in New Zealand is now considered uneconomic.

At Kaingaroa, there were twelve unpruned PSPs, four (two in each of two replicates) at each of the post-thinning stand densities of either 250, 500, or 750 stems ha^{-1}. Six were selected for study as detailed below.

At Ribbonwood, there were four unpruned treatments for each of the initial stand densities[3], an unthinned control and three thinned treatments. There was one PSP for each of the nominal post-thinning stand densities of either 250, 500 or 750 stems ha^{-1}. For the unthinned control treatment, there were four PSPs at an initial stand density of 833 stems ha^{-1}, and five PSPs when the initial stand density was either 1333 or 2000 stems ha^{-1}. This gave a total of 23 PSPs of which 15 were selected for study as detailed below.

Where there was more than one unpruned PSP per treatment, sample trees were selected from each of two PSPs. The two PSPs selected were ones where the stand density remained close to the prescribed initial or post-thinning stand density. This approach was used to minimise variation from the prescribed treatment. At Kaingaroa, an additional selection criterion was that one PSP was selected from each replicate. At Ribbonwood, where the experiment is on a hillside, an additional selection criterion was that one

PSP was near the top and one PSP was near the bottom of the slope. Selected PSP summary data considered relevant for interpreting the branch diameter data are given for Kaingaroa in Table 2, and for Ribbonwood in Table 3 and Fig. 1.

Selection of sample trees

Twelve sample trees were selected from each treatment using the historical PSP data. All 12 trees came from a single PSP when there was only one that was unpruned. Where there was more than one unpruned PSP, six sample trees were selected from each of two PSPs (see above).

Douglas-fir has a higher incidence of stem malformation in New Zealand compared with some other countries (Ledgard et al. 2005), and large branches are often associated with stem malformation. Therefore, the decision was made to avoid trees with obvious malformations because the underlying branching pattern has been disrupted by the malformation, and consequently were not considered representative of 'normal growth'.

The following approach was then used to select trees that consistently had a small, medium or large DBH. At each re-measurement of a PSP, the minimum, mean and maximum DBH were calculated. Two boundaries were then calculated using Equations 1 and 2.

$$B_{low} = DBH_{min} + (DBH_{mean} - DBH_{min}) \times 2.0/3.0 \quad (1)$$

$$B_{high} = DBH_{max} - (DBH_{max} - DBH_{mean}) \times 2.0/3.0 \quad (2)$$

The ith tree was assumed to have a small DBH at a given stand age (re-measurement) if:

$$DBH_{min} \le DBH_i < B_{low} \quad (3)$$

a medium DBH at a given stand age if:

$$B_{low} \le DBH_i \le B_{high} \quad (4)$$

or a large DBH at a given stand age if:

$$B_{high} < DBH_i \le DBH_{\max} \quad (5)$$

Where

DBH_{min} is the minimum DBH for the PSP at a given stand age

DBH_{mean} is the mean DBH for the PSP at a given stand age

DBH_{max} is the maximum DBH for the PSP at a given stand age

B_{high} and B_{low} are the two boundaries at a given stand age

Sample trees were then selected based on the following three criteria:

1. all, or if not possible, the majority of DBH measurements for a given tree fell into the same DBH class;
2. trees that had recorded stem malformation (i.e., been assigned a stem description code[4]) were avoided; and
3. sample trees were selected using a random number generator if there were more than the required number of sample trees in a DBH class.

Branch measurement

As these are long-term experiments, it was not feasible to fell trees to measure branch diameter. Branching data were collected using TreeD, a photogrammetrically based imaging methodology for capturing tree and branch dimensions from high-resolution digital images of standing trees (Brownlie et al. 2007). This methodology was selected because it provides a permanent record of the tree for future reference, and additionally avoids the health and safety risks associated with tree climbing. Data were collected in January to February 2013 when the Kaingaroa stand was 30.7-30.8 years old and the Ribbonwood stand was 29.7-29.8 years old.

Each pre-selected sample tree was located in the field and examined to determine the optimal direction to get the clearest view of the stem and branches. Generally

Table 2 Height, crown height and DBH data for the selected PSPs at Kaingaroa Forest

Nominal initial stand density (stems ha^{-1})	Nominal post-thinning stand density (stems ha^{-1})	Actual post-thinning stand density (stems ha^{-1})	Actual stand density (stems ha^{-1}) at age 29 years	Mean green crown height (m) at time of thinning (10.9 years)	Mean tree height (m) at time of thinning (10.9 years)	Mean green crown height (m) at age 29 years	Mean tree height (m) at age 29 years	Mean DBH (cm) at age 29 years
1666	250	250	250	0.8	10.2	10.6	26.9	46.4
1666	250	250	250	1.6	10.1	7.9	26.6	43.9
1666	500	500	500	1.2	9.9	14.6	25.2	38.3
1666	500	500	500	1.7	10.1	15.8	27.0	35.3
1666	750	738	725	1.3	10.2	14.1	28.2	30.4
1666	750	750	750	1.4	9.8	14.4	26.6	32.2

Table 3 Height, crown height, and DBH data for the selected PSPs at Ribbonwood Station

Nominal initial stand density (stems ha^{-1})	Nominal post-thinning stand density (stems ha^{-1})	Actual post-thinning stand density (stems ha^{-1})	Actual stand density (stems ha^{-1}) at age 29 years	Mean green crown height (m) at 12 years	Mean tree height (m) at time of thinning (10.1 years)	Mean green crown height (m) at age 29 years	Mean tree height (m) at age 29 years	Mean DBH (cm) at age 29 years
833	250	250	250	0.1	7.3	8.4	25.2	51.4
1333	250	250	250	0.1	7.0	6.8	24.7	46.9
2000	250	250	250	0.1	8.3	8.3	25.4	47.7
833	500	500	500	0.1	6.7	10.5	24.5	41.4
1333	500	500	500	0.1	7.3	11.6	24.9	42.5
2000	500	500	500	0.1	7.5	12.0	24.8	38.9
833	750	738	738	0.4	7.0	12.6	22.8	34.1
1333	750	750	750	0.2	8.3	12.2	24.1	34.8
2000	750	750	750	0.1	8.2	14.0	24.8	35.8
833	-	788	750	0.2	6.7	12.2	24.4	35.9
833	-	813	775	0.4	6.9	12.4	22.9	33.7
1333	-	1113	1025	0.2	6.5	12.7	23.5	30.8
1333	-	1263	1038	0.4	6.5	12.8	23.1	28.5
2000	-	1838	1688	1.4	7.4	15.5	23.0	26.4
2000	-	1950	1800	1.2	7.1	14.2	22.1	25.3

The earliest green crown measurements in this experiment were at 12 years

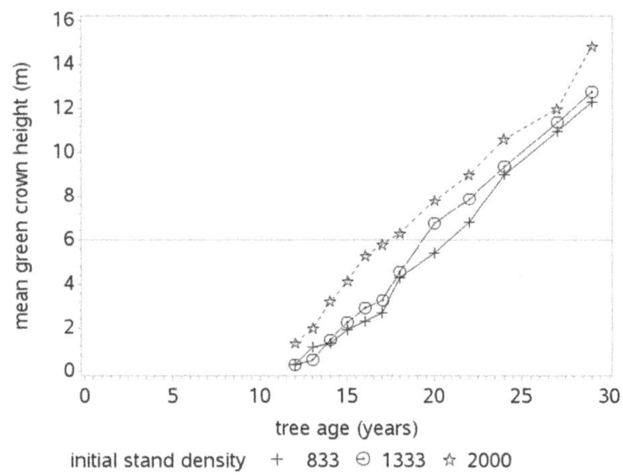

Fig. 1 Mean green crown height at Ribbonwood for the unthinned treatments with initial stand densities of 833, 1333 or 2000 stems ha^{-1}

the crowns were symmetrical but in some instances crown asymmetry was apparent; for example due to sloping ground and/or gaps in the canopy. In these cases, the camera position was selected to ensure that the stem and the larger branches were clearly visible in the image. Dead branches were pruned to enable a scaling pole to hang vertically close to the stem. From the chosen sample point, two images of the stem were taken with approximately 1 m separation. The 1 m separation enables the tree image to be viewed in stereo. The horizontal distance from the camera position to the stem was measured and the lean of the stem, if any, towards or away from camera was measured (see Brownlie et al. 2007 for further details on TreeD methodology). Four variables were extracted from the images: height to base and to top of a branch cluster; diameter of the largest visible branch in the cluster; and stem diameter below the branch cluster. Whilst all branches are not visible in the image, the likelihood of a large branch being missed is minimised through the selection of an appropriate camera position.

The data extracted from the images were screened to determine if there were any 'extra-large' branches. For each tree the ratio of maximum branch diameter measured to the mean of measured branch diameters was calculated. For those trees where the ratio was greater than two, the images were re-examined to determine whether the branch with the maximum diameter was part of the regular branching pattern or whether it might have grown in response to stem damage such as a broken top. Trees considered to have stem damage were excluded from further analysis as such trees were not considered representative of 'normal growth'.

Model development

Selected PSP and TreeD data were combined to form the dataset used to develop site-specific models. Decisions were made as to which branches to include, which section of stem to include and the mathematical equations to consider.

Tree data (from PSP system)

Tree diameter, total height and green crown height were measured during routine monitoring of the overall experiments when the trees were aged 29 years. These data were retrieved from the PSP system.

Branch diameter data (from TreeD)

As it is the larger branches that will influence log quality, it was considered appropriate to develop a model that would predict the diameter of the larger branches, rather than the average branch diameter.

Initially a subset of TreeD images were examined to determine whether or not it would be feasible to identify nodal branches (larger branches), but it was concluded that any such classification would be unreliable.

To select larger branches, the stem of each tree was divided into zones of equal length based on the estimated individual tree height at age 29 years as calculated within the PSP system. Twenty zones were selected. This was a compromise between having sufficient data points for model development and minimising the likelihood of a zone not containing a nodal cluster as measured annual height data were not available for each tree. The largest branch diameter in each zone was determined. The height assigned to the branch diameter was the height of the branch or the mean height of the branches if there were two or more branches with the same diameter.

The TreeD data were collected up to 2 years later than the PSP data (see previous section) but this timing difference was considered acceptable on the basis that minimal branch diameter growth on the section of stem measured would have occurred during the intervening period. For the PSPs with post-thinning stand densities of 500 and 750 stems ha^{-1}, the majority of the measured branches were below the mean green crown height (Fig. 2), so branch diameter growth would have been negligible. For the PSPs with a post-thinning stand density of 250 stems ha^{-1}, approximately half the branches measured were above the mean green crown height (Fig. 2). Comparing the maximum height to which branches were measured using TreeD with the PSP mean height data, it was estimated that the branches measured would be 11 years of age or older, and would occur in the lower half of the green crown. The diameters of branches used for model development (Fig. 2) were generally over 40 mm. By examining figures in Weiskittel et al. (2007b), it was estimated that annual branch diameter increment was likely to have been about 1 mm per year, well within the accuracy of TreeD measurements (Brownlie et al. 2007).

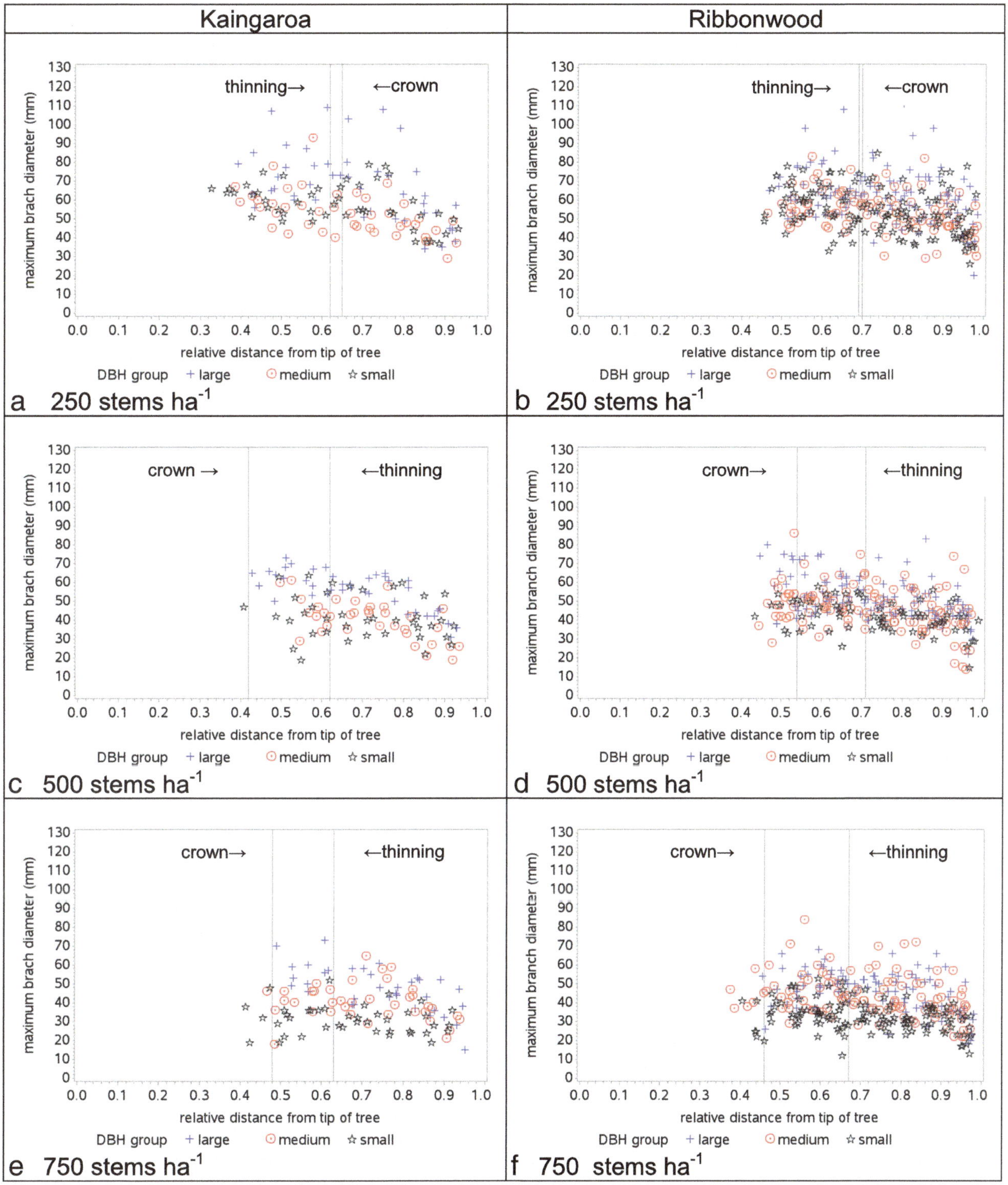

Fig. 2 Maximum branch diameter in a stem zone versus relative distance from the tip of the tree for the Kaingaroa (**a**, **c**, **e**) and Ribbonwood (**b**, **d**, **f**) experiments at nominal post-thinning stand densities of 250 stems ha^{-1} (**a**, **b**), 500 stems ha^{-1} (**c**, **d**) and 750 stems ha^{-1} (**e**, **f**). The vertical lines represent relative distance from tip of tree to mean green crown height at age 29 years (crown), and to mean height at time of thinning (thinning)

Stem section to be included in model

Two options were considered. One option was to predict branch diameter as a function of height above ground. A model developed on this basis is open to extrapolation beyond the range of heights measured. The second option was to assume branch diameter was zero at the tip

Table 4 Model fit statistics from fitting variants of Equation 7 to the Kaingaroa dataset

Model fitted	Mean residual (mm)	Standard deviation	Range of residuals (mm)	Error mean square
1. Equation 7: individual trees	−0.050	8.49	−31.0 to 24.6	
2. Equation 7, 8, and 9: all trees	−0.040	10.61	−26.8 to 40.0	113.5
3. Equation 7 and 8: all trees	−0.025	10.63	−27.0 to 37.6	113.7

There were 364 individual observations in the dataset

of the tree and predict branch diameter as a function of relative distance from the tip of the tree using a curvilinear equation that would predict a maximum branch diameter at some point along the stem as observed in other studies (Maguire et al. 1994, 1999; Roeh and Maguire 1997; Ishii et al. 2000; Weiskittel et al. 2007a; Hein et al. 2008a, b).

Even though there were few measurements from the upper third of the stem, the second option was preferred because relative height is constrained to be between 0 and 1.

The model profile was developed for the whole stem length because dead branches below the base of the green crown also influence log and timber value.

Model selection

Model selection was an iterative process with the results from the previous iteration influencing the methodology in the next iteration. Ratkowsky (1990) proposed that models should be as simple as possible. Based on this thinking, nonlinear regression functions were examined to determine an equation that would: enable branch diameter to be predicted as a function of relative distance from the tip of the tree; provide a good fit to the data; and be logical from a biological perspective.

For the first iteration, the same methodology was applied to both sites. Functions were fitted to the data for each tree individually, and the relationship between the predicted coefficients and the independent tree-level variable, 'growing space' (Equation 6) examined.

$$S_{ij} = \sqrt{\frac{10000}{SPH_j} \times \frac{DBH_{ij}}{\overline{DBH_j}}} \tag{6}$$

Where:

S_{ij} is the 'growing space' for tree i in plot j in m

DBH_{ij} is the current (stand age 29 years) DBH for tree i in plot j in cm

$\overline{DBH_j}$ is the current plot mean DBH (stand age 29 years) for plot j in cm

SPH_j is the current post-thinning stand density (stems ha^{-1}) for plot j

'Growing space' was selected as a 'tree-level' variable to be included in the model, as branch diameter is likely to be larger where there is a canopy gap (Maclaren 2009), suggesting that on average, branch diameter is likely to be related to the 'growing space' available to the tree.

A non-linear function which was appropriate on an individual tree basis for both sites and for which the predicted coefficients c_1 and c_2 were related to 'growing space' was:

$$BD = c_1 \times DIST_R \times e^{(-c_2 \times DIST_R)} \tag{7}$$

Where:

BD is the branch diameter

$DIST_R$ is the relative distance from the tip of the tree (with zero being the tip of the tree and one being ground level)

c_1 and c_2 are predicted coefficients

Kaingaroa

When Equation 7 (Model 1) was fitted to each tree individually, the overall mean residual was −0.050 mm (Table 4), and 97 % of residuals were in the range ± 2 cm (Table 5). The coefficients, c_1 and c_2, increased linearly with increasing 'growing space'. In the second iteration, Model 2 (Equation 7 with c_1 and c_2, replaced by Equation 8 and 9 respectively) was then fitted to the whole dataset.

$$c_1 = b_1 + b_2 \times S \tag{8}$$

$$c_2 = b_3 + b_4 \times S \tag{9}$$

Where

S is the growing space for the tree in m

b_1, b_2, b_3, b_4 are the predicted coefficients

Table 5 Distribution of residuals for models fitted at individual tree level, and selected site-specific model

	Kaingaroa Model 1 (individual trees)	Kaingaroa Model 3
Number of observations	364	364
Percentage of residuals in range −1.0 to 1.0 cm	78 %	68 %
Percentage of residuals in range −2.0 to 2.0 cm	97 %	93 %

Table 6 Predicted model coefficients and approximate standard errors for most appropriate model for Kaingaroa

Coefficient	Kaingaroa – Model 3
b_1	137.9 ± 13.8
b_2	40.47 ± 2.79
c_2	2.24 ± 0.07

When Model 2 was fitted, the mean residual was -0.040 mm (Table 4). As the coefficient, b_4, was not statistically different from zero, Model 3 (Equation 7 with c_1 replaced by Equation 8) was fitted in the third iteration. For Model 3, the mean residual was -0.025 mm (Table 4).

The residuals from fitting Model 3 (Table 4) were not correlated with predicted values, relative distance from the tip of the tree, individual tree 'growing space', or nominal post-thinning stand density. Consequently Model 3 was selected as the site-specific model for Kaingaroa. This model predicted 68 % of branch diameters to ≤ 1 cm, and 93 % of residuals to ≤2 cm (Table 5). The predicted coefficients are shown in Table 6.

Ribbonwood

When Equation 7 was fitted to each tree individually the overall mean residual was −0.019 mm (Table 7), and 98 % of residuals were in the range ± 2 cm (Table 8). The coefficient, c_1, increased linearly with increasing 'growing space', but c_2 was not correlated with 'growing space', nominal initial stand density, nominal post-thinning stand density, or the difference between nominal initial and post-thinning stand densities. For the second iteration, Model 3 (Equation 7 with c_1 replaced by Equation 8) was then fitted to the whole dataset. The residuals from fitting this model were correlated with initial, and nominal change in stand density due to the thinning. For the third iteration Model 4 (Equation 7 with c_1 replaced by Equation 8 and c_2 replaced by Equation 10) was then fitted to the whole dataset.

$$c_2 = a_3 + a_4 \times (SPH_{init} - SPH_{fin}) \qquad (10)$$

Where:
SPH_{init} is the nominal initial stand density (stems ha^{-1})
SPH_{fin} is the nominal post-thinning stand density (stems ha^{-1}) in the case of thinned PSPs or the stand density at age 29 year for unthinned PSPs (See Table 3)
a_3, a_4 are model coefficients.

Using Model 4, the residuals were not correlated with predicted values, relative distance from the tip of the tree, 'growing space', initial stand density, nominal post-thinning stand density or change in stand density as defined above. Consequently, Model 4 was selected as the site-specific model for Ribbonwood. This model predicted 73 % of branch diameters to ≤ 1 cm, and 96 % of residuals to ≤2 cm (Table 8). The predicted coefficients are shown in Table 9.

Results and discussion

The maximum height to which it was practical to measure branch diameter from the TreeD images was 17.7 m at Kaingaroa, and 15.4 m at Ribbonwood. As these heights are less than 20 m, branches are anticipated to have been measured to within ±1 cm (Brownlie et al. 2007). That few branch diameters were measured on the upper third of the stem (Fig. 2) is not considered critical in terms of model development, as most of the stand value comes from the lower part of the stem. For example, when radiata pine is grown for structural timber without pruning, 80 % of the stand value falls within the first 19 m of tree height (New Zealand Forest Owners Association 2013).

None of the sample trees selected at Kaingaroa were excluded from the dataset due to 'extra-large' branches. At Ribbonwood, there were 12 trees that were excluded from the data set as they were considered to have incidences of stem damage. One or two trees were affected in each of nine different treatments. There were two other trees with an 'extra-large' branch, but these were left in the data set as it was not obvious that the branches resulted from stem damage.

There was a large variation in measured branch diameter at a given relative distance from the tip of the tree at a given nominal post-thinning stand density even with trees with obvious stem damage excluded (Fig. 2). While branch diameter tended to be larger for the trees from the large DBH group, and smaller for trees from the small DBH group, there was a degree of overlap between the three DBH groups. Visually, branch diameter was

Table 7 Model fit statistics from fitting variants of Equation 7 to the Ribbonwood dataset

Model fitted	Mean residual	Standard deviation	Range of residuals	Error mean square
1. Equation 7: individual trees	−0.019	7.44	−22.5 to 44.4	
3. Equation 7, and 8: all trees	−0.016	10.26	−32.8 to 76.7	105.3
4. Equation 7, 8 and 10: all trees	−0.013	9.96	−32.4 to 75.5	99.5

Note: there were 1286 data points in the Ribbonwood dataset

Table 8 Distribution of residuals for models fitted at individual tree level, and selected site-specific model

	Ribbonwood Model 1 (individual trees)	Ribbonwood Model 4
Number of observations	1286	1286
Percentage of residuals in range −1.0 to 1.0 cm	83 %	73 %
Percentage of residuals in range −2.0 to 2.0 cm	98 %	96 %

similar for trees with a similar growing space at the two sites (Fig. 3). Of interest is the difference in branch form. At Kaingaroa, branches generally grew upwards, whereas at Ribbonwood branches often curved downwards. The exact reason for this difference is not known but there are several factors that may contribute. These include: different seedlots (Table 1); the fact that more cones are produced at Ribbonwood adding more weight to branches (C. Low, Scion, pers. comm.); and /or that Ribbonwood often receives snow in winter, whereas snow is a rare event in Kaingaroa. The site-specific models also indicated that branch diameters are similar at the two sites (compare Fig. 4a with Fig. 4b), but with slightly larger branch diameters at the base of the stem at Ribbonwood.

At Kaingaroa, all PSPs were planted at the same initial stand density and the only thinning occurred when the green crown height was still close to ground level (Table 2). Therefore, current 'growing space' was sufficient to explain the difference in branch diameter profile between individual trees. The predicted branch diameter profile for a tree of average DBH at various stand densities using Model 3 (Table 6) are shown in Fig. 4a. The profile for a stand density of 1666 stems ha^{-1} represents an unthinned stand. The fact that the predicted branch diameter profile for a tree of average DBH in a stand decreases with increasing post-thinning stand density (i.e., 250 >, 500 > 750 stems ha^{-1}) (Fig. 4a) and are all higher than that for the initial stand density of 1666 stems ha^{-1} indicates that branches at the base of the tree have been influenced by the severity of the thinning. This result is logical given that the green crown height was close to ground level (Table 2) when the thinning occurred.

Table 9 Predicted model coefficients and approximate standard errors for most appropriate model for Ribbonwood

Coefficient	Ribbonwood – Model 4
b_1	67.96 ± 6.23
b_2	45.38 ± 1.87
a_3	1.85 ± 0.04
a_4	0.00015 ± 0.00002

Fig. 3 TreeD image for two trees with a similar 'growing space'. Left image, tree from Kaingaroa with a 'growing space' of 4.4 m. Right image, tree from Ribbonwood with a 'growing space' of 4.5 m

At Ribbonwood, where PSPs were planted at a range of initial stand densities, current 'growing space' was not sufficient to explain the difference in branch diameter profile between individual trees. It was necessary to include an additional term in the model to account for the difference between initial and current stand density. The predicted branch diameter profile for a tree of average DBH in a stand with an initial density of 1666 stems ha^{-1} and post-thinning stand densities of 250, 500, 750 and 1666 stem ha^{-1} using Model 4 (Table 9) is shown in Fig. 4b. The predicted profiles for a stand at the post-thinning stand densities of 250, 500 and 750 stems ha^{-1} are higher than the profile for an unthinned stand at 1666 stems ha^{-1}, which indicates that branches at the base of the tree have been influenced by the severity of the thinning. This result is logical given that the green crown height was close to ground level at the time of thinning (Table 3). Even though the green crown height was close to ground level at the time of thinning, initial stand density has influenced the branch diameter profile in the lower two-thirds of the stem (Fig. 5a-c), but not in the upper part of the stem. The upper part of the crown would have been formed at the post-thinning stand

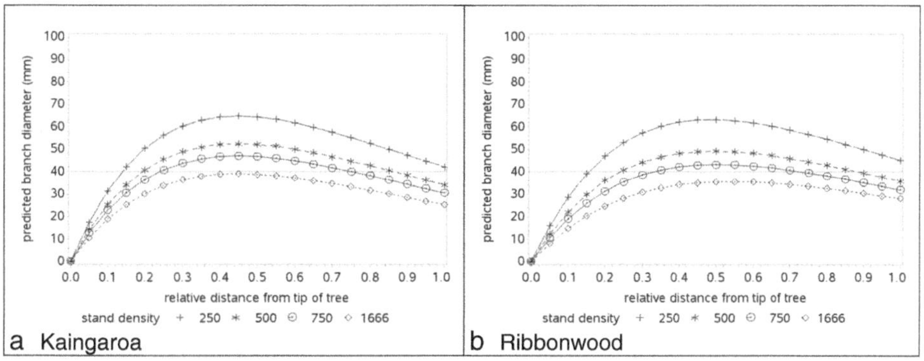

Fig. 4 The predicted crown profile for a tree of average DBH in a stand at Kaingaroa (**a**) where the initial stand density was 1666 stems ha^{-1} and the stand density after any thinning was 250, 500, 750 or 1666 stems ha^{-1}. For Ribbonwood (**b**), an initial stand density was 1666 stems ha^{-1} was used for comparative purposes and the stand density after any thinning was 250, 500, 750 or 1666 stems ha^{-1}

density, but considering the position of mean tree height at the time of thinning (Fig. 2b, d, f), it appears that the initial stand density may have had an impact on branch diameter development for branches formed after thinning.

For a tree of average DBH, some branches are predicted to be larger than the 40 mm maximum suggested by Ledgard et al. (2005) at all three post-thinning stand densities, namely 250, 500 and 750 stems ha^{-1} at both Kaingaroa (Fig. 4a) and Ribbonwood (Fig. 5a-c). Larger branch diameters were predicted at lower post-thinning stand densities. This suggests that post-thinning stand density of at least 750 stems ha^{-1} would be desirable to reduce the occurrence of branches with a diameter larger than 40 mm.

The predicted branch profiles, at a stand age of 29 years, for a tree of average DBH in unthinned stands with different initial stand densities (Fig. 5d) indicate that an initial stand density of at least 1333 stems ha^{-1} would be necessary to ensure branches in the lower stem are below 40 mm.

If carrying out a single thinning operation, a method for controlling branch diameter in the lower stem would

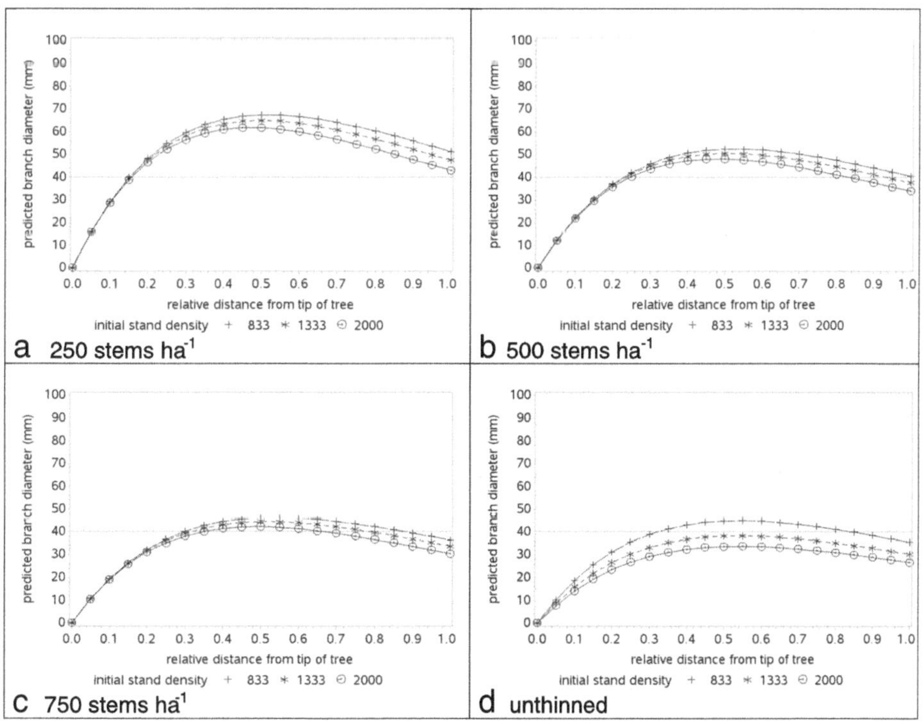

Fig. 5 The predicted crown profiles for a tree of average DBH in a Ribbonwood stand with an initial stand density of 833, 1333 or 2000 stems ha^{-1} and a nominal post-thinning stand density of 250 stems ha^{-1} (**a**), 500 stems ha^{-1} (**b**), 750 stems ha^{-1} (**c**) and unthinned (**d**)

be to delay the operation to an age where branches in the lower crown would be unlikely to take advantage of the increase in 'growing space'. For the unthinned treatments at Ribbonwood, the variation in green crown height with age depends on the initial stand density (Fig. 1). If the lowest log was cut from between stump height and 6 m, then thinning would need to be delayed until a stand age of approximately 17 years, 19 years or 21 years when the initial stand density was 2000, 1333, or 833 stems ha^{-1} respectively. This would ensure that the green crown height was at the top of the log at the time of thinning (Fig. 1).

Neither the timing nor severity of thinning should be based on predictions of branch diameter alone. Deciding on such variables will need to be a compromise between tree growth and branch growth as well as minimising the risk of wind damage since stem damage leads to larger than expected branches. A different approach to that taken in this study would be appropriate if one wished to predict stand value and/or create links to sawing simulators.

Conclusions

The TreeD methodology proved useful for collecting branch diameter data. The data collected from two different sites in New Zealand indicate that both initial and post-thinning stand density influence branch diameter for Douglas-fir. Branch diameters at the base of the tree were affected by the thinning, which occurred between 10 and 11 years of age. At post-thinning stand densities of less than or equal to 750 stems ha^{-1}, some branches on a tree of average DBH will be greater than 40 mm.

Endnotes

[1]A follower is a tree that is not pruned as severely as the 'crop' trees and is usually removed prior to stand maturity.

[2]In New Zealand forestry, a stand age of zero is assigned at the time of planting.

[3]For this study, the block at 1250 stems ha^{-1} was ignored as 1250 stems ha^{-1} is similar to 1333 stems ha^{-1}.

[4]Tree features such as a broken top may be noted using a stem description code.

Competing interests

The authors declare that they have no competing interests.

Authors' contributions

SK initiated the study, selected the experiments to use and reviewed the paper. RB collected and analysed the TreeD images and reviewed the paper. JG selected the sample trees, analysed the data and wrote the paper. All authors read and approved the final manuscript.

Acknowledgements

The authors wish to thank P. Hodgkiss, Scion for help with the TreeD field data collection; C. Low and J. Moore, Scion for providing helpful comments on the draft manuscript; and M. Heaphy, Scion for providing the environmental data. This research was supported by Future Forests Research Ltd.

Author details

[1]Scion, Private Bag, 3020 Rotorua, New Zealand. [2]Coillte, Kilmacurra Park, Kilbride, County Wicklow, Ireland.

References

Brownlie, RK, Carson, WW, Firth, JG, & Goulding, CJ. (2007). Image-based dendrometry system for standing trees. *New Zealand Journal of Forestry Science, 37*(2), 153–168.

Hayes, JD, & Andersen, C. (2007). The Scion permanent sample plot (PSP) database system. *New Zealand Journal of Forestry, 52*(1), 31–33.

Hein, S, Weiskittel, AR, & Kohnle, U. (2008a). Effect of wide spacing on tree growth, branch and sapwood properties of young Douglas-fir [*Pseudotsuga menziesii* (Mirb.) Franco] in south-western Germany. *European Journal of Forest Research, 127*(6), 481–493.

Hein, S, Weiskittel, AR, & Kohnle, U. (2008b). Branch characteristics of widely spaced Douglas-fir in south-western Germany: Comparisons of modelling approaches and geographic regions. *Forest Ecology and Management, 256*(5), 1064–1079.

Ishii, H, Clement, JP, & Shaw, DC. (2000). Branch growth and crown form in old coastal Douglas-fir. *Forest Ecology and Management, 131*(1–3), 81–91.

James, RN. (1990). Evolution of silvicultural practice towards wide spacing and heavy thinning in New Zealand. In RN James & GL Tarlton (Eds.), *New approaches to spacing and thinning in plantation forestry. Proceedings of a IUFRO symposium held at the Forest Research Institute, Rotorua, New Zealand, 10–14 April, 1989 (FRI Bulletin 151; Pp 13–20).* Rotorua, New Zealand: IUFRO, Forest Research Institute, N.Z. Ministry of Forestry.

Jensen, EC, & Long, JN. (1983). Crown structure of a codominant Douglas-fir. *Canadian Journal of Forest Research, 13*, 264–269.

Larson, PR. (1962). A biological approach to wood quality. *Tappi, 45*(6), 443–8.

Larson, PR. (1969). *Wood formation and the concept of wood quality. (School of Forestry Bulletin No. 74).* New Haven, CT, USA: Yale University.

Ledgard, N, Knowles, L, & de la Mare, P. (2005). Douglas-fir - the current New Zealand scene. *New Zealand Journal of Forestry, 50*(1), 13–16.

Maclaren, JP. (2009). *Douglas-fir manual. (FRI Bulletin No. 237).* Rotorua, New Zealand: Forest Research Institute.

Maguire, DA, Johnston, SR, & Cahill, J. (1999). Predicting branch diameters on second-growth Douglas-fir from tree-level descriptors. *Canadian Journal of Forest Research, 29*(12), 1829–1840.

Maguire, DA, Moeur, M, & Bennett, WS. (1994). Models for describing basal diameter and vertical distribution of primary branches in young Douglas-fir. *Forest Ecology and Management, 63*(1), 23–55.

Megraw, RA. (1985). *Wood quality factors in loblolly pine.* Atlanta, GA, USA: Tappi Press.

Miller, JT, & Knowles, FB. (1994). *Introduced forest trees in New Zealand: Recognition, Role and Seed Source. 14. Douglas-fir. (FRI Bulletin No. 124 Part 14).* Rotorua, New Zealand: Forest Research Institute.

Ministry for Primary Industries (MPI). (2013). *National exotic forest description as at 1 April 2013.* Wellington, New Zealand: Ministry for Primary Industry.

National Institute of Water and Atmospheric Science Ltd. (NIWA). (2012). *30-year climate normal period 1981–2010.* Wellington, New Zealand: National Institute of Water and Atmospheric Science Ltd.

Newton, M, Lachenbruch, B, Robbins, JM, & Cole, EC. (2012). Branch diameter and longevity linked to plantation spacing and rectangularity in young Douglas-fir. *Forest Ecology and Management, 266*, 75–82.

New Zealand Forest Owners Association. (2013). *New Zealand plantation forest industry, facts and figures 2012/13.* Wellington, New Zealand: New Zealand Forest Owners Association Inc.

Phillips, GE, Bodig, J, & Goodman, JR. (1981). Flow-Grain Analogy. *Wood Science, 14*(2), 55–64.

Ratkowsky, DA. (1990). *Handbook of nonlinear regression models.* New York, USA: Marcel Dekker.

Roeh, RL, & Maguire, DA. (1997). Crown profile models based on branch attributes in coastal Douglas-fir. *Forest Ecology and Management, 96*(1–2), 77–100.

Tustin, JR, & Wilcox, MD. (1978). *The relative importance of branch size and wood density to the quality of Douglas fir framing timber. (Symposium No. 15, 267–272).* New Zealand: New Zealand Forest Service, Forest Research Institute.

Weiskittel, AR, Maguire, DA, Monserud, RA, Rose, R, & Turnblom, EC. (2006). Intensive management influence on Douglas fir stem form, branch characteristics, and simulated product recovery. *New Zealand Journal of Forestry Science, 36*(2–3), 293–312.

Weiskittel, AR, Maguire, DA, & Monserud, RA. (2007a). Modeling crown structural responses to competing vegetation control, thinning, fertilization, and Swiss needle cast in coastal Douglas-fir of the Pacific Northwest, USA. *Forest Ecology and Management, 245*(1–3), 96–109.

Weiskittel, AR, Maguire, DA, & Monserud, RA. (2007b). Response of branch growth and mortality to silvicultural treatments in coastal Douglas-fir plantations: Implications for predicting tree growth. *Forest Ecology and Management, 251*(3), 182–194.

Whitehead, D, Grace, JC, & Godfrey, MJS. (1990). Architectural distribution of foliage in individual *Pinus radiata* D.Don crowns and the effects of clumping on radiation interception. *Tree Physiology, 7*, 135–155.

4

Use of UAV or remotely piloted aircraft and forward-looking infrared in forest, rural and wildland fire management: evaluation using simple economic analysis

B. R. Christensen

Abstract

Background: Investment in emerging technologies may contribute to a reduction in the suppression costs of wildfires, and is thus worth careful consideration and trialling by researchers and managers. This investigation looked at the potential incorporation of a newly emerging remote sensing technology, remotely piloted aircraft and forward-looking infrared investigated using a cost-benefit analysis approach.

Methods: An online survey-based approach targeting subject matter experts in wildland fire management and unmanned aircraft was used to evaluate the percentage change in the effectiveness of five different management scenarios using remotely piloted aircraft and forward-looking infrared. The most commonly accepted economic model in wildland fire management was adapted to determine the net value change between the five scenarios.

Results: The benefits of incorporated unmanned aerial vehicles (UAV) or remotely piloted aircraft and forward-looking infrared were measured as the reduction in cost-opportunity of helicopter use, and were estimated as an average of \$548 per fire (where the average cost of fires was \$1767) or a 31 % cost saving on total suppression costs. The return on investment of such technology was estimated at 24 fires, whereas the return of investment of the existing thermal camera/helicopter use occurred after 160 fires.

Conclusions: The incorporation of remotely piloted aircraft and forward-looking infrared if implemented and managed appropriately, could well improve the cost effectiveness of the current forest, rural and wildland fire fighting efforts. The inherent uncertainty on such unevaluated technology (the combination of remotely piloted aircraft and forward-looking infrared) was addressed by adding stochastic variability and a triangular probability distribution approximation. Key management issues and recommendations are identified such as; greater use of current thermal camera equipment where possible, taking a project management based approach for further testing of remotely piloted aircraft and other emerging technologies, and the improved financial recording and reporting of fire management efforts.

Keywords: Emerging technology; Forest fire; Wildland fire; Rural fire; Remotely piloted aircraft (RPA); "Drones"; Forward Looking Infrared (FLIR); Cost benefit analysis

Correspondence: bchristensen@doc.govt.nz
(Threats)–Conservation-Related Fire Research, Department of Conservation, 99 Sala Street, Rotorua, New Zealand

Background

Use of remotely piloted aircraft (RPA[1]), also known as unmanned aerial vehicles (UAV)/radio-controlled aircraft, 'drones' or semi/fully autonomous vehicles, is a rapidly developing area that lends itself to wildland rural fire management (Rango et al. 2006; Kumar et al. 2011; Royo et al. 2011; Merino et al. 2012). Royo et al. (2011) went further in stating that such systems provide significant advantages over current aerial means of fire fighting in terms of flexibility, precision, cost, and safety. The quantification of RPA cost and comparisons to established methods such as helicopter use, and modelling the return on investment (including reducing safety risks) for wildland rural fire management is an area that requires investigation. Similar to the growth in information technology, this is an area useful to consider in terms of organisational capability and potential uptake.

Economic theory has played an important role in establishing wildland fire management budgets in the USA (Donovan and Rideout 2003; Ganewatta 2008) and in Australia (Florec et al. 2012). As concern over growing mega-fires and the trend of suppression expenditures has increased in the USA and Australia, so has the use of economic theory to maximise social benefits by optimally allocating resources in wildland fire management (Ganewatta 2008). Laupacis et al. (1992) noted that the ideal time to evaluate the cost-effectiveness of a technology is before its widespread introduction. Such economic analysis has been done previously with the established geospatial technologies for wildfire management, producing a more efficient use of resources results (Hesseln et al. 2010). The industry is now in a similar situation with emerging technologies such as RPAs. The specific research question is as follows:

1. *What is the potential cost-opportunity of using an emerging technology in comparison to the current wildland rural fire management approach used in New Zealand?*

Methods

Specialist (expert elicitation) viewpoints

An expert elicitation method was used to retrieve and synthesise the opinions from members of relevant Linked-in™ networks that would hold a degree of experience and knowledge in RPAs, technology, their systems and wildland fire management. Expert elicitation, a method of coalescing knowledge and combining judgements (Runge et al. 2011) is often used where data is expensive, limited or unreliable, such as environmental health management, nuclear waste production and its management, and emerging energy production technology (Knol et al. 2010; Anadón et al. 2012). A survey request on December 9th 2013 was put to the members of Linked-in™ networks (Wildland Fire Management/Firewise/Fire Adapted Communities, Fire Behaviour Analysts, the Wildland Fire Network, and the Unmanned Aviation Group). It was assumed that the membership of these networks would have the greatest accumulation of expert knowledge in the field of wildland fire management, RPA use etc. As of January 6th 2014 a total of 11 individuals responded to this survey request. They were asked to give their opinion on expected potential changes in forest and rural fire management (see Table 2) in accordance with the technological investment scenarios outlined in Table 1. For example one of the questions was as follows;

- Please consider the suppression reduction from using (or potentially using) thermal or infrared cameras in fire management (for example its use in mop-ups, cold trailing, etc.). What in your considered opinion is the respective percentage cost-benefits occurring from using a Thermal or IR (Infrared) imaging device compared to the total suppression costs? (any number including negative values can also be used). For example:
  1. Highest value (highest percentage cost saving in using thermal or IR imaging devices)
  2. Lowest value (lowest percentage cost saving in using thermal or IR imaging devices)
  3. Most likely value (most likely cost saving in using thermal or IR imaging devices)

Table 1 Summary of scenario parameters

Scenario/parameters	Values
Scenarios:	0–Pre-thermal camera costs, 1–Thermal camera/helicopter costs, 2–Potential RPA use–25 %, 3–Potential RPA use–50 %, 4–Potential RPA use–100 %
Parameters:	Suppression reduction using (or potentially using) thermal or infrared cameras in fire management, Area burned, fire suppression cost, fatalities, injuries
Time:	0–10 years projection
Discount rate:	5 % (with random distribution parameters; high–0.03, low 0.009, most likely–0.01)
Fires/year:	1–20, 25
Numbers of units:	1, 2, 3, 5, 10

Cost-benefit analysis approach

Cost-benefit analysis (CBA) is used to evaluate the known or perceived costs and benefits of a number of scenarios including the *status quo*–or current approach. It is a monetary-based framework that identifies the scenario with the best return on investment, or best Net Present Value (NPV). NPV is the sum of the present value of future cash flows minus the purchase price, and/or management cost for a given scenario (Hyytiäinen and Haight 2010). The CBA approach broadly followed the framework as described by the New Zealand Treasury (The Treasury 2005) and Gould et al. (2013), although for wildland fires, quantifying all of the benefits or reduced losses is impossible (Gould et al. 2013). The available pre-suppression costs (i.e. equipment purchase costs) and known suppression costs were based on Wu et al. (2009) analysis.

C + NVC model

Donovan and Rideout's (2003) *C + NVC* model[2] has become the standard or at least most commonly accepted model in evaluating wildland rural fire management programmes (Ganewatta 2008; Gebert et al. 2008; Florec et al. 2012). Lankoandé (2005) used this model and estimated that every (US) dollar invested in pre-suppression efforts reduces suppression expenditures by 3.76 dollars. The *C + NVC* model is particularly useful for production forestry (including native forest that is available for logging). The *C + NVC* model (as utilised by Wu et al. 2009) was used to evaluate the incorporation of thermal camera and RPA technology into the New Zealand fire management 'toolkit'.

For the purposes of this article, the costs and benefits of using new technology focused on the fire suppression cost component, e.g. equipment cost and use, and the respective opportunity costs. These values were detailed in terms of a current baseline situation, i.e. monitoring of fire management conditions using helicopters and manned-thermal cameras. The specific *C + NVC* model used (see below) follows Wu et al. (2009) as the primary economic evaluation of wildfire management costs, benefits and damages within New Zealand to date. They investigated the wider costs of wildfire management, and specifically compared the net present values of resource outputs and management costs between fires occurring and not occurring. For example:

- *C* is the total sum of the technology (i.e. equipment) costs.
- *NVC* is the total sum and net present value of all specific technology costs associated with wildfires management within New Zealand, and equals the net value change (i.e. opportunity cost) between different scenarios.

Thus the specific model as used in this paper is;

$$NVC = NPVa - NPVb$$

Where
NVC net value change
NPVa net present value for scenario *a*
NPVb net present value for scenario *b*

For the purpose of this analysis, the cost and benefits of thermal cameras, RPA, helicopter use related to wildland fire activities, were compared with the context of the baseline situation (i.e. no thermal camera use). Variable suppression costs are a subset of those identified by Gould et al. (2013), and are as follows:

- fire suppression costs,
- area burned (ha),
- numbers of fatalities,
- number of injuries.

Costs were considered to be number of dollars (NZD at 2013 third quarter value (2013: 3Q)) to purchase thermal cameras, potential costs for RPAs and FLIR systems or miniaturised thermal cameras and helicopter running costs. It was assumed that the FLIR capabilities were effectively equivalent in terms of rural fire management use to that of the current Department of Conservation (DOC) thermal cameras. The baseline was then compared against the cost and benefit of thermal camera use currently either by using a helicopter or on foot, miniaturised FLIR thermal cameras in three different scenarios. It is expected that increased hot spot detection for example will reduce the "mop-up" costs for organisations, claims on the National Rural Fire Authority rural fire fund, and hence cost to the taxpayers of New Zealand. A Monte Carlo simulation method of 1000 repetitions was used to determine a distribution of Net Present Value (NPV) for each scenario.

The fire suppression costs are of key interest in terms of specific case study analysis for this paper. Safety is a key concern in wildland rural fire management, though the quantification of safety costs is subject to significant debate. Sixteen people have died in wildland rural fire management within New Zealand over last 40 years (data from http://www.ruralfirehistory.org.nz/fire_fighters.htm). Of these, four have died in helicopter crashes. The value of human life in the literature is generally noted as the Value of Statistical Life (VoSL). While also subject to significant debate, this does give a monetary value to a human life that can be utilised for economic analysis (MOT 2008). New Zealand's Ministry of Transport has produced a "willingness to pay" value for road safety (injury and death prevention) within the Transport Sector (comprising Road, Aviation and Maritime sectors). This was listed as $2 million in 1991 dollar terms; which equates $3.71 m in 2013: 3Q. Injuries to wildland rural fire fighters, and the details and costs of these have been published, e.g. within the Buckland Crossing Fire burnover 24 March 1998 (Pearce et al. 2004), though up-to-date data is difficult to source. Injuries were identified as 0.4–10 % of VoSL (MOT 2008; Wren and Barrell 2010).

Wildfire management costs

The wildfire management costs were compiled from a variety of sources, chiefly from the annual reports from New Zealand Fire Service (NZFS) and DOC, and DOC financial records. The DOC annual expenditure on fire management averaged around 7 % ($11.4 M) of the total natural heritage expenditure ($164.0 M). The fire-related DOC helicopter costs for the time period 2009/10–2012/13 averaged around 4 % ($367 k) of the total fire control expenditure.

Technological investment scenarios

The investment of new technology has both costs and benefits. In order to examine the potential benefit in dollars for using RPAs for monitoring, a current "status quo" scenario was compared with three other "alternative" scenarios. The status quo described the current use of helicopters and thermal cameras, whereas the alternative scenarios looked at three percentages (25, 50, and 100 %) of potential RPA incorporation into fire management. All scenarios utilized specialist (expert) opinions and cost information sourced from DOC financial information (for "status quo" scenario 1). The thermal camera use over the last 3 years (February 2011 onwards) was sourced from the thermal camera worksheets (DOC unpubl. data). These also gave the number of fires, and amount of time used. This equated to 2.5 h per unit per year at active fire fighting, a total of 22 h for the three units in the last 3 years since they were purchased, or 7.9 h use per year in total). Further cost information was sourced from specific RPA and FLIR websites (http://conservationdrones.org/, http://www.flir.com/NZ/, and http://multiwicopter.com/).

Cost information sourced from DOC financial reports, and RPA-FLIR costs. DOC currently has three thermal cameras (P620, I60 models) with total purchase costs of approx $106 k.

The key assumption is that the FLIR technology is comparable in terms of hotspot detection as the existing thermal camera capabilities. The potential outcomes in using RPA and FLIR technology may include quicker response times, greater aerial water and resources deployment, less repeat "mop-up" work, quicker demoblisation of resources, etc.–all of which reduce suppression expenditure on wildland rural fires.

Modelling of costs

Fire management costs are quite variable over time, with the average percentage expenditure for DOC of fire management costs (e.g. pre-suppression, suppression) is generally 7 % (of the total natural heritage expenditures of DOC for each of the previous six financial years). As some costs occurred over different years (e.g. Thermal imaging camera costs, total fire expenditures, current RPA cost, etc.), in order to reduce inflation-based bias, all costs are expressed in 2013 Q3 figures using the reserve bank inflation calculator (http://www.rbnz.govt.nz/monetary_policy/inflation_calculator/) in order to minimise any inflation ambiguity. A summary of the parameters tested for each scenario is presented in Table 1. As the cost data was not considered to be as detailed as possible (e.g. there could be more appropriate or cheaper RPA-FLIR systems available–such as the recent iphone/FLIR app/add-on (http://www.gizmag.com/flir-iphone-thermal-imaging/30447/ accessed 20/01/2014). unknown actual thermal/FLIR returns, unknown actual discount rates), variability was included into each cost benefit analysis, by running the NPV calculations using a Monte Carlo simulation model (a total of 5000 simulations).

As indicated above, there is general variability on the annual DOC expenditure relating to fire management. As the actual benefits relating to incorporating any new technology is generally unknown, the combination of different annual costs (expenditure) and the indefinite benefits creates a clear degree of uncertainty for an evaluation. This uncertainty was addressed using a stochastic process, where the key cost and benefit values were transformed using a randomisation process (triangular distribution model), and then run under a Monte Carlo simulation model. As the input information was limited, a subjective probability distribution was chosen to approximate the economic costs. A triangular distribution model (following Gould et al. 2013) was chosen as an approximation of the variables relating to the above scenarios. The following issues supported the choice of the triangular distribution model:

1. There is currently limited detailed and assessable data (Garthwaite et al. 2005) on the exact costs of wildfire management in New Zealand,
2. Unknown actual benefits resulting from thermal camera and RPA use,
3. It is a relatively simple elicitation method (Garthwaite et al. 2005), and is intuitive for survey participants (van Dorp and Kotz 2002).

In addition to the above identified issues, the triangular distribution model is often applied in business decision making, risk analysis and project management.

Table 2 Elicitation summary of the experts' average lowest, highest and most likely (bold) mean and range (in italics) estimates (to 1dp) of cost-benefit percentage change in wildland fire management activities for status quo (i.e. helicopters) and potential RPA use in monitoring of forest and rural fires

Parameter	"Status quo", i.e. solely helicopter use			"Potential RPA use"								
	Scenario 1			Scenario 2			Scenario 3			Scenario 4		
	0 % RPA use			25 % RPA use			50 % RPA use			100 % RPA use		
	lv	mlv	hv	lv	mlv	hv	lv	mlv	hv	lv	mlv	hv
Cost–Benefit in terms of area burned	−0.7	**6.3**	14.9	4.7	**8.6**	14.1	10.2	**15.2**	21.0	8.5	**17.4**	26.0
	−10–5	*0–15*	*5–25*	*0–25*	*2–25*	*4–28*	*3–50*	*5–52*	*7–55*	*−20–53*	*−10–62*	*0–67*
Cost–Benefit in terms of fire suppression costs	1.4	**8.0**	14.5	5.9	**12.4**	17.7	11.2	**22.8**	35.8	7.9	**19.8**	29.3
	0–5	*0–15*	*0–25*	*0–25*	*6–25*	*9–28*	*0–50*	*10–52*	*10–52*	*−20–53*	*−10–62*	*0–67*
Cost–Benefit in terms of fatalities	0.1	**1.6**	3.1	2.8	**5.0**	8.3	3.5	**7.7**	12.0	2.5	**7.0**	10.7
	−2–5	*0–5*	*0–7.5*	*0–10*	*1–15*	*2–20*	*0–10*	*1.5–30*	*2–40*	*0–10*	*1–30*	*2–50*
Cost–Benefit in terms of injuries	0.5	**2.2**	5.2	2.0	**4.7**	7.9	4.2	**7.0**	11.8	3.5	**7.1**	10.5
	−2–5	*0–10*	*0–20*	*0–10*	*1–25*	*2–30*	*0–20*	*1.3–30*	*2.5–45*	*0–20*	*1–30*	*2–45*

lv lowest value, *mlv* most likely value, *hv* highest value. Table adapted from Gould et al. (2013)

These methods are often used in evaluations of cost and benefits in incorporating new technology.

A Microsoft Exel™-based approach developed by Hesse (2000) was used to quantify the triangular distribution model from the three inputs for each parameter (see Table 2); Lowest value, Most likely value and Highest value. A cumulative probability curve (frequency distribution) is given in Fig. 1 (adapted from Gould et al. 2013) showing how the Most likely value influences the shape of the distribution relative to the minimum and maximum values. The output of these (means and standard deviations) were then put through a Monte Carlo simulation approach, to give 10,000 values for use in the economic evaluation.

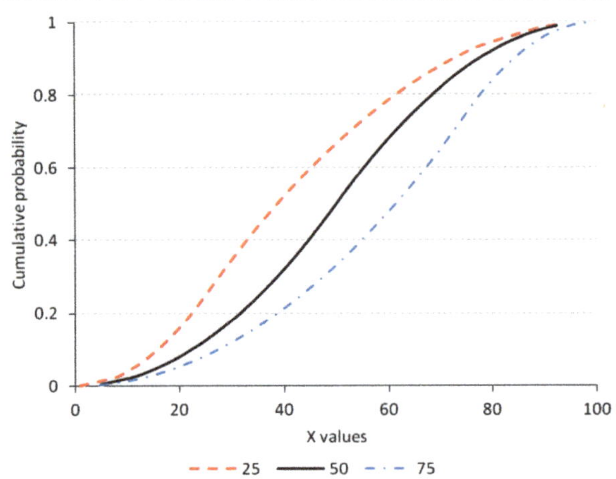

Fig. 1 Triangular distribution of a random variable (*x*). Minimum and maximum values are 1 and 100. Three cumulative frequency distributions are given for the "Most likely" values of 25, 50, and 75 respectively. Adapted from Gould et al. (2013)

Results

Incorporation of new technology into wildfire management

A survey investigated the perceived potential benefits of RPA use. In general, the perceptions from the Fire and RPA specialists were positive, with only few negative values ascribed to the lowest value resulting from RPA use (Table 2). The initial question attempted to identify participants' views on the benefit using current technology, i.e. suppression reduction using (or potentially using) thermal or infrared cameras in fire management. This was used as a baseline for the cost benefit analysis.

Of all scenarios including the current approach or status quo, scenario 3 (50 % incorporation of RPA use into aerial fire monitoring) was perceived by the expert participants' to have the greatest potential overall scenario benefits, and was marginally greater than the benefits perceived for scenario 4 (100 % RPA use). This scenario identified a most likely perceived reduction potential of fire suppression costs of 22.8 %. This was followed by area (extent) burnt (15.2 %), with reduction in fatalities and injuries (while still of greater benefit than all other scenarios) substantially less at 7.1 % and 7.0 % respectively. The total of percentage reductions for each scenario across all parameters (in decreasing order) was as follows; scenario 3 (50 % RPA use)–52.7 %, scenario 4 (100 % RPA use)–51.3 %, scenario 2 (25 % RPA use)–30.7 %, and the scenario 1 (status quo)–17.1 %.

An example of the cumulative probability of the perceived percentage changes for the different scenarios is shown in Fig. 2. In terms of fire suppression costs, the status quo (scenario 1) had a wide distribution of values, hence initially the incorporation of 25 % RPA use was considered to have in approximately 27 % of cases a

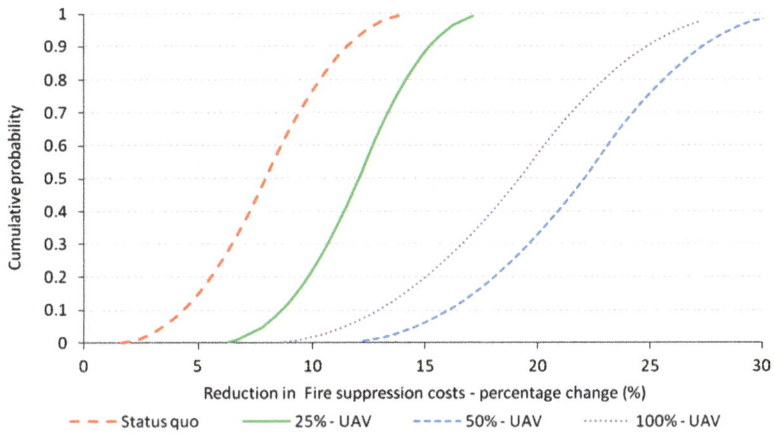

Fig. 2 Cumulative probability distributions for Fire suppression costs as elicited from the combined experts' estimation of percentage change within four scenarios (status quo, 25, 50 or 100 % RPA use) for monitoring wildfires

reduction of the area burnt by at least approximately 5 %, at 50 % RPA use costs could be further reduced by 10.5 %, and at 100 % RPA use such cost could be further reduced by 7 %.

Cost-benefit analysis of the incorporation of new technology for forest and rural fire management

The simulated NPV of the different technologies showed that for wildfire management cost reduction the most cost-beneficial approach was the incorporation of RPA-FLIR. NPV becomes positive for RPAs at a use rate of three fires per year by the 8th year. While the thermal camera use did clearly show a positive effect on reducing suppression costs, the NPV only became positive for thermal camera use, when used for at least 17 fires per year by the 10th year, or; at 18 and 19 fires per year (at year 9), or at 20 fires per year (at year 8). The estimated return on investment of RPA-FLIR was estimated at 24 fires, whereas the return on investment of thermal camera/helicopter use occurred after 160 fires.

Summary of the average difference in the Net Present Values (NPV) for all fire suppression (not counting the initial purchase costs of RPA-FLIR or Thermal cameras) costs at year 1 between the scenarios 2, 3 and 4 from scenario 1 is given in Table 3. This specifically considers the cost saving opportunity of helicopters at fires considering the first stage survey participants' elicitation percentages. The greatest return in terms of fire suppression per hectare comes from the incorporation of scenario 3, the utilisation of RPA in 50 % of all potential fire monitoring. Scenario 4, however, resulted in the greatest overall reduction in costs per fire.

The total potential suppression savings (including mortality and injury costs) in terms of the average of costs/fire (using previous 11 years NZFS data source: NZFS Annual reports) are listed below. As the average fire suppression activities (e.g. salaries, machinery & equipment usage) as per Wu et al. (2009) cost $1796 per fire, the following potential dollar savings were identified;

- Scenario 2–25 % RPA use–7 %,
- Scenario 3–50 % RPA use–23 %, and
- Scenario 4–100 % RPA use–31 %.

These potential savings include faster mop-up of fires using RPA-FLIR use. Even though expert elicitation put had the best percentage change for scenario 3–50 % RPA use, the analysis indicated the cost saving opportunity of helicopter suppression is potentially higher in Scenario 4. While Scenario 4 had the highest return, the participants noted that having people in helicopters for some of the monitoring has a clear benefit, thus the most optimal RPA-FLIR use is somewhere in between these two scenarios, i.e. 50–100 % of all monitoring efforts.

Table 3 Summary of average difference in scenarios from status quo Scenario 1: Net present value at year 1

Research scenario	Cost/fire	Cost/area burnt ha	Mortality and Injury costs/fire	Total savings/fire
Scenario 2	$ 109	$ 172	$ 13	$ 122
Scenario 3	$ 385	$ 406	$ 27	$ 412
Scenario 4	$ 494	$ 368	$ 54	$ 548

Discussion

Financial management issues and concerns are crucial in the incorporation of new technology. Through the use of CBA, and assuming that the imaging capabilities are effectively equivalent between DOC's current thermal cameras and present FLIR technology, the RPA-FLIR platform would offer a far greater return on investment. The cost-opportunity offered by freeing-up helicopter time would transfer a direct average cost saving of up to 31 % in costs for fighting rural and wildland fires. Depending on the availability and implementation of such technology, and assuming that the assumptions made are valid, this could represent an annual saving between \$450,000–\$2,260,000 NZD for rural and wildland fire management in New Zealand. This assumes that both a RPA-FLIR unit and helicopter is available for all fires. While not quantified here, perhaps the most advantageous use of RPA-FLIR is the potential time saved in the support of "mopping-up" tasks at smaller fires where no helicopter support is present.

On average, the return on investment of RPA-FLIR was estimated at 24 fires, whereas the return on investment of thermal camera/helicopter use occurred after 160 fires. RPA-FLIR systems would repay any investment far quicker that the current DOC thermal cameras, and in terms of cost-benefit are clearly worth consideration for further testing by agencies. As the current DOC thermal cameras are sunk costs, and will have an approximate lifetime, it is recommended that these be utilised as often as feasibly possible. While CBA is relatively simple and inexpensive to conduct, it can have unintended consequences if investment decisions are made solely on it alone. More testing and analysis requirements should be established and met prior to any implementation of RPA technology.

The economics of monitoring rural and wildland fires has changed over the last couple of decades. Previous costs for both thermal cameras and aerial monitoring of wildland rural fires costs were in the many tens of thousands of dollars. Both FLIR and RPA technology have brought this down to a few to several thousand dollars just for the initial purchase costs. FLIR technology has recently been linked with mobile phone technology and software applications, with costs currently down to less than \$500. Thus the CBA figures with the best ROI calculated for this paper will soon be "out-of-date", and thus be just a snap-shot in time, or at least represent a conservative estimate of returns. There is a clear need for active cost-return, and ROI analysis for new emerging technologies to see if they merit further testing.

New technology–and its implementation

Change is constant and its speed appears to be increasing, perhaps indicative if not of the decreasing costs of technology then certainly its accessibility (Shanks and Tay 2001; Ederington and McCalman 2008). Groen and Walsh (2013) noted that the science and technology of emerging technologies (such as RPA robotics) are clearly far ahead of any due consideration of their commercial, policy, environmental, ethical and societal implications. Emerging technologies are, by definition, not fully developed, and thus arguably are primarily appropriate for operational agencies only for trialling or testing. While the general focus by rural and wildland fire management is not on emerging technology, it is contended that just as thermal cameras were incorporated into operations, there is a clear need to consider trialling of key emerging technology such as RPAs.

The implementation and roll-out of an updated technology would have additional costs, thus a key recommendation would be to test capabilities in further trial studies, with a small number of units. It is suggested that a standard protocol for testing and the incorporation of new technology is used and incorporated under a project management framework.

Use of remote sensing technology

The current use of the DOC thermal cameras with an average of 2.95 fires per unit per year indicates that on perceived cost reduction alone for wildfire management, it is unlikely that they confer an effective return on their investment if considered in purely monetary terms. The professional rural fire managers indicated that 6.3 % of potential helicopter time is in use for thermal monitoring, compared with 13.7 % for all forms of visual monitoring of fires while using helicopters (Christensen 2015, in submittance). While such values are relatively small, they are a perhaps a useful indicator of the longer-term wildfire management efforts as opposed from the initial attack efforts. Helicopters are a hugely useful aerial platform for wildland rural fire work, though it is now clearly important to consider the potential integration of RPAs into wildland rural fire management.

There is a noted trade-off between pre-suppression and suppression (including 'mop-up') costs (Mercer et al. 2008; Florec et al. 2012), with reductions in one stage likely to be associated with increases in the other (Wu et al. 2009). This paper recognises such a relationship, though targets the implementation in technology that have both an initial reduced cost of investment as well as a released cost-opportunity value. By the addition of RPAs, greater and more cost-effective allocation of key resource capabilities can occur. This specialisation can only be enabled by the implementation of new technology.

Conclusion

RPAs are currently not used for wildland fire fighting in New Zealand, though are being tested by the New

Zealand Fire Service for urban use. Thermal cameras and imaging, however, have been incorporated in wildland fire monitoring from the 1970s (Fulton and Mason 1982), though only established in New Zealand within the previous decade. Plucinski and Pastor (2013) noted that there is an immense value of ortho-rectified aerial IR imagery for the management of wildland fires. The comparison of the cost-effectiveness of aircraft in fire suppression in New Zealand, has shown that improvement has occurred with the introduction of new aircraft (Fogarty and Smart 1996). While the DOC fire-related helicopter costs are relatively small in comparison to the overall fire-related expenditure, RPAs if implemented and managed appropriately, could well improve the cost effectiveness of such fire fighting efforts. While more aerial fire-fighting resources do not necessarily result in faster fire containment (Plucinski et al. 2012), the introduction of specific aerial monitoring by RPAs, e.g. in "mop-up" and "cold-trailing" activities is likely to obviate the need for longer ground operations or repeat visits.

Stephenson et al. (2013) noted the need to develop consistent values for environmental damage done by wildfires. The $C + NVC$ model is primarily used to examine the relationship of prescribed burning practices relative to future suppression costs, and shows that prescribed burning (within the current or recent fire seasons) decreases wildfire intensity, and thus increases the probability of successful suppression (Fernandes and Botelho 2003; Florec et al. 2012). The $C + NVC$ model with the incorporation of Wu et al. (2009) outputs was adapted for use in this paper, it was shown to give a useful approximation of the difference between scenarios, and could be further utilised with specific wildfire management data (i.e. pre-suppression efforts–such as prescribed burning) if and where available. The use of this model by agencies could lead to clearer business decision making and project management approaches, which is likely to have a direct cost saving in this field.

It is recommended that:

- Further testing of RPA and FLIR application in wildland rural fire fighting is warranted. It is also suggested that a project management approach is used in the testing of new technologies. Further research questions arise such as: *is the quality of data useful for real-time monitoring of wildfires*?
- The existing (DOC's and other agencies') thermal cameras are utilised as often as logistically and feasibly possible, to return as much of their investment costs.
- In order that clearer analysis on costs and benefits can occur, improved financial and operational accounting and access to data relating to wildland rural fire fighting efforts is warranted.
- Central capture of all New Zealand wildland rural fire data, including spatial information, costs, and resources used is made available for public use.

Endnotes

[1]The International Civil Aviation Organization (ICAO) uses the terms Remotely Piloted Aircraft (RPA) and Remotely Piloted Aircraft System (RPAS), as does the New Zealand Civil Aviation Authority (CAA).

[2]C = cost, NVC = Net Value Change.

Competing interests
The author declares that he has no competing interests.

Acknowledgements
Thanks go to the following people; my academic supervisors–Dr Michaela Balzarova, and Dr David Cohen for the review of the work as a dissertation. Thanks also to Dr Jim Gould for discussion and his approval to utilise and adapt the evaluation approach developed for research benefiting the Canadian Forest Fire Danger Rating System. I acknowledge and appreciate the support of my employer; the Department of Conservation, and specifically my Manager Simon Kelton. Dawn Smith was instrumental in obtaining fire-related costs for this project. Bruce Janes, Tom Barr, and Dean Strachan provided useful discussion on thermal camera use and capabilities. Tom Barr, Brenton Wilson and Emma Griffen provided the thermal camera use data. Members of the RPA and Wildland fire management online fraternity; Eric Lien, Doug Campbell, Joachim Dreibach, David Cant, F. Javier Garcia Garcia, Patrick "Lucidity" of the RPA Roswell Test Flight Team, and DOC staff; Dean Strachan and Earl Rewi for discussion on RPA and FLIR use and potential at wildland fires. I am grateful for suggestions from two anonymous reviewers which improved this article. Finally, I would like to thank all the survey participants, their input was critical to the value of this article and the completion of my dissertation. I acknowledge the previous work completed by researchers and compilers of fire data, which allowed the analyses within this article to take place.

References
Anadón, LD, Bosetti, V, Bunn, M, Catenacci, M, & Lee, A. (2012). Expert judgments about RD and D and the future of nuclear energy. *Environmental Science and Technology, 46*(21), 11497–11504.

Christensen, BR. (2015). Technological advances in rural and wildland fire management as determined using organisational knowledge. *New Zealand Journal of Forestry*, 60(2), 29–32.

Donovan, GH, & Rideout, DB. (2003). A reformulation of the Cost Plus Net value Change (*C + NVC*) model of wildfire economics. *Forest Science, 49*(2), 318–323.

Ederington, J, & McCalman, P. (2008). Endogenous firm heterogeneity and the dynamics of trade liberalization. *Journal of International Economics, 74*, 422–440.

Fernandes, PM, & Botelho, HS. (2003). A review of prescribed burning effectiveness in fire hazard reduction. *International Journal of Wildland Fire, 12*, 117–128.

Florec, V, Pannell, D, Burton, M, Kelso, J, Mellor, D, & Milne, G. (2012). Economic analysis of bushfire management programs: a Western Australian perspective. In RP Thornton & LJ Wright (Eds.), *Proceedings of bushfire CRC and AFAC conference research forum, 28 August 2012*. Bushfire CRC: Perth Australia.

Fogarty, L, & Smart, P. (1996). *Comparison of the cost-effectiveness of some aircraft used for fire suppression* (Fire Technology Transfer Note No. 8, p. 9).

Fulton, RW, & Mason, GF. (1982). Night vision electro-optics technology transfer: a decade of activity. *The Journal of Technology Transfer, 7*(1), 35–53.

Ganewatta, G. (2008). The economics of bushfire management. In J Handmer & K Haynes (Eds.), *Community bushfire safety* (pp. 151–159). Victoria: CSIRO Publishing. Collingwood.

Garthwaite, PH, Kaldane, JB, & O'Hagan, A. (2005). Statistical methods for eliciting probability distribution. *Journal of the American Statistical Association, 100*, 680–701.

Gerbert, K, Calkin, DE, Huggett, RJ, & Abt, K (2008). Toward a unified economic theory of fire program analysis with strategies for empirical modelling. In TP Holmes, JP Prestemon, & K Abt (Eds.), The economics of forest disturbances: wildfires, storms, and invasive species. (pp. 295–322). Dordrecht: Springer.

Gould, JS, Patriquin, MN, Wang, S, McFarlane, BL, & Wotton, M. (2013). Economic evaluation of research to improve the Canadian forest fire danger rating system. *Forestry, 86*, 317–329.

Groen, AJ, & Walsh, ST. (2013). Introduction to the field of emerging technology management. *Creativity and Innovation Management, 22*(1), 1–5.

Hesse, R. (2000). *Triangle distribution: Mathematica link for Excel* (Decision Line, May 2000, pp. 12–14).

Hesseln, H, Amacher, GS, & Deskins, A. (2010). Economic analysis of geospatial technologies for wildfire suppression. *International Journal of Wildland Fire, 19*(4), 468–477.

Hyytiäinen, K, & Haight, RG. (2010). Evaluation of forest management systems under risk of wildfire. *European Journal of Forest Research, 129*, 909–919.

Knol, AB, Slottje, P, van der Sluijs, JP, & Lebret, E. (2010). The use of expert elicitation in environmental health impact assessment: a seven step procedure. *Environmental Health, 9*, 19. doi:10.1186/1476-069X-9-19.

Kumar, M, Cohen, K, & Homchaudhuri, B. (2011). Cooperative control of multiple uninhabited aerial vehicles for monitoring and fighting wildfires. *Journal of Aerospace Computing, Information, and Communication, 8*(1), 1–16.

Lankoandé, MD. (2005). *Three essays on wildfire economics and policy*. Washington State University: Unpublished Doctor of Philosophy thesis. School of Economic Sciences.

Laupacis, A, Feeny, D, Detsky, AS, & Tugwell, PX. (1992). How attractive does a new technology have to be to warrant adoption and utilization? Tentative guidelines for clinical and economic evaluations. *Canadian Medical Association Journal, 146*(4), 473–481.

Mercer, DE, Haight, RG, & Prestemon, JP. (2008). Analyzing trade-offs between fuels management, suppression, and damages from wildfire. In TP Holmes (Ed.), *The economics of forest disturbances; wildfires, storms and invasive species* (pp. 247–272). Dordrecht: Springer Science + Business Media.

Merino, L, Caballero, F, Martinez-de-Dios, JR, Maza, I, & Ollero, A. (2012). An unmanned aircraft system for automatic forest fire monitoring and measurement. *Journal of Intelligent and Robotic Systems, 65*, 533–548.

MOT (Ministry of Transport). (2008). *The social cost of road crashes and injuries: June 2008*. Wellington: Ministry of Transport. New Zealand Government.

Pearce, HG, Hamilton, RW, & Millman, RI. (2004). *Fire behaviour and firefighter safety implications associated with the Bucklands Crossing Fire burnover of 24 March 1998* (Forest Research Bulletin No. 197, Forest and Rural Scientific and Technical Series, Report No. 4, p. 62). Forest Research, Rotorua, in association with the New Zealand Service Commission and National Fire Authority, Wellington.

Plucinski, MP, & Pastor, E. (2013). Criteria and methodology for measuring aerial wildfire suppression. *International Journal of Wildland Fire, 22*(8), 1144–1154.

Plucinski, M, McCarthy, GJ, Hollis, JJ, & Gould, JS. (2012). The effect of aerial suppression on the containment time of Australian wildfires estimated by fire management personnel. *International Journal of Wildland Fire, 21*, 219–229.

Rango, A, Laliberte, A, Steele, C, Herrick, JE, Bestelmeyer, B, Schmugge, T, Roanhorse, A, & Jenkins, V. (2006). Using remotely piloted aircraft for rangelands: current applications and future potentials. *Environmental Practice, 8*(3), 159–168.

Royo, P, Pastor, E, Solé, M, Lema, JM, López, J, & Barrado, C. (2011). *UAS architecture for forest fire remote sensing* (International symposium on remote sensing of environment. ISPRS Proceedings 2011. Sydney, pp. 1–4). http://upcommons.upc.edu/e-prints/bitstream/2117/16549/1/211104015final00187.pdf.

Runge, MC, Converse, SJ, & Lyons, JE. (2011). Special issue article: adaptive management for biodiversity conservation in an uncertain world. Which uncertainty? Using expert elicitation and expected value of information to design an adaptive program. *Biological Conservation, 144*, 1214–1223.

Shanks, G, & Tay, E. (2001). The role of knowledge management in moving to a customer-focused organization. In *Global co-operation in the new millennium. The 9th European conference on information systems, June 27–29, 2001, Bled, Slovenia*.

Stephenson, C, Handmer, J, & Betts, R. (2013). Estimating the economic, social and environmental impacts of wildfires in Australia. *Environmental Hazards, 12*(2), 93–111.

The Treasury. (2005). *Cost benefit analysis primer*. The Treasury. New Zealand Government. http://www.treasury.govt.nz/publications/guidance/planning/costbenefitanalysis/primer/cba-primer-v12.pdf.

van Dorp, JR, & Kotz, S. (2002). A novel extension of the triangular distribution and its parameter estimation. *The Statistician, 51*(1), 63–79.

Wren, J, & Barrell, K. (2010). *The costs of injury in New Zealand methods for prioritising resource allocation: a background briefing paper to inform the evaluation of the New Zealand injury prevention strategy*. Wellington, New Zealand: New Zealand Injury Prevention Secretariat, ACC.

Wu, J, Kaliyati, W, & Sanderson, K. (2009). *The economic cost of wildfires* (Business and Economic Research Ltd (BERL). New Zealand Fire Service Commission Research Report Number 99). Wellington: New Zealand Fire Service.

5

Observations of "coarse" root development in young trees of nine exotic species from a New Zealand plot trial

Chris J. Phillips[1*], Michael Marden[2] and Suzanne M. Lambie[3]

Abstract

Background: Forests and wide-spaced trees are used widely in New Zealand to control erosion from shallow landslides. Species that offer similar or better levels of protection to those currently used are sought to meet future needs. Determining what plants to use and when they become effective is important for developing guidelines and policy for land management. This study aimed to obtain data on above- and below-ground plant growth for young exotic tree species considered potential candidates for future 'erosion control forests'.

Methods: The above- and below-ground growth of nine exotic tree species was assessed annually for 3 years from planting in a randomised block field trial. Whole trees were excavated and destructively sampled and several below-ground metrics (total root length of all roots > 1 mm in diameter, lateral root spread, total root biomass) assessed.

Results: Differences between species for most metrics at the time of planting carried through to Year 3. The best performing species across most metrics was alder, followed by blackwood, cherry, and cypress. Allometric models relating total root length and below-ground biomass to root collar diameter were established.

Conclusion: Top performers with regard to root metrics were alder, cherry, and cypress followed by blackwood, radiata, and redwood. Root information contributes to improving our understanding of how and when, and at what planting density, plants become effective for controlling erosion in New Zealand.

Keywords: Above-ground biomass; Below-ground biomass; Root development; Shallow landslide erosion; Soil conservation

Background

The New Zealand plantation forest estate is dominated by *Pinus radiata* D.Don (radiata pine) with this one species occupying 90 % of the plantation estate (New Zealand Forest Owners Association 2013). Outside plantation forests, *Populus* and *Salix* species (often planted as single trees or wide-spaced groups of trees) are the most common exotic species and are used extensively for soil conservation and river control across New Zealand (Douglas et al. 2011). In spite of the widespread use of exotic species in New Zealand's managed (*i.e.* non-conservation) lands, there are a number of factors that are likely to shape future New Zealand production forests. These include: the sustainability of the New Zealand forest industry, in particular its resilience in the face of climate change (Ministry of Agriculture and Forestry 2008); the increasing demand for land for dairying (New Zealand Forest Owners Association 2013); and a move towards species diversification to reduce potential threats from disease or insect damage. There is also likely to be an increasing focus on non-wood production benefits (values or ecosystem services) from forests (plantations or natural) in the future (O'Loughlin 2005; Smaill et al. 2014). Of those services (amenity value, bioenergy production, carbon capture, biodiversity, and erosion control), the use of forests for controlling or reducing the impacts of rain events (storms) on erosion-susceptible land has continuing importance for many regions of New Zealand. Forests, together with wide-spaced

* Correspondence: phillipsc@landcareresearch.co.nz
[1]Landcare Research, PO Box 69040, Lincoln 7640, New Zealand
Full list of author information is available at the end of the article

erosion control or soil conservation trees, also continue to be actively promoted as erosion mitigation tools integral to sustainable land management in many regions (Ministry of Agriculture and Forestry 2010).

Forests are generally recognised to provide protection from hydrogeomorphic hazards such as floods, landslides, debris flows, snow avalanches and rockfalls (Sidle et al. 1985; Brang et al. 2001; Sidle and Ochiai 2006; Blaschke et al. 2008). An extensive literature exists on the benefits of trees, shrubs and grasses for reinforcing soils to control or reduce both surficial and shallow mass-movement erosion (*e.g.* Phillips et al. 1990; Marden & Rowan 1993; Phillips & Marden 2005; Stokes et al. 2008, 2009). Root reinforcement is now recognised as an important factor affecting several hydro-mechanical processes that give rise to the hazards mentioned above (Schwarz et al. 2013). However, while it is acknowledged that vegetation imparts additional strength to soil via its reinforcing root network, the amount of information recording the development and growth of a species' root system is sparse, particularly for larger trees (*e.g.* Mao et al. 2013). This paucity of quantitative data is largely because obtaining below-ground plant information is difficult, time-consuming, and expensive. Thus data sets for individual species over time are not abundant, the sample sizes per age class where these data exist are generally small or limited to one age class only (*e.g.* Danjon et al. 1999; Mulatya et al. 2002; Docker and Hubble 2008), and thus the statistical power of many relationships are generally limited by small sample sizes. Nevertheless, such data, be they from replicated trials or from single tree observations, are necessary to both improve our general understanding of a species' performance and to develop, calibrate, and validate either conceptual or quantitative models that incorporate root information for use in predicting the effects of vegetation on slope stability (Ekanayake & Phillips 1999, 2002; Schwarz et al. 2010; Phillips et al. 2011; Stokes et al. 2014) and hence the effectiveness of revegetation policies (Phillips et al. 2013b).

The on-site benefits of planted radiata pine (*Pinus radiata* D.Don) forests, on-farm plantings of *Populus* species, and some New Zealand native species for controlling erosion are relatively well understood even though the amount of quantitative data on their root system development is small (*e.g.* Phillips & Watson 1994; Phillips et al. 2013a, 2014). These benefits include a reduction in shallow landsliding (Marden & Rowan 1993; Phillips and Marden 2005; Douglas et al. 2011; Basher 2013), reduced rates of earthflow movement (O'Loughlin & Zhang 1986; Marden et al. 2008), reduced gully erosion (Parkner et al. 2007; Marden et al. 2012) and the retention of soil. Depending on factors such as geology, species, tree density, and slope steepness, plantation forests can be effective in controlling erosion as soon as 5–8 years after planting (O'Loughlin & Zhang 1986; Phillips et al. 2012).

In recent years, there has been increasing interest among some land owners and forestry stakeholders that species other than *Pinus radiata* might be better suited to the wider provision of ecosystem services (*e.g.* Trotter et al. 2005). However, in a recent survey of foresters and their stakeholders, strong opinions were revealed regarding the relative performance and suitability of various tree species to deliver those ecosystem services from New Zealand's plantation forests (Smaill et al. 2014). This study demonstrated that many of the opinions held by stakeholders were inaccurate when compared with available data from growth, physiological and ecological studies (Smaill et al. 2014). These data where they exist, are largely from studies of above-ground growth performance but our understanding of many species' development of below-ground traits required for root reinforcement and erosion control is severely lacking both in New Zealand and more generally internationally (*e.g.* Stokes et al. 2009; Phillips et al. 2011). However, although limited in many cases, data on root spread, root depth, and total root length, can and have been used to compare, at least in a general sense, the performance of a species in terms of effectiveness for erosion control (Phillips et al. 2011; Stokes et al. 2009; Phillips et al. 2013a, 2014). These data are required to underpin needs identified by both researchers and practitioners alike concerning the ecological mitigation of hillslope instability (Stokes et al. 2014).

To address this lack of basic below-ground information, a field plot trial was established to provide early growth data and observations on the "coarse"[1] root development of nine exotic tree species (Table 1). Species selection was not based on any assessment using rigorous criteria. Rather, the species selected are common, are in widespread use by farmers and farm foresters across New Zealand, and were readily obtainable from New Zealand nurseries. Researcher consensus narrowed potential candidates down to the final nine. Data were collected in such a way that they could be directly compared with existing data (mostly for radiata pine).

This paper reports on the observations and results of that trial, compares results with other species for which below-ground New Zealand root data exist, and then briefly discusses the implications for their use in erosion control. Our focus for the latter is on the use of trees to mitigate shallow landsliding (*i.e.* shallow landslides (soil slips), debris avalanches, debris flows, earthflows) rather than on surface erosion (slopewash, dry ravelling, etc.) because the former are the main erosion processes of concern in New Zealand's hill country and steeplands (Crozier 2005; Basher 2013).

Table 1 Seed sources and nursery treatment

Species	Common name[a]	Source and treatment	Physiological age at planting (years)	Mean seedling height at planting (m)
Quercus robur	Oak	Seedlings sourced from a local tree where seed had fallen and germinated in leaf litter/bark/garden soil. 2009 germination.	2	0.27
Eucalyptus fastigata	Eucalypt	Direct open-ground-sown into commercial nursery bed, October 2009.	1	0.36
Pseudotsuga menziesii	Douglas-fir	Direct open-ground-sown into commercial nursery bed, October 2009.	1	0.15
Sequoia sempervirens	Redwood	Direct open-ground-sown into commercial nursery bed, October 2009.	1	0.37
Prunus serrulatus	Cherry	Transplanted from under a local tree where the seed had fallen and germinated in the leaf litter. 2009 germination.	2	0.79
Acacia melanoxylon	Blackwood	Direct open-ground-sown into commercial nursery bed, October 2009.	1	0.72
Alnus rubra	Alder	Direct open-ground-sown into commercial nursery bed, October 2009.	1	0.83
Cupressus lusitanica	Cypress	Direct open-ground-sown into commercial nursery bed, October 2009.	1	0.42
Pinus radiata	Radiata	GF 19 improved seed. Direct open-ground-sown into commercial nursery bed, October 2009.	1	0.43

[a]Common name used in this paper

Methods

A randomised field plant trial with three "replicates" in two blocks of each of nine species grown over 3 years (81 plants) was established in July 2010 near Gisborne, North Island, New Zealand. Species names, common names, and nursery treatments are listed in Table 1. The trial was established on the same site as earlier trials that examined root growth of both native and exotic species (Marden et al. 2005; Phillips et al. 2014). The field site is located on a low-lying, even-surfaced near-flat alluvial terrace adjacent to the Taraheru River (Fig. 1). The soil is free-draining, Te Hapara soil (Typic Sandy Brown; Hewitt 2010) and may require irrigation in summer to ensure plant survival. The soil has no physical or chemical impediments to root development to at least about 1.2 m depth, other than a variable-depth water table. Across the site, the soil appeared to be relatively uniform, with a sandy loamy texture. The topsoil was homogenised through a history of cultivation, with no obvious development of tillage pans, though soil density appeared to be greater at depths beyond 0.5 m. The climate is also favourable to tree growth, being warm temperate with average winter low temperatures of about 5 °C and average summer highs of 25 °C and about 1000 mm annual rainfall. The trial plot (50 × 20 m) was fenced to exclude grazing animals and tilled before plants were established in two side-by-side blocks at 2.5 m spacing according to the trial design.

Seedlings were either grown in-house from locally collected seed or supplied by a commercial/research nursery in Rotorua (Scion pers. comm.) (Table 1) and were typical of what would be supplied to farmers or foresters (Fig. 2). Thus, the physiological age of each species was not consistent and results are presented as "years since planting" (*i.e.* Year 0 being material as supplied from the nursery, Year 1 after one year's growth in the trial plot, etc.) and species like alder, cherry, and oak were considerably larger at time of planting than radiata and Douglas-fir (Table 1; Figs. 2 & 3). As part of commercial nursery practice, seedling roots were conditioned/undercut (*i.e.* root pruned). To help ensure the survival of seedlings in the trial plot, weed mat was laid to reduce competition from weeds, trickle irrigation was installed as a precaution against extended dry periods (not required), and wire cages were placed over plants initially to limit browsing damage by hares (*Lepus europaeus*) and pūkeko (*Porphyrio melanotus*), a native bird known to pull out small plants.

Measurement methods of both above- and below-ground metrics followed well-established procedures used in earlier studies (Watson et al. 1999; Czernin & Phillips 2005; Marden et al. 2005; Phillips et al. 2014). Various metrics or traits have also been used to help assess species performance in terms of erosion-control effectiveness (Stokes et al. 2009; Phillips et al. 2011). In this study, above-ground metrics included tree height, canopy spread (*i.e.* canopy diameter), root-collar diameter (RCD – measured at ground level) and diameter at breast height where applicable (DBH – measured 1.4 m above ground level). Above-ground biomass was measured

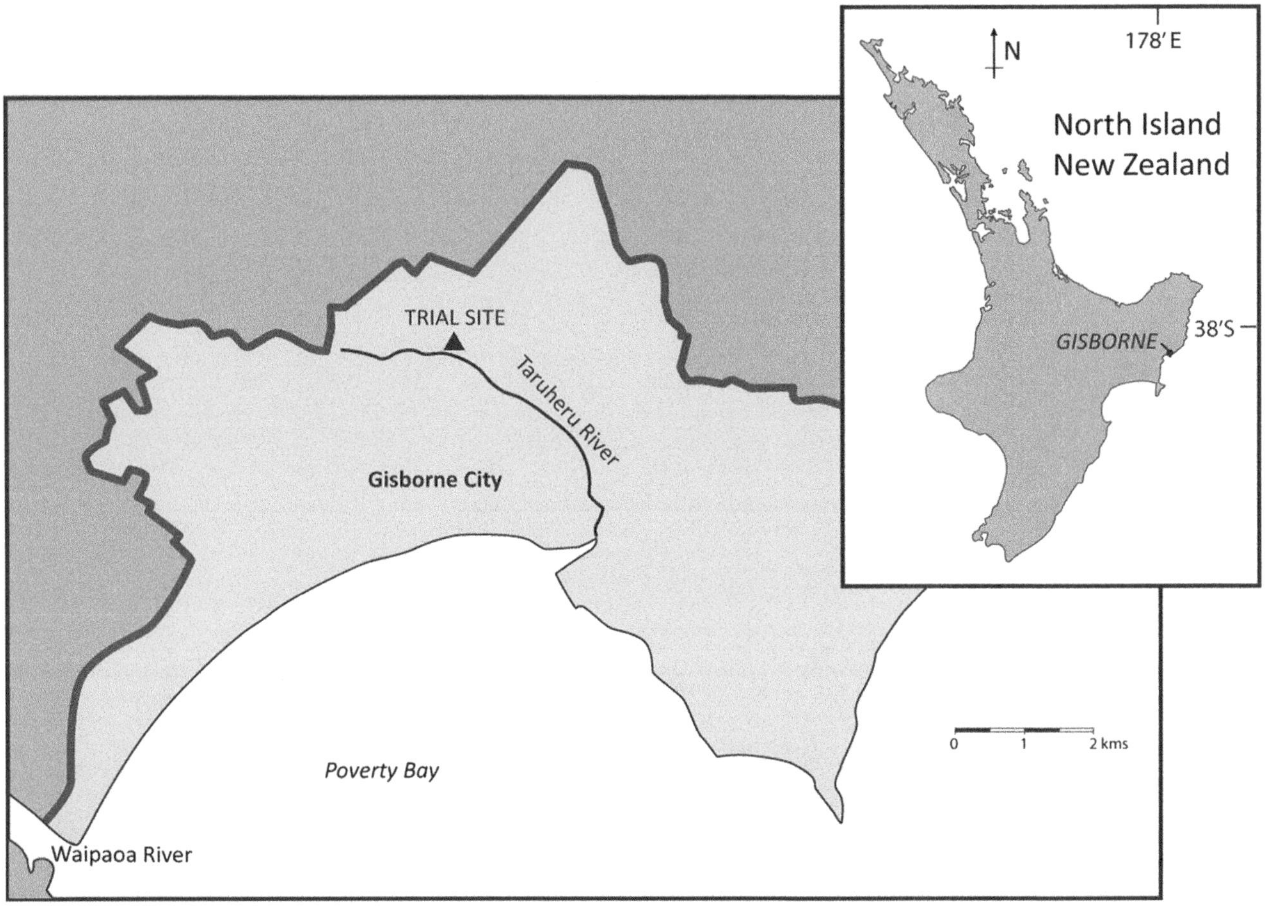

Fig. 1 Map showing trial location

separately for foliage, branches and stem. Foliage was not present at the time of extractions in the deciduous species. Maximum depth, spread of lateral roots, total root length of all roots > 1 mm in diameter, root diameter size class distribution, and total below-ground biomass were assessed. Lateral root spread together with canopy spread, was taken as the average of the maximum diameters measured by tape measure in two directions at right angles, approximating a N-S and E-W line. The long axis of the field plot was oriented approximately N-S. Plants were extracted and measured each year about the anniversary of the planting date (Year 1, Year 2, and Year 3). Trees of each species were extracted in July 2011 after 1 year in the ground. Year-2 extractions were conducted in late June and early July 2012 after 2 years in the ground. Year-3 and final extractions took place in late May and early June 2013.

Root systems were extracted using an air spade (an ultrasonic high pressure device to disaggregate and remove soil from around the roots) or by hand, particularly in the first year (s) when plants were small. Once removed from the ground, the plants were destructively sampled to determine the parameters outlined above, all of which could then be related to over bark root-collar diameter (*i.e.* groundline diameter or basal diameter at root collar) and over bark diameter at breast height (1.35 m) (when plants grew tall enough). Ten additional sample trees of each species were also destructively sampled at the time of planting (Year 0, *i.e.* seedlings as delivered from the nursery, see Fig. 2).

The complete root system of each plant was photographed before the plant was destructively partitioned into 0.5 m radial and depth segments (see Czernin & Phillips 2005). Roots within each radial segment were further partitioned into diameter size-classes (<1 mm (fibrous), 1–2, >2–5, >5–10, >10–20, >20–50, >50–100 mm) (Watson & O'Loughlin 1990). Over-bark root diameters of all roots were measured with digital callipers. For each radial segment, total root length for each diameter size-class was determined by placing roots end to end (excluding fibrous roots < 1mm in diameter). Roots were then oven-dried for biomass determination. Both above- and below-ground components were oven-dried at 80° C for 24 h then weighed to the nearest 0.1 g.

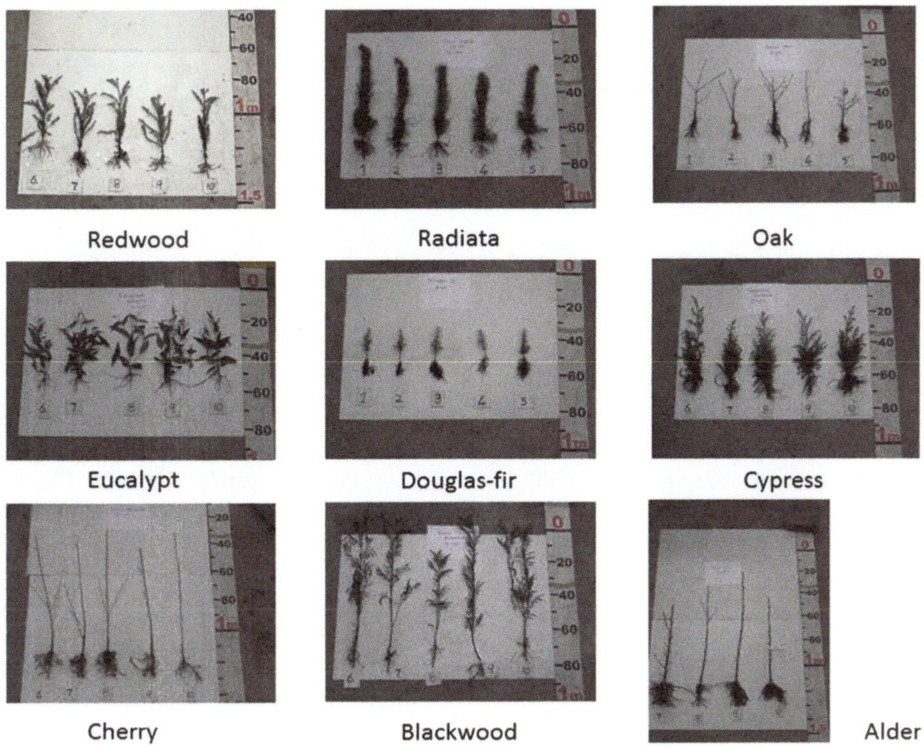

Fig. 2 Photographs of seedlings at time of planting to illustrate size and form differences. Not to scale

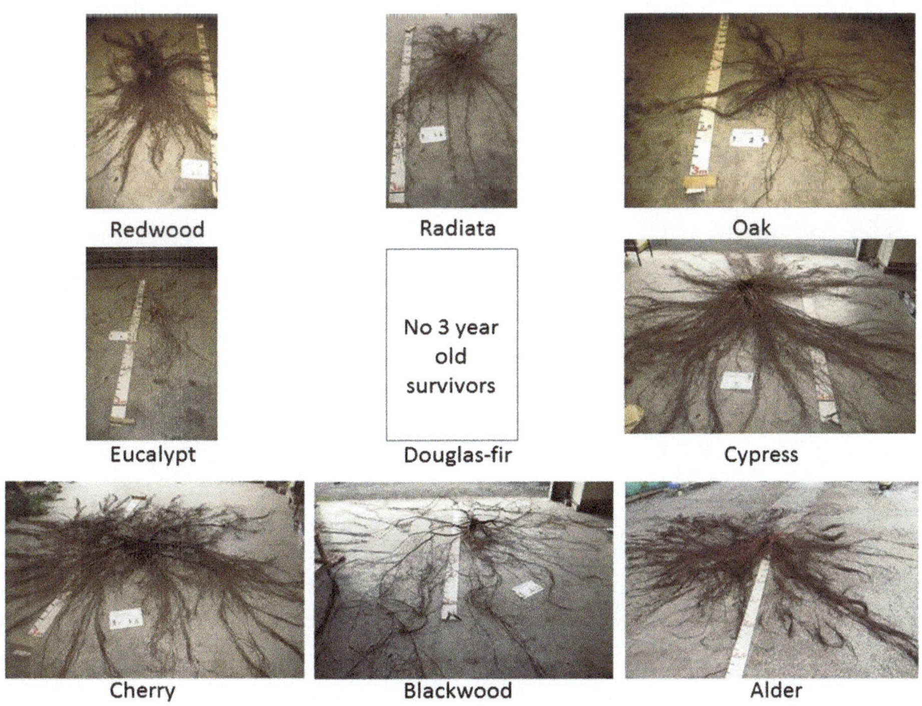

Fig. 3 Photographs at Year 3 of representative root systems illustrating differences in root architecture. Not to scale. Tape is 3 m

Though three alders survived through to Year 3, one tree only was randomly selected for full excavation and processing as resources were insufficient to deal with the remaining two due to their large size and the time to extract and process them (on average more than 3–4 days per tree for a team of four people).

While it is acknowledged that obtaining powerful statistical relationships for many root parameters due to small sample sizes is difficult, it is possible to move beyond simple empirical observations where sufficient sample numbers exist. Differences in growth characteristics among tree species within each year of the trial were determined using 2×2 factorial analysis of variance (ANOVA) blocked by planting blocks (GenStat 12; VSN International, Hemel Hempstead, UK). Analysis of differences within each tree species over the course of the trial was also undertaken using ANOVA. Analyses were considered significant if $P < 0.05$. Some of the data were transformed (log10) for ANOVA following residual normality analysis. Nevertheless, all results are presented as untransformed means and standard errors of the means, unless otherwise stated.

Results

Survival and general description of root systems

There were minor losses of blackwood, redwood and eucalypt soon after planting due, we suspect, to browsing by animals (rabbits - *Oryctolagus cuniculus*; hares; and pūkeko) or from the water table rising into the rooting zone over the winter months and causing subsoil waterlogging. Douglas-fir showed initial signs of stress due to a post-planting 'wet period' but losses at Year 1 were not severe enough to compromise the objectives of the trial. However, at Year 2, all Douglas-fir had died. Other trees (5 of 24), also sustained 'damage' in the second year of the trial, exhibiting broken branches or toppling likely caused by wind. At Year 3, 19 of the 27 trees marked for extraction had survived. Of these, blackwood, alder and radiata had three surviving replicates with the other species having only two each.

The nine trialled species were a mix of evergreen and deciduous trees and exhibited a range of tree form. Canopy shape and density were variable with some species having a bushy shrub-like appearance and others more streamlined canopies and forms. In a similar manner, the below-ground appearance of the root systems was also variable. Several species had low root density (*i.e.* low numbers of roots per m^2 of soil) and elongated "spidery" lateral roots with few small side branches (low root branching order) (blackwood, eucalypt, oak) (Fig. 3). Alder and cherry had dense branching lateral root systems with high root numbers (high root density and high root order) while redwood and cypress had similar dense root systems, but many fine long branching roots. Radiata had long laterals with limited branching order, similar to blackwood but with a greater number of short branch roots on the main laterals. Alder, redwood and cypress roots were reddish in colour, radiata orange, blackwood and oak pale brown, cherry pale brown to white, and eucalypt dark brown.

Growth performance parameters

Visual and quantitative differences among species for most metrics at the time of planting carried through to Year 3. At both 2 and 3 years from planting, the best performing species across most metrics was alder, followed by blackwood, cherry, and cypress (Tables 2, 3, 4 & 5, Figs. 3, 4, 5, 6 & 7). Alder was the tallest tree at Year 3 (6.4 m maximum tree height) with cypress (4.9 m) and eucalypt (4.0 m) also performing well (Fig. 4).

Largest root spread was in a Year-3 blackwood at 10.3 m with one Year-2 blackwood having 9.0 m root spread. The Year-3 alder had a root spread of 7.7 m, though one Year 2 tree had a spread of 8.5 m. Blackwood had a mean root spread of 6.9 m, eucalypt 4.5 m, and cherry was 3.2 m. Overall, blackwood and alder had significantly greater root spread than the other species at Year 2 ($P < 0.001$) although there was no statistically significant difference between any of the tree species in Year 3 due to large variation in growth within a particular species, even though visibly the Year-3 root systems looked larger than those in year 2 (Fig. 5).

Root depth was variable (total mean range at Year 2 was 0.18 – 0.45 m) among species and trends established in Year 1 continued through to Year 3. Alder roots were significantly deeper than the other species in Year 3, and roots did not go deeper than about 0.5 m in any year of the trial.

Mean above-ground biomass (AGB) of Year-3 cypress was 29 kg with the largest of the two trees having 46 kg above- and 8 kg below-ground (including root bole or stump). Alder was the next best performing species in mean total biomass and had the greatest below-ground biomass (BGB) of all species at both Year 2 and 3, closely followed by cherry. However, statistically there was no difference between species in both above- and below-ground biomass at Year 3, largely because of the wide range in values for cypress and cherry (big error bars) ($P = 0.130$ for AGB and $P = 0.365$ for BGB) (Fig. 6).

Alder had the greatest total root length (all roots >1 mm in diameter) of the trial species at Year 2, ranging from 1347 to 2262 m. A Year-2 alder had the greatest total root length of all trees excavated at 2262 m, considerably more than the random Year-3 alder chosen for excavation (1268 m). At Year 3, the excavated alder had a significantly greater root length than the other species ($P = 0.006$) but was not significantly different from cypress (Fig. 7).

Table 2 Attributes for 0-year-old trialled species

Attribute	Oak	Eucalypt	Douglas-fir	Redwood	Cherry	Blackwood	Alder	Cypress	Radiata
Trees sampled	10	10	10	10	10	10	10	10	10
Tree height (m)	0.27 (0.02)	0.36 (0.01)	0.15 (0.00)	0.37 (0.01)	0.79 (0.02)	0.72 (0.03)	0.83 (0.04)	0.42 (0.03)	0.43 (0.03)
Canopy spread (m)	0.14 (0.02)	0.24 (0.01)	0.06 (0.00)	0.23 (0.01)	0.13 (0.02)	0.19 (0.02)	0.11 (0.03)	0.19 (0.01)	0.15 (0.01)
DBH[a] (mm)	-	-	-	-	-	-	-	-	-
RCD[b] (mm)	9 (0.6)	6 (0.3)	3 (0.7)	5 (0.2)	9 (0.5)	8 (0.5)	13 (0.6)	8 (0.3)	7 (0.3)
Max. root diameter (m)	0.12 (0.01)	0.17 (0.02)	0.09 (0.00)	0.16 (0.01)	0.22 (0.01)	0.15 (0.02)	0.39 (0.05)	0.21 (0.02)	0.14 (0.01)
Max. root depth (m)	0.15 (0.001)	0.12 (0.001)	0.12 (0.001)	0.12 (0.001)	0.17 (0.001)	0.13 (0.001)	0.18 (0.001)	0.13 (0.001)	0.10 (0.001)
Total AGB[c] (g)	4 (0.4)	6 (0.7)	1 (0.1)	5 (0.4)	11 (0.8)	13 (1.9)	18 (2.5)	14 (1.0)	13 (1.0)
Total BGB[d] (g)	7 (1.0)	1 (0.1)	1 (0.1)	1 (0.3)	11 (1.1)	3 (0.5)	20 (3.0)	3 (0.3)	2 (0.2)
Total root mass[e] (g)	7 (1.0)	1 (0.1)	1 (0.1)	1 (0.3)	11 (1.1)	3 (0.5)	20 (3.0)	3 (0.3)	2 (0.2)
Total root length > 1 mm (m)	0.64 (0.11)	0.59 (0.08)	0.29 (0.05)	0.71 (0.08)	1.63 (0.19)	0.53 (0.14)	4.11 (0.43)	1.34 (0.18)	0.77 (0.11)

Data are presented as means, numbers in brackets are standard error of the mean. [a]diameter at breast height, [b]root collar diameter, [c]total above-ground biomass, [d]total below-ground biomass including stump/root bole, [e]total root mass excluding the stump/root bole

The distribution of total root length by root diameter class about the stem shows two distinct patterns (Table 6). Firstly, oak, redwood and radiata have greater than 50 % of their total root length within 1 m of the stem, and show a rapid linear decline in total root length with radial distance. The remaining species, apart from eucalypt, have less than 40 % of their total root length within 1 m of the stem, and show a pattern of increasing total root length with distance from the stem to a maximum at about 1.5–2.0 m and then declining (Fig. 8). Greatest total root length was found in the finest root diameter class (1–2 mm) independent of radial distance from the stem, except for alder at 3.5 m distance from the stem where there was more total root length in the >2–5 mm diameter size class than the 1–2 mm class (Table 6). Total root length declined as roots increased in diameter and the thickest roots were found close to the stem.

In terms of growth differences within species between years, for most trialled species and for most metrics, there were significant differences between years. The metrics that tended to show no difference were root length and root:shoot ratio for most species although alder and blackwood were significantly different between Year 0, 1 and 2 but Year 2 and 3 were the same ($P < 0.001$).

For individual metrics, Year-0 data tended to show the most differences between species, although for each metric, different species grouped differently. Across all metrics, alder by being larger at the time of planting tended to be the best, or in the group of best performers for each year.

Allometric relationships

Allometric relationships are used to predict tree growth on the basis of easy-to-measure above-ground parameters such as diameter at breast height

Table 3 Attributes for 1-year-old trialled species

Attribute	Oak	Eucalypt	Douglas-fir	Redwood	Cherry	Blackwood	Alder	Cypress	Radiata
Trees sampled	3	3	1	3	3	3	3	3	3
Tree height (m)	0.80 (0.25)	1.01 (0.41)	0.18	1.00 (0.15)	1.51 (0.17)	1.57 (0.17)	2.86 (0.58)	1.21 (0.11)	0.92 (0.17)
Canopy spread (m)	0.56 (0.02)	0.61 (0.28)	0.05	0.81 (0.08)	0.55 (0.25)	0.55 (0.05)	0.97 (0.50)	0.54 (0.03)	0.37 (0.15)
DBH[a] (mm)	-	4 (N/A)	-	-	8 (N/A)	5 (N/A)	20 (4.7)	-	-
RCD[b] (mm)	21 (1.2)	24 (10.2)	6	24 (0.4)	27 (0.7)	25 (1.4)	49 (5.6)	24 (2.0)	19 (2.9)
Max. root diameter (m)	0.97 (0.24)	0.73 (0.45)	0.09	0.93 (0.03)	1.69 (0.41)	2.23 (0.39)	2.60 (0.80)	0.93 (0.07)	0.87 (0.16)
Max. root depth (m)	0.32 (0.02)	0.11 (0.01)	0.09	0.22 (0.04)	0.21 (0.02)	0.27 (0.03)	0.28 (0.02)	0.20 (0.02)	0.17 (0.03)
Total AGB[c] (g)	86 (37)	218 (151)	4	147 (54)	232 (56)	199 (51)	996 (240)	239 (60)	117 (42)
Total BGB[d] (g)	71 (21)	28 (19)	1	37 (11)	197 (43)	57 (10)	731 (201)	39 (12)	18 (5)
Total root mass[e] (g)	40 (10)	20 (13)	1	28 (8)	149 (39)	30 (4)	596 (182)	23 (3)	13 (3)
Total root length > 1 mm (m)	9.0 (2.5)	6.0 (3.8)	0.2	19.4 (3.7)	29.6 (18.7)	10.6 (2.1)	140 (40.8)	16 (3.5)	16 (10)

Data are presented as means (with the exception of Douglas-fir), numbers in brackets are standard error of the mean. [a]diameter at breast height, [b]root collar diameter, [c]total above -ground biomass, [d]total below-ground biomass including stump/root bole, [e]total root mass excluding the stump/root bole

Table 4 Attributes for 2-year-old trialled species

Attribute	Oak	Eucalypt	Douglas-fir	Redwood	Cherry	Blackwood	Alder	Cypress	Radiata
Trees sampled	3	3	-	3	3	3	3	3	3
Tree height (m)	0.63 (0.03)	2.51 (0.56)	-	1.47 (0.38)	2.60 (0.36)	3.53 (0.52)	4.73 (2.30)	2.27 (0.26)	1.30 (0.47)
Canopy spread (m)	0.39 (0.12)	1.43 (0.36)	-	1.05 (0.34)	1.08 (0.17)	1.63 (0.29)	3.73 (0.19)	1.67 (0.18)	0.38 (0.14)
DBH[a] (mm)	-	21 (3.3)	-	8 (1.4)	19 (5.0)	39 (7.5)	60 (8.3)	16 (0.1)	19 (N/A)
RCD[b] (mm)	19 (1.4)	73 (15.9)	-	45 (13.8)	63 (10.9)	71 (9.0)	144 (11.7)	43 (0.1)	29 (10.6)
Max. root diameter (m)	0.57 (0.06)	2.10 (0.20)	-	1.32 (0.34)	3.33 (0.54)	6.33 (1.45)	6.70 (0.99)	3.00 (0.25)	1.10 (0.44)
Max. root depth (m)	0.45 (0.03)	0.18 (0.03)	-	0.20 (0.03)	0.22 (0.01)	0.39 (0.04)	0.42 (0.04)	0.21 (0.01)	0.23 (0.12)
Total AGB[c] (g)	46 (17)	2454 (1032)	-	669 (313)	2180 (830)	4347 (1146)	11106 (377)	3833 (759)	301 (171)
Total BGB[d] (g)	65 (28)	523 (250)	-	138 (65)	2088 (910)	1675 (1403)	8438 (1212)	1037 (582)	52 (22)
Total root mass[e] (g)	49 (22)	350 (166)	-	81 (40)	1699 (748)	1275 (244)	7091 (1279)	801 (510)	24 (10)
Total root length > 1 mm (m)	6.0 (1.2)	43.4 (18.1)	-	33.3 (16.2)	180.9 (86.6)	119.5 (21.9)	1508.2 (397.2)	147 (31.3)	11.3 (4.4)

Data are presented as means, numbers in brackets are standard error of the mean. [a]diameter at breast height, [b]root collar diameter, [c]total above-ground biomass, [d]total below- ground biomass including stump/root bole, [e]total root mass excluding the stump/root bole

or root-collar diameter. Because the trial plants were young and the number of replicates low, fitting trend lines with limited data may not be appropriate. However, in other root studies, power-law curves have been fitted to limited data sets to provide some predictive ability (*e.g.* McIvor et al. 2009; Phillips et al. 2013a). Using the trial data, it was possible to generate simple relationships between root collar diameter and growth attributes with reasonable r^2 values as shown in Table 7 for total root length and below-ground biomass.

In a similar manner, allometric relationships can be generated if all the data are combined (Equations 1 and 2).

$$\text{Total root length (m)} = 0.009\ \text{root-collar diameter}^{2.22}\text{(mm)} \quad R^2 = 0.83 \tag{1}$$

$$\text{Below-ground biomass (g)} = 0.012\ \text{root-collar diameter}^{2.63}\text{(mm)} \quad R^2 = 0.89 \tag{2}$$

Discussion

The differences in initial height among species is probably the biggest factor in interspecies variation in the parameters measured for the early years of growth, though the physiological age of all species was roughly similar (Table 1). Species like alder, cherry, and oak were considerably larger at time of planting than radiata and Douglas-fir (Fig. 2), and this continued to show for the first year or two of growth in the trial plot. Rather than having planting materials of the same size and grade, our trial used "nursery-raised" planting grades as would normally be used by the forestry or "land management" sectors. This may have been a shortcoming in trial

Table 5 Attributes for 3-year-old trialled species

Attribute	Oak	Eucalypt	Douglas-fir	Redwood	Cherry	Blackwood	Alder	Cypress	Radiata
Trees sampled	2	2	-	2	2	3	1	2	3
Tree height (m)	1.85 (0.45)	3.65 (0.35)	-	2.50 (0.60)	2.70 (0.50)	3.37 (0.08)	6.43	4.00 (0.90)	1.32 (0.38)
Canopy spread (m)	1.95 (0.40)	2.03 (0.27)	-	2.53 (0.53)	1.93 (0.87)	2.22 (0.42)	4.4	3.23 (0.73)	2.67 (0.73)
DBH[a] (mm)	12 (N/A)	40 (1.5)	-	34 (24.0)	28 (6.0)	52 (1.5)	80	84 (39.0)	71 (3.0)
RCD[b] (mm)	82 (0.25)	94 (11.5)	-	91 (32.0)	101 (32.5)	107 (9.3)	129	158 (32.0)	103 (25.2)
Max. root diameter (m)	1.50 (1.10)	4.50 (0.30)	-	3.10 (0.50)	3.20 (1.70)	6.93 (1.69)	7.70	5.60 (0.40)	3.27 (1.11)
Max. root depth (m)	0.35 (0.05)	0.30 (0.05)	-	0.28 (0.03)	0.28 (0.08)	0.32 (0.02)	0.50	0.35 (0.05)	2.27 (0.03)
Total AGB[c] (g)	1098 (628)	4743 (876)	-	3921 (2468)	4552 (3625)	6834 (2055)	12071	28942 (17326)	6848 (3384)
Total BGB[d] (g)	1022 (711)	1561 (260)	-	1129 (506)	6200 (5807)	2520 (770)	8959	5088 (2596)	2046 (945)
Total root mass[e] (g)	483 (370)	1022 (160)	-	605 (253)	4780 (4549)	1595 (456)	8117	2954 (1296)	1268 (617)
Total root length > 1 mm (m)	94.7 (51.8)	83.0 (38.7)	-	208.3 (70.0)	403.2 (357.4)	209.3 (81.8)	1268.7	837.4 (431.7)	128.1 (46.3)

Data are presented as means (with the exception of Alder), numbers in brackets are standard error of the mean. [a]diameter at breast height, [b]root collar diameter, [c]total above-ground biomass, [d]total below-ground biomass including stump/root bole, [e]total root mass excluding the stump/root bole

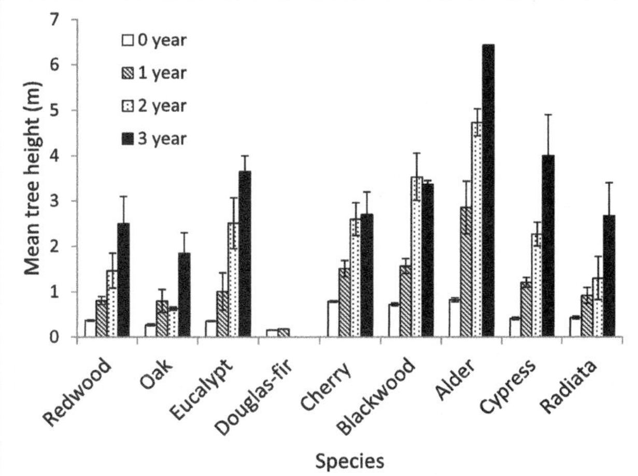

Fig. 4 Mean tree heights for nine trial species 0–3 years after planting. Vertical bars are one standard error of the mean

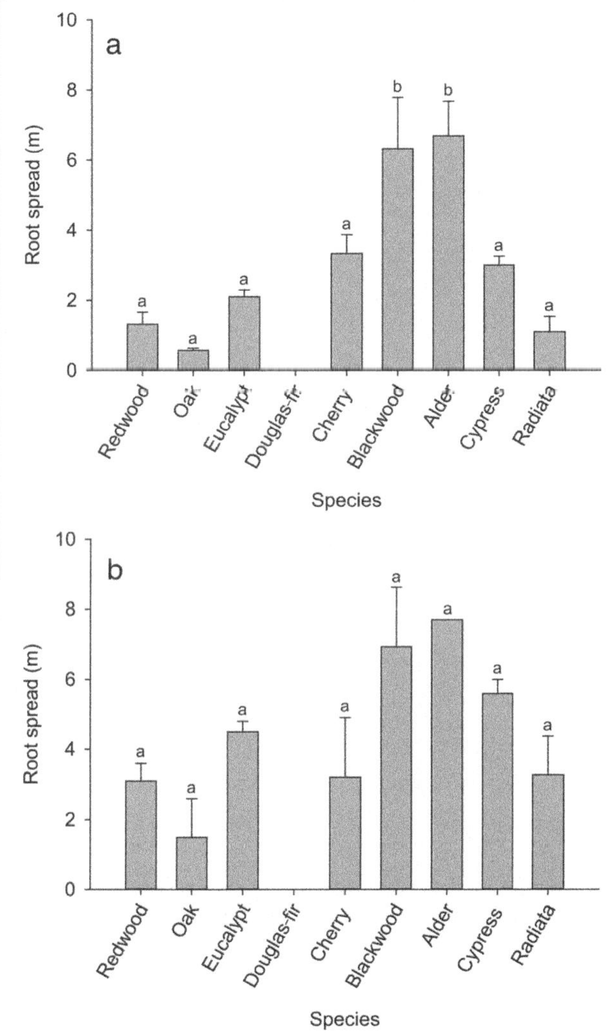

Fig. 5 Mean root spread in Year 2 (**a**) and Year 3 (**b**) for the nine trial species. Error bars represent one standard error of the mean; bars with different letters were significantly different ($P < 0.001$)

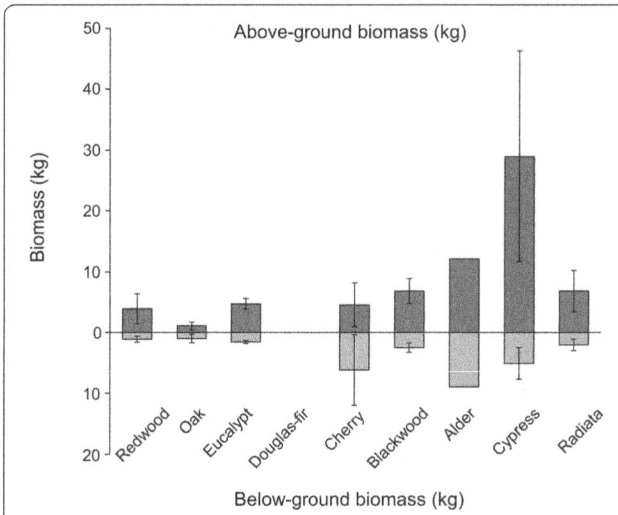

Fig. 6 Plant biomass above- (dark bars) and below-ground (light bars) in Year 3. Error bars represent one standard error of the mean for each of above- and below-ground components of biomass. Mean values except for alder

design. In hindsight, it would have been preferable for seedlings to have been of the same size/grade, though the growth in early years probably reflects the actual performance and practice of nursery-raised and supplied stock. However, in another trial but using native species, where size at planting and nursery treatment were highly variable, in the first 3–5 years no difference was found in survival or growth between plants raised in open-ground compared with those grown in larger containers for the shrub hardwood and monocot species trialled (Cole et al. 2014).

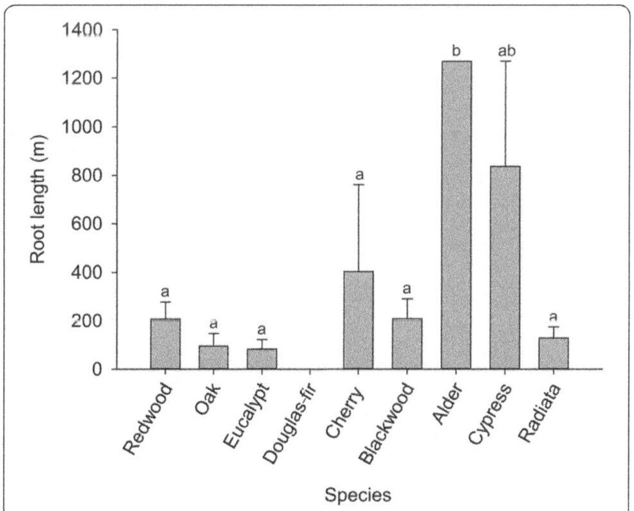

Fig. 7 Mean total root length (roots >1 mm in diameter) of trial species in Year 3. Error bars represent one standard error of the mean; bars with different letters were significantly different ($P = 0.006$)

Table 6 Mean total root length (m) by root diameter size class (mm) with radial distance from stem (m)

		Radial distance from stem (m)									
Species	Root diameter class (mm)	0-0.5	>0.5-1.0	>1.0-1.5	>1.5-2.0	>2.0-2.5	>2.5-3.0	>3.0-3.5	>3.5-4.0	>4.0-4.5	>4.5-5.0
Oak	1-2	22.0	22.2	12.1	5.5						
	>2-5	16.2	8.7	2.7	0.4						
	>5-10	4.8	1.8								
	>10-20	1.7	0.3								
	>20-50	0.6									
	Total root length (m)	45.2	33.0	14.8	5.9						
Eucalypt	1-2	9.8	20.6	7.9	2.9	1.9	1.0				
	>2-5	5.5	8.9	6.4	4.2	1.8	1.0				
	>5-10	3.0	2.0	1.6	0.6	0.3					
	>10-20	0.8	1.0	0.4							
	>20-50	1.0	0.4								
	>50-100	0.2									
	Total root length (m)	20.3	32.9	16.3	7.8	4.0	2.0				
Douglas-fir	No data at age 3										
Redwood	1-2	74.3	65.5	28.6	8.6						
	>2-5	22.7	8.0	1.6							
	>5-10	3.1	0.7								
	>10-20	0.5									
	>20-50	0.3									
	Total root length (m)	100.9	74.2	30.2	8.6	0.0					
Cherry	1-2	42.3	80.8	182.7	111.4	22.2	0.8				
	>2-5	18.4	29.8	52.7	26.9	4.0					
	>5-10	7.2	8.4	6.8	0.6						
	>10-20	3.7	8.2	1.1							
	>20-50	3.5	3.2	0.2							
	>50-100	1.0									
	Total root length (m)	76.2	130.3	243.5	138.9	26.2	0.8				
Blackwood	1-2	7.7	23.6	31.7	26.1	21.2	13.6	9.9	7.7	0.3	0.5
	>2-5	6.2	11.8	13.6	8.0	8.7	5.3	6.3	0.7		
	>5-10	4.0	4.5	2.8	0.7	0.7					
	>10-20	3.5	2.4								
	>20-50	1.5	0.3								
	>50-100	0.2									
	Total root length (m)	23.2	42.6	48.1	34.9	30.6	18.9	16.2	8.4	0.3	0.5
Alder	1-2	64.0	127.7	131.0	154.0	101.1	95.6	52.0	22.7	5.1	2.7
	>2-5	36.1	56.5	54.5	61.2	43.3	34.8	22.8	73.8	1.7	0.6
	>5-10	18.8	23.0	17.1	14.5	12.2	6.0	2.5	0.6		
	>10-20	7.0	7.8	4.1	2.1	3.5	1.4	0.7			
	>20-50	5.2	0.9								
	>50-100	0.4									
	Total root length (m)	131.5	215.8	206.7	231.8	160.1	137.7	78.0	97.1	6.7	3.3

Table 6 Mean total root length (m) by root diameter size class (mm) with radial distance from stem (m) ***(Continued)***

Cypress	1-2	75.3	106.9	139.8	160.1	98.8	49.7
	>2-5	48.5	64.4	49.1	14.8	3.1	
	>5-10	21.8	15.8	3.2	0.0		
	>10-20	8.2	0.5				
	>20-50	1.9					
	>50-100	0.1					
	Total root length (m)	155.8	187.7	192.1	175.0	102.0	49.7
Radiata	1-2	30.7	26.3	29.9	19.7	8.0	1.4
	>2-5	13.2	11.0	7.7	5.0	1.2	
	>5-10	3.1	7.9	3.6	0.8		
	>10-20	1.3	2.0	0.4			
	>20-50	2.2	0.2				
	Total root length (m)	50.4	47.4	41.6	25.4	9.3	1.4

While differences in performance existed in the early years post-planting, by Year 3, some of the smaller seedlings such as radiata, redwood, and cypress had "caught up" to the better early performers and exhibited strong growth between Years 2 and 3. While general observations can be made about plant performance in early years as above, it is accepted that real differences between species are unlikely to emerge until the plants are considerably older – at least beyond 4 years from planting.

The differences observed in plant growth may also be attributed to environmental factors such as the presence of a high water table at the time of establishment, which for some species such as Douglas-fir may have been an inhibiting factor and ultimately was probably responsible for their non-survival, while in others such as alder it may have enhanced growth. The rapid growth of alder may also be due to it being both a natural coloniser species and its ability to fix nitrogen. Edaphic factors such as soil type are often cited as the major controlling factor on plant growth (*e.g.* Phillips and Watson 1994). However, the trial site was reasonably uniform and the soil being a sandy loam without stones provided an ideal growing medium that did not impede the lateral root growth of the plants. However, the presence of the seasonal water table and apparent increase in soil density beyond 0.5 m depth may have limited the development of vertical root growth.

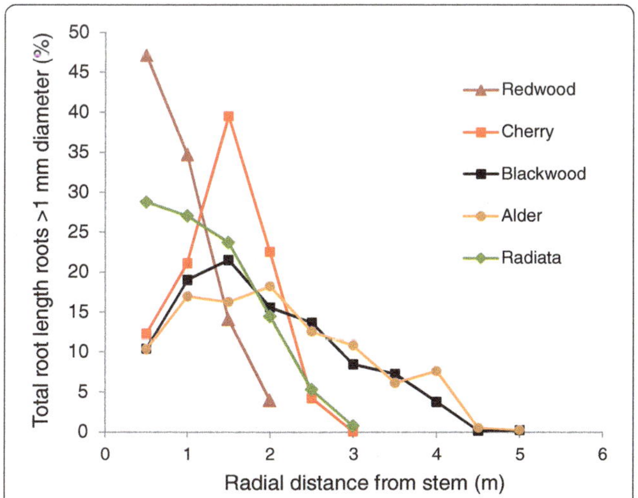

Fig. 8 Selected trial species showing percent total root length > 1mm diameter in each half-metre radial segment from the stem. Values are plotted at the outer value of the radial segment, *i.e.* 1.0-1.5 m value plotted at x = 1.5 m

After 3 years' growth, it is clear that some species outperformed others in terms of the metrics deemed important for soil reinforcement and hence erosion control (*e.g.*, total root length, root depth, root spread, etc.) (Marden et al. 2005). Whether these levels of performance persist beyond 3 years is unknown. Notwithstanding the limited sample sizes or the fact that the trial site may have been more or less favourable than others for establishment and juvenile growth, it may still be possible to compare "performance" metrics with other species for which root data exists. Comparing species grown in earlier trials on the same site with those from this study suggest that in terms of total root length, the best performing species from this trial exceeded those of native woody species (Marden et al. 2005) but are less than poplar and willows grown from poles after 3 years (Phillips et al. 2014 and from unpublished data).

The primary objective of this study was to assess potential candidates for future 'erosion control forests' so comparisons with the root characteristics of the most commonly used species in New Zealand forestry (radiata

Table 7 Allometric relationships for trialled species

Species	Root-collar diameter (x, mm) vs total root length (y, m)	R^2	Root-collar diameter (x, mm) vs below-ground biomass (y, g)	R^2
Redwood	$y = 0.02x^{1.97}$	0.72	$y = 0.02x^{2.36}$	0.94
Oak	$y = 0.03x^{1.81}$	0.97	$y = 0.19x^{1.93}$	0.94
Eucalypt	$y = 0.002x^{2.31}$	0.99	$y = 0.002x^{2.87}$	0.98
Douglas-fir	No data		No data	
Cherry	$y = 0.026x^{2.05}$	0.80	$y = 0.04x^{2.5}$	0.83
Blackwood	$y = 0.017x^{2.01}$	0.91	$y = 0.012x^{2.66}$	0.95
Alder	$y = 0.027x^{2.19}$	0.89	$y = 0.09x^{2.31}$	0.93
Cypress	$y = 0.063x^{1.90}$	0.87	$y = 0.04x^{2.41}$	0.84
Radiata pine	$y = 0.064x^{1.62}$	0.88	$y = 0.014x^{2.46}$	0.96

Relationships are shown between root collar diameter and total root length >1 mm diameter (m) and below-ground biomass (g)

pine) are important. However, direct comparisons for many attributes are not possible largely because earlier studies of radiata root systems only examined "structural roots", *i.e.* those greater than 2 mm in diameter. In addition, these earlier studies did not examine or report the radial distribution of root information. To enable comparison with previous studies, the 1–2 mm diameter class information has to be removed from the current results. For example, three-year-old radiata mean total root length > 2mm diameter from this study was 42.5 m (n = 3, range 9 to 62 m). This is similar to that reported for a Tauranga trial (56 m, n = 5) and more than from a Taranaki trial (15 m, n = 5) (Watson and Tombleson 2002, 2004). It was, however, considerably less than from a hillslope growth factor trial of 3-year-old radiata plantings, which had a mean root length of 95.5 m (n = 25) (Marden unpublished data). Further, mean total root lengths > 2 mm diameter for alder (512 m, n = 1), cypress (232 m, n = 2) and cherry (121 m, n = 2) exceed all reported values for radiata total root lengths at age 3. It is not possible to compare total root length by root diameter and distance from the stem for radiata as these data are not available from those earlier studies. As might be expected, the coarsest roots are located closest to the stem in all the species trialled. Lateral roots closest to the stem tend to increase diameter growth in response to an abiotic stress such as wind or slope (Telewski 1995). Cherry had the greatest length of roots in the largest diameter class (50–100 mm). All species had some roots in the 20–50 mm diameter class. Alder (5.2 m), cherry (3.5 m), and radiata (2.2 m) had the greatest length of roots in this diameter class.

The mean maximum root spread for each trialled species shows a similar pattern to that for total root length > 1 mm, with alder (7.7 m), blackwood (6.9 m) and cypress (5.6 m) having the greatest lateral root spread. Alder and cypress also have greater root density and a more even proportion of total root length from the stem as well as occupying more soil around the stem than the other species tested (Fig. 3, Table 6). The mean maximum root spread of radiata from this trial at Year 3 (3.3 m) was close to that recorded from a growth factor seedling trial on a hillslope near Gisborne (3.8 m) (Marden unpublished data).

Use of trialled species for erosion control

Several of the trialled species have specifically been recommended for, or are commonly used for, erosion control in New Zealand (*e.g.* Van Kraayenoord and Hathaway 1986) and elsewhere (*e.g.* USDA Natural Resource Conservation Centre United States Department of Agriculture and Natural Resources Conservation 2014). Alder species, for example, are used in many countries as parts of bioengineering control programmes (*e.g.* Naghdhi *et al.* 2013; Stokes et al. 2009). Further, some of the trialled species have uses beyond soil stabilisation, such as for timber production (*e.g.* blackwood, redwood) while others are often used for amenity or aesthetics (cherry, oak). Overall, in terms of their potential for root reinforcement and erosion control, the top performers from this trial were alder, cherry, blackwood, radiata, and redwood, with eucalypt and oak showing the least potential largely because their total root lengths were the lowest. It is not possible, however, to provide details on factors that might limit either the establishment of, or use of any particular trialled species across New Zealand as the variation in site requirements are wide in relation to optimum performance. However, the fact that many of these species can be seen growing in a wide range of climatic, soil type, and slope positions across New Zealand suggests that they may be able to be established.

Metrics such as total root length can be used to infer the potential soil reinforcement capability of species through either direct comparison or through modelling site occupancy (Phillips et al. 2011). Results from this trial when compared with those from earlier studies grown on the same site under similar climatic conditions as outlined above, indicate a wide range of performance (Fig. 9). The similar root attributes of alder to the currently preferred soil conservation species poplar and willow (rapid growth, extensive lateral spread) suggest it is a prime candidate for consideration as an erosion control species (notwithstanding potential issues such as 'wildling' spread; *e.g.* Stanley 2002). This supports the view that some introduced species, many of which may be colonising species in their country of origin, may make excellent candidates for use in soil conservation and for controlling shallow landslide erosion (van Kraayenoord and Hathaway 1986). For example, alders have been widely used for revegetating and ameliorating infertile industrially disturbed sites in Europe, the U. S. and Japan and Italian alder (*Alnus cordata* (Lois.) Desf.) has been used to consolidate landslips (ibid).

Because tree root excavations are time-consuming, only a limited number of investigations of whole tree/ plant root systems have been carried out in New Zealand and internationally. Further, time-series growth performance trials in which below-ground tree growth attributes such as biomass, root architecture, and total root length are assessed, are also limited. Many studies, including this one, involve assessment of only one, or a few specimens, of different ages and usually from only one site. This restricts the data available to make inter-species comparisons and also limits the capability to develop statistically-reliable predictive models/tools. However, by using simple allometric relationships such as described above, it may be possible to begin to model more complex relationships between plant numbers, plant performance, and time since planting to assess when forests or groups of trees begin to contribute significantly to soil reinforcement and slope stabilisation (*e.g.* Schwarz et al. 2013). This would help meet a need identified by Stokes et al. (2014) to allay fears in the civil and geotechnical engineering communities, who are still wary about using soft engineering structures and associated vegetation.

Conclusions

This trial of nine exotic tree species has provided important new information and expanded the knowledge base of root growth and performance of young tree species in New Zealand. When compared with similar age *Pinus radiata* root information, all the trialled species other than Douglas-fir (which did not survive) showed some potential for use in erosion control based on their root attributes. However, root-system development beyond the term of this trial and its limitation to only one site and soil type, precludes more definitive statements on the widespread use of the trialled species for erosion control across New Zealand. The best overall performer was alder with cherry, cypress, blackwood, and redwood also performing well across the range of metrics assessed.

Though based on limited data, regression relationships established for individual species between above- and below-ground growth parameters at different stages of growth, are useful for determining planting density requirements for situations where trees are the preferred soil conservation practice to prevent shallow landslides on New Zealand's erosion-prone hill country. These relationships assist in determining when canopy closure or root site occupancy will occur (Phillips et al. 2011) which in turn helps determine when that species might become "effective" in terms of providing an erosion control function. However, the paucity of below-ground tree root data globally still limits our ability to develop and refine predictive models for different species and to refine or

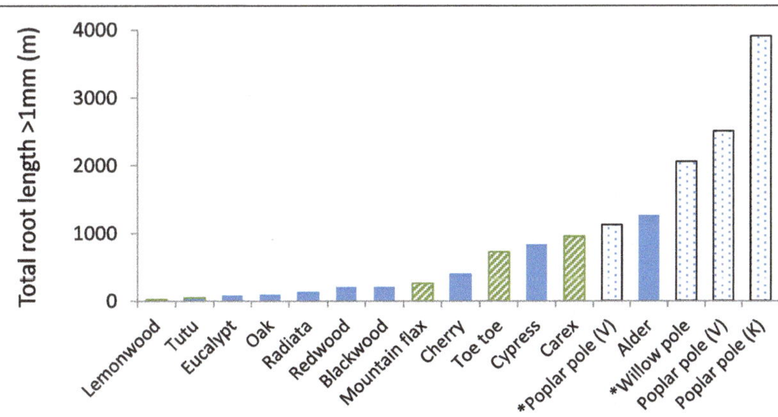

Fig. 9 Total root length (roots > 1 mm diameter) from this trial (solid blue bars) compared with existing 3-year-old data from other species trialled at the same site. Mean values except for toe toe, carex, alder, poplar, and willow. Native species (green diagonal shading - Marden et al. 2005). Poplars (V = 'Veronese'; K = 'Kawa') and willow (lightly stippled bars – C. Phillips unpublished data). * = 2-year-old data

enhance recommendations on tree density and pattern to achieve erosion control. Root observations such as those presented here make a contribution to improving our understanding of how and when, and at what density, plantings become effective for controlling erosion. However, the statistical power of relationships reported here, are acknowledged as being weak, and reflect the generally low sample numbers obtained in most international root system studies for trees beyond a few years old.

If similar trials are contemplated in future to obtain accurate interspecies comparisons in the early years of plant growth, it is recommended that plant materials should be of a similar size at the time of establishment where possible and of similar physiological age, even though nursery and planting practices result in different sized materials being sold and planted.

Endnote

[1]"Coarse" is used here to mean roots greater than 1 mm diameter.

Competing interests
There are no competing interests.

Authors' contributions
CP and MM conceived the idea for the trial during discussions with Scion and contributed to plant extractions and processing. CP drafted the initial manuscript and MM refined it. SL contributed to plant extractions and processing, data management and performed the statistical analysis. All authors read and approved the final manuscript.

Acknowledgements
We thank Richard Hemming, Stephen Hough, interns Sophie Michel, Georg Richter and volunteer Alex Watson for assisting with plant extraction and processing during this trial; Guy Forrester for help in designing the trial; and Christine Bezar for editing services. The Eastland Institute of Technology (formerly Tairawhiti Polytechnic) is thanked for providing the land on which this and earlier plant trials were located. Comments and suggestions from two reviewers have significantly improved the final paper. This research was funded from Future Forests Research and Scion via MBIE contract C04X0806 "Forestry and Environment".

Author details
[1]Landcare Research, PO Box 69040, Lincoln 7640, New Zealand. [2]Landcare Research, PO Box 445, Gisborne 4010, New Zealand. [3]Landcare Research, Private Bag 3127, Hamilton 3240, New Zealand.

References
Basher, LR. (2013). Erosion processes and their control in New Zealand. In JR Dymond (Ed.), *Ecosystem services in New Zealand – conditions and trends* (pp. 363–374). Lincoln, New Zealand: Manaaki Whenua Press.

Blaschke, P, Hicks, D, & Meister, A. (2008). *Quantification of the flood and erosion reduction co-benefits, and co-costs, of climate change mitigation measures in New Zealand*. Wellington: Ministry for the Environment.

Brang, P, Schonenberger, W, Ott, E, & Gardner, B. (2001). Forests as protection from natural hazards. In J Evans (Ed.), *Applying forest science for sustainable management* (pp. 53–81). Oxford: Blackwell Science.

Cole, C, Bergin, D, Kimberley, M. (2014). Establishment performance of native shrubs – a comparison of container and open-ground plants.Technical Article No. 5.4. Tane's Tree Trust. http://maxa.maf.govt.nz/sff/about-projects/search/L12-098/ttt-technical-article-2.pdf. Accessed 16 April 2014.

Crozier, MJ. (2005). Multiple-occurrence regional landslide events in New Zealand: hazard management issues. *Landslides, 2*, 247–256. doi:10.1007/s10346-005-0019-7.

Czernin, A, & Phillips, CJ. (2005). Below-ground morphology of *Cordyline australis* (New Zealand cabbage tree) and its suitability for riverbank stabilisation. *New Zealand Journal of Botany, 43*, 851–864. doi:10.1080/0028825X.2005.9512995.

Danjon, F, Bert, D, Godin, C, & Trichet, P. (1999). Structural root architecture of 5-year-old *Pinus pinaster* measured by 3D digitising and analysed with AMAPmod. *Plant and Soil, 217*(1–2), 49–63. doi:10.1007/978-94-017-3469-1_6.

Docker, BB, & Hubble, TCT. (2008). Quantifying root-reinforcement of river bank soils by four Australian tree species. *Geomorphology, 100*, 401–418. doi:10.1016/j.geomorph.2008.01.009.

Douglas, GB, McIvor, IR, Manderson, AK, Koolaard, JP, Todd, M, Braaksma, S, & Gray, RAJ. (2011). Reducing shallow landslide occurrence in pastoral hill country using wide-spaced trees. *Land Degradation & Development, 24*, 103–114. doi:10.1002/ldr.1106.

Ekanayake, JC, & Phillips, CJ. (1999). A method for stability analysis of vegetated hillslopes: an energy approach. *Canadian Geotechnical Journal, 36*, 1172–1184. doi:10.1139/t99-060.

Ekanayake, JC, & Phillips, CJ. (2002). Slope stability thresholds for vegetated hillslopes: a composite model. *Canadian Geotechnical Journal, 39*, 849–862. doi:10.1139/t02-026.

Hewitt, AE. (2010). *New Zealand soil classification. 3nd edn. Landcare research science series 1*. Lincoln: Manaaki Whenua Press.

Mao, Z, Jourdan, C, Bonis, M, Pailler, F, Rey, H, Saint-André, L, & Stokes, A. (2013). Modelling root demography in heterogeneous mountain forests and applications for slope stability analysis. *Plant and Soil, 363*, 357–382. doi:10.1007/s11104-012-1324-2.

Marden, M, & Rowan, D. (1993). Protective value of vegetation on tertiary terrain before and during Cyclone Bola, east coast, North Island, New Zealand. *New Zealand Journal of Forestry Science, 23*, 255–263.

Marden, M, Rowan, D, & Phillips, CJ. (2005). Stabilising characteristics of New Zealand indigenous riparian colonising plants. *Plant and Soil, 278*, 95–105. doi:10.1007/s11104-004-7598-2.

Marden, M, Rowan, D, & Phillips, C. (2008). Recurrent displacement of a forested earthflow and implications for forest management, East Coast Region, New Zealand. In J Schmidt, T Cochrane, C Phillips, S Elliott, T Davies, & L Basher (Eds.), *Sediment dynamics in changing environments* (pp. 491–501). Wallingford, UK: International Association of Hydrological Sciences. Publication 325.

Marden, M, Arnold, G, Seymour, A, & Hambling, R. (2012). History and distribution of steepland gullies in response to land use change, East Coast region, North Island. *Geomorphology, 153*(154), 81–90. doi:10.1016/j.geomorph.2012.02.011.

McIvor, IR, Douglas, GB, & Benavides, R. (2009). Coarse root growth of Veronese poplar trees varies with position on an erodible slope in New Zealand. *Agroforestry Systems, 76*, 251–264. doi:10.1007/s10457-009-9209-y.

Ministry of Agriculture and Forestry. (2008). *Future drivers for New Zealand forestry*. Wellington, New Zealand: MAF.

Ministry of Agriculture and Forestry (2010). *Sustainable land management hill country erosion programme*. Wellington, New Zealand: MAF. http://www.mpi.govt.nz/funding-and-programmes/natural-resources/sustainable-land-management-and-hill-country-erosion-programme/. Accessed 1 July 2015.

Mulatya, JM, Wilson, J, Ong, CK, Deans, JD, & Sprent, JI. (2002). Root architecture of provenances, seedlings and cuttings of *Melia volkensii*: implications for crop yield in dryland agroforestry. *Agroforestry Systems, 56*, 65–72. doi:10.1023/A:1021165830511.

Naghdi, R, Maleki, S, Abdi, E, Mousavi, R, & Nikooy, M. (2013). Assessing the effect of Alnus roots on hillslope stability to use in soil bioengineering. *Journal of Forest Science, 59*(11), 417–423.

New Zealand Forest Owners Association (2013). http://www.nzfoa.org.nz/plantation-forestry Accessed 1 July 2015

O'Loughlin, CL. (2005). The protective role of trees in soil conservation. *N Z J For, 49*(4), 9–15.

O'Loughlin, CL, & Zhang, X. (1986). *The influence of fast-growing conifer plantations on shallow landsliding and earthflow movement in New Zealand steepland* (pp. 217–226). Ljublijana, Yugoslavia: Proceedings 18th IUFRO World Congress. Division 1, Volume 1.

Parkner, T, Page, M, Marden, M, & Marutani, T. (2007). Gully systems under undisturbed indigenous forest, East Coast Region, New Zealand. *Geomorphology, 84*, 241–253. doi:10.1016/j.geomorph.2006.01.042.

Phillips, CJ, & Marden, M. (2005). Reforestation schemes to manage regional landslide risk. In T Glade, M Anderson, & MJ Crozier (Eds.), *Landslide hazard and risk* (pp. 517–547). Chichester: John Wiley.

Phillips, CJ, & Watson, AJ. (1994). *Landcare research science series 7*. Lincoln, New Zealand: Manaaki Whenua Press. *Structural tree root research in New Zealand, a review*.

Phillips, CJ, Marden, M, & Pearce, AJ. (1990). Effectiveness of reforestation in prevention and control of landsliding during large cyclonic storms. *In Proceedings of International Union of Forest Research Organisations XIX World Congress, 1*, 340–350.

Phillips, CJ, Ekanayake, JC, & Marden, M. (2011). Root site occupancy modelling of young New Zealand native plants: implications for soil reinforcement. *Plant and Soil, 346*, 201–214. doi:10.1007/s11104-011-0810-2.

Phillips, C, Marden, M, & Basher, L. (2012). Plantation forest harvesting and landscape response – what we know and what we need to know. *New Zealand Journal of Forestry Science, 56*(4), 4–12.

Phillips, CJ, Marden, M, Lambie, S, Watson, A, Ross, C, & Fraser, S. (2013a). Observations of below-ground characteristics of young redwood trees (*Sequoia sempervirens*) from two sites in New Zealand – implications for erosion control. *Plant Soil, 363*, 33–48. doi:10.1007/s11104-012-1286-4.

Phillips, CJ, Rey, F, Marden, M, & Liebault, F. (2013b). Revegetation of steeplands in France and New Zealand: geomorphic and policy responses. *New Zealand Journal of Forestry Science, 43*, 14. doi:10.1186/1179-5395-43-14.

Phillips, CJ, Marden, M, & Lambie, S. (2014). Observations of root growth of young poplar and willow planting types. *New Zealand Journal of Forestry Science, 44*, 15. doi:10.1186/s40490-014-0015-6.

Schwarz, M, Lehmann, P, & Or, D. (2010). Quantifying lateral root reinforcement in steep slopes from a bundle of roots to tree stands. *Earth Surface Processes and Landforms, 35*, 354–367. doi:10.1002/esp.1927.

Schwarz, M, Giadrossich, F, & Cohen, D. (2013). Modeling root reinforcement using root-failure Weibull survival function. *Hydrology and Earth System Sciences. Discussion, 10*(3), 3843–3868. doi:10.5194/hessd-10-3843-2013.

Sidle, RC, & Ochiai, H. (2006). *Landslides: processes, prediction, and land use. water resources monograph 18*. Washington, DC: American Geophysical Union.

Sidle, RC, Pearce, AJ, & O'Loughlin, CL. (1985). *Hillslope stability and land use. water resources monograph 11*. Washington, DC: American Geophysical Union.

Smaill, SJ, Bayen, KM, Coker, GWR, Paul, TSH, & Clinton, PW. (2014). The right tree for the job? Perceptions of species suitability for the provision of ecosystem Services. *Environmental Management, 53*, 783–799. doi:10.1007/s00267-014-0239-5.

Stanley, M. (2002). *Alternatives to willows for riverbank protection: more weeds? Protect, (summer), 22–24*.

Stokes, A, Norris, JE, van Beek, LPH, Bogaard, T, Cammeraat, E, Mickovski, SB, Jenner, A, Di Iorio, A, & Fourcaud, T. (2008). How vegetation reinforces soil on slopes. In JE Norris, A Stokes, SB Mickovski, E Cammeraat, R van Beek, BC Nicoll, & A Achim (Eds.), *Slope stability and erosion control: ecotechnological solutions (pp. 65–118)*. Dordrecht: Springer.

Stokes, A, Atger, C, Bengough, A, Fourcaud, T, & Sidle, R. (2009). Desirable plant root traits for protecting natural and engineered slopes against landslides. *Plant and Soil, 324*, 1–30. doi:10.1007/s11104-009-0159-y.

Stokes, A, Douglas, GB, Fourcaud, T, Giadrossich, F, Gillies, C, Hubble, T, Kim, JH, Loades, KW, Mao, Z, McIvor, IR, Mickovski, SB, Mitchell, S, Osman, N, Phillips, C, Poesen, J, Polster, D, Preti, F, Raymond, P, Rey, F, Schwarz, M, & Walker, LR. (2014). Ecological mitigation of hillslope instability: ten key issues facing researchers and practitioners. *Plant and Soil, 377*, 1–23. doi:10.1007/s11104-014-2044-6.

Telewski, FW. (1995). Wind-induced physiological and developmental responses in trees. In MP Coutts & J Grace (Eds.), *Wind and Trees* (pp. 237–263). Cambridge, UK: Cambridge University Press.

Trotter, C, Tate, K, Scott, N, Townsend, J, Wilde, H, Lambie, S, Marden, M, & Pinkney, T. (2005). Afforestation/reforestation of New Zealand marginal pasture lands by indigenous shrublands: the potential for Kyoto forest sinks. *Annals of Forest Science, 62*, 865–871.

United States Department of Agriculture, Natural Resources Conservation Service (2014) http://plants.usda.gov/java/ Accessed 1 July 2015.

Van Kraayenood, CWS, & Hathaway, RL. (1986). *Plant Materials Handbook for Soil Conservation. Volume 2: Introduced Plants. Water and Soil Miscellaneous Publication No. 94*. Wellington, New Zealand: Ministry of Works and Development.

Watson, A, & O'Loughlin, C. (1990). Structural root morphology and biomass of three age-classes of *Pinus radiata*. *New Zealand Journal of Forestry Science, 20*(1), 97–110.

Watson, AJ, & Tombleson, JD. (2002). Toppling in juvenile pines: a comparison of the root system characteristics of direct-sown seedlings, and bare-root seedlings and cuttings. *Plant and Soil, 239*, 187–196.

Watson, AJ, & Tombleson, JD. (2004). Toppling in young pines: temporal changes in root system characteristics of bare-rooted seedlings and cuttings. *New Zealand Journal of Forestry Science, 34*(1), 39–48.

Watson, AJ, Phillips, CJ, & Marden, M. (1999). Root strength, growth, and rates of decay: root reinforcement changes of two tree species and their contribution to slope stability. *Plant and Soil, 217*, 39–47. doi:10.1023/A:1004682509514.

6

Detection of asymptomatic fungal microorganisms in *Pinus radiata* tissue culture material

Rebecca J Ganley*, Cathy L Hargreaves and Lloyd A Donaldson

Abstract

Background: Tissue culture has been viewed as a relatively safe method of transporting conifer germplasm as the risk of pathogens associated with tissue cultured material has been assumed to be low. Despite these assumptions, it is unknown whether tissue cultured propagation material contains microorganisms or whether such microorganisms could grow out onto the media used.

Methods: To determine whether asymptomatic tissue culture material contains fungal microorganisms, 30 different *Pinus radiata* D.Don genotypes from four different types of tissue culture material were analysed for the presence of fungal DNA or mycelium. In addition, thirteen fungal and oomycete isolates were cultured on standard tissue propagation media.

Results: Fungal DNA was detected in all samples tested. No fungal or bacterial microorganisms were able to be grown from any of the tissue culture material tested. However, confocal microscopy showed the presence of filaments that could have been fungal mycelium. Growth of thirteen fungal and oomycete isolates on standard tissue propagation media indicated these media can severely limit the growth of some of these microorganisms.

Conclusion: The results from this study show that tissue culture material does contain asymptomatic fungal microorganisms but the exact risk from these microorganisms is unclear.

Keywords: Biosecurity risk; Fungal endophytes; Confocal microscopy; Cryptic infections; *Phytophthora*

Background

The perceived threat that the fungus *Fusarium circinatum*, the causal agent of pitch canker disease, poses to the plantation forest industry has rigorously curtailed the importation of some conifer material into New Zealand. There is currently a complete ban on the importation of any live plant material and restrictions on the importation of seed from *Pseudotsuga menziesii* (Mirb.) Franco (Douglas-fir) or any *Pinus* species (Ministry for Primary Industries 2010). However, new breeding material is required for genetic improvement of existing stocks, to enhance health traits, including resistance to disease, as well as improving productivity and wood quality. Thus, restrictions on the importation of conifer germplasm are of concern to the New Zealand forest industry.

In an effort to find alternative methods to import new *Pinus* spp. or *Pseudotsuga menziesii* material into New Zealand, the importation of live material in tissue culture has been suggested. Tissue culture has been viewed as a relatively safe method of transporting conifer germplasm for a variety of reasons. Firstly, there are very few viruses or bacterial pathogens associated with *Pinus* spp. or *Pseudotsuga menziesii* and those that are known are not considered to be serious pathogens (Sinclair and Lyon 2005; Veliceasa et al. 2006). Secondly, it has been assumed that any fungal or bacterial pathogens present in tissue culture material would grow out onto the growth media. Thirdly, the likelihood that fungi could survive the intensive nature of many currently used tissue culture procedures is thought to be low.

Despite these assumptions, it is unknown whether tissue cultured propagation material contains microorganisms or whether the tissue propagation media would

* Correspondence: Rebecca.Ganley@scionresearch.com
Scion, New Zealand Forest Research Institute Ltd., 49 Sala Street, Rotorua 3046, New Zealand

contain the required nutrients to support the growth of such microorganisms. Furthermore, the discovery of unculturable microorganisms in plant cultures grown *in vitro* is becoming increasingly common (Thomas 2011). Although these reports have mainly dealt with bacteria that are able to survive antibiotic cleansing treatments, it is possible that similar asymptomatic associations could also occur with fungi in conifer tissue culture. In addition to the possibility of undesirable fungi, *Pinus* spp. are known to contain a wide diversity of fungal endophytes (Deckert and Peterson 2000; Ganley and Newcombe 2006) and total sequencing approaches indicate the number of unculturable fungi present may exceed initial diversity estimates (Fröhlich and Hyde 1999; Hawksworth 2001). The function of fungal endophytes in their hosts is largely unclear (Rodriguez et al. 2009). It is likely that many are involved in beneficial symbiotic associations and it is possible that some may be so intricately linked that they are required for host survival.

The purpose of this study was to determine whether asymptomatic *Pinus radiata* D.Don (radiata pine) tissue culture material contains fungal microorganisms. This species was chosen for study because of its importance as a commercial species in New Zealand. Also, if microorganisms are present, what species are there, what might be their ecological significance, are they viable and do they pose a biosecurity risk? In addition, selected tissue propagation media were tested to see if they could support growth of a selection of fungi and one *Phytophthora* species associated with *Pinus radiata*.

Methods

Tissue culture preparation

Four types of tissue culture material were prepared from *Pinus radiata*: embryogenic tissue derived from immature zygotic embryos; adventitious-axillary organogenesis shoots derived from cotyledons dissected from mature zygotic embryos; epicotyl-axillary organogenesis shoots derived from the epicotyls of mature zygotic embryos; and mature organogenesis shoots derived from field grown material and established as *in-vitro* shoot cultures. The same ten genotypes were used for both the adventitious-axillary and epicotyl-axillary organogenesis shoots. Ten different genotypes were used for the embryogenic tissue and another 10 for the mature organogenesis shoots so 30 different genotypes were tested in total. Embryogenic tissue was initiated using techniques described by Walter and Grace (2000) and Walter et al. (2005). The adventitious-axillary organogenesis shoots were prepared from excised cotyledons that were cryopreserved at – 196 °C and then had shoots initiated adventitiously. Non- germinated hypocotyls (including epicotyls) from the embryos that provided the cotyledons for cryopreservation were retained and these provided the non-cryopreserved epicotyl-axillary shoots, thus material was available from two methodology origins using the same genotypes (Hargreaves et al. 2004, 2005). Mature material was established using techniques described by Horgan and Holland (1989).

DNA extractions

Total DNA was extracted from each genotype/cell line for the four different types of *Pinus radiata* tissue culture material prepared. Approximately one gram of embryonic callus material from each cell line was ground to a fine powder in liquid nitrogen using a mortar and pestle. For the adventitious-axillary, epicotyl-axillary and mature organogenesis shoots, one shoot from each cell line was ground to a fine powder in liquid nitrogen using a mortar and pestle. Extraction of DNA was then performed using a DNeasy Plant Mini Kit (Qiagen, California, USA), according to the manufacturer's instructions. For genotypes/cell lines with tissue or shoots that weighed more than 100 mg, the tissue culture powder obtained was separated into samples less than 100 mg and individual DNA extractions were performed on each sample.

PCR amplification

To confirm that the DNA extractions worked, the total extracted DNA was subjected to PCR amplification using plant specific primers ConiferF (5′-GGG AGG TTT TAC TGC AAA TAC TTC-3′) and ConiferR (5′-CAT CCC CAA AGA TCT CGG TCA GAG CAG GC-3′) (Marler et al. 1999). The PCR reaction mixture was prepared in a total volume of 15 μL using 10 mM Tris-hydrochloric acid at pH 8.3, 50 mM potassium chloride, 1.5 mM magnesium chloride, 0.2 mM dNTPs, 0.2 mg/mL bovine serum albumin, 5 pmol of each primer, 1.5 U *Taq* DNA polymerase and 1 μL total extracted DNA. All PCR amplifications were performed according to Marler et al. (1999). After PCR amplification, products were run on 1–2 % agarose gels and examined for bands of the expected size to confirm the DNA extractions were successful.

To determine if fungal DNA was present in the tissue culture material, total extracted DNA was tested with fungal-specific nested primers designed to amplify the internal transcribed spacers (ITS) of the fungal ribosomal RNA gene region (rDNA). The fungal-specific primary PCR primers used were NSI1 (5′-GATTGAATGGCT TAGTGAGG) and NLB4 (GGATTCTCACCCTCTAT GAC-3′) (Martin and Rygiewicz 2005), and the nested PCR primers were ITS1-F (5′-CTTGGTCATTTAGAG GAAGTAA) and ITS4 (TCCTCCGCTTATTGATATGC-3′) (White et al. 1990). The PCR reaction mixture was prepared as described above. For the primary PCR, 1 μL total extracted DNA was added to the reaction mixture, whereas 1 μL of purified PCR product was used for the nested PCR. The purified PCR product was prepared by adding 5 μL of the primary round PCR product to 10 U Exonuclease I (Fermentas) and 2 U Shrimp Alkaline Phosphatase (USB) and incubating on a thermal cycler at 37 °C for 15 mins then 80 °C 15 mins. The primary and nested PCR reactions were incubated in a thermal cycler at 95 °C for 6 min followed by 35 cycles of 95 °C for 30 sec, 60 °C for 40 sec, 72 °C for 40 sec, and finally 72 °C for 5 min. A high annealing temperature was used so that single band products, if present, could be later sequenced. After PCR amplification, products were run on 1–2 % agarose gels and inspected for bands of the expected size.

Each of the ten cell lines/genotypes for the four different tissue culture types were tested for the presence of fungal DNA. In addition, a negative control (water only) was included. For the nested PCR reactions, the negative control from the previous PCR round was PCR amplified as well as a new negative control.

Fungal DNA identification

Products amplified from the nested PCR reactions using the fungal-specific primers were sequenced. Only nested PCR products with single bands were sequenced, and these were purified using Exonuclease I and Shrimp Alkaline Phosphatase as described above. The resultant purified products were sequenced by Macrogen (Seoul, Korea). The sequences obtained were then compared with those present in the National Center for Biotechnology Information (NCBI) GenBank sequence database using the BLAST search program to confirm the products were fungal in origin and to determine closest sequence-based match. Only sequences that showed single peaks were used; sequences that had double peaks, indicating more than one species present, were discarded.

Growth of fungi from tissue culture material

To determine whether any culturable fungi were present, embryogenic tissue and shoots of *Pinus radiata* were plated onto four different general fungal media. The shoot material of these species originated from field collected material that was sterilised and used to establish organogenic cultures. Five genotypes were tested that were different from those used in the DNA analysis. Shoot bases with associated callus from organogenic cultures of *Pinus radiata* were used on the test media. The shoot bases were chosen as test material as they have had the longest contact with the growing shoots. The four types of general fungal media used were 4 % (w/v) potato dextrose agar (PDA), 2 % (w/v) malt extract agar (MEA), carrot agar (CA) and prune agar (PA). The CA was prepared by liquidising 200 g carrots in 1 L distilled water with 15 g agar and the PA was made using 50 mL prune juice, 2.5 g lactose, 0.5 g yeast extract, 10 g agar and 450 mL distilled water, adjusted to pH 5.8–6.0. Embryogenic tissue and shoots were plated onto each type of media, the embryogenic callus was simply placed on the test media surface. The shoot bases were cut length-wise to increase contact of the tissue with the test media. The tissue culture material was incubated in standard growth room conditions (Hargreaves et al. 2005) and monitored for 10 days for any sign of bacterial or fungal growth.

Confocal microscopy

Ten samples of *Pinus radiata* somatic embryogenic tissue growing in sterile culture were examined by a range of light microscopy techniques. Tissue was examined by differential interference contrast (DIC), and by confocal fluorescence microscopy after staining with Alexa labelled-wheat germ agglutinin (WGA), a lectin which binds to chitin in some types of fungal hyphae, or by Acriflavin, a general purpose fluorescent stain. Cotton blue and toluidine blue stains were also used on some samples.

Growth of microorganisms on standard tissue propagation media

All isolates were obtained from the Forest Research Culture Collection at Scion. Twelve fungal (*Amylostereum aerolatum* NZFS 2534, *Armillaria novae-zelandiae* NZFS 1973, *Colletotrichum acutatum* NZFS 7 T/3, *Cyclaneusma minus* NZFS 682, *Cylindrocladium scoparium* NZFS 1828, *Dothistroma septosporum* NZFS 2728, *Fusarium oxysporum* NZFS 2952, *Junghuhnia vincta* NZFS 2473, *Leptographium procerum* NZFS 626, *Lophodermium conigenum* NZFS 709, *Pestalotiopsis* sp. NZFS 3259, *Strasseria geniculata* NZFS 2998) and one oomycete (*Phytophthora cinnamomi* NZFS 2968) were selected to encompass the wide array of microorganisms known to be associated with *Pinus radiata*.

All isolates were initially grown on MEA or CA for *Phytophthora cinnamomi*. For each isolate, a 5 mm plug of mycelium obtained from these plates was placed in the centre of a 90 mm diameter Petri dish containing one of the five nutrient media. The four standard tissue culture media were: shoot initiation medium ½LP5 (Hargreaves and Menzies 2007), shoot multiplication medium LPch (Hargreaves and Menzies 2007), embryogenesis medium EDM6 (Hargreaves et al. 2009) and embryogenesis medium Glitz (Hargreaves et al. 2009). The Petri dishes were sealed and placed in a culture cabinet at 20 °C in the dark. Three replicates were completed for each of the five nutrient media. The diameter of each colony was measured in two directions at 7, 14, 21, 28 and 35 days.

The average colony diameter ± standard deviation was calculated from the three replicates on each of the five nutrient media using SAS software. Analyses of variance were also performed to determine any significant difference in colony growth on the five nutrient media for each of the 13 microorganisms at 7, 14, 21, 28 and 35 days. Fisher's Least Significant Difference (LSD) ($\alpha = 0.05$) was used to determine significant differences between the five nutrient media when sample sizes were equal ($n = 30$). When sample sizes were unequal ($n < 30$), due to contamination, the Student-Newman-Keuls procedure ($\alpha = 0.05$) was used.

Results

Fungal DNA in tissue culture

Total DNA was successfully extracted from all the tissue culture samples. Amplification of PCR using the general nested fungal primers NSL1/NLB4 and ITS1-F/ITS4 detected DNA in all the tissue culture material tested and sequencing of single band PCR products confirmed the DNA was fungal (Table 1). The number and intensity of the bands amplified varied among the samples (Fig. 1). All four tissue culture types produced multiple and single band PCR products, including the cryopreserved (adventitious-axillary) shoot cultures. Analysis of DNA sequences obtained from single band PCR products showed a variety of different fungal microorganisms present (Table 1).

Growth on general fungal media

No fungal or bacterial growth was observed from any of the tissue culture material plated onto the four general fungal media (PDA, MEA, CA and PA) tested (data not shown).

Confocal microscopy

There was no clear unambiguous evidence of fungal mycelium in any of the tissue culture samples analysed. None of the stains used were useful in highlighting any low level fungal infection and the embryogenic tissue reacted with the WGA reagent reducing the potential for specific labelling of hyphae amongst the mass of tissue. However, filaments were observed in four out of the 10 samples examined. It is possible that these filaments were fungal hyphae. The cytoplasmic appearance of these filaments would suggest they were alive. There were two types of structures observed that might represent fungal mycelium. The first were bundles of filaments 1–5 μm wide with granular cytoplasmic contents that appeared relatively stiff compared with suspensors (Fig. 2). These structures were non-pigmented, and WGA negative. No evidence of clamp connections could be seen. The second were minute filaments attached to the outer surface of suspensor cells (Fig. 3). These filaments were definitely outside of the cell wall of the suspensor and had a cytoplasmic appearance. The cell wall was very thin so was difficult to assess but appeared to be WGA positive.

Growth on standard tissue propagation media

Growth of the microorganisms on standard tissue propagation and general fungal/oomycete media differed for each of the 13 isolates tested (Table 2, data only shown for six of the isolates tested). There were significant differences in growth on the five different media for all of the isolates except one, *Junghuhnia vincta*. There was no difference in growth for *J. vincta* on any of the media tested and by 21 days this isolate had grown to the edge on all Petri dishes. For two of the isolates, *Cylindrocladium scoparium* and *Dothistroma septosporum*, there was no significant difference in growth at 7 days, but after 21 days growth differences were significant (Table 2). Conversely, *Colletotrichum acutatum* showed significant differences in growth until 35 days when the isolate grew to the edge of the Petri dish on all plates. The remainder of the isolates showed significant differences in growth over the entire 35 days.

The majority of isolates grew better on the five nutrient media in the following order: MEA (CA for *Phytophthora cinnamomi*), Glitz, EDM6, ½LP5, LPch. The only two isolates that did not grow best on the general fungal medium (MEA) were *Colletotrichum acutatum* and *Lophodermium conigenum*, which both grew best on Glitz. Conversely, the isolates tended to grow slowest on LPch, with 11 of the isolates growing significantly slower on LPch than on MEA (CA for *Phytophthora cinnamomi*) and two of these isolates (*Cyclaneusma minus* and *L. conigenum*) did not grow on the LPch medium beyond the original plug. The only isolates with no significant difference in growth between these two media were *Cylindrocladium scoparium* and *J. vincta*. *Colletotrichum acutatum* grew significantly faster on MEA than LPch from 7–21 days, although there was no difference in growth after 28 days.

Discussion

The results from this study show that *Pinus radiata* tissue culture material is able to carry cryptic infections of asymptomatic fungal microorganisms. All of the tissue culture material tested was found to carry fungal DNA and multiple bands were found in many of the DNA samples tested, indicating more than one species present. Overall, this suggests that fungal DNA is present in the embryo and is propagated throughout the tissue culture process, also indicating that the fungal microorganisms present are viable. If the fungal microorganisms present were non-viable then fungal DNA would be expected to be found only in the embryonic tissue and not be propagated through to the adventitious- or epicotyl-axillary organogenesis shoot material. Hence the presence of fungal DNA, and the implied viability of the microorganisms, in the adventitious-axillary shoots indicates that cryopreservation has no effect on the ability of this material to harbour cryptic fungal microorganisms and cannot

Table 1 Fungal DNA present in *Pinus radiata* tissue culture material

Tissue culture type and genotype number	PCR product amplified[a]	Closest GenBank match of sequenced single-band PCR products		
		Fungal microorganism	Accession number	Sequence identity (%)
Embryogenic tissue				
1	yes	-	-	-
2	yes	-	-	-
3	yes	-	-	-
4	yes	Uncultured ectomycorrhiza	EF218819	98
5	yes	-	-	-
6	yes	*Mycena rubromarginata*	EF530939	91
6	yes	*Botryosphaeria parva*	AM410965	99
7	yes	-	-	-
8	yes	-	-	-
9	yes	-	-	-
10	yes	*Resincium bicolor*	AF518763	99
Adventitious-axillary organogenesis shoots				
11	yes	*Epicoccum nigrum*	AF455403	99
12	yes	*Trametes* sp.	AY968077	75
13	yes	-	-	-
14	yes	-	-	-
15	yes	*Oligoporus placentus*	AJ249267	79
16	yes	*Drechslera erythrospila*	AY004782	99
17	yes	*Cladosporium* sp.	EF432298	99
18	yes	*Oligoporus placentus*	AJ249267	79
19	yes	Basidiomycete	AF519892	79
20	yes	-	-	-
Epicotyl-axillary organogenesis shoots				
11	yes	Fungal sp.	AM231388	98
12	yes	-	-	-
13	yes	-	-	-
14	yes	*Armillaria hinnulea*	AF394918	97
15	yes	-	-	-
16	yes	*Coriolaceae*	AY840564	93
16	yes	*Botryobasidium subcoronatum*	AJ389785	95
17	yes	*Cyclaneusma minus*	KM216324	100
18	yes	-	-	-
19	yes	*Phellinus* sp. or *Trametes* sp.	AF251437/ AY840588	99
20	yes	*Rhodotorula yarrowii*	AF444628	98
Mature organogenesis shoots				
21	yes	*Trametes versicolor*	AF13996	95
22	yes	-	-	-
23	yes	-	-	-
24	yes	*Petriella sordida*	AY882360	98
25	yes	-	-	-
26	yes	-	-	-
27	yes	-	-	-

Table 1 Fungal DNA present in *Pinus radiata* tissue culture material ***(Continued)***

28	yes	*Resinicium bicolour*	AF518763	99
29	yes	-	-	-
30	yes	-	-	-

[a]PCR products generated from nested PCR amplification using primers NSL1/NLB4 and ITS1-F/ITS4 (see Methods and materials)

be relied on as a sanitising method for tissue culture material.

Mature organogenesis shoot material can have a high level of fungal contamination, which grows out onto the tissue culture media used during the first stages of the tissue propagation process and is discarded as contaminated. The majority of the contaminants are likely to be fungal endophytes, which are known to reside within mature needle tissue (Ganley and Newcombe 2006; Johnson and Whitney 1989). The remainder of the fungal microorganisms retained in the asymptomatic mature shoot material, as identified in this study, are likely to represent non-culturable fungal assemblages present in mature needle tissue. A high annealing temperature was used in this study so single-band products could be obtained. If a lower annealing temperature was used it is expected more species would have been amplified, if present in the tissue culture material. To get a better understanding of the fungal biodiversity present, the use of next-generation technologies would be recommended, and the use of quantitative PCR methods could be informative in understanding the extent of colonisation for specific species.

A BLAST search analysis of the single-band PCR products generated revealed a variety of fungal species present. The majority of the GenBank taxa with close sequence identity matches to the tissue culture sequences were fungi known to have conifer associations, had been isolated as fungal endophytes from other plant species or had been obtained from environmental samples from forest settings, according to their GenBank accession details. There was a wide variation in the sequence identity to the closest GenBank match, ranging from 79 to 100 % identity. The *Cyclanuesma minus* sequence that was 100 % identical to sequences on GenBank matches shows known fungal associates of *Pinus radiata* are able to survive the tissue culture process. *Cyclaneusma minus* is a pathogen of *Pinus radiata* that causes a disease known as cyclaneusma needle cast (Gadgil 1984). However, it can be readily isolated from healthy *P. radiata* needles suggesting it also has an endophytic phase (unpublished data). Molecular analysis has indicated that *C. minus sensu lato* contains multiple species (Prihatini et al. 2014) and potentially the species within this complex may encompass a range of pathogenic, saprophytic or endophytic behaviours.

For the remainder of the putative fungal isolates identified, whether the tissue culture sequences represent fungal microorganisms that are the same species or genus as their closest GenBank match is unclear. Likewise, the possible function of these microorganisms within *Pinus radiata* tissue is unknown. Fungal

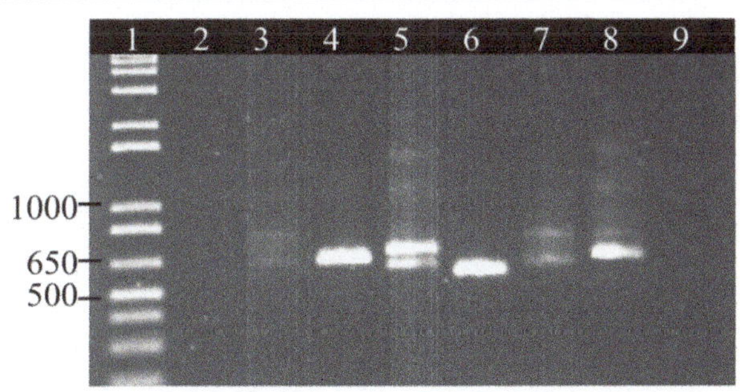

Fig. 1 PCR products generated from nested PCR amplification using primary primers NSL1/NLB4 followed by the nested primers ITS1-F/ITS4, visualised on a 1 % agarose gel. Lane 1: 1 kb ladder, band sizes (bp) are indicated on the left. Lane 2: primary negative control, NSL1/NLB4 negative control re-amplified with nested primers ITS1-F/ITS4. Lane 3–8: adventitious-axillary shoot PCR products, Lane 3: genotype number 19; Lane 4: 16; Lane 5: 13; Lane 6: 11; Lane 7: 15; and Lane 8: 18. Lane 9: secondary negative control, amplified with primers ITS1-F/ITS4 only

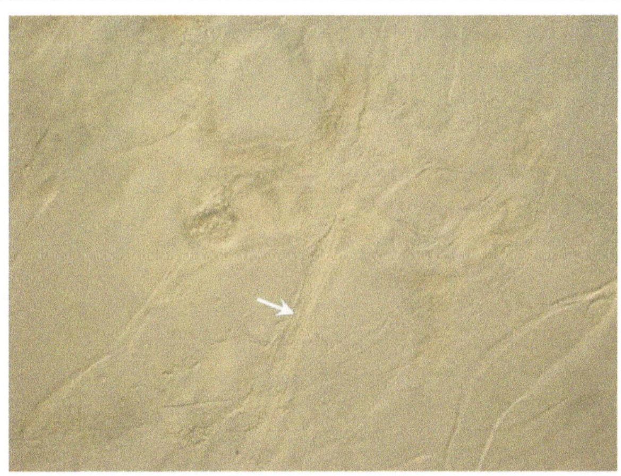

Fig. 2 Potential strands of 2–3 mycelia amongst the embryo tissue (*arrow*). Field of view 1000 × 750 μm

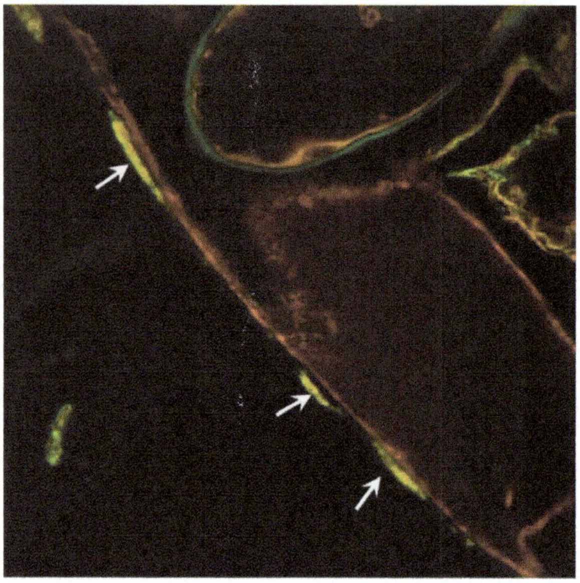

Fig. 3 Potential mycelium (*arrows*) adhering to the outside surface of a suspensor cell. Confocal fluorescence image of WGA stained tissue. Field of view 159 × 159 μm

organisms that live entirely endophytic lifestyles are unlikely to pose a biosecurity risk, in contrast to fungi such as *F. circinatum* that are pathogenic but can live endophytically within their hosts (Storer et al. 1998). Although host switching or the discovery of unrelated host species of pathogens does occur, for example the recent discovery that corn is a hosts of the pine pathogen *F. circinatum* (Swett and Gordon 2015), the frequency of such occurrences are low.

Fungal DNA was found in all tissue culture samples tested but no fungal microorganisms grew out of the tissue culture material plated on the general fungal media tested. The lack of fungal outgrowth was not surprising as most of the closely related GenBank taxa were listed as uncultured or were from DNA-based assessments of environmental samples. Others which are known to be readily culturable would have the potential to grow out on general fungal media. However, it is not known if these fungal microorganisms are capable of outgrowth onto general fungal media during the endophytic or latent phase of their lifecycle. The abundance of these fungal microorganisms in the tissue culture material is also likely to be very low based on the low number of filaments observed microscopically and the lack of amplification of fungal DNA using single primer sets. It is possible that the sample of tissue culture material tested did not contain culturable fungal microorganisms.

The results from this study show that standard tissue culture propagation media can severely limit the growth of some fungal and oomycete microorganisms. In view of these results, it cannot be assumed that all pathogens present in tissue culture material would be able grow out and be detectable. For the isolates that had severely restricted growth on some of the tissue propagation media, especially LPch, it is likely that the original agar plug of mycelium was able to support limited growth of the isolates. It is unknown whether these isolates would have been able to grow at all on these media in the absence of the original agar plug, such as would occur under normal tissue propagation conditions.

In general, the length of time that tissue culture material is grown on EDM6, Glitz or ½LP5 should be sufficient for any culturable fungi present to grow. Callus tissue derived from zygotic embryonic tissue is usually grown on EDM6 or Glitz for 10–14 days before it is sub-cultured and can be grown on these media for up to 12 months until plants are formed. For most of the fungi tested, and *Phytophthora cinnamomi*, a detectable colony should have grown within this timeframe as growth tended not to be as restricted on these media. Likewise for ½LP5, a medium used for between 7–21 days for shoot initiation, colonies should be detectable if left on this medium for longer than seven days. As a shoot multiplication medium which contains charcoal, LPch is generally used for between 28–42 days; growth on this medium would be dependent on the microorganisms present.

Conclusions

The exact risk of importation of asymptomatic fungal microorganisms in tissue culture material is unclear. This study indicates that fungi do occur on *Pinus radiata* tissue culture material and that these microorganisms are alive. However, there is no indication of long-term survival of these microorganisms, whether they would be functional (either positively or negatively) in the host plant or whether they could become established outside of their host in a new environment. For example, current scientific understanding of the symbiotic relationship between endophytic microorganisms and their hosts is poor so it is possible these microorganisms could be intricately involved in plant survival, and without some of them the plant may not survive. Likewise, the transmission of endophytes is poorly understood; fungal endophytes frequently do not sporulate in culture and observations of fruiting bodies under natural conditions are uncommon. Finally, the risk of these microorganisms becoming pathogenic in their host or host switching under new environmental conditions is expected to be low. Further research is required to confirm whether these microorganisms are alive in tissue culture material and to elucidate whether they pose a biosecurity risk.

Table 2 Average colony diameter for fungal and oomycete isolates grown on standard tissue propagation and general fungal or oomycete media

		Average colony diameter ± standard deviation on nutrient media (mm) [significant difference][a]							
Taxa	Day	MEA	Glitz	EDM6	½LP5	LPch	n	F	P
Amylostereum aerolatum	7	26.7 ± 2.2 a	21.0 ± 3.1 b	10.8 ± 1.0 c	10.0 ± 2.8 c	12.3 ± 1.0 c	26	54.7	<0.0001
	14	76.3 ± 2.0 a	67.3 ± 3.3 b	52.5 ± 2.9 c	33.3 ± 8.4 d	16.0 ± 0.8 e	26	142.4	<0.0001
	21	85.0 ± 0.0 a	83.5 ± 1.5 a	75.5 ± 2.6 b	52.2 ± 3.3 c	17.3 ± 1.0 d	26	874.3	<0.0001
	28	85.0 ± 0.0 a	85.0 ± 0.0 a	78.5 ± 3.7 b	54.3 ± 2.3 c	21.0 ± 2.9 d	26	787.1	<0.0001
	35	85.0 ± 0.0 a	85.0 ± 0.0 a	82.8 ± 1.0 a	54.8 ± 2.3 b	24.0 ± 2.6 c	26	1389.2	<0.0001
Colletotrichum acutatum	7	34.3 ± 1.6 c	42.5 ± 1.4 a	38.5 ± 2.7 b	40.3 ± 2.5 ab	17.2 ± 1.8 d	30	143.3	<0.0001
	14	68.0 ± 5.4 a	69.8 ± 3.8 a	64. 8 ± 4.3 a	69.8 ± 4.2 a	44.5 ± 3.1 b	30	38.8	<0.0001
	21	84.2 ± 2.0 a	84.2 ± 1.3 a	75.2 ± 6.7 b	74.5 ± 2.4 b	71.5 ± 6.1 b	30	11.0	<0.0001
	28	85.0 ± 0.0 a	85.0 ± 0.0 a	81.0 ± 3.1 b	85.0 ± 0.0 a	83.7 ± 1.2 a	30	8.2	0.0002
	35	85.0 ± 0.0	85.0 ± 0.0	84.0 ± 1.5	85.0 ± 0.0	85.0 ± 0.0	30	2.5	0.0681
Cyclaneusma minus	7	16.8 ± 0.8 a	9.3 ± 1.5 c	10.5 + 0.5 b	6.2 ± 0.4 d	6.3 ± 0.5 d	30	158.5	<0.0001
	14	32.5 ± 2.0 a	15.7 ± 2.7 c	18.7 ± 1.2 b	9.0 ± 1.3 d	6.2 ± 0.4 e	30	224.1	<0.0001
	21	49.5 ± 1.4 a	26.8 ± 3.8 b	25.3 ± 1.6 b	14.7 ± 2.1 c	6.3 ± 0.5 d	30	341.0	<0.0001
	28	59.8 ± 3.5 a	32.8 ± 3.4 b	32.3 ± 1.8 b	18.7 ± 2.4 c	6.5 ± 0.5 d	30	354.9	<0.0001
	35	65.7 ± 3.7 a	37.3 ± 3.6 b	37.7 ± 2.6 b	23.0 ± 2.6 c	6.8 ± 0.8 d	30	349.0	<0.0001
Dothistroma septosporum	7	8.0 ± 2.4	7.5 ± 1.5	9.2 ± 2.9	6.2 ± 0.8	6.5 ± 0.5	30	2.5	0.0674
	14	13.7 ± 1.8 a	10.3 ± 1.2 b	10.8 ± 1.2 b	8.3 ± 1.2 c	8.0 ± 0.9 c	30	19.1	<0.0001
	21	19.3 ± 2.1 a	13.5 ± 1.5 b	15.0 ± 2.4 b	13.0 ± 1.3 b	8.8 ± 1.0 c	28	24.6	<0.0001
	28	25.8 ± 0.5 a	15.5 ± 1.2 c	19.2 ± 2.0 b	15.5 ± 2.1 c	12.0 ± 1.4 d	28	48.5	<0.0001
	35	32.5 ± 1.7 a	18.8 ± 1.2 c	23.5 ± 2.3 b	19.8 ± 2.4 c	15.2 ± 1.3 d	28	57.2	<0.0001
Junghuhnia vincta	7	85.0 ± 0.0	85.0 ± 0.0	85.0 ± 0.0	83.0 ± 3.6	81.8 ± 3.8	30	2.4	0.0787
	14	85.0 ± 0.0	85.0 ± 0.0	85.0 ± 0.0	85.0 ± 0.0	83.8 ± 2.9	30	1	0.4261
	21	85.0 ± 0.0	85.0 ± 0.0	85.0 ± 0.0	85.0 ± 0.0	85.0 ± 0.0	30	-	-
Phytophthora cinnamomi	7	82.0 ± 0.0 a	71.0 ± 6.9 b	74.3 ± 2.9 b	58.3 ± 2.7 c	16.5 ± 1.9 d	30	303.5	<0.0001
	14	85.0 ± 0.0 a	83.3 ± 2.0 ab	82.0 ± 1.9 b	85.0 ± 0.0 a	18.8 ± 2.5 c	30	1829.6	<0.0001
	21	85.0 ± 0.0 a	85.0 ± 0.0 a	84.5 ± 0.9 a	85.0 ± 0.0 a	19.8 ± 2.0 b	30	5215.9	<0.0001
	28	85.0 ± 0.0 a	85.0 ± 0.0 a	85.0 ± 0.0 a	85.0 ± 0.0 a	23.2 ± 3.9 b	30	1532.8	<0.0001
	35	85.0 ± 0.0 a	85.0 ± 0.0 a	85.0 ± 0.0 a	85.0 ± 0.0 a	40.0 ± 24.3 b	30	20.6	<0.0001

[a]Average colony diameters with the same letter were not significantly different; those with different letters were significantly different

Competing interests
The authors declare that they have no competing interests.

Authors' contributions
All authors have contributed substantially to this manuscript and were involved in planning, data analyses and manuscript writing. RG carried out the DNA and growth media experiments; CH prepared all tissue culture material; and LD carried out the microscopy experiments. All authors read and approved the final manuscript.

Acknowledgments
Dolina Skudder and Keiko Gough are thanked for their support with initiation, maintenance and preparation of the plant tissue culture material; Christina Maher and Rita Tetenburg for preparation of the media; Matt Power and Tracey Bates for their assistance with colony measurements. This work was funded by New Zealand's Foundation for Research, Science and Technology through the Better Border Biosecurity (B3) Programme.

References
Deckert RJ, & Peterson, L. (2000). Distribution of foliar fungal endophytes of *Pinus strobus* between and within host trees. *Canadian Journal of Forest Research, 30*, 1436–1442.

Fröhlich J, & Hyde, KD. (1999). Biodiversity of palm fungi in the tropics: are global fungal diversity estimates realistic? *Biodiversity Conservation, 8*, 977–1004.

Gadgil, PD. (1984). *Cyclaneusma* (*Naemacyclus*) needle-cast of *Pinus radiata* in New Zealand. *New Zealand Journal of Forestry Science, 14*, 179–196.

Ganley RJ, & Newcombe, G. (2006). Fungal endophytes in seeds and needles of *Pinus monticola*. *Mycological Research, 110*, 318–327.

Hargreaves CL, Grace, LJ, van der Maas, SA, Menzies, MI, Kumar, S, Holden, DG, Foggo, MN, Low, CB, & Dibley, MJ. (2005). Comparative *in vitro* and early shoot performance of adventitious shoots from cryopreserved cotyledons and axillary shoots from epicotyls of the same zygotic embryo of control-pollinated *Pinus radiata*. *Canadian Journal of Forest Research, 35*, 2629–2641.

Hargreaves CL, Grace, L, van der Maas, S, Reeves, C, Holden, G, Menzies, MI, Kumar, S, & Foggo, M. (2004). Cryopreservation of *Pinus radiata* zygotic embryo cotyledons: effect of storage duration on adventitious shoot formation and plant growth after two years in the field. *Canadian Journal of Forest Research, 34*, 600–608.

Hargreaves CL, & Menzies, MI. (2007). Organogenesis and cryopreservation of juvenile radiata pine. In SM Jain & H Haggman (Eds.), *Protocols for Micropropagation of Woody Trees and Fruits* (pp. 51–65). New York: Springer.

Hargreaves CL, Reeves, CB, Gough, K, Josekutty, P, Skudder, DB, van der Maas, SA, Sigley, MR, Menzies, MI, Low, CB, & Mullin, TJ. (2009). Improving initiation, genotype capture and family representation in somatic embryogenesis of *Pinus radiata* D.Don. by a combination of zygotic embryo maturity, media and explant preparation. *Canadian Journal of Forest Research, 39*, 1566–1574.

Hawksworth DL. (2001). The magnitude of fungal diversity: the 1.5 million species estimate revisited. *Mycological Research, 105*, 1422–1432.

Horgan K, & Holland, L. (1989). Rooting micropropagated shoots from mature radiata pine. *Canadian Journal of Forest Research, 19*, 1309–1315.

Johnson JA, & Whitney, NJ. (1989). A study of fungal endophytes of needles of balsam fir (*Abies balsamea*) and red spruce (*Picea rubens*) in New Brunswick, Canada, using culture and electron microscope techniques. *Canadian Journal of Botany, 67*, 3513–3516.

Marler M, Pedersen, D, Mitchell-Olds, T, & Callaway, RM. (1999). A polymerase chain reaction method for detecting dwarf mistletoe infection in Douglas-fir and western larch. *Canadian Journal of Forest Research, 29*, 1317–1321.

Martin KJ, Rygiewicz PT .(2005). Fungal-specific PCR primers developed for analysis of the ITS region of environmental DNA extracts. *BMC Microbiol*, 5. doi:10.1186/1471-2180-5-28

Ministry for Primary Industries. (2010). *Import Requirements*. http://www.biosecurity.govt.nz/pests/pine-pitch-canker. Accessed 17 November 2014.

Prihatini I, Glen, M, Wardlaw, TJ, & Mohammed, CL. (2014). Multigene phylogenetic study of *Cyclaneusma* species. *Forest Pathology, 44*, 299–309.

Rodriguez RJ, White, JF, Jr, Arnold, AE, & Redman, RS. (2009). Fungal endophytes: diversity and functional roles. *New Phytologist, 182*, 314–330.

Sinclair WA, & Lyon, HH. (2005). *Diseases of Trees and Shrubs*. New York: Cornell University Press.

Storer AJ, Gordon, TR, & Clark, SL. (1998). Association of the pitch canker fungus, *Fusarium subglutinans* f. sp. *pini* with Monterey pine seeds, and seedlings in California. *Plant Pathology, 47*, 649–656.

Swett CL, & Gordon, TR. (2015). Endophytic association of the pine pathogen *Fusarium circinatum* with corn (*Zea mays*). *Fungal Ecology, 13*, 120–129. doi:10.1016/j.funeco.2014.09.003.

Thomas P. (2011). Intense association of non-culturable endophytic bacteria with antibiotic-cleansed in vitro watermelon and their activation in degenerating cultures. *Plant Cell Reports, 30*, 2313–2325.

Veliceasa D, Enünlü, N, Kós, PB, Köster, S, Beuther, E, Morgun, B, Deshmukh, SD, & Lukács, N. (2006). Searching for a new putative cryptic virus in *Pinus sylvestris* L. *Virus Genes, 32*, 177–186.

Walter C, & Grace, L. (2000). Genetic engineering of conifers for plantation forestry: *Pinus radiata* transformation. In SM Jain & SC Minocha (Eds.), *Molecular Biology of Woody Plants* (pp. 79–104). Dortrecht: Kluwer.

Walter C, Find, J, & Grace, LJ. (2005). Somatic embryogenesis and genetic transformation in *Pinus radiata*. In SM Jain & PK Gupta (Eds.), *Protocols for Somatic Embryogenesis in Woody Plants* (pp. 11–24). Netherlands: Springer.

White TJ, Bruns, TD, Lee, S, & Taylor, JW. (1990). Amplification and direct sequencing of fungal ribosomal RNA genes for phylogenetics. In MA Innis, DH Gelfand, JJ Sninsky, & TJ White (Eds.), *PCR Protocols: A guide to Methods and Applications* (pp. 315–322). San Diego: Academic.

Post-harvest respiration of *Pinus radiata* logs under different temperature and storage conditions

Jinquan Feng[1*], Allan Woolf[1], Xingyun Yang[1], Shane Olsson[1], Don Brash[2], Alistair Hall[2] and Bruce Bycroft[2]

Abstract

Background: Radiata pine (*Pinus radiata* D.Don) log exports are a growing and important trade for New Zealand. Oxygen (O_2) depletion and carbon dioxide (CO_2) accumulation in ship holds caused by log respiration may affect the efficacy of fumigation treatments and safety to people working around the holds. Currently very little is known about changes in respiration rates of harvested logs. This study examined the respiration rate of harvested logs after temperature equalisation overnight to a range of temperatures and when stored at typical indoor or outdoor conditions to identify key factors affecting log respiration rate.

Results: The results indicate that the respiration rate of logs increased as a function of temperature following Arrhenius' equation, with activation energy (*Ea*) value of 27708 J mol^{-1}. The respiration rate at a reference temperature of 20 °C decreased from 25.1 mg CO_2 kg DW^{-1} h^{-1} one week after harvest to 16.2 and 11.5 mg CO_2 kg DW^{-1} h^{-1} three and six weeks after harvest, respectively. The dominant respiration tissue shifted from cambial and parenchymatous tissues in 'fresh' logs towards the internal sapwood over a six-week period. This shift may be attributed to severe dehydration of the outer layers. Microbial activity also contributed to the respiration of logs when stored at humid conditions.

Conclusion: Irrespective of storage conditions, the rates of respiratory CO_2 release of log sections remained high during the first two weeks post harvest and declined rapidly within the next a few weeks as logs become dehydrated. Respiration rates at a given storage time increased nearly 50 % for every 10 °C increase in temperature from 10 to 40 °C. Drying logs for 4–6 weeks before fumigation treatment at moderate temperature may help to achieve consistent fumigation results, prevent life-threatening conditions and to reduce transportation cost.

Keywords: Temperature; Moisture; Storage condition; Wood tissue; Microbial growth

Background

Radiata pine (*Pinus radiata* D.Don) log exports are a growing and important trade for New Zealand. New Zealand exported 16 million m^3 of logs in the year to June 2013, worth $NZ2.3 billion (MPI 2013). Logs are usually exported within six weeks of harvest and are transported by ships. Approximately 70 % of the log cargo is stowed in closed holds and the remainder are top-stowed on the open deck. Typical in-transit transport time from when loading commences to markets is 20–25 days. Fumigation to control insect pests is usually carried out either in-transit for within-hold stowed logs or on dockside for top deck-stowed logs. Fumigation of radiata pine logs in the past heavily depended on the use of the insecticide methyl bromide but the use of methyl bromide is due to stop in the future because of its ozone depleting effect. Therefore, research has been set up to replace methyl bromide with more environmentally acceptable pesticides, such as phosphine and other alternatives. However, the efficacy of these alternative treatments has been found to be impaired by high CO_2 and low O_2 concentrations in log holds caused by log respiration (Brash and Page, 2009).

Svedberg et al. (2009) found that oxygen (O_2) concentration inside ship holds loaded with logs and wood chips declined to a mean of 10 ± 7.3 % over a mean period of 46 h

* Correspondence: ringo.feng@plantandfood.co.nz
[1]The New Zealand Institute for Plant & Food Research Limited, 120 Mt Albert Rd, Auckland 1142, New Zealand
Full list of author information is available at the end of the article

after loading. Concomitantly, CO_2 concentration increased to a mean of 7.5 ± 5.2 %. This indicates that the atmosphere inside holds may become life-threatening within two to three days after loading. The changes in atmosphere were thought to be a consequence of wood respiration, biomass decomposition and microbiological activities (Svedberg et al. 2009; Fan & Bi 2013). Literature on the post-harvest respiration rate of logs is rare. Most studies focus on the respiration rate of living stems (Ryan et al. 1994; Damesin et al. 2002; Guidolotti et al. 2013; Rodriguez-Calcerrada et al. 2014) or coarse wood litter (Progar et al. 2000; Chambers et al. 2001). For both living stems and coarse wood litter, respiration rate increases as temperatures rise within a day or throughout the year, although discrepancies exist because of genetic, developmental and other environmental factors, such as shelter, which affects moisture content, and heterotrophic community composition (Progar et al. 2000; Rodriguez-Calcerrada et al. 2014). Therefore, it is expected that the respiration rates of harvested radiata pine logs also respond to temperature, storage condition and time.

We evaluated the respiration rate of harvested radiata pine logs at a range of temperature and storage conditions. The aims were to identify factors affecting log respiration and to use this information to predict the potential modification of the atmosphere in a sealed headspace.

Methods

Three typical logs, 4.8 m long and 0.2-0.3 m in diameter were commercially harvested on 17 January 2013 from a forest approximately 25–28 years old in Wanganui, New Zealand. The logs were commercially de-limbed as an industry practice, which retained about 20 % bark by surface area. A total of 48 sections, each 0.29 m long, were cut from three replicate logs. Each section was paraffin wax-sealed at both ends as described by Pranamornkith et al. (2014). These log sections were then designated into four sets of twelve log sections for experiments started one week after harvest. Each set consists of four log sections from each replicate logs.

Effect of temperature on respiration rate

This first experiment aimed to determine the effect of temperature on respiration rate using three sets of twelve log sections. Each of these was assessed one, three and six weeks after harvest, respectively. The log sections were stored at ambient temperature in a ventilated room before respiration measurements commenced. At each assessment time, twelve log sections were equilibrated to four different temperatures of 10, 20, 30 and 40 °C (three log sections for each temperature) overnight before measuring the respiration rate (record of temperature at the centre of log sections indicated that temperature equilibration overnight is sufficient to get internal temperature to within in 1 °C from the targeted temperature). The four temperatures cover the temperature range that logs may possibly experience in transit.

Each log section was then sealed in a 20-L food grade plastic bucket. Headspace CO_2 concentration was measured at different intervals until a concentration of 0.5-1.0 % CO_2 was reached; this concentration range was selected to ensure the measured value was sufficiently higher than detection limit of the analyser (0.01 % CO_2). The CO_2 concentration of the headspace was measured using 1 mL of gas sampled from each sealed container using the method described by Hertog et al. (2003). Briefly, gas samples were analysed using a miniature infrared CO_2 analyser (Analytical Development Company, Hoddesdon, UK). Output signals were linear over the range applied and analysed using HP integrators (Hewlett-Packard, model 3396A). The time to reach 0.5-1.0 % CO_2 in the headspace typically took around 40–100 min, depending on temperature and storage time.

The weight of each log section was measured at the beginning of the experiment (initial weight one week after harvest) and again after respiration measurements, allowing the calculation of moisture content. The dry weight of each log section was calculated based on the initial weight and an initial moisture content of 54 % as measured on additional logs of the same harvest (Dry weight = Initial weight x (1 – 0.54)). Respiration rate was calculated on a dry weight basis.

Spatial variation in respiration rate

The second experiment investigated the spatial variation in respiration rate from the bark/cambium layer towards the centre of the log. The purpose of this was to show which part of the log was most actively respiring at a given interval postharvest.

Three log sections used for respiration measurement at 20 °C in the temperature experiment described above were used here also. Each log section was cut into eight blocks approximately 90 mm in length, 90 mm in width and 10 mm in thickness across the log section (Fig. 1). Blocks one and eight represent the outermost zone containing bark and cambium; blocks two and seven represent the outer part of the log; blocks three and six represent sapwood; and blocks four and five represent the heartwood of the log section. These blocks were weighed and then sealed in 1-L respiration jars at 20 °C for up to four hours before headspace gas measurements were taken to determine respiration rate. The moisture content of each block was measured by weighing before and after a 72-h drying period at 105 °C.

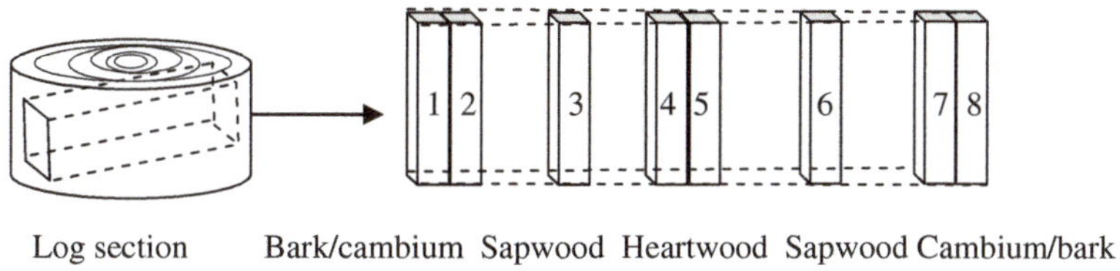

Fig. 1 Eight blocks of wood tissue cut from each radiata pine log section for the measurement of spatial variation in respiration rate

Effect of storage condition on respiration rate

The third experiment evaluated the effect of storage conditions on respiration rate over time. Twelve log sections were stored in four different environmental conditions with three replicate sections in each:

A. Indoor dry air; three log sections stored in a 360-L plastic container flushed with dry air with approximately 50 % relative humidity
B. Indoor humid air; three log sections stored in a 360 L plastic container flushed with humidified air with approximately 98 % relative humidity
C. Outdoor covered; three log sections stored in a plastic crate located outside with cover using a plastic film and a cardboard sheet to prevent direct exposure to rain and sunshine
D. Outdoor uncovered; three log sections stored in a plastic crate located outside without cover.

The respiration rate was measured on each log section at seven times during storage: one, two, three, five, six, twelve and twenty four weeks after harvest to reflect time when logs are transported. Log sections were equilibrated at 20 °C overnight before being sealed in 20-L plastic buckets for respiration measurement at 20 °C.

The weight of each log section was recorded on each measurement date to allow calculation of respiration rate and weight loss over time. The volumes of the three log sections stored in humidified air were measured after the final respiration measurement by submerging each log section into a 20-L plastic bucket full of water. The volume of water that overflowed from the bucket was taken as the volume of the log section. Log density at the beginning of the experiment was calculated as the ratio between initial weight and log volume (volume change should be negligible for log sections stored in humidified air). The average density of these three log sections (0.90 kg L^{-1}) was used to calculate initial volume of all log sections based on initial weight.

The experimental data were analysed using a nonlinear regression procedure for temperature effects and a general linear model procedure for tissue zone difference or the effects of storage treatments. All the analyses were carried out using Statistical Analysis System, version 9.4 (SAS Inc., USA).

Results and discussion

Effect of temperature on respiration rate

Respiration rates (rCO_2) measured at week one increased 3.2-fold with a temperature increase from 10 to 40 °C (16.3 vs. 51.4 mg CO_2 kg DW h^{-1}). Respiration rates measured at week three and week six showed a similar response to temperature, and a clear decline in overall respiration rate was obvious, with up to 50 % lower rCO_2 towards the end of the measurement period (Fig. 2). The respiration rates increased 3.2-fold with a temperature increase from 10 to 40 °C (16.3 vs. 51.4 mg CO2 kg DW h^{-1}).

The increase in respiration rate was described as a function of temperature following Arrhenius' equation (Equation 1).

$$r_{CO_2} = r_{CO_2 ref} e^{\left(\left(\frac{Ea}{R}\right)\left(\frac{1}{T_{ref}} - \frac{1}{T}\right)\right)} \qquad (1)$$

where rCO_{2ref} is the rCO_2 measured at a reference temperature of 20 °C; Ea is the activation energy; R is the universal gas constant (8.314 J mol^{-1} K^{-1}); T_{ref} =293K (20 °C); and T is temperature (K) of log sections during respiration measurement. A high value of Ea indicates high temperature sensitivity.

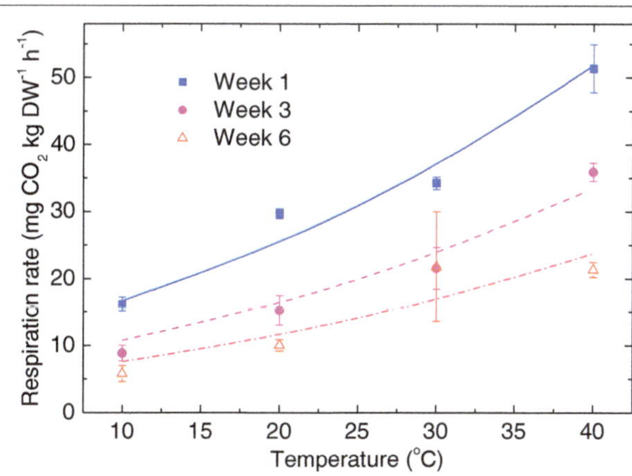

Fig. 2 Effect of temperature on respiration rate of radiata pine logs on a dry weight basis. Each point is the means of three replicate log sections. Vertical bars are standard errors of the means. Lines represent the Arrhenius' model fitted to the data, with parameters listed in Table 1

A generic model fitted to combined data from three assessments made one, three and six weeks after harvest had a R^2 value of 0.87 and a root mean squared error (RMSE) of 5 mg CO_2 kg FW^{-1} h^{-1}. The estimated value for reference respiration rate at 20 °C decreased from 25.1 mg CO_2 kg DW^{-1} h^{-1} one week after harvest to 11.5 mg CO_2 kg DW^{-1} h^{-1} six weeks after harvest (46 % decrease). The activation energy estimated was 27708 J mol^{-1} (Table 1 and Fig. 2). These results are consistent with the exponential increase in CO_2 flux observed for tree trunks as a function of increasing temperature (Rodriguez-Calcerrada et al. 2014). The parameters listed in Table 1 could be used to estimate respiration rates at any temperature between 10 and 40 °C based on respiration rate measured at a 20 °C. Based on the *Ea* value of 27708 J mol^{-1}, respiration rate will increase nearly 50 % for every 10 °C increase in temperature.

Spatial variation in respiration rate

At the first assessment made one week after harvest, the respiration rate measured on wood blocks showed a clear pattern indicating a reduction of respiration rate from the bark/cambium layer towards the heartwood. However, this pattern became less pronounced three weeks after harvest, and the wood blocks cut off half way between the bark/cambium layer and the heartwood had the highest rate of respiration rate six weeks after harvest (Fig. 3a). The significant reduction in respiration rate of outer layers over time may have been a consequence of severe water loss in the layers (Fig. 3b). It was also noticed that the two bark samples (Blocks one and eight) from opposite sides of the log have significantly different respiration rates at weeks 1 and 3 (Fig. 3a). This difference could be attributed to the difference in moisture content (Fig. 3b), which could be the consequence of stacking log sections horizontally (the side facing down and touching ground or other logs beneath may have lost less water than the opposite side. The average respiration rates across different zones (65.6, 27.3 and 11.9 mg CO_2 kg DW^{-1} h^{-1} respectively as measured one, three and six weeks after harvest) were higher than the r_{CO_2} values measured for log sections at the same temperature (Fig. 3), particularly for week one (1.6 times higher) and week three (0.7 times higher). These differences could be attributed to the reduction in resistance to gas diffusion within wood tissue compared with that in log sections (CO_2 from respiration was more freely released to the respiration jar when the wood was cut into small pieces than from the whole log section because of the increased surface to mass ratio and the decreased distance from inside to outside). Therefore, the respiration rate measured on small blocks of log tissues in the current study serves as an indication of potential respiration activity of different tissue zones. However, prediction of ship-hold atmospheres should be based on the respiration rate measured on log sections rather than that measured on small blocks of log tissue. The respiration rates measured on log sections should be useful for predicting initial changes in ship-hold atmospheres before O_2 depletion and consequent CO_2 accumulation becomes a limiting factor for full expression of respiration potential. The impacts of O_2 depletion and CO_2 accumulation on respiration rate need to be investigated further for accurate prediction of ship hold atmospheres.

Table 1 Parameters of equation 1 estimated using combined data from three assessment times (N=36)

Parameter	Week	Estimate	se	R^2	RMSE
$r_{CO_2 ref}$	1	25.08	1.60		
(mmol CO_2 kg^{-1} h^{-1})	3	16.22	1.32	0.87	5.00
	6	11.52	1.20		
Ea (J mol^{-1})	All	27708	2780		

Effect of storage condition on respiration rate

The respiration rate of log sections was stable during the first two weeks after harvest. The respiration rate then decreased rapidly between weeks two through five, possibly because of depletion of soluble sugars in the harvested logs. Following this rapid decrease in respiration, low respiration rates were measured for log sections stored outdoors. However, there was an unexpected increase in respiration between twelve and twenty four weeks after harvest, particularly for log sections stored outdoors under ambient conditions (Fig. 4a). This increase may be explained by rehydration during autumn rains and consequent microbial growth. Indeed, a decline in moisture content was observed between twelve and twenty four weeks after harvest for log sections stored in outdoor uncovered conditions (Fig. 4b). Log sections stored indoors respired similarly to those stored outdoors until five weeks after harvest. Respiration of log sections stored indoors increased rapidly between five and twelve weeks after harvest compared with respiration of those stored outdoors. The respiration rate of log sections stored in humidified air increased continuously after twelve weeks, but that of log sections stored in dry air declined significantly between twelve and twenty four weeks (Fig. 4a). The increase in respiration rate of the log sections stored indoors after five weeks of storage could be attributed to the noticeable increase in mould growth (Fig. 5b). The decrease in respiration rate of log sections stored in dry air twenty four weeks after harvest was associated with moisture content declining to less than 30 % (Fig. 4b). The lower moisture content

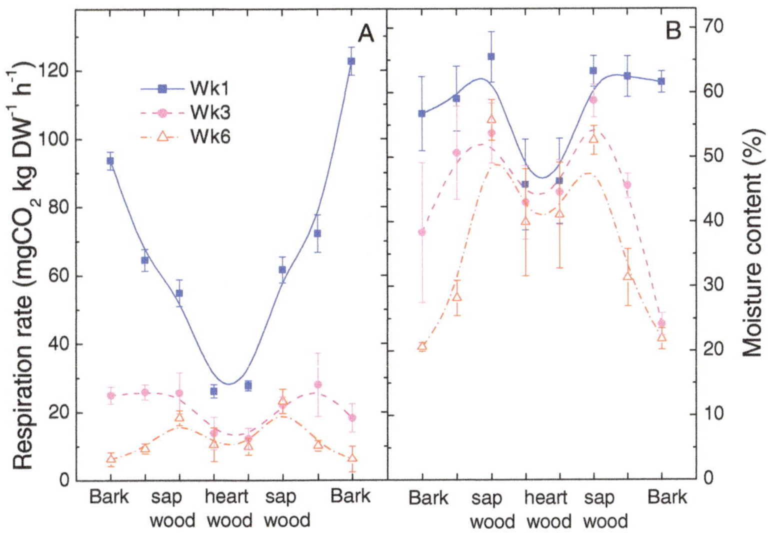

Fig. 3 Respiration rate (**a**) and water content (**b**) measured on wood tissue cut across radiata pine log sections. Each point is the means of three replicates. Vertical bars are standard errors of the means

could have limited both microbial activities and the respiration rate of the log sections. This is consistent with the lowest respiration rate recorded twelve weeks after harvest for log sections stored outdoors, when moisture content dropped to a similar range. If a drying process were introduced to reduce moisture content from 54 % at harvest to about 30 % prior to export, the total weight of logs would reduce to two thirds of the original. This may result in a significantly less cost for transporting logs overseas (less fuel consumption and reduced carbon footprint). However, the feasibility of such a drying process for the forest industry also depends on the cost of building and operating appropriate drying facilities. Achieving consistent moisture loss could be practically difficult due to the impact of seasonal changes in climate on the drying process (Visser et al., 2014). Without a consistent drying process, it might be necessary to adjust fumigation protocols based on assessment of moisture and microbial activity.

Conclusions

Irrespective of storage conditions, the rates of respiratory CO_2 release of log sections remained high during the first two weeks post harvest and declined rapidly

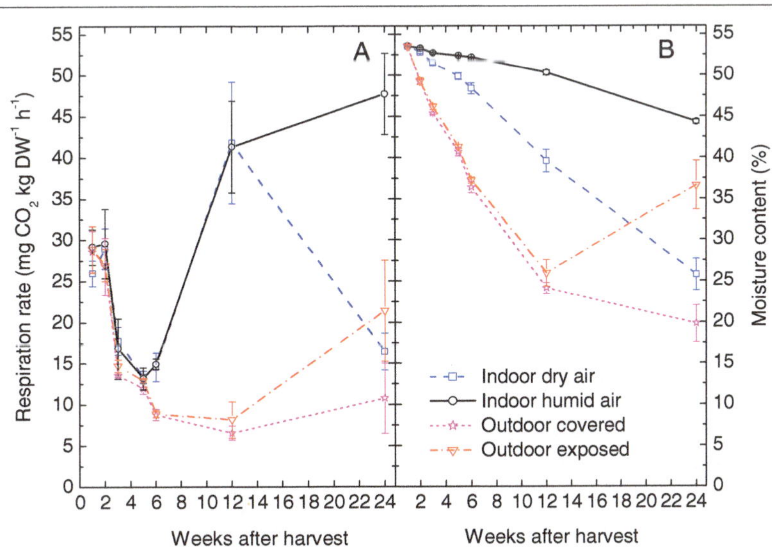

Fig. 4 Changes in respiration rate (**a**) and moisture content (**b**) of radiata pine log sections stored under different storage conditions. Each point is calculated from the means of three replicates. Vertical bars are standard errors of the means

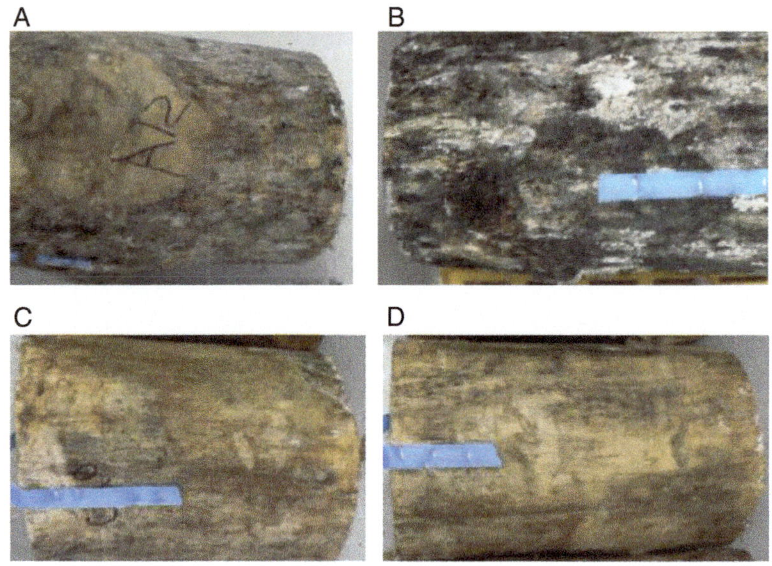

Fig. 5 Mould growth observed after 6 weeks of storage on radiata pine log sections stored indoors in dry air (**a**), humidified air (**b**), outdoors covered (**c**), outdoors uncovered (**d**)

within the next a few weeks as logs become dehydrated. A logical recommendation for the forestry industry would be to apply fumigation treatments after 4–6 weeks post harvest to reduce the interference with insecticides. The respiration rate increased nearly 50 % for every 10 °C increase in temperature so the CO_2 issue is exacerbated during the warm summer months when a temperature-driven stimulation of respiratory processes will result in very high CO_2 concentrations in ship holds. Also, because respiratory processes are strongly dependent on moisture availability, any actions that can be taken to accelerate moisture loss will mitigate the CO_2 issue.

Competing interests
The authors declare that they have no competing interests.

Authors' contributions
JF was lead author and project manager and carried out data analyses and drafted the main manuscript. AW had a supervisory role in project planning and resourcing. XY contributed to the collection of data. SO contributed to the collection of data and contributed to data analysis and the preparation of the manuscript. DB sourced the materials for trials and commented on the manuscript. AH assisted with data analyses and commented on the manuscript. BB sourced materials for trials. All authors read and approved the final manuscript.

Acknowledgements
We are very grateful to STIMBR (Stakeholders in Methyl Bromide Reduction) for funding this research project.

Author details
[1]The New Zealand Institute for Plant & Food Research Limited, 120 Mt Albert Rd, Auckland 1142, New Zealand. [2]The New Zealand Institute for Plant & Food Research Limited, Batchelar Rd, Palmerston North 4474, New Zealand.

References
Brash, DW and Page, BBC. (2009). Review of phosphine research for control of timber quarantine pests. Plant & Food Research Confidential Report No. 2370 prepared for STIMBR (Stakeholders in Methyl Bromide Reduction). Retrieved 29 May 2015 from http://maxa.maf.govt.nz/sff/about-projects/search/08-061/phosphine-research-review.pdf

Chambers, JQ, Schimel, JP, & Nobre, AD. (2001). Respiration from coarse wood litter in central Amazon forests. *Biogeochemistry, 52*, 115–131.

Damesin, C, Ceschia, E, Le Goff, N, Ottorini, J-M, & Dufrêne, E. (2002). Stem and branchrespiration of beech: from tree measurements to estimations at the stand level. *New Phytologist, 153*, 159–172.

Fan, CG, & Bi, XTT. (2013). Development of Off-Gas Emission Kinetics for Stored Wood Pellets. *Annals of Occupational Hygiene, 57*, 115–124.

Guidolotti, G, Rey, A, D'Andrea, E, Matteucci, G, & De Angelis, P. (2013). Effect of environmental variables and stand structure on ecosystem respiration componentsin a Mediterranean beech forest. *Tree Physiology, 33*, 960–972.

Hertog, MLATM, Nicholson, SE, & Whitmore, K. (2003). The effect of modified atmospheres on the rate of quality change in 'Hass' avocado. *Postharvest Biology and Technology, 29*, 41–53.

MPI (New Zealand Ministry for Primary Industries). (2013). Annual forestry export statistics. http://www.mpi.govt.nz/news-resources/statistics-forecasting/forestry/annual-forestry-exportstatistics.aspx.

Pranamornkith, T, Hall, MKD, Adlam, AR, Somerfield, KG, Page, BBC, Hall, AJ, et al. (2014). Effect of fumigant dose, timber moisture content, end-grain sealing, and chamber load factor on sorption by sawn timber fumigated with ethanedinitrile. *New Zealand Plant Protection, 67*, 66–74.

Progar, RA, Schowalter, TD, Freitag, CM, & Morrell, JJ. (2000). Respiration from coarse woody debris as affected by moisture and saprotroph functional diversity in Western Oregon. *Oecologia, 124*, 426–431.

Rodriguez-Calcerrada, J, Martin-StPaul, NK, Lempereur, M, Ourcival, JM, del Rey, MD, Joffre, R, et al. (2014). Stem CO_2 efflux and its contribution to ecosystem CO_2 efflux decrease with drought in a Mediterranean forest stand. *Agricultural and Forest Meteorology, 195*, 61–72.

Ryan, MG, Linder, S, Vose, JM, & Hubbard, RM. (1994). *Dark respiration of radiata pines* (*Ecological Bulletins*, pp. 50–63).

Svedberg, U, Petrini, C, & Johanson, G. (2009). Oxygen Depletion and Formation of Toxic Gases following Sea Transportation of Logs and Wood Chips. *Annals of Occupational Hygiene, 53*, 779–787.

Visser, R., Berkett, H., & Spinelli, R. (2014). Determining the effect of storage conditions on the natural drying of radiata pine logs for energy use. *New Zealand Journal of Forestry Science*, 44:3. http://www.nzjforestryscience.com/content/44/1/3

8

Relative persistence of commonly used forestry herbicides for preventing the establishment of broom (*Cytisus scoparius*) seedlings in New Zealand plantations

Hop Tran[1], Kerry C Harrington[1*], Alastair W Robertson[1] and Michael S Watt[2]

Abstract

Background: Many of the herbicides used in site preparation of forests and during initial selective weed management can persist in the soil for several months, killing or stunting seedlings of broom (*Cytisus scoparius* (L.) Link) as they establish. The objective of this research was to determine the relative persistence and effect over time of the herbicides most widely applied within radiata pine (*Pinus radiata* D.Don) plantations in New Zealand on broom survival and growth.

Methods: Eleven herbicide treatments were applied in early summer to a Tokomaru silt loam soil. Soil samples were collected from each treatment at fortnightly intervals for up to a year. The samples were placed into pots and immediately sown with scarified viable broom seeds in a heated glasshouse. Relative growth and biomass of seedlings were determined as the ratio of growth and survival in each treatment relative to growth and survival in an untreated control. Logistic curves were fitted to recorded values of relative growth and survival over time and the time to reach 50 and 95% of the vigour of the plants in the untreated control was determined.

Results: The treatments can be approximately ranked in the following order from most to least persistent: triclopyr/picloram > high rate of clopyralid > high rate of hexazinone > terbuthylazine/hexazinone > low rate of hexazinone > low rate of clopyralid > high rate of terbuthylazine > triclopyr > high rate of metsulfuron-methyl > low rate of terbuthylazine > low rate of metsulfuron-methyl.

Conclusion: As the rate of triclopyr/picloram used in this work is too phytotoxic to be used selectively over newly planted radiata pines, this treatment would only be suitable prior to tree planting so residual activity would control seedlings that develop after scrub clearance. For post-plant weed management operations, results show hexazinone and high rates of clopyralid to be useful for long-term management of broom as both these herbicides have low phytotoxicity to radiata pine. When combined with low rates of triclopyr and picloram, clopyralid could provide an effective treatment that could be used in combination with oversown or naturally occurring grasses to manage broom.

Keywords: Bioassay; Broom; Clopyralid; Herbicide; Hexazinone; Metsulfuron-methyl; Picloram; Residue; Terbuthylazine; Triclopyr

* Correspondence: K.Harrington@massey.ac.nz
[1]Institute of Agriculture and Environment, Massey University, Palmerston North 4442, New Zealand
Full list of author information is available at the end of the article

Background

During plantation establishment, competition from weeds is often an important constraint for the growth and survival of young trees (White et al. 1990; Balandier et al. 2006; Wagner et al. 2006). Weed species compete and reduce crop tree growth by using water, nutrients and light (Nambiar and Sands 1993; Löf 2000; Wang et al. 2000; Richardson and Whitehead 2002). The main reason for practising weed management during plantation establishment is to improve survival and growth rates of young trees by decreasing competition from weed species (Balneaves 1982; Walstad and Kuch 1987).

Broom (*Cytisus scoparius* L. (Link)) is a large, deciduous leguminous shrub that can grow rapidly to form dense stands, excluding native plants and altering community structure of prairies, woodlands and young forests (Bossard and Rejmánek 1994; Wearne and Morgan 2004; Caldwell 2006; Prévosto et al. 2006; Wearne and Morgan 2006). Abundant seed production in combination with seed longevity of up to 81 years (Turner, 1933) results in the accumulation of persistent seed under forest stands, making the long-term management of this weed difficult (Oneto et al. 2010).

As broom is fast growing and competes strongly for both water and light, it is one of the most serious competitors of radiata pine (*Pinus radiata* D.Don) plantations in New Zealand (Richardson and Whitehead 2002). Studies carried out on both moist (Richardson et al. 1996; Richardson et al. 1999) and dry sites (Clinton et al. 1995; Richardson et al. 1997; Richardson and Whitehead 2002; Watt et al., 2003) have shown that broom can have a large negative influence on radiata pine growth during the establishment phase. These effects persist in stands that are slow to close canopy. Understorey broom has been shown to compete strongly with radiata pine for water and retard tree growth at stand ages between 9–12 years (McCracken 1995; Richardson and Whitehead 2002).

Although a number of management measures have been used to reduce competition from broom, few studies report on the efficacy of these methods (Oneto et al. 2010). Manual and mechanical methods, such as cutting or hand pulling, may be effective in controlling broom seedlings and small shrubs but are labour intensive and costly over large areas as repeated actions are required (Hoshovsky 1986; Bossard 1990). Until viable non-chemical methods of management are determined, the routine application of herbicides during the plantation establishment phase (Richardson 1993) will continue to play an important part in successful management of broom.

Chemical weed management is preferred over manual weeding or over-sowing of cover crop species, as herbicides are less expensive to apply and more effective than other methods (Gous 2005). Herbicides with appropriate selective activity suppress competing weeds so that the desired trees can grow more quickly (Gratkowski 1975; Morash and Freedman 1989). Some herbicides, such as glyphosate, kill weeds on application and leave no residue in the soil. Others, called residual herbicides, can continue to be effective for some time after application because they persist in the soil. Some of these residual herbicides, such as clopyralid, are used more for their knockdown effect at application, whereas others, such as terbuthylazine, are used more for their persistence than their knockdown ability. An important determinant of the success of residual herbicides is their persistence in the soil and consequently the post-application duration over which the herbicide will continue to prevent seed germination, seedling emergence and/or development (Ketchum and Rose 2003).

Herbicide persistence is usually compared by estimating a half-life (DT_{50}), which is the time taken for its activity to be reduced by 50% through dissipation from the soil (Helling 2005). Half-life can be determined directly by analysing changes in residual herbicide levels in soil samples using chromatographic methods (Roy et al. 1989), or indirectly with bioassays (Sahid and Teoh 1994) which measure the ability of indicator species to successfully grow within treated soil (Günther et al. 1989). Persistence of herbicides within soil varies depending on herbicide chemical structure and mechanism of action, soil properties (soil type and soil organic matter), meteorological factors (temperature and rainfall) and microbial degradation (Helling 2005).

A recent survey of New Zealand plantation owners has shown that the major herbicides used by the forest industry are terbuthylazine, glyphosate, hexazinone, clopyralid, metsulfuron-methyl, triclopyr and picloram (Rolando et al. 2013). Two of these (terbuthylazine and hexazinone) were prohibited temporarily for use in forests certified by the Forest Stewardship Council (FSC). Criteria for prohibition relate to the persistence and/or toxicity of specific herbicides in the environment. It was possible for the owners of certified forests to obtain a derogation from the FSC during this period of prohibition that allowed them to use limited amounts of prohibited herbicides in a controlled way if they supported research into alternative chemical and non-chemical weed management options.

As the prohibition of terbuthylazine and hexazinone was under review by the FSC when this trial was conducted, studying the effectiveness of these herbicides against broom was deemed worthwhile. Hexazinone is known to provide good knockdown and residual management of a range of perennial weeds. This herbicide is typically applied in the spring, rather than in the winter soon after planting, and can continue to inhibit weed development for up to two years when applied at 4.0 kg ai (active ingredient) ha^{-1} (Cameron and Stokes 1978). In Washington and Oregon (USA), hexazinone applied at 2.2 kg ai ha^{-1} was found to provide both effective and lasting management of grasses and forbs after pre-planting and post-planting applications (Dimock et al. 1983). Research conducted in southern New Mexico with applications of hexazinone at 0.56 and 1.12 kg ai ha^{-1} showed early season grass and forb control of 65 and 90%, respectively (Fisher et al. 1986). Although these rates were safe for conifer growth, higher rates (6.0–8.0 kg ai ha^{-1}) were found to increase mortality of loblolly pine (*Pinus taeda* L.) (Fitzgerald and Fortson 1979). In New Zealand, hexazinone is registered for killing weeds during site preparation and for the release of radiata pine during the post-establishment phase at rates as high as 7.5 kg ai ha^{-1} though label recommendations state that broom is a less susceptible species and will only be controlled while seedlings are very young even at this high rate (Young 2013).

Hexazinone is degraded by light and microbial activity to various metabolites and is also readily leached in soil (USDA-FS 1984a). The persistence of hexazinone varies with soil type, environmental conditions (Rhodes 1980; Rahman 1981; Harrington et al. 1982; USDA-FS 1984a), and timing of application (Coackley and Moore 1977) with a half-life ranging from 10–180 days (Gaskin and Zabkiewicz 1986; Khan and Liang 1989; Michael and Neary 1993). For example, hexazinone was found to be more persistent under cold and dry conditions than under warm and wet conditions and was more mobile in a sandy loam soil than in a silt loam soil (Rahman 1981). These results are in agreement with other studies in North America (Rhodes 1980; Feng 1987; Roy et al. 1989) and New Zealand (Close et al. 2008; Sarmah et al. 2009). Coackley and Moore (1977) assessed the persistence of hexazinone in various radiata pine plantations under different climatic conditions throughout New Zealand and suggested that the optimum time of application was September (early spring) compared with other times assessed due to less rainfall after this time and thus less loss from leaching.

Terbuthylazine is a residual herbicide used either by itself or in combination with hexazinone, and is absorbed mainly by roots but also through the foliage (Lavy et al. 1989; Gous 2005). Its tolerance by coniferous trees makes it a popular choice in forestry (Lavy et al. 1989) and it is one of the most commonly used herbicides in radiata pine plantations in New Zealand (Wang et al. 2010) for pre- and post-emergent management of annual and perennial grass and broadleaf weeds (James et al. 1998; Gous 2005). The half-life of terbuthylazine in the soil has been reported to vary between 5 and 116 days after treatment (DAT) depending on soil characteristics and temperature (Sahid and Teoh 1994; James et al. 1998; Watt et al. 2010).

Metsulfuron-methyl is not tolerated by radiata pine so this herbicide is only used for site preparation, and trees cannot be planted until 2 months after application to allow for persistence of the herbicide within the soil (Young 2013). It is commonly applied for clearance of woody weeds such as gorse (*Ulex europaeus* L.) and blackberry (*Rubus fruticosus* agg.) prior to planting radiata pine either alone or, more commonly, in combination with glyphosate (Gous 2005). Metsulfuron-methyl is considered to have high soil activity and a soil half-life that varies between 5 and 63 days (Ismail and Lee 1995; Günther et al. 1989; Trabue et al. 2006), with the principal modes of degradation being microbial breakdown and chemical hydrolysis (Beyer et al. 1987). Metsulfuron-methyl can have high persistence and mobility depending on the environmental conditions (Sarmah et al. 2000), with a negative relationship between sorption and soil pH (Walker et al. 1989; Walker and Welch 1989). Rahman et al. (1991)) reported that the persistence of metsulfuron-methyl was shorter in soils with lower pH (below 5.5) and/or high organic matter (above 10% organic carbon), and phytotoxic residues differed between time of applications (autumn > spring). James et al. (2004) found that when 0.012 kg ai ha^{-1} metsulfuron was applied to a silt loam soil, between 75% and 90% of the herbicide had degraded 56 DAT, and they reported that similar work at four sites around New Zealand found the persistence ranged from 28 days to more than 63 days.

Triclopyr is used for clearing scrub weeds during site preparation, either by itself or in combination with picloram (Davenhill 1997). When used prior to planting pine seedlings, rates as high as 6.0 kg ai ha^{-1} are recommended, with lower rates of up to 1.2 kg ai ha^{-1} when spraying over newly planted (or young) radiata pine seedlings for the selective control of broom (Young 2013). Triclopyr is a phytotoxic pyridine compound that has been developed to control broadleaf weeds and brush (Jotcham et al. 1989). The soil persistence of triclopyr is variable with the half-life ranging from 10 to 100 days (Cox 2000). Triclopyr applied at 2.64 kg ai ha^{-1} was moderately persistent in some Canadian sandy and clay forestry soils with times to 50% and 90% disappearance at 14 and 20 DAT, respectively (Stephenson et al. 1990). Residual triclopyr (applied at 3.5 kg ai ha^{-1} to control infestations of the exotic vine, kudzu (*Pueraria montana* (Lour.) Merr.) in loblolly pine plantation establishment sites) was more persistent during a drier period (170 DAT) following spot treatment of regenerating kudzu plants (applied only to the kudzu plants and soil in the immediate vicinity) than during a wet period (29 DAT) (Berisford et al. 2006), probably because this herbicide is degraded quickly through microbial activity which is more likely to occur under warm, moist conditions (USDA-FS 1984b).

The two pyridine herbicides, clopyralid and picloram, are used within New Zealand radiata pine forests for management of broom and some other weeds such as gorse, woolly nightshade (*Solanum mauritianum* Scop.), tree lucerne (*Chamaecytisus palmensis* (Christ) F.A.-Bisby&K.W.Nicholls), wattles (*Racosperma* spp.) and ragwort (*Jacobaea vulgaris* Gaertn.). However, picloram is normally used in combination with triclopyr for both site preparation and (at lower rates) for selective management, while clopyralid is used with triclopyr and picloram for selective management of young broom after planting during plantation establishment (Davenhill 1997; Young 2013). Picloram aerially sprayed at 1.1 kg ai ha^{-1} on gorse can remain active in the soil for up to 450 days (MacDiarmid 1975) but residual triclopyr was not detected at 120 days after the application of 12.0 kg ai ha^{-1} (MacDiarmid 1977). Clopyralid can remain active for 60–420 days depending on the rate applied and environmental conditions (Pik et al. 1977; Bovey and Richardson 1991; Cox 1998).

The persistence of these herbicides, whether used during site preparation or for spraying over newly planted (or young) radiata pine seedlings, may provide management of new broom seedlings in the weeks following their application, and this has been shown by Harrington (2014) for clopyralid and aminopyralid. Published information related to the residual activity of the herbicides discussed is available but often provides broad ranges and usually gives limited indication of how soon after application that broom seedlings will no longer be affected. The objective of this research was to determine the effect over time on broom survival and growth of the herbicides most widely applied within radiata pine plantations in New Zealand.

Methods

Study site and preparation

The experiment was conducted at a flat site located at the Moginie Pasture and Crop Research Unit of Massey University, Palmerston North, New Zealand (40°23′S, 175°36′E). The site was previously in pasture grazed by sheep and was cleared on 19 November 2008 (two weeks before the start of the experiment) by spraying with glyphosate (Roundup Transorb, Nufarm Ltd; 2.2 kg ai ha^{-1}) to kill all pasture species present. Thirty-six plots of 1 m^2 (1 × 1 m) were then pegged out with 1 m buffers between each plot to eliminate contamination from adjacent plots.

Average monthly temperature and rainfall for the experimental period were obtained from the Grasslands AgResearch weather station (approximately 1 km away from the study site).

The soil was a Tokomaru silt loam, and 40 samples (2.5 cm diameter cores to a depth of 7.5 cm) were taken from across the trial site on 18 November 2008, air dried, passed through a 2 mm aperture sieve then mixed together. Characteristics of this soil were then ascertained using methods outlined by Blakemore et al. (1987), or by Hesse (1971) for CEC. Soil pH was measured in water at a soil:water ratio of 1:2.5. Phosphate available for plant absorption (expressed as “Olsen P”) was determined colorimetrically after extraction in 0.5 M sodium bicarbonate at a soil:water ratio of 1:20. Cation concentrations (specifically Na, Mg, K and Ca) were quantified by leaching samples with 1M ammonium acetate (pH 7.0) and determining concentrations in soil:-water slurries of 1:50 using atomic absorption spectroscopy. Cation exchange capacity (CEC) was determined by the summation of the extractable cations and the extractable acidity, with the extractable acidity being determined by the decrease in pH of the extraction solution. Soil organic matter was estimated by measuring total carbon, which was determined by heating soil in a stream of high purity oxygen in a Leco furnace to produce CO_2 and measuring this with an infrared detector.

Field treatments

The experiment was conducted from December 2008 to December 2009. A randomised block design (blocked depending on proximity of plots to each other within the field) was used that consisted of three replicated blocks of 12 treatments including an untreated control (Table 1). Herbicides were applied either at the recommended rate for managing broom or gorse (metsulfuron, clopyralid, triclopyr, triclopyr/picloram mixture) or at the highest rate of the recommended range (hexazinone, terbuthylazine, terbuthylazine/hexazinone mixture), and some herbicides were also applied at half this rate (metsulfuron, clopyralid, hexazinone, terbuthylazine).

Each herbicide was applied to the soil within the allocated plot using a hand-held sprayer with a solid cone nozzle on 4 December 2008 at a rate equivalent to 1,000 L ha^{-1} of herbicide solution. A back-and-forth spray pattern was used until all the measured amount of solution had been applied to ensure uniform application of the herbicide. The two treatments that were herbicide mixtures are available commercially already mixed together so were applied as a mixture.

Not surprisingly, some weeds established in the untreated plots during the experiment. Weeds also emerged in some herbicide-treated plots once the herbicide residues began dissipating. These weeds had the potential to adversely affect subsequent bioassays as the roots could regrow in the bioassay pots described below. Therefore, weeds present in the plots were sprayed twice with herbicides containing active ingredients that are deactivated on contact with soil (Young 2013) reducing any potential influence on the treatment effects being measured. This

Table 1 Details of the herbicide treatments used in the trial

Active ingredient (rate, kg ai ha $^{-1}$)	*Trade name* (rate)*	*Treatment code*
metsulfuron-methyl (0.10)	Escort (0.17 kg ha^{-1})	metsul-L
metsulfuron-methyl (0.20)	Escort (0.33 kg ha^{-1})	metsul-H
clopyralid (1.5)	Versatill (5 L ha^{-1})	clopyr-L
clopyralid (3.0)	Versatill (10 L ha^{-1})	clopyr-H
hexazinone (3.75)	Velpar DF (5 kg ha^{-1})	hexa-L
hexazinone (7.5)	Velpar DF (10 kg ha^{-1})	hexa-H
terbuthylazine (5.0)	Gardoprim 500 (10 L ha^{-1})	terbu-L
terbuthylazine (10.0)	Gardoprim 500 (20 L ha^{-1})	terbu-H
terbuthylazine/hexazinone (8.5/1.5)	Agpro Valzine 500 (20 L ha^{-1})	terbu/hexa
triclopyr/picloram (3.0/1.0)	Tordon Brushkiller (10 L ha^{-1})	tric/pic
triclopyr (6.0)	Grazon (10 L ha^{-1})	triclo
untreated control	NA	untreated

*Manufacturers of the products were DuPont (NZ) Ltd (Escort, Velpar DF), DowAgroSciences (NZ) Ltd (Versatill, Grazon, Tordon Brushkiller), Orion Crop Production Ltd (Gardoprim 500) and Agpro (NZ) Ltd (Agpro Valzine 500).

spraying was carried out on 22 December 2008 using glyphosate (Roundup Transorb; 9.2 g ai L^{-1}) and 22 January 2009 using glufosinate (Buster, Bayer (NZ) Ltd; 2.0 g ai L^{-1}).

Soil sampling

Soil samples were taken at regular intervals from the plots. The first soil samples were taken one day after the herbicide treatments were applied and then at 2-week intervals until 361 days after treatment (DAT). Three cores were randomly extracted from each treatment plot at each sampling time using a soil corer (2.5 cm in diameter and 5 cm in depth). The soil samples from each plot were combined, mixed and partitioned into three pots to give three sub-samples of residual herbicide levels for each field plot. Pots were each 5 cm × 5 cm × 5 cm and were 30% filled (as a bottom layer) with vermiculite to reduce the soil requirement to fill each pot, to assist with drainage of pots and also help adsorb any herbicide leaching out of the soil which might contaminate neighbouring pots.

Bioassay

A bioassay was developed to determine the time at which the residual herbicide was no longer at a high enough concentration to prevent the establishment of broom seeds. Broom seeds were obtained on 8 August 2008 by removing soil from beneath mature broom plants growing at the Veterinary Large Animal Teaching Unit of Massey University (40°23′S, 175°37′E). The seeds extracted from this soil were laid out on filter paper overnight to dry at room temperature then stored in paper bags at 5°C until the experiment commenced. To maximise the germination of the seed, mature broom seeds were selected on the basis of hardness. All soft seeds were discarded as they were considered less viable than hard ones. The selected hard seeds were scarified using a Forsberg scarifier (a drum lined with P100-grit sand paper) and rotated for 20 seconds at 1425 revolutions per minute to abrade the seed coat. Scarified seeds were treated with thiram powder (800 g ai kg^{-1} as Thiram 80W, Yates (NZ) Ltd) to prevent fungal infection, and six seeds were sown 1 cm deep into each pot of soil within a day of taking the soil samples.

Pots were arranged in a randomised layout on a bench in a glasshouse in which the mean monthly temperatures ranged from 16.1°C in October 2009 to 20.3°C in January 2010. Pots were watered three times per day by a subsurface irrigation system to ensure that the soil remained moist, with water being added to felt mats under the pots from where it moved into pots by capillary flow.

Data collection and measurements

The number of live seedlings per pot was recorded fortnightly. Each set of soil samples was kept for *ca.* 60 days to allow time for broom seeds to germinate and grow. After this time, all living broom seedlings from each treatment pot were cut at the base of their stems and the above-ground components were oven dried at 80°C to a constant mass then weighed.

Once there was no visible difference in seedling health between a herbicide treatment and the corresponding untreated control treatment, the soil sampling for that treatment was terminated. Sampling for most herbicide treatments had been terminated by 253 DAT but sampling of triclopyr/picloram and terbuthylazine/hexazinone treatments continued until 361 DAT.

Data analysis

All statistical analyses were conducted using SAS 9.2 (SAS Institute Inc 2008). For each herbicide treatment sample, both survival to 60 days (the number of seedlings that emerged and survived in each pot) and biomass were expressed as a fraction of the untreated control treatment for the appropriate block. The data were checked for homogeneity and normality before analysis. They were found to be normally distributed so no transformations were necessary. The following two-parameter logistic model (Equation 1) was fitted to the repeated measures (batches) of residual herbicide,

$$y = \exp(a + bt)/(1 + \exp(a + bt)) \quad (1)$$

where y is the seedling survival or biomass, expressed as a fraction of the untreated control for each period, a and b are empirical parameters, and t is the time (in days) after treatment that each batch was sampled. This logistic model was selected as it is a flexible functional form with only two parameters and has the useful property that predictions are constrained between zero and one. Biomass data in which treated pots exceeded the biomass of untreated pots were assumed to have a value of one for this analysis. The accuracy of the models was determined by the coefficient of determination (R^2). Plots of residuals against predicted values showed little apparent bias for any of the fitted models.

Through rearranging Equation 1 to output time, T_i, the estimated time (T_i, days) to 50 and 95% survival (i.e. Ts_{50} and Ts_{95}, respectively) and to 50 and 75% biomass (i.e. Tb_{50} and Tb_{75}, respectively) of the untreated control were derived from each nonlinear regression equation.

Relative survival and biomass at 113, 211 and 253 DAT were analysed with one-way analysis of variance (ANOVA). The logistic parameters a, b and the obtained T_i values were also subjected to ANOVA to test the hypothesis of equality between herbicide treatments using the model shown in Equation 2 that accommodated the underlying randomised block design:

$$Y_{ij} = \mu + \tau_i + B_j + \varepsilon_{ij} \quad (2)$$

where Y_{ij} is the value observed for treatment i and block j, μ is overall mean, τ_i is a fixed effect associated with the ith treatment ($i = 1–11$), B_j is a random effect associated with jth block ($j = 1–3$) and ε_{ij} is the random variation. Where there were significant differences among the treatment least square means, a Tukey's honestly significant difference (HSD) test was used to undertake multiple comparison testing.

Results

Soil at the site had a mean pH of 5.2 and an Olsen-P of 43 μg mL^{-1}. The Na, Mg, K and Ca levels averaged 0.16, 0.74, 0.77 and 3.2 me 100 g^{-1}, respectively. Mean soil organic matter was 5.3% and mean CEC was 13 me 100 g^{-1}. The maximum daily temperature ranged from 22.6°C in January to 11.8°C in June (Figure 1).

Several techniques were assessed prior to the bioassay experiment to determine the most appropriate method for maximising the germination of broom seeds (data not shown). The technique chosen (see Methods section) provided the best germination rate (86.5% healthy seedlings, mean of four replicates each with 50 seeds).

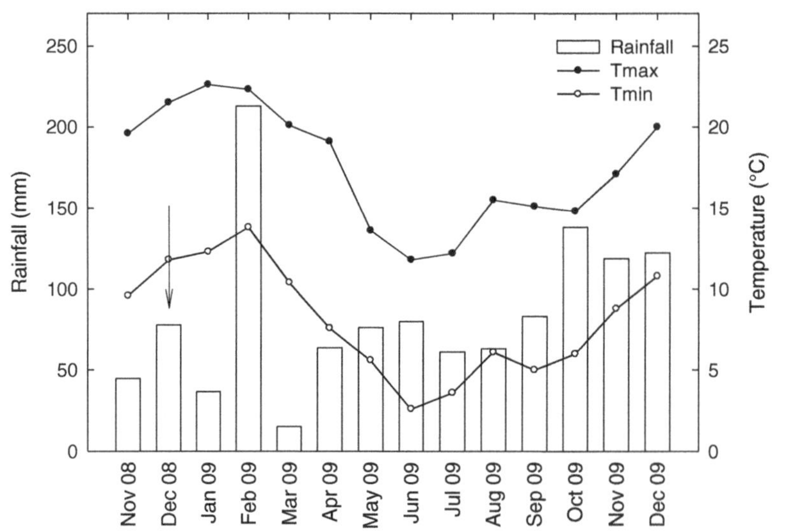

Figure 1 Climate data for the field plots in which residual herbicides were applied. Data are mean total monthly rainfall (mm) and mean monthly maximum (T_{max}) and minimum (T_{min}) temperature (°C) over the duration of the experiment. The arrow shows the time when herbicide treatments were applied.

Survival and biomass of broom seedlings for each herbicide treatment increased over time when expressed as a proportion of the untreated control treatment. Although the relationships followed the shapes of sigmoid curves, variability among treatments meant the fit of the curves to the data varied in precision and shape (Figure 2) with R^2 values ranging from 0.80 to 0.98 (survival) and from 0.61 to 0.95 (biomass).

Residual triclopyr/picloram herbicide mixture in the soil adversely affected the survival of establishing broom seedlings for the longest period, with few surviving in the first 180 DAT (Figure 2). Soil treated with the high rate of hexazinone contained sufficient residue to prevent survival of establishing seedlings for 160 days, compared with 150 days for the hexazinone/terbuthylazine mixture, 105 days for the high rate of clopyralid and 90 days for the high rate of terbuthylazine. Soil treated with the high rate of metsulfuron-methyl and the triclopyr treatment did not contain sufficient residual activity to prevent survival of broom seedlings germinating beyond 70 DAT.

Seedling survival was significantly different among treatments until 211 DAT. Treatments were categorised into three groups based on treatment means at 113 DAT. Group 1 was defined as total prevention of seedling survival, Group 2 was defined as survival <65% and Group 3 was defined as survival >65% (Table 2). The triclopyr/picloram mixture and the high rate of hexazinone were in Group 1. Seedling survival increased from 31 to 53% and 71 to 86% for these treatments at 211 and 253 DAT, respectively. Both rates of metsulfuron-methyl and the low rates of terbuthylazine and triclopyr were classed as Group 3 because seedling survival increased rapidly with 86 and 78% of seedlings surviving under the high rate of metsulfuron-methyl and low rate of terbuthylazine at 113 DAT, respectively. The intermediate group of treatments (Group 2) showed a high variation in seedling survival 113 DAT with 65 and 14% of seedlings surviving the high rate of terbuthylazine and low rate of hexazinone, respectively. No significant difference in seedling survival was found between treatments within Groups 2 and 3 at either 211 or 253 DAT.

Seedling biomass showed a similar pattern to seedling survival with significant differences again between the least and most effective treatments at 113 DAT (Table 2). The same groupings applied to biomass as to survival. Seedling biomass for the triclopyr/picloram mixture and the high rate of hexazinone (Group 1) was reduced by 86% and 27-37% at 211 and 253 DAT, respectively. For the least persistent herbicides (Group 3), seedling biomass was only slightly lower than untreated plants 113 DAT, and only reduced by 14 and 31% under the low and high rates of metsulfuron-methyl, respectively. As occurred with the seedling survival data, the intermediate group showed a high variation in relative biomass. Relative differences among treatments changed over time as the high rate of clopyralid appeared to dissipate less rapidly than the low rate of hexazinone after 211 days. Again, no significant difference in seedling biomass was found between treatments within Groups 2 and 3 at 211 and 253 DAT (Table 2).

Estimates of relative persistence in Table 2 (i.e. Groups 1–3) relied on measurements from a few specific dates, whereas estimates based on the logistic model used data from all of the harvest dates (Table 3). Although the logistic model parameters *a* and *b* did not significantly differ among the treatments ($P > 0.05$) for the relative seedling survival and biomass, the estimated time (days) of residual activity did differ significantly among treatments (Table 3). Estimates of the time to reach 50 and 95% survival, or 50 and 75% biomass, showed similar trends of relative herbicide dissipation between treatments (Table 2), but with a few differences. Using all four of these estimates, the combined triclopyr/picloram treatment was the most persistent and the two metsulfuron-methyl treatments, triclopyr and the two terbuthylazine treatments were the least persistent.

Discussion

The bioassay used here as a measure of herbicide persistence involves seeds that were scarified and treated with fungicide. The reason for selecting hard seeds and artificially scarifying them was to ensure that a consistent level of germination would be obtained from each pot. This probably resulted in higher establishment of seedlings than normally occurs in the field although some seeds in the soil would probably be naturally scarified through such mechanisms as fluctuating temperature and moisture within soil breaking the seed coat (Moreno-Casasola et al. 1994). However, the broom seedlings in this study were used as indicators of the long-term effectiveness of various herbicides rather than to simulate typical seedling dynamics.

The persistence of herbicides estimated from germination and biomass of broom seedlings was compared with an untreated control. There were some fluctuations within the actual data so the predicted values from the logistic model are likely to provide a more robust estimate of persistence for specific treatments over time. However, it was possible to compare the relative persistence of the various treatments studied here, and the discussion below is based on the modelled data.

Two of the herbicides assessed in this trial, hexazinone and terbuthylazine, were both prohibited for a time from use in certified forests where forest owners subscribe to the principles set out by the FSC. Hexazinone was considered too mobile (so likely to move to surrounding areas), and terbuthylazine was liable to bioaccumulate in

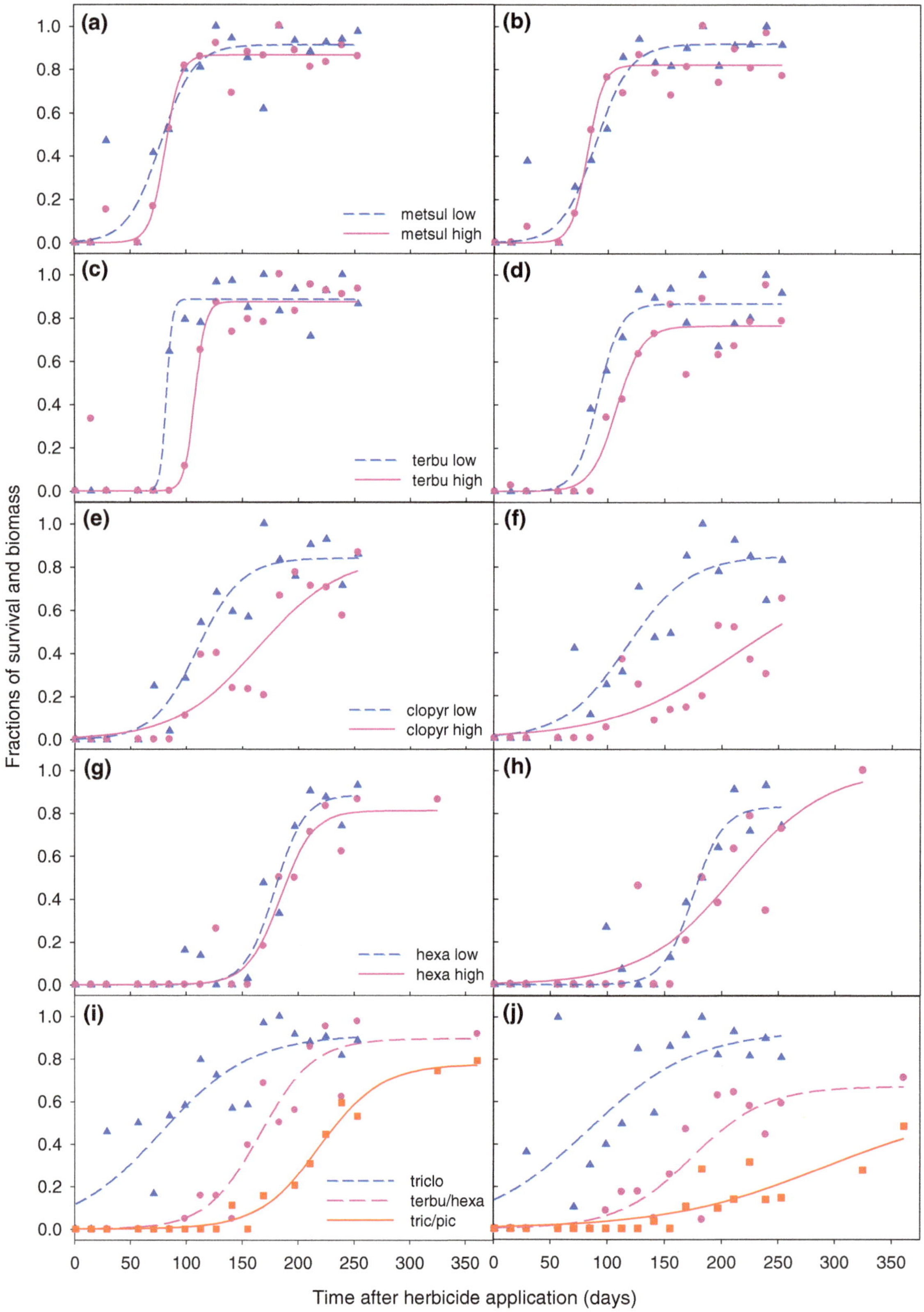

Figure 2 (See legend on next page.)

(See figure on previous page.)
Figure 2 Survival (left) and biomass (right) of broom seedlings grown in soil previously treated with herbicides. Treatments were **(a, b)** metsulfuron, **(c, d)** terbuthylazine, **(e, f)** clopyralid, **(g, h)** hexazinone, **(i, j)** triclopyr (triangle), terbuthylazine/hexazinone mixture (filled circle) and triclopyr/picloram mixture (squares). Data are expressed as a proportion of untreated control plants, plotted against the time after herbicide application. The symbols show the measurements (triangles for low rates in a-h, filled circles for high rates in a-h). Curves were fitted to the data using the logistic model described in Equation 1.

soils (Rolando and Watt 2012). A recent decision made by the FSC has resulted in terbuthylazine and hexazinone both being reinstated as acceptable active ingredients for use in certified forests (FSC 2015).

The higher rate of hexazinone used in this trial (7.5 kg ai ha^{-1}) equates to the maximum rate recommended for pre-plant site preparation (Young 2013) while the lower rate tested here is the highest rate of hexazinone recommended for use in release spraying of radiata pine. Its high cost means that this herbicide is seldom used for site preparation with the more affordable glyphosate/metsulfuron-methyl mix preferred instead. Both rates (3.75 and 7.5 kg ai ha^{-1}) suppressed germination of the majority of broom seedlings for ca. 160 DAT. Differences between these treatments became more evident after this time. The growth of broom seedlings that established 14–20 days after that date were stunted by the remaining herbicide residual activity resulting in less than half the biomass of untreated seedlings for approximately another 20–50 days for low and high rates respectively. The persistence of hexazinone in the soil compared with the other herbicides tested in this experiment provides evidence that this herbicide is useful for managing broom.

Estimates of half-life for hexazinone determined here are broadly consistent with previous published research (Sarmah et al. 2009). The differences in persistence between the two rates of herbicide (7.5 kg ai ha^{-1} and 3.75 kg ai ha^{-1}) used in the current trial provide an approximate indication of the half-life of herbicides with one half-life being the time taken for 7.5 kg to reduce to 3.75 kg within the soil. The fitted models for these two rates are ca. 20 days apart. Sarmah et al. (2009) determined that the half-life for hexazinone was either 8 or 45 days at 20°C for two New Zealand silt loam soils, which is consistent with our result given the temperatures over December and January. They also found the half-life in these two soils increased to 36 or 96 days respectively at 7.5°C. Their result indicates that the persistence of hexazinone would have been longer in the current study if it had been applied at a cooler time of the year.

Table 2 Effects of residual herbicide activity on survival and biomass of broom seedlings

Treatment	*Survival*						*Relative biomass*					
	113 DAT		*211 DAT*		*253 DAT*		*113 DAT*		*211 DAT*		*253 DAT*	
Group 1: most persistent												
tric/pic	0.0	d	30.7	b	53.0	a	0.0	c	13.9	b	14.5	b
hexa-H	0.0	d	71.2	ab	86.5	a	0.0	c	63.4	ab	72.8	a
Group 2: intermediate persistence												
hexa-L	13.7	cd	90.5	a	93.2	a	7.2	bc	91.3	a	74.3	a
terbu/hexa	15.6	bcd	85.7	ab	83.3	a	17.2	bc	64.4	ab	59.1	a
clopyr-H	39.3	abcd	71.2	ab	86.7	a	36.8	abc	51.7	ab	65.2	a
clopyr-L	54.3	abcd	90.5	a	86.0	a	31.1	abc	92.5	a	83.1	a
terbu-H	65.2	abc	95.2	a	93.3	a	42.4	abc	67.2	ab	78.7	a
Group 3: least persistent												
terbu-L	77.9	ab	71.7	ab	86.5	a	71.2	ab	77.6	a	91.7	a
triclo	79.7	a	88.1	a	88.9	a	49.7	abc	93.3	a	80.9	a
metsul-L	81.0	a	88.1	a	97.6	a	85.7	a	90.8	a	91.2	a
metsul-H	85.7	a	81.0	ab	85.9	a	68.9	ab	89.2	a	76.9	a
Significance of one-way ANOVA ($F_{10,32}$)												
Treatment	7.48^{***}		2.68^{*}		1.50^{ns}		5.43^{***}		4.51^{**}		6.41^{***}	

Data are expressed as a percentage of mean untreated control 113, 211 and 253 days after treatment (DAT). Treatments are sorted in an increasing order of survival for 113 DAT. Means within each column followed by the same letter are not significantly different at $\alpha = 0.05$. For the analysis of variance (ANOVA), *F*-values are shown, followed by the *P*-category. Values with one, two and three asterisks are significant at $P < 0.05$, 0.01 and 0.001, respectively and ^{ns}denotes no significance at $P \geq 0.05$.

Table 3 Estimated time (days) after herbicide application before broom seedlings reached 50 and 95% survival and 50 and 75% of biomass relative to untreated seedlings

Treatment	*Time to reach % survival (day)*				*Time to reach % relative biomass (day)*			
	Ts_{50}		Ts_{95}		Tb_{50}		Tb_{75}	
tric/pic	246	a	383	a	418	a	524	a
clopyr-H	180	abc	310	ab	243	b	316	ab
hexa-H	198	ab	246	abc	210	bcd	245	b
terbu/hexa	176	bcd	245	abc	228	bc	299	b
hexa-L	180	abc	207	bc	185	bcd	197	b
clopyr-L	122	cde	178	bc	131	bcd	169	b
terbu-H	107	de	167	bc	131	bcd	176	b
triclo	91	e	221	abc	100	cd	133	b
metsul-H	90	e	170	bc	94	d	132	b
terbu-L	83	e	116	c	98	cd	126	b
metsul-L	73	e	159	bc	87	d	118	b
Significance of one-way ANOVA ($F_{10,32}$)								
Treatment	18.02^{***}		5.55^{***}		14.39^{***}		8.92^{***}	

Treatments are sorted in a descending order of herbicide persistence, based both on Ts_{50} and Tb_{50}. Means within each column followed by the same letter are not significantly different at $\alpha = 0.05$. For the analysis of variance, *F*-values are shown, and values with three asterisks are significant at $P < 0.001$.

Terbuthylazine was one of the less persistent herbicides tested. Residual activity was able to kill most broom seedlings for about 90 days at 10 kg ai ha^{-1}, the highest rate recommended for release spraying in New Zealand pine plantations. Suppression of seedlings to less than half their normal biomass continued until about 130 DAT.

Results from the mixture of hexazinone and terbuthylazine used were intermediate between the two components alone, killing most broom seedlings for about 150 days. Applied in combination at the rate tested here, a terbuthylazine and hexazinone mixture has been the most widely used herbicide for post-plant management of broom (Rolando et al. 2013). This mixture is usually applied during early to late spring so trees can effectively grow weed-free for almost a year. Ideally, grass species would then establish but both terbuthylazine and hexazinone provide management of a broad spectrum of woody and herbaceous weeds, including grasses, so more aggressive weeds than grasses are likely to establish once residual effects dissipate. Thus, broom seeds that do successfully germinate following dissipation of residual herbicide activity are likely to dominate the site as this species generally out-competes other weed species during the establishment of pine plantations.

In contrast, clopyralid, triclopyr and picloram have little effect on grass species and so can be used to promote a shift in vegetation type to either naturally occurring or artificially sown grasses. Watt and Rolando (2014) found that the application of clopyralid on sites dominated by broom promoted a post-control shift in regenerating vegetation to grass species that compete less with trees than broom compared with use of a terbuthylazine/hexazinone mixture. Clopyralid, triclopyr and picloram can also be used in over-sowing operations that predominantly use grass ground covers such as Yorkshire fog (*Holcus lanatus* L.). Such operations typically occur during early autumn prior to tree planting. Once established, these grasses can reduce germination and establishment of broom seedlings. Moreover, should any broom germinate and establish, clopyralid, triclopyr and picloram can be used to kill the broom during spring with little effect on the grass species. These herbicides can also be used for broom management when establishing trees directly onto pasture sites where there is existing grass cover. Competition from grass immediately around the tree can be removed through spot application of hexazinone/terbuthylazine mixture; otherwise haloxyfop (not tested here) can be used (Rolando et al. 2011). The use of ground covers is an example of a non-chemical weed management strategy encouraged by the FSC (Rolando and Watt 2012).

The most persistent treatment was the triclopyr/picloram mixture. However, this was applied at a rate equivalent to that recommended for killing mature broom plants and other scrub species during pre-plant site preparation (Young 2013). When applied alone, triclopyr was one of the less persistent treatments, indicating that it was the 1.0 kg ai ha^{-1} of picloram in the triclopyr/picloram mixture that was responsible for killing most broom seedlings in later weeks. Picloram is well known for being persistent in the soil although its persistence alone was not tested here. MacDiarmid (1975) found residual activity remaining in soil up to 450 DAT when picloram was applied at 1.7 kg ai ha^{-1} as an aerial spray for the management of gorse. However, this rate of picloram is not safe for selective use over young pine trees. Rates between 10 and 20% of that used in this trial are recommended for use over young trees to manage establishing broom, and even then, there is a risk of tree distortion (Young 2013). Therefore, the triclopyr/picloram treatment tested is only suitable for site preparation, which is normally almost exclusively undertaken using a mixture of the less expensive glyphosate/metsulfuron-methyl (Gous 2005). On sites to be planted with radiata pine that have a pasture cover with some broom present, the triclopyr/picloram mixture could be quite effective as it would conserve the grass cover and leave residual activity in the soil to discourage any new broom germination. However, the long residual action of this herbicide may prolong the period following application before trees can be safely planted without suffering damage.

In contrast, clopyralid can be applied over young radiata pine trees for selectively managing broom as a release spray treatment at rates as high as 3 kg ai ha^{-1}. Overall, this rate of clopyralid had the second highest duration of residual soil activity, continuing to kill 50% of seedlings for 175 DAT, with surviving seedlings reaching less than 50% of the control biomass until 245 DAT, though some seedlings were beginning to establish successfully after 100 days (Figure 2). Clopyralid at half of this rate (1.5 kg ai ha^{-1}) was less effective (Figure 2). Tran et al. (2010) showed that this lower rate of clopyralid (when applied with low rates of triclopyr) could kill 12-month-old broom plants without damaging young radiata pine so further research should investigate whether the higher rate of clopyralid has phytotoxic effects on radiata pine. Harrington (2014) has also identified clopyralid as a useful herbicide for managing emerging broom seedlings for several months after application to soil.

The individual metsulfuron-methyl, triclopyr and terbuthylazine treatments were the least persistent of the herbicides tested as they were not significantly different. Even so, both metsulfuron-methyl rates reduced broom germination and/or growth for about 70 to 90 DAT, which would be useful for killing any broom that germinated soon after spraying. Metsulfuron-methyl is commonly used in combination with glyphosate for site preparation in New Zealand forests, though it is not suitable for release spraying as it is not sufficiently tolerated by radiata pine (Young 2013). Harrington and He (2010) used similar bioassay techniques to that used here in a number of soil types to determine how many months after metsulfuron-methyl application that lotus (*Lotus pedunculatus* Cav.), another legume species, could be safely sown within forests as a ground cover. For autumn sowing of lotus, metsulfuron-methyl applied over the previous summer months had safely dissipated if applied more than two months before sowing of the lotus, a similar result to that found with broom in the present work. However, applications during the cooler months of the year meant a delay of more than eight months prior to a spring sowing of lotus to ensure safety. This result illustrated increased persistence under cooler conditions. As with the hexazinone studies cited above, the application of herbicides in December may underestimate the persistence of these herbicides if applied under cooler conditions.

Overall, a combination of clopyralid with appropriate rates of triclopyr and/or picloram has the most potential for managing broom within radiata pine plantings. Such a mixture will not damage grass ground cover so will promote a shift towards naturally occurring grass species. Broom seedlings that do germinate (along with desirable grass seedlings during establishment of ground cover) can be killed within the first year using this herbicide mixture, and residual activity will prevent further broom establishment for about 100 days, by which time grass ground cover should be growing strongly. The broom seedlings that do establish when residual activity declines will be deformed and stunted, and thus will be more susceptible to competition from grass.

Conclusion

Several of the herbicides routinely used in radiata pine forests can remain active in the soil for a period of time after application to kill or stunt broom seedlings that germinate later. The triclopyr/picloram mixture tested provided the best performance but can only be used at this application rate if applied prior to pine planting. Also, it is not as broad-spectrum or inexpensive as the glyphosate/metsulfuron-methyl mixture normally applied at pre-planting. Both rates of hexazinone also controlled broom seedlings for ca. 160 days after application but will also stop competitive grass swards from forming to discourage broom germination once residual activity dissipates. Clopyralid on its own was less persistent than these other treatments tested but a combination of clopyralid with low rates of triclopyr and/or picloram could provide effective management of broom in New Zealand forests and should be tested using the bioassay system applied here.

Chemical names of herbicides mentioned in text

Clopyralid (3,6-dichloropyridine-2-carboxylic acid), glufosinate ((2*RS*)-2-amino-4-[hydroxy(methyl)phosphinoyl] butyric acid), glyphosate (*N*-(phosphonomethyl)glycine), haloxyfop ((*RS*)-2-{4-[3-chloro-5-(trifluoromethyl)-2-pyridyloxy]phenoxy}propionic acid), hexazinone (3-cyclohexyl-6-dimethylamino-1-methyl-1,3,5-triazine-2,4-dione), metsulfuron-methyl (methyl 2-(4-methoxy-6-methyl-1,3,5-trazine-2-ylcarbamoylsulfamoyl)benzoate), picloram

(4-amino-3,5,6-trichloro-2-pyridinecarboxylic acid), terbuthylazine (N-tert-butyl-6-chloro-N'-ethyl-1,3,5-triazine-2,4-diamine), triclopyr ([(3,5,6-Trichloro-2-pyridinyl)oxy] acetic acid).

Competing interests
The authors declare that they have no competing interests.

Authors' contributions
HT was the primary author, applied treatments, conducted bioassays and analysed the data. KCH designed the trial, assisted with treatments and wrote parts of the paper. AWR and MSW helped with data analysis and writing. All authors have read and approved the final manuscript.

Acknowledgements
We wish to thank the Foundation for Research, Science and Technology (FoRST) and New Zealand's International Aid and Development Agency (NZAID) for financial assistance, the Fertiliser and Lime Research Centre (Massey University) for analysing characteristics of the soil prior to the trial, chemical companies for providing their products, and Mark Osborne and Plant Growth Unit (Massey University) staff for technical assistance. The comments and suggestions of referees and editors on earlier drafts of the paper helped to significantly improve its clarity.

Author details
[1]Institute of Agriculture and Environment, Massey University, Palmerston North 4442, New Zealand. [2]Scion, Fendalton, PO Box 29237, Christchurch 8540, New Zealand.

References
Balandier, P, Collet, C, Miller, JH, Reynolds, PE, & Zedaker, SM. (2006). Designing forest vegetation management strategies based on the mechanisms and dynamics of crop tree competition by neighbouring vegetation. *Forestry, 79*(1), 3–27.

Balneaves, JM. (1982). Grass control for radiata pine establishment on drought sites. *New Zealand Journal of Forestry, 27*(2), 259–276.

Berisford, YC, Bush, PB, & Taylor, JW, Jr. (2006). Leaching and persistence of herbicides for kudzu (*Pueraria montana*) control on pine regeneration sites. *Weed Science, 54*(2), 391–400.

Beyer, EM, Brown, HM, & Duffy, MJ. (1987). *Sulfonylurea herbicide soil relations* (Proceedings of the 1987 British Crop Protection Conference – Weeds, 2, pp. 531–540).

Blakemore, LC, Searle, PL, & Daly, BK. (1987). *Methods for chemical analysis of soils* (New Zealand Soil Bureau Scientific Report 80).

Bossard, CC. (1990). *Secrets of an ecological interloper: ecological studies on Cytisus scoparius (Scotch broom) in California*. Davis, USA: University of California.

Bossard, CC, & Rejmánek, M. (1994). Herbivory, growth, seed production and resprouting of an exotic invasive shrub *Cytisus scoparius*. *Biological Conservation, 67*(3), 193–200.

Bovey, RW, & Richardson, CW. (1991). Dissipation of clopyralid and picloram in soil and seep flow in the blacklands of Texas. *Journal of Environmental Quality, 20*(3), 528–531.

Caldwell, BA. (2006). Effects of invasive Scotch broom on soil properties in a Pacific coastal prairie soil. *Applied Soil Ecology, 32*(1), 149–152.

Cameron, JN, & Stokes, KG. (1978). *Efficacy and selectivity of hexazinone in radiata pine* (Proceedings of the New Zealand Weed and Pest Control Conference, 31, pp. 58–65).

Clinton, PW, Payn, TW, Sun, OJ, & Richardson, B. (1995). Impacts of understorey vegetation on radiata pine growth and nutrition. In *Second International Conference on Forest Vegetation Management, 20–24 March 1995* (pp. 237–239).

Close, ME, Lee, R, Sarmah, AK, Pang, L, Dann, RL, Magesan, GN, Watt, JPC, & Vincent, JW. (2008). Pesticide sorption and degradation characteristics in New Zealand soils—a synthesis from seven field trials. *New Zealand Journal of Crop and Horticultural Science, 36*(1), 9–30.

Coackley, A, & Moore, RW. (1977). *DPX 3674 - A broad spectrum herbicide for weed control in forestry* (Proceedings of the New Zealand Weed and Pest Control Conference, 30, pp. 233–237).

Cox, C. (1998). Herbicide factsheet: clopyralid. *Journal of Pesticide Reform, 18*(4), 15–19.

Cox, C. (2000). Herbicide factsheet: triclopyr. *Journal of Pesticide Reform, 20*(4), 12–19.

Davenhill, NA. (1997). *Forest weed control manual - a guide to herbicide use in forests* (*FRI Bulletin* (Vol. 180)). Rotorua: New Zealand Forest Research Institute Ltd.

Dimock, EJ, II, Beebe, TF, & Collard, EB. (1983). Planting-site preparation with herbicides to aid conifer reforestation. *Weed Science, 31*(2), 215–221.

Feng, JC. (1987). Persistence, mobility and degradation of hexazinone in forest silt loam soils. *Journal of Environmental Science and Health, Part B: Pesticides, Food Contaminants, and Agricultural Wastes, 22*(2), 221–233.

Fisher, JT, Dudoich, DJ, & Fancher, GA. (1986). Efficacy of pre- and postemergent herbicides in field-planted *Pinus eldarica*. *Forest Ecology and Management, 16*(1–4), 253–258.

Fitzgerald, CH, & Fortson, JC. (1979). Herbaceous weed control with hexazinone in loblolly pine (*Pinus taeda*) plantations. *Weed Science, 27*(6), 583–588.

FSC (2015). FSC list of "highly hazardous" pesticides. Bonn, Germany: Forestry Stewardship Council. http://ic.fsc.org/pesticides.88.htm

Gaskin, RE, & Zabkiewicz, JA. (1986). *Comparison of effectiveness and persistence of three hexazinone formulations* (Proceedings of the New Zealand Weed and Pest Control Conference, 39, pp. 81–84).

Gous, S. (2005). Herbicides. In M Colley (Ed.), *Forestry Handbook* (pp. 112–113). Christchurch: New Zealand Institute of Forestry Inc.

Gratkowski, H. (1975). *Silvicultural use of herbicides in Pacific Northwest forests* (USDA Forest Service General Technical Report, p. PNW-37).

Günther, P, Rahman, A, & Pestemer, W. (1989). Quantitative bioassays for determining residues and availability to plants of sulphonylurea herbicides. *Weed Research, 29*(2), 141–146.

Harrington, TB. (2014). Synthetic auxin herbicides control germinating Scotch broom (*Cytisus scoparius*). *Weed Technology, 28*(2), 435–442.

Harrington, KC, & He, XZ. (2010). *Determining how soon ground covers can be planted in forests after applying metsulfuron* (Proceedings of the 17th Australasian Weeds Conference, pp. 340–343).

Harrington, KC, Rolston, MP, & Ivens, GW. (1982). *Movement of hexazinone spots applied to hill slopes* (Proceedings of the New Zealand Weed and Pest Control Conference, 35, pp. 162–165).

Helling, CS. (2005). The science of soil residual herbicides. In RC Acker (Ed.), *Soil Residual Herbicides: Science and Management* (Vol. 3, pp. 3–22). Sainte-Anne-de-Bellevue, Canada: Canadian Weed Science Society.

Hesse, PR. (1971). *A Textbook of Soil Chemical Analysis*. London, UK: John Murray.

Hoshovsky, M. (1986). *Element stewardship abstract for Cytisus scoparius and Genista monspessulanus: Scotch broom, French broom*. Arlington, VA, USA: The Nature Conservancy.

Ismail, BS, & Lee, HJ. (1995). Persistence of metsulfuron-methyl in two soils. *Journal of Environmental Science and Health, Part B: Pesticides, Food Contaminants, and Agricultural Wastes, 30*(4), 485–497.

James, TK, Rahman, A, Holland, PT, McNaughton, DE, & Heiermann, M. (1998). *Degradation and movement of terbuthylazine in soil* (Proceedings of the New Zealand Plant Protection Conference, 51, pp. 157–161).

James, TK, Rahman, A, Mellsop, JM, & Trolove, M. (2004). Effect of rainfall on the movement and persistence of metsulfuron-methyl and clopyralid applied to pasture. *New Zealand Plant Protection, 57*, 271–276.

Jotcham, JR, Smith, DW, & Stephenson, GR. (1989). Comparative persistence and mobility of pyridine and phenoxy herbicides in soil. *Weed Technology, 3*(1), 155–161.

Ketchum, JS, & Rose, R. (2003). Preventing establishment of exotic shrubs (*Cytisus scoparius* (L.) Link. and *Cytisus striatus* (Hill)) with soil active herbicides (hexazinone, sulfometuron, and metsulfuron). *New Forests, 25*(2), 83–92.

Khan, MA, & Liang, T. (1989). Mapping pesticide contamination potential. *Environmental Management, 13*(2), 233–242.

Lavy, TL, Mattice, JD, & Kochenderfer, JN. (1989). Hexazinone persistence and mobility of a steep forested watershed. *Journal of Environmental Quality, 18*(4), 507–514.

Löf, M. (2000). Establishment and growth in seedlings of *Fagus sylvatica* and *Quercus robur*: influence of interference from herbaceous vegetation. *Canadian Journal of Forest Research, 30*(6), 855–864.

MacDiarmid, BN. (1975). *Soil residues of picloram applied aerially to New Zealand brushweeds* (Proceedings of the New Zealand Weed and Pest Control Conference, 28, pp. 109–114).

MacDiarmid, BN. (1977). *Properties and activity of triclopyr on some New Zealand brushweeds* (Proceedings of the New Zealand Weed and Pest Control Conference, 30, pp. 66–70).

McCracken, IJ. (1995). *The effect of understorey broom (Cytisus scoparius) on soil moisture and growth of Pinus radiata* (Second International Conference on Forest Vegetation Management, 20–24 March, pp. 208–220).

Michael, JL, & Neary, DG. (1993). Herbicide dissipation studies in southern forest ecosystems. *Environmental Toxicology and Chemistry, 12*(3), 405–410.

Morash, R, & Freedman, B. (1989). The effects of several herbicides on the germination of seeds in the forest floor. *Canadian Journal of Forest Research, 19*(3), 347–350.

Moreno-Casasola, P, Grime, JP, & Martinez, ML. (1994). A comparative study of the effects of fluctuations in temperature and moisture supply on hard coat dormancy in seeds of coastal tropical legumes in Mexico. *Journal of Tropical Ecology, 10*(2), 67–86.

Nambiar, E, & Sands, R. (1993). Competition for water and nutrients in forests. *Canadian Journal of Forest Research, 23*(10), 1955–1968.

Oneto, SR, Kyser, GB, & DiTomaso, JM. (2010). Efficacy of mechanical and herbicide control methods for Scotch broom (*Cytisus scoparius*) and cost analysis of chemical control options. *Invasive Plant Science and Management, 3*(4), 421–428.

Pik, AJ, Peake, E, Strosher, MT, & Hodgson, GW. (1977). Fate of 3,6-dichloropicolinic acid in soils. *Journal of Agricultural and Food Chemistry, 25*(5), 1054–1061.

Prévosto, B, Dambrine, E, Coquillard, P, & Robert, A. (2006). Broom (*Cytisus scoparius*) colonization after grazing abandonment in the French Massif Central: impact on vegetation composition and resource availability. *Acta Oecologica, 30*(2), 258–268.

Rahman, A. (1981). *Persistence and movement of hexazinone in some New Zealand soils* (Proceedings of the 8th Asian Pacific Weed Science Society, 8, pp. 475–479)

Rahman, A, James, TK, Martin, P, & Fullerton, DK. (1991). *Persistence of metsulfuron in different soils* (Proceedings of the New Zealand Weed and Pest Control Conference, 44, pp. 99–104).

Rhodes, RC. (1980). Soil studies with ^{14}C-labeled hexazinone. *Journal of Agricultural and Food Chemistry, 28*(2), 311–315.

Richardson, B. (1993). Vegetation management practices in plantation forests of Australia and New Zealand. *Canadian Journal of Forest Research, 23*(10), 1989–2005.

Richardson, B, & Whitehead, D. (2002). Root-zone water storage and growth of *Pinus radiata* in the presence of a broom understorey. *New Zealand Journal of Forestry Science, 32*(2), 208–220.

Richardson, B, Vanner, A, Ray, J, Davenhill, N, & Coker, G. (1996). Mechanisms of *Pinus radiata* growth suppression by some common forest weed species. *New Zealand Journal of Forestry Science, 26*(3), 421–437.

Richardson, B, Vanner, A, Ray, J, & Balneaves, J. (1997). *Effect of some common weed species on Pinus radiata growth at a dry South Island site* (Proceedings of the New Zealand Plant Protection Conference, 50, pp. 373–376).

Richardson, B, Kimberley, MO, Ray, JW, & Coker, GW. (1999). Indices of interspecific plant competition for *Pinus radiata* in central North Island of New Zealand. *Canadian Journal of Forest Research, 29*(7), 898–905.

Rolando, CA, & Watt, MS. (2012). Perspectives on Forest Stewardship Council certification and vegetation management in New Zealand. *New Zealand Journal of Forestry, 56*(4), 37–40.

Rolando, CA, Gous, S, & Watt, MS. (2011). Preliminary screening of herbicide mixes for the control of five major weed species on certified *Pinus radiata* plantations in New Zealand. *New Zealand Journal of Forestry Science, 41*, 165–175.

Rolando, CA, Garrett, LG, Baillie, BR, & Watt, MS. (2013). A survey of herbicide use and a review of environmental fate in New Zealand planted forests. *New Zealand Journal of Forestry Science, 43*: 17.

Roy, DN, Konar, SK, Charles, DA, Feng, JC, Prasad, R, & Campbell, RA. (1989). Determination of persistence, movement, and degradation of hexazinone in selected Canadian boreal forest soils. *Journal of Agricultural and Food Chemistry, 37*(2), 443–447.

Sahid, I, & Teoh, S. (1994). Persistence of terbuthylazine in soils. *Bulletin of Environmental Contamination and Toxicology, 52*(2), 226–230.

Sarmah, AK, Kookana, RS, & Alston, AM. (2000). Leaching and degradation of triasulfuron, metsulfuron-methyl, and chlorsulfuron in alkaline soil profiles under field conditions. *Australian Journal of Soil Research, 38*(3), 617–631.

Sarmah, AK, Close, ME, & Mason, NWH. (2009). Dissipation and sorption of six commonly used pesticides in two contrasting soils of New Zealand. *Journal of Environmental Science and Health, Part B: Pesticides, Food Contaminants, and Agricultural Wastes, 44*(4), 325–336.

SAS Institute Inc. (2008). *SAS/STAT 9.2 User's Guide*. Cary, NC, USA: SAS Institute Inc.

Stephenson, GR, Solomon, KR, Bowhey, CS, & Liber, K. (1990). Persistence, leachability, and lateral movement of triclopyr (Garlon) in selected Canadian forestry soils. *Journal of Agricultural and Food Chemistry, 38*(2), 584–588.

Trabue, SL, Palmquist, DE, Lydick, TM, & Singles, SK. (2006). Effects of soil storage on the microbial community and degradation of metsulfuron-methyl. *Journal of Agricultural and Food Chemistry, 54*(1), 142–151.

Tran, H, Harrington, KC, Robertson, AW, & Watt, MS. (2010). *Assessment of herbicides for selectively controlling broom (Cytisus scoparius) growing with Pinus radiata* (Proceedings of 17th Australasian Weeds Conference, p. 372).

Turner, JH. (1933). The viability of seeds. *Bulletin of Miscellaneous Information (Royal Gardens, Kew), 1933*(6), 257–269.

USDA-FS (1984a). Pesticide Background Statements. Vol. 1. Herbicides. Hexazinone. Agriculture Handbook 633: Washington, DC: U.S. Department of Agriculture. Pp. H1–H86.

USDA-FS (1984b). Pesticide background statements. Vol. 1. Herbicides. Triclopyr. Agriculture Handbook 633: Washington, DC: U.S. Department of Agriculture. Pp. T1–T62

Wagner, RG, Little, KM, Richardson, B, & Mcnabb, K. (2006). The role of vegetation management for enhancing productivity of the world's forests. *Forestry, 79*(1), 57–79.

Walker, A, & Welch, SJ. (1989). The relative movement and persistence in soil of chlorsulfuron, metsulfuron-methyl and triasulfuron. *Weed Research, 29*(5), 375–383.

Walker, A, Cotterill, EG, & Welch, SJ. (1989). Adsorption and degradation of chlorsulfuron and metsulfuron-methyl in soils from different depths. *Weed Research, 29*(4), 281–287.

Walstad, JD, & Kuch, PJ. (1987). *Forest Vegetation Management for Conifer Production*. New York, USA: John Wiley.

Wang, GG, Su, J, & Wang, JR. (2000). Height growth of planted black spruce seedlings in response to interspecific vegetation competition: a comparison of four competition measures at two measuring positions. *Canadian Journal of Forest Research, 30*(4), 573–579.

Wang, H, Lin, K, Hou, Z, Richardson, B, & Gan, J. (2010). Sorption of the herbicide terbuthylazine in two New Zealand forest soils amended with biosolids and biochars. *Journal of Soils and Sediments, 10*(2), 283–289.

Watt, MS, & Rolando, CA. (2014). Alternatives to hexazinone and terbuthylazine for the chemical control of *Cytisus scoparius* in *Pinus radiata* plantations in New Zealand. *Weed Research, 54*(3), 265–273.

Watt, MS, Whitehead, D, Richardson, B, Mason, EG, & Leckie, AL. (2003). Modelling the influence of weed competition on the growth of young *Pinus radiata* at a dryland site. *Forest Ecology and Management, 178*(3), 271–286.

Watt, MS, Wang, H, Rolando, CA, Zaayman, M, & Martin, K. (2010). Adsorption of the herbicide terbuthylazine across a range of New Zealand forestry soils. *Canadian Journal Forest Research, 40*(7), 1–10.

Wearne, LJ, & Morgan, JW. (2004). Community-level changes in Australian subalpine vegetation following invasion by the non-native shrub *Cytisus scoparius*. *Journal of Vegetation Science, 15*(5), 595–604.

Wearne, LJ, & Morgan, JW. (2006). Shrub invasion into subalpine vegetation: implications for restoration of the native ecosystem. *Plant Ecology, 183*(2), 361–376.

White, DE, Witherspoon-Joos, L, & Newton, M. (1990). Herbaceous weed control in conifer plantations with hexazinone and nitrogen formulations. *New Forests, 4*(2), 97–105.

Young, S. (2013). *New Zealand Novachem Agrichemical Manual*. Christchurch, New Zealand: Agrimedia Limited.

9

Application of LiDAR data to maximise the efficiency of inventory plots in softwood plantations

Gavin Melville[1*], Christine Stone[2] and Russell Turner[3]

Abstract

Background: Precision in describing plantation attributes is a key requirement for forestry managers and inventory surveys aim to extract the most precise information possible using the smallest number of plots. This paper quantifies the potential efficiencies to be gained by using Light Detection and Ranging (LiDAR) data as an aid to estimation of standing timber volume in softwood plantations. A range of inventory design and estimation methods were investigated in terms of their overall predictive efficiency.

Methods: Field measurements representing four different populations from two *Pinus radiata* D. Don plantations in New South Wales, Australia, were used to inform statistical models which were then employed to simulate populations of inventory plots. These plots were then "surveyed" using a variety of simulated sampling strategies to quantify the benefits from using LiDAR data as auxiliary information. Model-based and design-based methods were both investigated. Survey design options included stratification and plot selection strategies; estimation options included ratio estimation and regression modelling. Results were compared in terms of the relative bias and root mean squared error of the estimates.

Results: The study suggests that relative efficiencies of two-fold or better, are possible with either model-based or model-assisted estimators compared to traditional inventory surveys which use grid samples and simple design-based estimators. This would enable a halving in the required sample size for the same precision for field inventories in these plantations.

Conclusion: The use of LiDAR data as an aid to survey design produces marked efficiency gains compared to traditional inventory methods.

Keywords: Sampling efficiency; LiDAR; Inventory surveys; Model-based sampling; Design-based sampling

Background

Plantation managers are concerned with the establishment, health, inventory and growth trajectory of commercial forests. Greater precision in describing plantations should lead to more accurate forecasts and therefore better commercial decisions. As the gathering of information requires expenditure, these benefits must be balanced with the costs in order to determine the optimal amount of information that should be collected for making a decision (Gilabert and McDill 2010). Therefore, there is a strong motivation to improve sampling design efficiencies in field inventory, with the aim of minimising the number of labour intensively measured plots while maximising predictive performance and minimising bias (Junttila et al. 2008).

The increasing availability of high-resolution remotely sensed information, including Light Detection and Ranging (LiDAR) (also referred to as airborne laser scanning (ALS)) sensors, is now routinely being coupled with geo-referenced inventory plot measurements to improve large- and small-scale estimation of forestry parameters (e.g. White et al. 2013). Several studies have now demonstrated the utility of LiDAR data in the survey design phase for optimal sample selection (e.g. Gobakken et al. 2013; Maltamo et al. 2009a) as well as in the estimation

* Correspondence: gavin.melville@dpi.nsw.gov.au
[1]Trangie Agricultural Research Centre, Mitchell Hwy, Trangie 2823, Australia
Full list of author information is available at the end of the article

phase (e.g. McRoberts et al. 2012) to improve prediction accuracies and/or cost efficiencies. The benefits can be an outcome of differing efficiencies arising from either the sampling design and/or the estimation method. In both cases, efficiencies are gained when there is a strong relationship between the forest variable of interest and the derived LiDAR metrics (Maltamo et al. 2011). If this relationship is strong and the postulated model is able to capitalise on this relationship, then substantial gains in efficiency can be realised (Junttila et al. 2013). Junttila et al. (2013) also claimed that the most important factor improving sampling design efficiency occurred when the selected plots were well spread in feature space. This is supported by Grafström and Ringvall (2013) who reported, when examining several sampling designs, that well-spread samples not only improved model fitting but also design-based estimation.

Sampling strategies have been investigated across numerous forest-related activities through a range of modelling approaches within both design-based (probability) and model-based inference frameworks (e.g. Hansen et al. 1983: McRoberts 2010; Wulder et al. 2012 and references therein). Traditionally, forest inventories calculate estimates of growing stock volume using design-based simple random sampling, stratified or model-assisted estimators (McRoberts 2006). More recently, several design-based and model-based variance estimators have been investigated using data from LiDAR-assisted forest surveys (e.g. Gregoire et al. 2011; McRoberts et al. 2013; Ståhl et al. 2011). Efficient sampling strategies are attained by applying an optimal sampling design in addition to identifying the best variance estimators for that particular design as well as for the specified model.

In the design-based framework, the population is regarded as fixed whereas the sample is regarded as a realisation of a stochastic process (Gregoire 1998). Model-assisted estimators are a subclass of design-based estimators in which a model is used to describe the population (e.g. Gregoire et al. 2011; McRoberts 2010). Relative to stratified sampling using the same auxiliary data, model-assisted estimators can potentially make better use of each variable by using the complete range of values rather than dividing the variable into classes. More recently, model-based inference in LiDAR survey sampling has gained popularity (e.g. Ståhl et al. 2011). In model-based inference, the sampling method does not necessarily have to be random; that is, the sample plots can be treated as fixed, the population variables treated as random, and variation (and therefore inference) is derived from the model (McRoberts 2010). Both model-assisted and model-based approaches rely more heavily on models and auxiliary variables to produce estimates than traditional design-based approaches and therefore may produce more precise estimates for small areas (McRoberts et al. 2013). Also model-based procedures allow prediction and error estimation at the pixel level, and hence can result in attribute mapping (McRoberts et al. 2013). However, their main disadvantage is the potential for substantial bias (if the model is poorly specified) plus such models are often computationally intensive (McRoberts et al. 2013).

Within these statistical frameworks, the performance of various sampling designs incorporating LiDAR variables has been examined through a range of modelling techniques including k-Most Similar Neighbours (MSN) (e.g. Maltamo et al. 2009b), Bayesian regression (Junttila et al. 2008), nonlinear logistic regression (McRoberts et al. 2013), linear regression techniques (Dalponte et al. 2011; Ene et al. 2013) and nonlinear regression modelling (Næsset et al. 2011). A common approach is to compare the prediction accuracies of the estimation techniques under investigation after reducing the number of sample plots (Dalponte et al. 2011; Junttila et al. 2013). McRoberts et al. (2012), on the other hand, assessed the utility of LiDAR data as the basis for post-stratification when applied to three techniques for estimating mean growing stock volume: a multiple linear regression model, a nonlinear logistic regression model, and the k-Nearest Neighbour technique.

If appropriate LiDAR data are available prior to sampling, then they can be used directly to stratify the field survey (Gobakken et al. 2013; Hawbaker et al. 2009; Maltamo et al. 2011; van Aardt et al. 2006). Hawbaker et al. (2009), for example, improved sampling efficiency and prediction accuracies of mean diameter, basal area, stand height and biomass using gridded LiDAR data as *a priori* information to define additional strata within a mixed species coniferous forest in northeastern Wisconsin, USA. For each forest type, potential sampling locations were stratified into 10 x 3 strata according to mean LiDAR height (10 levels) and standard deviation (3 levels) of mean LiDAR height. This ensured that the entire data range of the predictor variables was sampled on the ground and resulted in better predictions by the LiDAR-derived regression models. If, however, the LiDAR data are acquired after the fieldwork, an increase in precision may still be achieved through incorporation of LiDAR-derived post-stratification within the estimation process (Dalponte et al. 2011; Ene et al. 2013; McRoberts et al. 2012). Note that stratification for the estimation of a single population attribute is generally advantageous but the situation is less clear for a multi-purpose inventory with many attributes of interest (Köhl et al. 2006; Næsset et al. 2013). After stratification, there exist several strategies for plot allocation that provide the optimal sample size for each stratum. For example, Neyman allocation is a method used to allocate samples based on the stratum variances and population sizes, assuming similar sampling costs in the strata, and given a fixed total sample size (Cochran 1977).

Recently, several studies have investigated a number of theoretical alternatives to stratification for LiDAR-assisted sample plot selection strategies (e.g. Grafström and Ringvall 2013; Grafström et al. 2014). Grafström et al. (2014), for example, introduced the local pivotal method, to select well-spread probability samples, in multidimensional spaces, from larger populations.

The scheduling of LiDAR campaigns is defined by the characteristics of the plantation and planning objectives, and schedules need to be adjusted annually. Efficiency gains from using LiDAR are dependent on strong correlations between the LiDAR variables and the variables to be estimated, and hence they benefit from minimum time delay between the acquisition of LiDAR data and the collection of field data. A number of researchers have investigated the effects of integrating historical and recently collected data. Rombouts et al. (2010) examined data from several campaigns separated by up to seven years and found that campaign was statistically significant in models relating stand volume to LiDAR data. However, they also concluded that the integration of prior data with current data produces gains in efficiency especially when the number of recent plots is limited. Junttila et al. (2010) employed earlier data to improve estimates using Bayesian regression. They also provided a method for correcting for time differences using linear correction mapping and stated that small changes in canopy height and forest structure and small changes in scanning conditions can be accommodated by their procedure.

There are two generic approaches to attribute estimation using LiDAR data. Firstly, "area-based" methods where the area-based prediction of forest variables is based on the statistical dependency between variables measured in field plots and the predictor features derived from LiDAR data (e.g. Corona and Fattorini 2008; White et al. 2013). Currently the area-based approach is operationally applied in several countries when carrying out standwise forest inventories (e.g. Hyyppä et al. 2012; White et al. 2013). Secondly, "tree-based" methods where individual trees are detected and tree-level variables are measured or predicted from the LiDAR data (e.g. Hyyppä et al. 2012). In this study, area-based methods were used where LiDAR data was summarised into virtual plots and which were co-located with inventory plots which were measured on the ground. It is generally acknowledged that if individual tree crowns can be recognised accurately in the LiDAR data then this approach tends to outperform the area-based methods (Yu et al. 2010). However, the area-based approach enables one to retrieve canopy height information by means of relatively coarse pulse density LiDAR data (~2 points m^{-2}) and hence cheaper acquisition costs (Jakubowski et al. 2013).

In the area-based approach, statistical metrics and other nonphysical distribution-related features of LiDAR height measurements can be extracted either directly from normalised LiDAR point clouds or from a normalised rasterised representation of laser hits (e.g. the normalised Digital Surface Model or Canopy Height Model (CHM) (Hyyppä et al. 2008; White et al., 2013). By using the CHM to derive metrics, the authors acknowledge the under-utilisation of the information content of the LiDAR point cloud. However this approach has proven to be operationally robust and requires less computational input (Hyyppä et al. 2008). Both area-based distributional metrics and individual tree level metrics have been extracted from CHMs in an operational workflow designed for automatic inventory estimates of *Pinus radiata* D. Don from LiDAR data, and this operationally orientated approach (Chen and Zhu 2013) was followed here.

In this study several survey design components were varied and compared using auxiliary variables derived from LiDAR data to improve efficiencies in the inventory of *Pinus radiata* plantations. Design-based and model-based methods were both investigated. Survey design options included stratification and plot selection strategies, and estimation options included ratio estimation and regression modelling. The methodology involved using simulations based on linear mixed-effects modelling of four plot based surveys. The variable of interest in the simulations was total stand volume (TSV) per plot. Specifically, the objective of this study was to determine whether it is possible to guide the choice of ground truth samples, using *a priori* knowledge, so as to reduce the number of field plots for *P. radiata* inventories, without compromising prediction accuracies.

Methods

Study site

Two sites were selected for this study. The first of these was the Nundle State Forest *P. radiata* plantation, approximately 7000 ha in size and located in the Northern Tablelands of New South Wales (NSW), Australia. The second site was the Green Hills State Forest, which comprises 5000 ha of *P. radiata,* and is located in the Southern Tablelands of NSW. Both plantations are managed by the Forestry Corporation of New South Wales (FCNSW). In the Nundle study area, two unthinned compartments were selected representing young and old age classes respectively. The young compartment was planted in 2002 (2002 age class (AC)) at a stand density of 1000 trees ha^{-1}, while the older compartment was planted in 1977 (1977 AC) at a stand density of 1200 trees ha^{-1} (Table 1). The Green Hills study site contained a full representation of age classes (from just planted to > 35 years old) and a

Table 1 Comparison of stand attributes

Survey	Stand density (stems ha^{-1})	Stand volume (m^3 ha^{-1})	DBH (cm)	Basal area (m^2 ha^{-1})	Mean dominant height (m)	Mean height (m)
Nundle 1977	279 ± 123	500 ± 151	45.6 ± 7.5	45.2 ± 13.7	35.2 ± 2.5	33.3 ± 1.0
	(48–676)	(205–855)	(34.7 - 74.1)	(18.6 - 78.0)	(27.8 - 39.4)	(31.1 - 35.9)
Nundle 2002	593 ± 289	57 ± 25	16.8 ± 3.0	13.5 ± 5.8	11.0 ± 1.0	9.8 ± 0.5
	(100–1250)	(11–135)	(11.6 - 24.5)	(2.7 - 31.7)	(8.8 - 12.9)	(8.9 - 10.9)
Green Hills Inventory (1979)	197 ± 34	249 ± 44	40.0 ± 2.1	24.6 ± 3.7	30.3 ± 2.0	28.0 ± 2.2
	(140–260)	(161–342)	(35.3 - 43.6)	(17.6 - 32.8)	(25.6 - 34.1)	(23.5 - 31.4)
Green Hills Research (30 + yrs)	274 ± 40	413 ± 46	42.1 ± 1.3	38.8 ± 4.7	33.4 ± 1.2	31.3 ± 1.1
	(187–305)	(322–463)	(40.3 - 43.9)	(29.5 - 43.6)	(31.2 - 35.0)	(30.0 - 33.6)

mean ± sd (min - max).

range of associated silvicultural thinning treatments (described in Stone et al., 2011).

Data collection

Airborne scanning LiDAR data

Discrete return airborne LiDAR data were acquired over the Nundle State Forest *P. radiata* plantation (Nundle) in June 2011 using a Trimble Harrier 68i system mounted on a Cessna U206G airplane. The technical specifications of the LiDAR data acquisition are provided in Table 2. The LiDAR points were processed, geo-referenced and classified by the service provider into ground and non-ground categories using their proprietary method and MARS software (Merrick & Company, Colorado, USA).

The discrete return LiDAR data were acquired over the Green Hills State Forest (SF) *P. radiata* plantation (Green Hills) in July 2008 using a Lite Mapper LMS-Q5600 ALS system (Riegl, Austria) using a mean swath width of 500 m (other technical specifications are listed in Table 2). The points in this dataset were also classified by the provider into ground and non-ground categories using their proprietary method and TerraScan software (TerraSolid Ltd., Helsinki, Finland).

Table 2 LiDAR acquisition specifications

LiDAR acquisition specification	Nundle S.F. study area	Green Hills S.F. study area
Swath width (m)	925	500
Scan overlap (%)	40	
Sensor system	Trimble Harrier	Lite Mapper
	68i-LMS-Q5600 (Riegal)	LMS-Q5600 (Riegal)
Measurement rate (KHz)	81	88
Field of view (sum of the two angles off nadir, degrees)	30	30
Flying height (metres above ground level)	1155	1100
Point density (points m^{-2})	3.6	2.0
Vertical/horizontal accuracy (σ in metres)	0.25 / 0.5	0.25 / 0.5
Average footprint size (m)	0.5	0.6
Beam divergence (milliradians)	0.5	0.5

σ = standard deviation.

For both the Nundle and Green Hills study areas, a 1 m spatial resolution Digital Elevation Model (DEM) was constructed from the ground point data using a standard linear triangulation surface modelling technique in Environment for Visualizing Images (ENVI 4.7) software (Research Systems Inc., USA). A similar triangulation approach was used to generate a Digital Surface Model (DSM) from all first returns (selecting the highest point in each grid cell). Then a 1 m pixel resolution CHM surface was constructed by subtracting the DEM from the DSM which represents the height of the canopy above ground level (ENVI 4.7).

For the Nundle stands, the CHM raster was initially masked to minimise potential edge effects by applying an internal 18 m buffer to the compartment boundary layer. In the older stand (1977 AC), circular 0.1 ha plots (17.84 m radius) were located at the intersection points of a 35 m grid to generate 1496 contiguous, virtual plots. The grid pattern ensured that the circular plots did not overlap and the plot dimensions were chosen so that each virtual plot would contain 19 trees on average, based on the LiDAR estimated stand density. In the younger stand (2002 AC), circular 0.04 ha plots (11.28 m radius) were located at the intersection points of a 23 m grid to generate 572 plots and each of these virtual plots was expected to contain 19 trees on average. When the stand densities were later measured on the ground they were somewhat higher than the LiDAR estimates so that the number of trees per virtual plot was actually 28 and 24 respectively for the 1977 and 2002 AC stands. LiDAR metrics were extracted for each virtual plot based on the 1 m CHM pixels contained within the plot (pixels on the circumference were included if they mostly lay inside the plot). The LiDAR metrics extracted are described in Table 3 and included mean height, mean above mean height, mean dominant height, estimated stand density

Table 3 Description of LiDAR metrics

LiDAR metric	Definition
Mean height	Mean height of pixels in virtual plot
Mean above mean height	Mean height of pixels which are above the mean height
Mean dominant height	Mean height of pixels identified as maxima using a 5 × 5 m window
Stocking density	Estimated stems per ha based on a 3 × 3 m local maxima (the window size most closely related to stand density)
Canopy cover	Percentage of pixels above 3 m in height
Occupied volume	Sum of all pixel heights in virtual plot
Variance	Variance of pixels in virtual plot
Skewness	Skewness of pixels in virtual plot

per hectare percentage canopy cover (CC), occupied volume (OV), height variance and height skewness (Turner et al. 2011).

At the second study site, also using the 1 m CHM, the Green Hills plot centre coordinates were imported into ENVI 4.2 software and buffered to a distance of 35 m. These vectors were then used as sampling templates to extract the same suite of LiDAR metrics as for the Nundle study.

Ground survey design

At the time of measurement, both of the Nundle stands were relatively heterogeneous in terms of stand density and tree spacing due to the absence of any pre-harvest thinning, as well as historic exposure to browsing and drought which affected tree establishment and survival. This heterogeneity was expected to make accurate estimates difficult to achieve using ground-based assessment based on a traditional systematic survey design. Because of this, stratification of the two Nundle stands was achieved using the LiDAR data as *a priori* information (e.g. Hawbaker et al. 2009; Maltamo et al. 2011), specifically mean height per virtual plot and estimated stand density. Both variables were divided into three ranges using the Dalenius-Hodges approach (Cochran 1977). This method resulted in nine non-contiguous strata, and each virtual plot was assigned to one of these strata. A random sample of 90 plots was then selected with an equal number of plots per stratum (10) using simple random sampling without replacement in each stratum. Subsequently, a subset of six plots per stratum was randomly selected for ground-based measurement (Table 4). The sample size was chosen so that the relative standard error of stand volume was 5% or better, using the design-based expansion estimator [3] which is given below.

At the Green Hills site, plot-based tree data from two surveys undertaken at approximately the same time were used for the analysis. One of the surveys comprised 100 conventional inventory plots with four strata (operational Planning Units), defined according to age class and thinning status (specifically the four strata were 1998 AC and unthinned; 1979 AC and thinned twice; 1983 AC and thinned twice; and 1983 AC and unthinned) (Table 4). These plots were established in accordance with standard FCNSW inventory procedures; they are based on a sampling intensity of one plot per four hectares, employ a systematic grid design with a random starting point and are designed to achieve a probable limit of error (PLE, as defined below) of 10% in terms of predicted stand volume, using the design-based expansion estimator [3] which is given below. Within the same study area, a further 63 plots were measured as part of a research study (Stone et al. 2011) with a sampling design comprising 16 strata defined according to age class, thinning history and ground slope categories. Four circular plots were randomly assigned to each stratum (one stratum had only three plots), bringing the total to 63 research plots. This sample size enabled models of tree height vs LiDAR height to be estimated with a relative root mean squared error (RMSE) of 5%.

Field data

In the Nundle 1977 AC stand, all of the 54 plots were used (ranging from 12.62 m to 19.95 m in radius), providing a cluster of selected trees containing approximately 22 trees per plot on average. In the Nundle 2002 AC stand, 52 of the plots were used (ranging from 7.98 m to 17.84 m in radius), providing a cluster of selected trees containing approximately 23 trees per plot on average (two plots were inaccessible). The use of "flexible radius" plots that is varying the plot radius according to the estimated stocking density, was employed to maintain an approximately constant workload throughout the survey (Melville et al. 2015). However the plots themselves were chosen with equal probability within strata. In the field, plot centres

Table 4 Comparison of survey data

Survey	Stratification	Number of strata	Number of plots	Sample size	R^2 values for model [1]	CV of timber volume (%)
Nundle 1977	LiDAR	9	1496	54	0.69	31
Nundle 2002	LiDAR	9	572	52	0.57	44
Green Hills inventory	Management	4	-	100	0.88	61
Green Hills research	Management	16	-	63	0.93	71

CV = coefficient of variation.

were located using two GPS units (Garmin Ltd.), and then validated through crown interpretation of open spaces between trees using printed copies of the LiDAR imagery. Every tree in each selected cluster was labelled and diameter at breast height over bark (DBH at 1.3 m) measured. At least five trees per selected cluster, representing the tallest trees, were selected and their heights measured using a hypsometer (Vertex III, Haglof, Sweden). These trees were used to predict the heights of the non-measured trees using the nonlinear dominant height vs DBH relationships which are used by FCNSW for this species. Plot standing volume was then estimated using the tapering equations employed by FCNSW for these plantations.

In the Green Hills inventory survey, plot measurements and tree assessments were undertaken by an independent local inventory crew using ATLAS Cruiser methodology (Atlas Technology, Rotorua, NZ) which provided a cluster of selected trees containing approximately 18 trees per plot on average. The inventory plot centres were located using a Scout Pak GPS (Juniper Systems, Logan, Utah, USA) with post-processing differential correction. In the Green Hills research survey the 63 plots were assessed using flexible radius plots, as described above, with radii ranging from 7.98 m to 17.84 m. This provided a cluster of selected trees containing approximately 15 trees per plot on average. The research plot centres were located using differential Global Positioning System (Trimble Navigation Ltd., Sunnyvale, California, USA) and a precision survey (with a laser theodolite). These plots were then measured using the same methodology as the Nundle plots.

Simulation approach

Overview

The approach used in the study was to construct simulation models using actual data, which comprised the data acquired using LiDAR together with the data collected from the ground-based surveys. The simulation models were then used to generate artificial plantations which were sampled using a range of different strategies. These strategies were then compared in terms of the precision of the resulting estimates. A flowchart illustrating the approach is provided in Figure 1. Using the Nundle 1977 AC as an example, the procedural steps were as follows:-

1. Sample 54 plots from a compartment containing 1496 plots. This sample was stratified using LiDAR height and stocking density (nine strata).
2. Conduct ground measurements of the sampled plots.
3. Estimate the parameters in simulation models (1) and (2).
4. Use models (1) and (2) to simulate data for 2700 plots in nine strata (300 plots per strata – same stratification as step 1), including stand volume, OV and CC.
5. Since the strata used to simulate the plot population is deemed to be "unknown" for the simulated sampling, the simulated population is re-stratified using the auxiliary variables OV and CC.
6. Select 24 plots using one of the sampling strategies under consideration.
7. Estimate the total volume of the simulated population and compare the estimate to the simulated "actual" value.
8. Repeat steps 4 to 7 a total of 10,000 times.
9. Calculate the relative bias and RMSE using the estimates and actual values.

Statistical modelling and plot simulation

The four datasets were examined separately by fitting linear mixed models to the variables of interest, namely timber volume and the LiDAR variables OV and CC (Turner et al. 2011). Although several LiDAR variables were available, the variables OV and CC were chosen because they were the best predictors of stand volume. It was also considered important to have separate variance terms for the strata which were defined in the original surveys; this is because each of the key variables changes markedly from one stratum to another, both in terms of the mean value and the variance. The use of mixed models allows for the estimation of separate stratum variances thus providing a more complete description of the data (Gilmour et al. 1995). The parameter estimates from the mixed models can then be used to generate plot data with characteristics very similar to the original data they were based on.

For each of the four plantation surveys, the following model was used for timber volume at the plot level

$$\boldsymbol{Y} = \boldsymbol{Z}^T\boldsymbol{\beta} + \boldsymbol{\varepsilon} \quad (1)$$

where $\boldsymbol{Y}$ represents total volume, $\boldsymbol{Z}$ is a design matrix consisting of stratum and OV and $\boldsymbol{\varepsilon} \sim \mathrm{N}(\boldsymbol{0}, \boldsymbol{R})$. The term $\boldsymbol{R} = \mathrm{diag}(\boldsymbol{V}_1, \ldots \boldsymbol{V}_H)$ represents the error structure, which is assumed to be a diagonal matrix with separate stratum variances, represented by V_h where h indexes strata and H is the total number of strata. In this model $\boldsymbol{V}_h$ may also be written as $\sigma_h^2 \boldsymbol{I}_{mh}$ where σ_h^2 and m_h represent the plot variance and number of plots respectively for stratum h and $\boldsymbol{I}_{mh}$ is the identity matrix of order m_h. A preliminary model which included off-diagonal correlation terms was subsequently replaced by the diagonal variance model because the correlation terms were very small.

A secondary linear mixed model was used to simultaneously model the LiDAR variables OV and CC at the plot level. This model has the same form as equation [1]

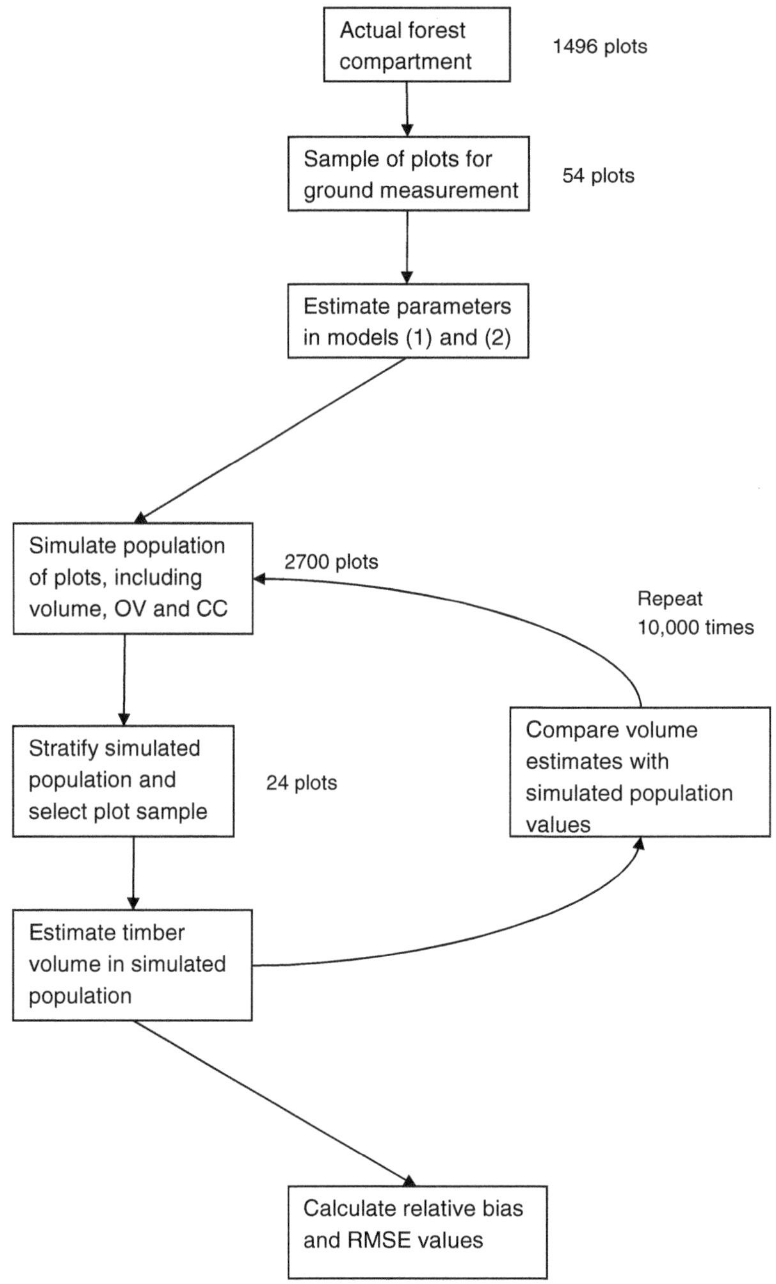

Figure 1 Flowchart illustrating simulation approach for the Nundle 1977 AC stand.

where $\boldsymbol{Y}$ is multivariate matrix with column vectors for OV and CC, $\boldsymbol{Z}$ is a design matrix consisting of stratum and the multivariate mean vector (with intercept terms for OC and CC) and $\boldsymbol{R} = \text{diag}(\boldsymbol{V}_1, \ldots \boldsymbol{V}_H)$ represents the error structure, which is assumed to be a block diagonal matrix with variance matrices, $\boldsymbol{V}_h$, given by

$$V_h = \begin{pmatrix} \sigma^2_{h1} & \sigma_{h12} \\ \sigma_{h12} & \sigma^2_{h2} \end{pmatrix} \quad (2)$$

In this model σ^2_{h1} and σ^2_{h2} are the within stratum variances of OV and CC respectively and σ_{h12} is the within stratum covariance term. As with the model for timber volume, preliminary analysis revealed that the off-diagonal correlation terms are sufficiently small to ignore.

For each of the four forestry surveys, a "population" of 2700 plots (each plot equal to 0.05 ha) was simulated using the parameter estimates from model [1]. The actual model parameters and stratum variances were estimated separately for each survey. Values for OV and CC were also simulated using the multivariate model for use as stratification variables. The R^2 values associated with model [1] are shown in Table 4 as well as the coefficients of variation (CV) for plot volume.

In the Nundle surveys, a measure of OV based on a fixed circular plot was used and timber volume was expressed in terms of cubic metres per hectare. In the two Green Hills surveys, a measure of OV related to the actual size of the measured plots was used and timber volume was expressed in terms of volume (m^3) per plot. In both cases, the 1 m pixel resolution enables calculation of the LiDAR variables for an approximately circular plot and both methods produce satisfactory correlations between OV and timber volume. Hence the decision to use either metrics calculated according to a fixed radius, or metrics calculated according to the actual measured radius was based on the available data. The LiDAR variables used for stratification in the Nundle surveys were mean canopy height and stand density. The management variables used for stratification in the Green Hills surveys were age class and thinning (operational inventory survey) and age class, thinning and slope (research survey). With the Green Hills stands, after simulating data for all strata, the oldest of these were used in comparing the various sampling strategies. In the Green Hills inventory plots these were the 1979 AC stands while in the Green Hills research plots they were the 30+ year AC compartments. The oldest plots were deemed to be most relevant to this study as they were the plots which were closest to harvest.

Simulated survey methods

Inferential framework After simulating the population of plots, both design-based and model-based strategies were investigated for survey design and estimation. Since stratification and Neyman allocation are typically used with design-based estimators, including model-assisted estimators (Foreman 1991), a number of different stratification options and two different types of estimation were explored. Stratification options included the use of different numbers of strata (two, three, four and six) and the use of one or two stratification variables (OV and OV + CC). Types of estimation included expansion and regression estimators, as described below. For the model-based strategies balanced sampling was used to obtain a representative sample of plots. Balanced sampling ensures that the Horvitz-Thompson estimators of the population totals of the design variables (OV and CC) equal the known totals of these variables and the method may be used without stratification. Since balanced samples are not strictly random they were not employed with the design-based estimators. They were used with the model-based methods because they were expected to be more efficient than a random sample. In these simulations the R-package "samplecube" was used which employs an algorithm developed by Deville (2004) for sample balance. Note that the selection probabilities themselves are not used in the model-based estimate even though the balanced sample is defined with respect to the Horvitz-Thompson estimator.

Stratification and plot selection A simple random sample of plots was chosen from the population of available plots to simulate sampling without stratification. To simulate stratified sampling, the primary variable OV was chosen as it is best correlated with stand volume (Turner et al. 2011). Where stratification was based on two variables, the secondary variable CC was chosen as it is also closely related to stand volume and the two variables together form the best prediction pair for stand volume. With a single stratification variable, the number of strata employed was two, three, four or six. With two stratification variables, the number strata employed was either four (2 × 2) or six (3 × 2). Stratum boundaries were calculated using the Dalenius-Hodges method (Cochran 1977). Neyman allocation was used to allocate plots to strata. Using this scheme plots are allocated according to $n_h \sim N_h\ S_h$, where h indexes strata, n_h represents stratum sample size and N_h and S_h represent stratum population size and standard deviation of timber volume respectively (Cochran 1977). We assumed that the standard deviation of timber volume, S_h, would be unknown for the purposes of survey design and used the standard deviation of the primary stratification variable instead. A minimum sample size constraint was also specified (two plots per stratum).

For the model-based methods, the sampling was based on either one or two balancing variables and stratification was not employed as a sampling tool. The design variable(s) used for plot selection was either OV (single balancing variable) or both OV and CC (two balancing variables). The target sample size was 24 plots throughout the study which is similar to the number of plots used by FCNSW for this sized population. Minor variation sometimes occurred due to minimum sample size constraints and the actual sample size was occasionally 23 or 25. RMSEs were adjusted where necessary to reflect the comparison size of 24 plots.

Estimation methods

Design-based estimates Where Y represents the total volume of timber in the stand of interest, then the basic design-based expansion estimator is given as

$$\hat{Y} = \frac{M}{m}\sum_{i\in q} Y_i, \tag{3}$$

where Y_i is the volume of timber in plot i, q represents the set of sampled plots, and M and m represent the total number of plots and the number of sampled plots respectively (Cochran 1977).

The stratified sampling method employed a stratified design-based expansion estimator which is given by (Cochran 1977)

$$\hat{Y} = \sum_{h=1}^{H} \frac{M_h}{m_h} \sum_{i \in q_h} Y_i, \tag{4}$$

where H represents the total number of strata, and M_h and m_h represent the total number of plots and the number of sampled plots in stratum h respectively.

In addition to the usual design-based estimator, a regression estimator (model-assisted) was also employed using strata and OV as the auxiliary variables. This estimator may be written as (Särndal et al. 1992)

$$\hat{Y} = \sum_{h=1}^{H} \sum_{i \in q_h} \frac{M_h}{m_h} Y_i - \hat{\boldsymbol{\beta}}^T (\hat{\boldsymbol{Z}} - \boldsymbol{Z}), \tag{5}$$

where $\boldsymbol{Z}$ represents the design matrix for the model covariates, $\hat{\boldsymbol{Z}}$ is a design-based expansion estimator for $\boldsymbol{Z}$ and $\boldsymbol{\beta}$ is a vector of model parameters who's design-based least squares estimate is given by $\hat{\boldsymbol{\beta}}$.

Model-based estimates The model-based predictions used either OV (single balancing variable) or OV and CC (two balancing variables) as auxiliary data. The model-based estimator for timber volume may be written as (Valliant et al. 2000)

$$\hat{Y} = \sum_{i \in q} Y_i + \sum_{j \notin q} \hat{\boldsymbol{\beta}}^T \boldsymbol{z}_j, \tag{6}$$

where $\boldsymbol{z}_j$ represents the covariate vector for plot j and $\hat{\boldsymbol{\beta}}$ is an appropriate model-based estimate for $\boldsymbol{\beta}$.

Comparisons between methods The various sampling methods were compared in terms of the relative bias, and the relative root mean squared error (RMSE%). The relative bias is defined as

$$RB = \frac{\sum_k (\hat{Y}_k - Y_k)}{\sum_k Y_k}, \tag{7}$$

where k indexes the set of realisations of the simulated plantations, $\hat{Y}_k$ is the estimate of total timber volume for the k'th realisation and Y_k is the actual total timber volume.

The relative root mean squared error is defined as

$$RMSE = \frac{\sqrt{\frac{1}{k} \sum_k (\hat{Y}_k - Y_k)^2}}{\frac{1}{k} \sum_k Y_k}. \tag{8}$$

The sampling strategies were also compared in terms of their relative efficiency with respect to the simple random sample. The relative efficiency for any method is calculated by squaring the ratio obtained by dividing the RMSE value for that method by the RMSE value for the design-based unstratified estimate (3). The PLE (probable limits of error; Goulding and Lawrence 1992) was also calculated, where the PLE is half of the 95% confidence interval expressed as a percentage of the mean and was calculated in this study as twice the RMSE

$$\text{PLE} = 2 * \text{RMSE}. \tag{9}$$

Results

The parameter estimates for model [1] are given in Table 5. The comparisons of survey methods for each of the four surveys are presented in Tables 6, 7, 8, 9. The different sampling strategies in terms of RMSE and bias were compared and the relative bias was found to be very small for all methods. In terms RMSE, all strategies outperformed the simple random sample. In the simulations based on these plantation stands, the performance of simple random sampling is similar to that of a grid sample, which is commonly used in inventory sampling. It could be argued that a grid sample should be more efficient than a simple random sample because it reduces the effect of the spatial variability in the plots. However, when the Nundle data were pre-stratified according to the LiDAR derived variables related to plot mean height and stand density, it was noticeable that the plots within

Table 5 Parameter estimates for model [1]

Parameter	Nundle 1977	Nundle 2002	Green Hills (30+ yrs)	Green Hills (1979)
Mean	54.7236	−0.4823	1.1713**	7.8131**
OV	0.0229***	0.0170***	0.0014***	0.0014***
Stratum 2	−12.8711	−4.1180	2.5097***	−2.5224
Stratum 3	47.3760	−6.3374	2.0857	−7.3314
Stratum 4	−56.9808	−1.7643	7.2840***	−2.7101
Stratum 5	43.0417	2.2229	−1.1264	
Stratum 6	10.3448	−24.734	2.1840**	
Stratum 7	−31.9621	6.2664		
Stratum 8	93.5574	−10.8385		
Stratum 9	113.1250**	−9.4114		

$^{**}p < 0.01$ $^{***}p < 0.001$.

Table 6 Nundle 1977 AC plots: comparison of methods

Inferential framework	Number of strata	Design variable	Estimation variables	Regression model	Relative bias	RMSE	PLE	Relative efficiency
Design-	-	-	Expansion	-	0.02	6.5	13.0	1.0
based	2	OV	Expansion	-	−0.02	5.3	10.6	1.5
	3	OV	Expansion	-	0.04	4.0	8.0	2.0
	4	OV	Expansion	-	0.01	4.4	8.8	2.2
	6	OV	Expansion	-	0.00	3.8	7.6	2.4
	2	OV	Regression	Strata + OV	0.06	4.4	8.8	2.2
	3	OV	Regression	Strata + OV	0.11	4.0	8.0	2.6
	4	OV	Regression	Strata + OV	0.04	4.0	8.0	2.6
	6	OV	Regression	Strata + OV	0.04	4.1	8.2	2.5
	4	OV + CC	Expansion	-	0.02	4.9	9.8	1.8
	6	OV + CC	Expansion	-	0.05	4.6	9.2	2.0
	4	OV + CC	Regression	Strata + OV	−0.14	4.4	8.8	2.2
	6	OV + CC	Regression	Strata + OV	−0.05	5.0	10.0	1.7
Model-	-	OV	Model	OV	−0.08	4.1	8.2	2.5
based	-	OV + CC	Model	OV + CC	−0.12	4.2	8.4	2.4

OV = the LiDAR metric 'occupied volume'; i.e. the sum of all pixel heights per plot.
CC = the LiDAR metric 'canopy cover'; i.e. the % of pixels above 3 m in height.
Expansion = estimator [4]; regression = estimator [5]; model = predictor [6].

the same stratum were not spatially contiguous which suggests that grid sampling and simple random sampling are likely to perform similarly in these plantations.

With the usual design-based expansion estimator, the number of strata is important. For example, better results were achieved using six rather than two strata in the Nundle and Green Hill research stands (Tables 6, 7 and 9). In addition to this, the number of stratification variables is also important with one single, LiDAR-derived variable (OV) leading to better estimates than two separate variables (OV + CC) for three of the four datasets and nearly equivalent estimates for the fourth.

Table 7 Nundle 2002 AC plots: comparison of methods

Inferential framework	Number of strata	Design variable	Estimation variables	Regression model	Relative bias	RMSE	PLE	Relative efficiency
Design-	-	-	Expansion	-	0.04	9.1	18.2	1.0
based	2	OV	Expansion	-	−0.05	7.8	15.6	1.4
	3	OV	Expansion	-	−0.10	8.0	16.0	1.7
	4	OV	Expansion	-	0.05	6.6	13.2	1.9
	6	OV	Expansion	-	0.01	6.5	13.0	2.0
	2	OV	Regression	Strata + OV	−0.04	6.4	12.8	2.0
	3	OV	Regression	Strata + OV	0.08	6.3	12.6	2.1
	4	OV	Regression	Strata + OV	0.02	6.3	12.6	2.1
	6	OV	Regression	Strata + OV	−0.07	6.4	12.8	2.0
	4	OV + CC	Expansion	-	−0.02	7.5	15.0	1.5
	6	OV + CC	Expansion	-	−0.04	7.4	14.8	1.5
	4	OV + CC	Regression	Strata + OV	−0.08	6.3	12.6	2.1
	6	OV + CC	Regression	Strata + OV	−0.10	6.6	13.2	1.9
Model-	-	OV	Model	OV	−0.12	6.3	12.6	2.1
based	-	OV + CC	Model	OV + CC	−0.14	6.5	13.0	2.0

OV = the LiDAR metric 'occupied volume'; i.e. the sum of all pixel heights per plot.
CC = the LiDAR metric 'canopy cover'; i.e. the % of pixels above 3 m in height.
Expansion = estimator [4]; regression = estimator [5]; model = predictor [6].

Table 8 Green Hills inventory plots (1979 AC): comparison of methods

Inferential framework	Number of strata	Design variable	Estimation variables	Regression model	Relative bias	RMSE	PLE	Relative efficiency
Design-	-	-	Expansion	-	0.04	4.4	8.8	1.0
based	2	OV	Expansion	-	0.06	3.8	7.6	1.3
	3	OV	Expansion	-	0.02	3.9	7.8	1.3
	4	OV	Expansion	-	0.01	4.0	8.0	1.2
	6	OV	Expansion	-	0.10	4.1	8.2	1.2
	2	OV	Regression	Strata + OV	0.05	3.5	7.0	1.6
	3	OV	Regression	Strata + OV	0.03	3.8	7.6	1.3
	4	OV	Regression	Strata + OV	−0.01	3.9	7.8	1.3
	6	OV	Regression	Strata + OV	0.10	4.1	8.2	1.2
	4	OV + CC	Expansion	-	0.02	3.8	7.6	1.3
	6	OV + CC	Expansion	-	−0.01	4.1	8.2	1.2
	4	OV + CC	Regression	Strata + OV	0.03	3.5	7.0	1.6
	6	OV + CC	Regression	Strata + OV	−0.01	4.0	8.0	1.2
Model-	-	OV	Model	OV	−0.01	3.5	7.0	1.6
based	-	OV + CC	Model	OV + CC	−0.01	3.5	7.0	1.6

OV = the LiDAR metric 'occupied volume'; i.e. the sum of all pixel heights per plot.
CC = the LiDAR metric 'canopy cover'; i.e. the % of pixels above 3 m in height.
Expansion = estimator [4]; regression = estimator [5]; model = predictor [6].

The relationship between stand volume and OV for the four stands is shown in Figure 2. These relationships form the basis of the model-assisted and model-based methods. With the regression estimator (model-assisted) fewer strata led to better estimates for the Green Hills datasets (Tables 8 and 9) but had minimal effect on the RMSE values for the Nundle datasets (Tables 6 and 7). A possible reason is that the regression models were more effective in the Green Hills stands, thus requiring fewer strata to obtain the same level of efficiency. The

Table 9 Green Hills research plots (30+ yrs AC): comparison of methods

Inferential framework	Number of strata	Design variable	Estimation variables	Regression model	Relative bias	RMSE	PLE	Relative efficiency
Design-	-	-	Expansion	-	0.01	6.9	13.8	1.0
based	2	OV	Expansion	-	0.00	4.9	9.8	2.0
	3	OV	Expansion	-	0.02	4.4	8.8	2.4
	4	OV	Expansion	-	−0.01	4.2	8.4	2.7
	6	OV	Expansion	-	−0.05	4.1	8.2	3.0
	2	OV	Regression	Strata + OV	0.00	3.3	6.6	4.4
	3	OV	Regression	Strata + OV	0.02	3.5	7.0	3.9
	4	OV	Regression	Strata + OV	0.00	3.6	7.2	3.7
	6	OV	Regression	Strata + OV	−0.07	3.8	7.6	3.3
	4	OV + CC	Expansion	-	−0.07	4.9	9.8	2.0
	6	OV + CC	Expansion	-	−0.04	4.4	8.8	2.4
	4	OV + CC	Regression	Strata + OV	−0.02	3.3	6.6	4.4
	6	OV + CC	Regression	Strata + OV	−0.02	3.5	7.0	3.9
Model-	-	OV	Model	OV	−0.01	3.6	7.2	4.4
based	-	OV + CC	Model	OV + CC	−0.05	3.3	6.6	4.4

OV = the LiDAR metric 'occupied volume'; i.e. the sum of all pixel heights per plot.
CC = the LiDAR metric 'canopy cover'; i.e. the % of pixels above 3 m in height.
Expansion = estimator [4]; regression = estimator [5]; model = predictor [6].

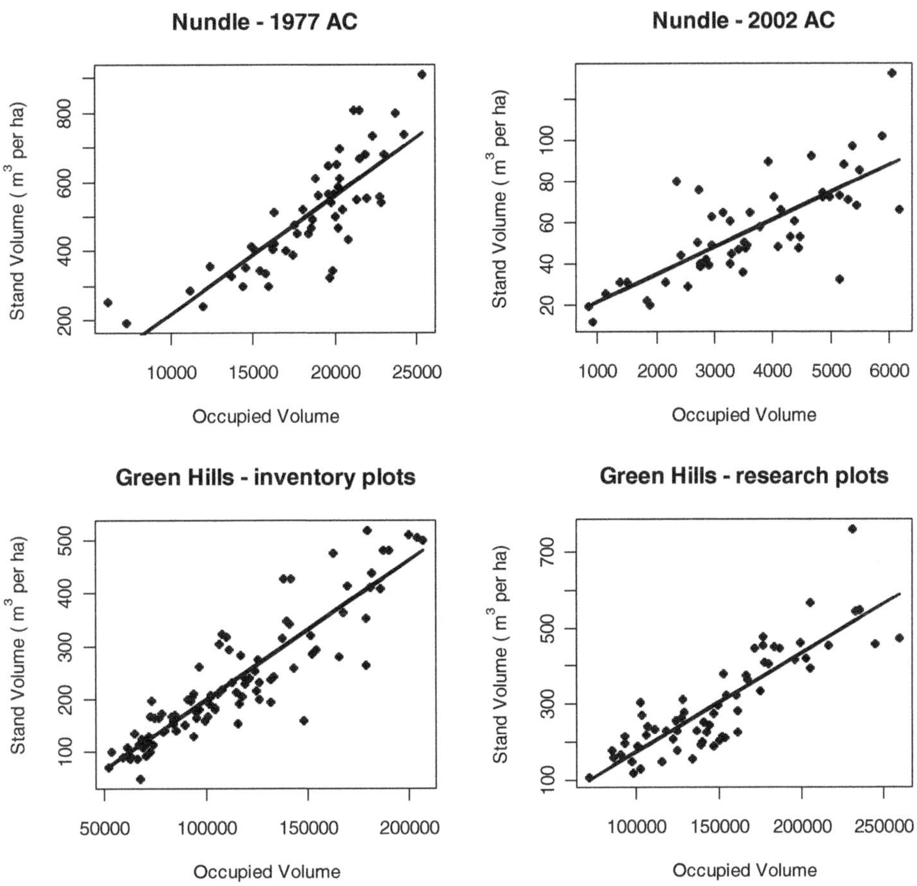

Figure 2 Mean stand volume vs occupied volume (OV) for the four stands used in the study.

preferred option with the Nundle stands was for a single stratification variable rather than two variables but this was not the case for the Green Hills stands.

With the model-based approach, the number of auxiliary variables (either one or two) had a minor impact on the prediction errors. Care is needed in interpreting the results for the Nundle stands because there was evidence of multi-collinearity in the auxiliary variables and the addition of a second variable to the model may lead to only minor improvement if the two variables are correlated. In the model-based approach the auxiliary variable/s (i.e. the OV and/or CC) was used both in the design phase, via balanced sampling, and in the regression model. Overall, there is not a substantial difference between the model-based and model-assisted methods, with the simulations slightly favouring the model-based approach (Tables 6, 7, 8, 9). The main difference between these two strategies is the use of stratification in the design-based approach (both for sample selection and as a regression variable) compared to use of balanced sampling *in lieu* of stratification with the model-based approach.

The efficiencies gained through using a regression model, either in a design-based or model-based framework, are pronounced in most of the stands which were studied. The best relative efficiency, of 4.4, was achieved in the Green Hills research stands (Table 9). A relative efficiency of 4.4 is equivalent to 4.4-fold reduction in sample size for the same precision. With the Nundle stands, the plots on which the simulations were based already utilised LiDAR data though a pre-stratification process (i.e. the initial Dalenius-Hodges stratification approach). With the Green Hills stands the initial stratification was based on management-level rather than LiDAR data and in both stands only one of these strata (the oldest) was selected for simulation. Therefore, the simulated plots display a broader range of auxiliary data in the Nundle stands.

Only in the Green Hills inventory plots is it possible to directly explore total recoverable volume (TRV), which is defined as tree stems assessed for product yield, and which was calculated by the inventory crew. However based on other (unpublished) work over several FCNSW regions, the correlation between TSV and TRV is very high. Therefore, none of the above results were expected to change substantially if TRV were considered instead of TSV. Looking at the sampling schemes and sample sizes currently being used it is obvious that the target of 10% PLE is not being met with most of these data (Tables 6, 7 and 9). The exception is the Green Hills inventory plots (Table 8) where, based on these simulations, a sample of 24 plots would result in a PLE of 8.8%.

Discussion

Several studies in the northern hemisphere have demonstrated the benefits of LiDAR data as *a priori* auxiliary information for improving the efficiency of field plot designs (e.g. Dalponte et al. 2011; Gobakken et al. 2013; Grafström and Ringvall 2013; Grafträm et al. 2014; Hawbaker et al. 2009; Junttila et al. 2013; Maltamo et al. 2011). There is now a consensus that LiDAR-assisted plot selection should lead to cost savings in operational inventories. More specifically, several recent studies have also investigated the utility of LiDAR data to predict stem volume in stands of planted *P. radiata* (Chen and Zhu 2013; Gonzàlez-Ferreiro 2012; Stone et al. 2011; Watt and Watt 2013; Watt et al. 2013). However, this study is the first to systematically compare the potential efficiencies gained through a series of LiDAR-assisted sampling designs for this important plantation species. Ultimately, the total number of plots will depend on the variation within the area of interest and how comprehensively the sample describes this variation (Junttila et al. 2008).

In this study, the relative efficiency achieved by the model-based sampling strategy ranged from 1.6 in the Green Hills inventory study area to 4.4 in the Green Hills research study area. It is possible to calculate that between 10 plots (Green Hills research study) and 38 plots (Nundle 2002 AC study) would have been required to achieve the 10% PLE target which is used by FCNSW. This compares with the 46 and 79 plots, respectively that would have been required to meet the same target using a grid-based sampling procedure. Even in the Green Hills inventory study, where the relative efficiency gains were least, the sample could have been reduced from 19 plots to 12 plots while still meeting the 10% PLE target.

For the purpose of this study, a distinction was drawn between the variables used to generate the simulated plot data and those used for the simulated survey designs and estimation. The simulation parameters were estimated from the observed data and the original stratification was based on the LiDAR variables mean height and estimated stocking. This original stratification was included in the models used to estimate the simulation parameters, with the covariance parameters estimated separately within each stratum. The strata that were used to generate the plot data were excluded from the simulated survey design and estimation because the underlying mechanisms which give rise to the plot data are usually unknown to the observer. The primary aim was to determine what efficiencies could be gained from using LiDAR variables even when these are different to those which were used to generate the data.

The plots in the simulated plantation stands were selected in accordance with the sampling method under consideration. The assumptions underlying the design-based approaches were observed in the simulated samples; specifically the plots were selected independently using known selection probabilities within strata. In the model-assisted methods the plots were also randomly selected and the known selection probabilities were employed in the estimates. As for the model-based approach, the models used for prediction were not the same as those used for simulating the plots. This is a common situation in model-based sampling since the underlying population model is usually unknown. Some of the model assumptions, such as constant variance, are not actually correct owing to the way that the data were simulated. Again this is a very common situation in actual surveys since the underlying variance structure is usually unknown. Model deficiencies can sometimes be revealed via a detailed residuals analysis but this is not always conclusive. What is important in these results is that the model-based predictions are still equal to or better than the design-based estimates in terms of the RMSE despite the known model deficiencies (and the associated bias).

Both the auxiliary variables used in this study, OV and CC, are easily derived from LiDAR data (Turner et al. 2011). The LiDAR variables chosen must capture the variability of the stand parameter being surveyed, which in turn is influenced by local topography, soils and silvicultural history. These variables were chosen because they were the best predictors of stand volume in the initial regression modelling of the data. Therefore they were used as auxiliary data, either for stratification and/ or within the models, using the model-assisted or model-based methods. Maltamo et al. (2011) used the LiDAR metrics VEG (the proportion of ground echoes vs canopy echoes using a threshold value of 2 m) and H90 (plot level 90 percentile for height based on first returns) as a basis for plot selection. Their VEG metric is a measure of canopy penetration which, although different to OV, is somewhat similar to the CC used in this study. Similarly, Gobakken et al. (2013) used the metric h_{70f} (related to canopy height) and d_{0f} (related to canopy density) to define their strata.

The simulations tend towards favouring the model-based approach and there are other advantages to using this approach such as the increased flexibility associated with model-based methods. For example, spatial prediction, small domain (small area) estimation and variance modelling are all model-based techniques which have no obvious design-based counterparts. Nevertheless, to limit bias it is essential that the model is well specified (e.g. Hansen et al. 1983) and captures the spatial variation presented by the population of trees in a Planning Unit. A common source of variation in *P. radiata* is stand height and disparity in growth is related to local topography (Saremi et al. 2014), especially in regions that may experience periods of drought (e.g. Álvarez et al. 2013). The determination of attributes associated with planted forests within these regions may benefit more from model-based (or model-assisted) sampling design strategies than plantations with relatively homogeneous stands.

A further advantage of model-based surveys is that they can be used with non-random strategies including balanced sampling. This permits greater flexibility in the choice of ground plots, with greater emphasis on achieving a good range of auxiliary variables as opposed to that which results from a purely random selection. It is a key requirement, however, that the models (which are estimated from the selected plots) are representative of the true models for the study area (Ståhl et al. 2011; Wulder et al. 2012). Therefore, plots which are relatively more accessible to forestry personnel could be used as substitutes for those which are inaccessible, provided they have similar LiDAR metrics. However, this plot selection strategy would need to be implemented carefully, since the integrity of the model needs to be maintained. Specifically the reason for the accessibility needs to be unrelated to the key variable of interest, such as timber volume.

Although LiDAR constitutes an additional cost, the results of this study have demonstrated that plot sampling intensity can be reduced yet still achieve the targeted level of precision for stand volume estimates. Typical softwood plantation inventory costs in Australia average around AU$100 per management plot. This equates to approximately AU$25 per hectare, depending on topography, weed occurrence and abundance, and time of measurement (Chen and Zhu 2012; D. Watt, Planning Manager, FCNSW, pers. comm.). Also, LiDAR survey costs can vary considerably depending on the cost of aircraft mobilisation and flight distance and the specifications of the data being acquired. The net benefits, therefore, need to be confirmed through a cost efficiency analysis taking into account local requirements, conditions and circumstances.

This paper is entirely concerned with sampling at a fixed point in time. The need to measure changes over time, for example to estimate growth rates, requires a sample which is designed to be efficient for this purpose. In most population surveys changes over time are best estimated by keeping the sampling units constant or at least maximising the degree of overlap between successive surveys (see for example Cochran 1977). However, some degree of rotation is usually required to remove sampling units that are no longer representative, or for other reasons (such as harvesting). In repeated surveys it will be necessary to replace some of the initial plots when management or environmental changes have had an adverse impact on the inventory plots and/or changes occur within the plantation such as the establishment of new compartments. If the initial sample was efficient in terms of estimating the initial population, then it is likely that it will also be efficient in terms of estimating changes over time, provided plot rotation is employed when it becomes necessary. It will be necessary to revisit inventory plots periodically, both to measure allometric changes and to recalibrate the regression models. However, the use of LiDAR data for survey design should result in a reduced number of inventory plots and also inform the plot selection process when older plots need to be rotated out of the survey and replaced by newer ones. It will also be necessary to obtain updated LiDAR data as the current data becomes obsolete however the exact requirements in this respect are still to be determined (see for example Junttila et al. 2010).

The model-based approach proved to be the preferred strategy in this study. If using a model-based approach then balanced sampling with no stratification gives a precision which is comparable with, or better than, any of the design-based samples and would normally be the method of choice. If using a design-based approach then a regression estimator using a single auxiliary variable is the preferred option. In the design-based approach the allocation of plots to strata should be in accordance with the Dalenius-Hodges approach, after specifying a minimum number of plots per stratum (a minimum of two plots per stratum is required for variance estimation). Since the intention is to have a small overall sample size the number of strata also needs to be small, at most two or three. The number of plots required to achieve a 10% PLE will depend on the variation in timber volume. However, the high correlation between timber volume and OV means that variation in OV (together with historical or pilot data) could be used to determine the number of plots required.

The model-based strategy uses a widely available package for balanced sampling ("samplecube" Deville (2004)) together with simple linear models to compute the predicted values. The variance estimates, which are not considered in this paper, utilise matrix computations which are available in any mathematical or statistical package, such as R or Matlab. The next phase of this study will be to examine other modelling procedures, in particular the non-parametric techniques of MSN and

Random Forests and to determine the optimal design strategy for product volume estimates.

Conclusion

This paper compared the efficiency gains which are achievable using LiDAR data as *a priori* auxiliary information in *P. radiata* inventory samples. The efficiency gains observed in the model-based strategy, compared to simple random sampling, were equivalent to reducing the sample size from 46 to 10 plots in the Green Hills study site and from 79 to 38 plots in the Nundle 2002 AC site. Although the simulations favoured the model-based approach, the model-assisted design-based estimators also achieved good levels of precision. Balanced sampling was used in the model-based strategies as an alternative to stratification and proved to be an effective option.

Competing interests

The authors declare that they have no competing interests.

Author's contributions

GM was the primary author and undertook all analyses and simulations. CS and RT collected the field data, co-wrote the manuscript and provided technical detail on the forestry issues. RT processed the LiDAR data, extracted metrics and produced the maps. All authors read and approved the final manuscript.

Acknowledgements

The authors thank the following people who assisted us in collating the data for this study: Bob Cooper, Tony Brown & Duncan Watt (Forestry Corporation NSW), Matt Nagel & Amrit Kathuria (NSW Dept Primary Industries), Murray Webster (ForeSence Pty Ltd), Hanieh Saremi (University of New England) and Cate MacGregor (University of New England). The LiDAR and inventory plot data were provided by the Forestry Corporation of NSW and this study was supported with funds from Forest & Wood Products Australia. Finally, we are grateful for the comments provided by Huiquan Bi (NSW DPI) and two anonymous reviewers which considerably improved the manuscript.

Author details

[1]Trangie Agricultural Research Centre, Mitchell Hwy, Trangie 2823, Australia. [2]NSW Department of Primary Industries, Forest Research, Level 12, 10 Valentine Ave, Parramatta 2124, Australia. [3]Remote Census Pty Ltd, PO Box 527, Morisset 2264, Australia.

References

Álvarez J, Allen HL, Albaugh TJ, Stape JL, Bullock BP, & Song C. (2013). Factors influencing the growth of radiata pine plantations in Chile. *Forestry, 86*, 13–26. doi:10.1093/forestry/cps072.

Chen Y, & Zhu X. (2012). Site quality assessment of a *Pinus radiata* plantation in Victoria, Australia, using LiDAR technology. *Southern Forests, 74*, 217–227. doi:10.2989/20702620.2012.741767.

Chen Y, & Zhu X. (2013). An integrated GIS tool for automatic forest inventory estimates of *Pinus radiata* from LiDAR. *GIScience & Remote Sensing, 50*, 667–689. doi:10.1080/15481603.2013.866783.

Cochran W. (1977). *Sampling Techniques*. New York: John Wiley and Sons.

Corona P, & Fattorini L. (2008). Area-based lidar-assisted estimation of forest standing volume. *Canadian Journal of Forest Research, 38*, 2911–2916. doi:10.1139/X08-122.

Dalponte M, Martinez C, Rodeghiero M, & Gianelle D. (2011). The role of ground reference data collection in the prediction of stem volume with LiDAR data in mountain areas. *ISPRS Journal of Photogrammetry and Remote Sensing, 66*, 787–797. doi:10.1016/j.isprsjprs.2011.09.003.

Deville JC, & Tillé Y. (2004). Efficient balanced sampling: the cube method. *Biometrika, 91*, 893–912.

Ene LT, Næsset E, Gobakken T, Gregoire TG, Ståhl G, & Holm S. (2013). A simulation approach for accuracy assessment of two-phase post-stratified estimation in large-area LiDAR biomass surveys. *Remote Sensing of Environment, 133*, 210–224. Available from http://dx.doi.org/10.1016/j.rse.2013.02.002 (Accessed 28 April 2015).

Foreman EK. (1991). *Survey Sampling Principles*. New York: Marcel Dekker, Inc.

Gilabert H, & McDill ME. (2010). Optimizing inventory and yield data collection for forest management planning. *Forest Science, 56*, 578–591.

Gilmour AR, Thompson R, & Cullis BR. (1995). Average Information REML, an efficient algorithm for variance parameter estimation in linear mixed models. *Biometrics, 51*, 1440–1450.

Gobakken T., Korhonen L. & Næsset E. (2013) Laser-assisted selection of field plots for an area-based forest inventory. *Silva Fennica, 47*(5), article id 943. http://dx.doi.org/10.14214/sf.943 (Accessed 28 April 2015)

González-Ferreiro E, Diéguez-Aranda U, & Miranda D. (2012). Estimation of stand variables in *Pinus radiata* D.Don plantations using different LiDAR pulse densities. *Forestry, 85*, 281–292. doi:10.1093/forestry/cps002.

Goulding CJ, & Lawrence ME. (1992). *Inventory practice for managed forests* (Forest Research Institute Bulletin 171, p. 52). New Zealand: Rotorua.

Grafström A. & Ringvall A.H. (2013) Improving forest field inventories by using remote sensing data in novel sampling designs. *Canadian Journal Forest Research, 43*, 1015–1022. dx.doi.org/10.1139/cjfr-2013-0123.

Grafström A., Saarela S. & Ene L.T. (2014) Efficient sampling strategies for forest inventories by spreading the sample in auxiliary space. *Canadian Journal of Forest Research, 44*, 1156–1164. dx.doi.org/10.1139/cjfr-2014-0202.

Gregoire TG. (1998). Design-based and model-based inference in survey sampling: appreciating the difference. *Canadian Journal of Forest Research, 28*, 1429–1447.

Gregoire TG, Ståhl G, Næsset E, Gobakken T, Nelson R, & Holm S. (2011). Model-assisted estimation of biomass in a lidar sample survey in Hedmark count, Norway. *Canadian Journal of Forest Research, 41*, 83–95. doi:10.1139/X10-195.

Hansen MH, Madow WG, & Tepping BJ. (1983). An evaluation of model-dependent and probability-sampling inferences in sample surveys. *Journal of the American Statistical Association, 78*, 776–793. Available from http://www.jstor.org/stable/2288182 (Accessed 28 April 2015).

Hawbaker TJ, Keuler NS, Lesak AA, Gobakken T, & Contrucci, K. (2009). Improved estimates of forest vegetation structure and biomass with a lidar-optimized sampling design. *Journal of Geophysical Research, 114*: G00E04. doi:10.1029/2008JG000870.

Hyyppä J, Hyyppä H, Leckie D, Gougeon F, Yu X, & Maltamo M. (2008). Review of methods of small-footprint airborne laser scanning for extracting forest inventory data in boreal forest. *International Journal of Remote Sensing, 29*, 1339–1366. doi:10.1080/01431160701736489.

Hyyppä J, Yu X, Hyyppä H, Vastaranta M, Holopainen M, Kukko A, Kaartinen H, Jaakkola A, Vaaja M, Koskinen J, & Alho P. (2012). Advances in Forest InventoryUsing Airborne Laser Scanning. *Remote Sensing, 4*, 1190–1207. doi:10.3390/rs4051190.

Jakubowski M.K., Guo Q. & Kelly M. (2013) Tradeoffs between lidar pulse density and forest measurement accuracy. *Remote Sensing of Environment*, 130, 245–253. http://dx.doi.org/10.1016/j.rse.2012.11.024 (Accessed 28 April 2015).

Junttila V, Maltamo M, & Kauranne T. (2008). Sparse Bayesian estimation of forest stand characteristics from airborne laser scanning. *Forest Science, 54*, 543–552.

Junttila V, Kauranne T, & Leppänen V. (2010). Estimation of forest stand parameters from airborne laser scanning using calibrated plot databases. *Forest Science, 56*, 257–270.

Junttila V, Finley AO, Bradford JB, & Kauranne T. (2013). Strategies for minimizing sample size for use in airborne LiDAR-based forest inventory. *Forest Ecology and Management, 292*, 75–85. Available from http://dx.doi.org./10.1016/j.foreco.2012.12.019 (Accessed 28 April 2015).

Köhl M, Magnussen SS, & Marchetti M. (2006). *Sampling methods, remote sensing and GIS multisource forestry inventory* (p. 373). Heidelberg: Springer-Verlag.

Maltamo M, Bollandsås OM, Næsset E, Gobakken T, & Packalén P. (2009a) Different sampling strategies for field training plots in ALS inventory. In: *Proceedings of the SilviLaser 2009: The 9th International Conference on Lidar Applications for Assessing Forest Ecosystems*. Editors: S. Popescu and K. Zhao. Independent Publisher, USA. ISBN 9781616239978.

Maltamo M, Peuhkurinen J, Malinen J, Vauhkonen J, Packalén P, & Tokola T. (2009b). Predicting tree attributes and quality characteristics. *Silva Fennica, 43*, 507–521. Available from http://www.metla.fi/silvafennica/full/sf43/sf433507.pdf (Accessed 28 April 2015).

Maltamo M, Bollandsås OM, Næsset E, Gobakken T, & Packalén P. (2011). Different plot selection strategies for field training data in ALS-assisted forest inventory. *Forestry, 84*, 23–31. doi:10.1093/forestry/cpq039.

McRoberts RE. (2006). A model-based approach to estimating forest area. *Remote Sensing of Environment, 103*, 56–66. doi:10.1016/j.rse.2006.03.005.

McRoberts RE. (2010). Probability- and model-based approaches to inference for proportion forest using satellite imagery and ancillary data. *Remote Sensing of Environment, 114*, 1017–1025. doi:10.1016/j.rse.2009.12.013.

McRoberts RE, Gobakken T, & Næsset E. (2012). Post-stratification of forest area and growing stock volume using lidar-based stratifications. *Remote Sensing of Environment, 125*, 157–166. doi:10.1016/j.rse.2012.07.002.

McRoberts RE, Næsset E, & Gobakken T. (2013). Inference for lidar-assisted estimation of forest growing stock volume. *Remote Sensing of Environment, 128*, 268–275. Available from http://dx.doi.org/10.1016/j.rse.2012.10.007 (Accessed 28 April 2015).

Melville GJ, Welsh AH and Stone C. (2015) Improving the efficiency and precision of tree counts in pine plantations using airborne LiDAR data and flexible-radius plots: model-based and design-based approaches. *Journal of Agricultural, Biological and Environmental Statistics 20*, 29pp.

Næsset E, Gobakken T, Solberg S, Gregoire TG, Nelson R, Ståhl G, et al. (2011). Model-assisted regional forest biomass estimation using LiDAR and InSAR as auxiliary data: A case study from a boreal forest area. *Remote Sensing of Environment, 115*, 3599–3614. doi:10.1016/j.rse.2011.08.021.

Næsset E, Bollandsås OM, Gobakken T, Gregoire TG, & Ståhl G. (2013). Model-assisted estimation of change in forest biomass over an 11 year period in a sample survey supported by airborne LiDAR: A case study with post-stratification to provide ``activity data". *Remote Sensing of Environment, 128*, 299–314. Available from http://dx.doi.org/10.1016/j.rse.2012.10.008 (Accessed 28 April 2015).

Rombouts J, Ferguson I, Leech J, & Culvenor D. (2010). *An evaluation of the field sampling design of the first operational LiDAR based site quality survey of radiata pine plantations in South Australia*. Freiburg, Germany: Proceedings of the 2011 Silvilaser conference, September 14–17, 2010.

Saremi H, Kumar L, Turner R, & Stone C. (2014). Airborne LiDAR derived canopy height model reveals a significant difference in radiata pine (*Pinus radiata* D.Don) heights based on slope and aspect of sites. *Trees, 28*, 733–744. doi:10.1007/s00468-014-0985-2.

Särndal C, Swensson B, & Wretman J. (1992). *Model Assisted Survey Sampling*. New York: Springer-Verlag.

Ståhl G, Holm S, Gregoire TG, Gobakken T, Næsset E, & Nelson R. (2011). Model-based inference for biomass estimation in a lidar sample survey in Hedmark County, Norway. *Canadian Journal of Forest Research, 41*, 96–107. doi:10.1139/X10-161

Stone C, Penman T, & Turner R. (2011). Determining an optimal model for processing lidar data at the plot level: results for a *Pinus radiata* plantation in New South Wales, Australia. *New Zealand Journal of Forestry Science, 41*, 191–205.

Turner R, Kathuria A, & Stone C. (2011). *Building a case for lidar-derived structure stratification for Australian softwood plantations*. Hobart, Tasmania, Australia: Proceedings of the SilviLaser 2011 conference, Oct. 16–19, 2011.

Valliant R, Dorfman A, & Royall R. (2000). *Finite population sampling and inference: a prediction approach*. New York: John Wiley.

van Aardt JAN, Wynne RH, & Oderwald RG. (2006). Forest volume and biomass estimation using small-footprint lidar-distributional parameters on a per-segment basis. *Forest Science, 52*, 636–648.

Watt P, & Watt MS. (2013). Development of a national model of *Pinus radiata* stand volume from lidar metrics for New Zealand. *International Journal of Remote Sensing, 34*, 5892–5904. Available at http://dx.doi.org/10.1080/01431161.2013.798053 (Accessed 28 April 2015).

Watt MS, Adams T, Marshall H, Pont D, Lee J, Crawley D, & Watt P. (2013). Modelling variation in *Pinus radiata* stem volume and outerwood stress-velocity from LiDAR metrics. *New Zealand Journal of Forestry Science, 43*, 1–7. Available from http://www.nzjforestryscience.com/content/43/1/1 (Accessed 28 April 2015).

White JC, Wulder MA, Varhola A, Vastaranta M, Coops NC, Cook BD, Pitt D, & Woods M. (2013). A best practices guide for generating forest inventory attributes from airborne laser scanning data using an area-based approach. Canadian Forest Service Canadian Wood Fibre Centre Information Report FI-X-010. *The Forestry Chronicle, 89*(6), 722–723.

Wulder MA, White JC, Nelson RF, Næsset E, Ørka HO, Coops NC, et al. (2012). Lidar sampling for large-area forest characterization: A review. *Remote Sensing of Environment, 121*, 196–209. doi:10.1016/j.rse.2012.02.001.

Yu, X, Hyyppä J, Holopainen M, & Vastaranta M. (2010). Comparison of area-based and individual tree-based methods for predicting plot-level forest attributes. *Remote Sensing, 2*, 1481–1495. doi:10.3390/rs2061481.

10

Inventory estimates of stem volume using nine sampling methods in thinned *Pinus radiata* stands, New Zealand

Andrew D Gordon* and David Pont

Abstract

Background: Simulation is an established tool for examining the efficacy of forestry sampling designs yet there is little empirical information on the effect that spatial layout of a sample has on stand-level inventory of managed, even-aged stands. This simulation study examines the performance of nine different sampling methods in terms of bias and reliability.

Methods: Data sets, derived from five stands of radiata pine, consisted of census lists of every stem, including the location of each stem, breast-height diameter over-bark (DBH), height and derived volume. In four small stands, stems had been geo-located using ground-based methods, whereas the data for a larger stand were derived from an Airborne Laser Scanning data set. Nine sampling methods (random, stand-boundary, quasi-random, Zigzag transects, grid-based plots of four sizes and single-point) were simulated and applied repeatedly to each stand, and the bias and reliability of the estimate of mean stem volume calculated.

Results: Sampling the stand boundary produced a biased estimate, averaging a 12% over-estimate for the four stands aged 22 years or more. The other sampling methods generally showed little bias with most estimates within ±0.5% of the population mean, although the Single-point method was considerably less accurate. The Stand-boundary, Grid-plot (>0.02 ha), and Single-point methods produced unreliable confidence intervals.

Conclusions: Most sampling methods showed little bias and good reliability when analysed as simple random samples. Sampling plots in the range of 0.02 to 0.04 ha, located systematically on a grid with random orientation and origin, produced some of the most unbiased and reliable estimates. However, the Zigzag method may be appropriate in small stands as it produces little bias, good reliability and is likely to be operationally efficient.

Keywords: Forest inventory; Sampling; Simulation; *Pinus radiata*

Background

Simulation studies have long been used to investigate forest sampling (O'Regan and Palley 1965), as simulation is the only viable technique for testing the validity of a sampling design over a large number of samples. For example, a recent study to evaluate empirically the efficacy of different estimators in complex designs for large-scale biomass inventories (Ene et al. 2013) showed large differences between the actual and standard analytical variances. It also highlighted other methods that produced satisfactory statements of the precision of the estimates.

Airborne laser scanning (ALS) is increasingly becoming more accessible as a standard forest management tool (Næsset et al. 2004, Watt et al. 2013). Adequate spatial coverage has long been considered a critical feature of good forest inventory design. Auxiliary information from remote sensing, especially ALS, is more frequently available at the design stage so it is possible to construct a sample that is balanced both spatially and across the auxiliary variables. The benefits of this approach have been demonstrated by simulation (Grafström and Hedström 2013).

In stand or woodlot inventory, simple designs can be used to provide adequate information provided they are practical and cost efficient. While the inventory planner aims to ensure systematic spatial coverage when locating

* Correspondence: andrew.gordon@scionresearch.com
Forest Science, Scion, Rotorua, New Zealand

samples in a forest stand (Goulding and Lawrence 1992, Gordon 2005), there is little empirical information as to the importance of coverage, or guidance on the best way to achieve it. This simulation study was designed to examine the consequences of different spatial methods of sampling for mean stem volume when applied to even-aged plantation stands (in the 1 – 30 ha size range). The number of stems in a stand can be determined by a tally in small stands, or via stem-counting methods where remote imagery or ALS data are available (Culvenor 2002, Pont et al. 2015). The combination of mean stem volume and number of stems provides stand-level volume estimates. Two measures of a sampling method's success were used for comparison: bias, the difference between the population mean and the average sample mean over a large number of samples, and reliability, the percentage of samples which contained the population mean within the calculated confidence interval.

Methods

Data

Data sets from five, even-aged stands of radiata pine (*Pinus radiata* D.Don) in New Zealand were assembled. Details of each stand are provided in Table 1. Stand 1 was approximately 10 times larger than each of the other four stands. Stand 3 was approximately half the age of the other four stands. These sets were census lists of every stem in a forest stand, including each stem's location, breast-height diameter over-bark (DBH) and height. In cases where height had not been measured on every stem, a Petterson height-DBH equation (Schmidt, 1967) was fitted and the stem assigned a conditional, average height. The volume under-bark of each stem was derived from the DBH and height using a volume function (Kimberley and Beets 2007).

In the four small stands (Stand 2, 3, 4, 5), stems had been geo-located using ground-based, surveying methods.

The data for the largest stand (Stand 1) was obtained from an ALS set supported by 371 stems measured in ground plots. The ALS point cloud was processed to create a canopy–height model (CHM), represented as a grayscale image. Stand density estimated from the ground plots was used to define a suitable range of smoothing levels to be applied to the CHM image. A smoothed image was generated that corresponded to each of ten smoothing levels covering the range. A tree-detection algorithm was used to process each image so that the location (Figure 1), crown radius and height of every stem could be retrieved.

Under the assumption that the crown radius and stem DBH are linearly related (Patton 1988, Madgwick 1994), the standardised distributions of DBH from the ground plots were compared with the standardised distributions of crown radius from the ten segmented CHM images. The Kolmogorov-Smirnof (KS) D statistic (Conover 1999) was used to select the best-matching distribution (corresponding to a smoothing level). A simple linear relationship was then used to estimate DBH for every stem from its standardised crown radius. The resulting distribution of DBH was strongly correlated with the actual distribution of DBH from the 371 measured stems (Pont et al. 2015).

Sampling methods

Computer programmes in C# were written to simulate the following sampling methods in such a way that they could be repeatedly applied to the list of stems from each stand:

Random sample

A pseudo-random number generator was used to select stems from the population with equal probability and without regard for their spatial location. A random-sampling method was included in this study as a benchmark because its characteristics are well known and conform to the requirements of sampling theory even though such methods often fail to provide good spatial coverage.

Stand-boundary sample

This approach was used to simulate a spatially biased sample by selecting stems based on their proximity to random points around the boundary of the stand. This method was included in the comparison to quantify any size difference between edge stems and interior stems as this would result in a biased estimate of stem volume.

Table 1 Details of Data Sets

Name	Location	Size (ha)	Topography	Stand Age	Basal Area (m^2 ha^{-1})	Mean top height (m)	Stem count (stems ha^{-1})	Volume (m^3 $stem^{-1}$)
Stand 1	Eastern Bay of Plenty	22.8	Very steep	26	40.20	38.18	212	2.350
Stand 2	Rotorua	1.93	Undulating	25	48.09	38.22	255	2.341
Stand 3	Nelson	1.34	Steep	12	25.04	18.98	255	0.657
Stand 4	Kaingaroa	1.47	Flat	23	32.31	29.65	247	1.272
Stand 5	Kaingaroa	2.29	Flat	22	23.70	27.31	194	1.117

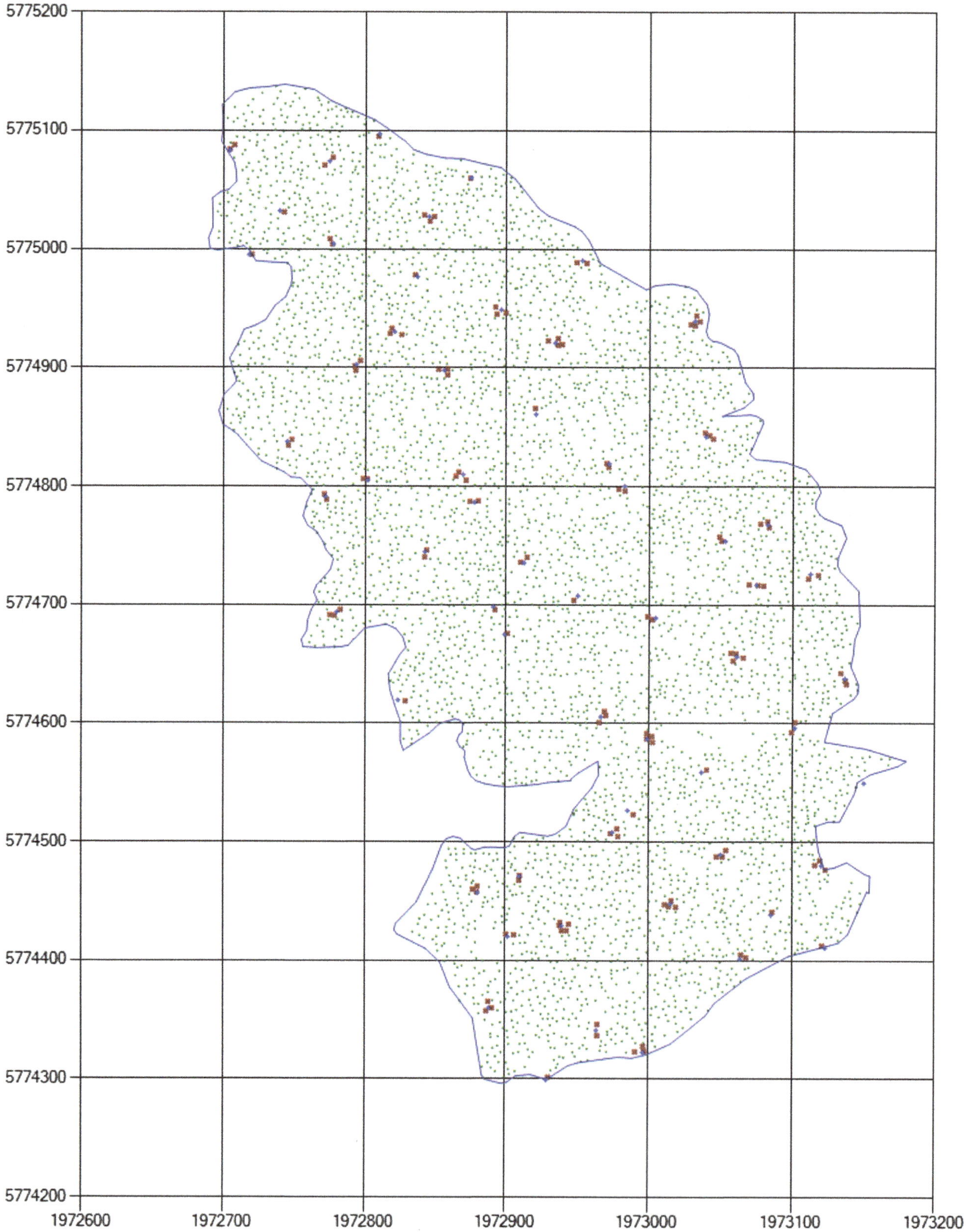

Figure 1 Stand 1 shown as a Quasi-random group (Sobol) sample. The stand boundary is shown as a blue line, stem locations as green dots, selected stems as red crosses, plot centres as blue plus signs. The underlying map grid is labelled in metres.

Quasi-random group sample

A quasi-random sequence is less random than a pseudo-random sequence but is still useful for tasks such as numerical integration and optimisation because it tends to sample n-dimensional space "more uniformly" than random numbers. Quasi-random sampling also allows for additional sample points to be added, with continual improvement in accuracy if, say, a target precision has not been met after measuring an initial sample. It provides good spatial coverage with "random" locations, without the risk of the locations coinciding with some periodic variation in the population, a risk which grid-based sampling always carries.

For each sample, a set of Sobol (Sobol 1967) points in two dimensions was found that was within the stand, or within a certain radius of the stand boundary. All stems that were within this radius of each point were selected. Radii were used to correspond to circular plots of 0.01 hectares. An example of this type of sample is shown in Figure 1.

Zigzag-line transects

Zigzag-line transects (also known as Z-plots) can be arranged with equal angles, but this introduces a coverage bias. The Zigzag-line approach used here was an equal-spaced sampler as this has satisfactory coverage properties (Strindberg and Buckland 2004). Each transect was 2 m wide and each sample contained all the stems that fell within all transects. Z-plots are an effective way of traversing a small stand or woodlot in a single pass, while ensuring good coverage in a repeatable procedure.

Grid-based plots

A grid (with random origin and orientation) that extended beyond the stand boundary was overlaid onto the stand. All stems within a certain radius of each grid point within the stand were selected along with stems within a certain radius of grid points at a set distance from the stand boundary. Radii were used that corresponded to circular areas (plots) of 0.01, 0.02, 0.04 and 0.08 hectares, which provided four separate samples. Each selected stem was labelled with its grid point reference to facilitate cluster analysis. Grids are commonly used when planning plantation forest inventory as they usually result in good coverage and are simple to navigate.

Single-point sample

A random point was selected within the stand and the closest stems (for the required sample size) were selected. This would be a valid procedure if it is assumed that there is no effect of location/stand edge on stem volume.

Several existing methods for adjusting for stand-edge sampling have been shown to provide improved estimates (West, 2013), but incorporating such methods into the simulations outlined above would increase their complexity. Instead, a simpler method was applied whereby sample plots whose centres lay outside the stand were included in any sample only if they intersected the stand boundary (Flewelling and Iles 2004, Schmid-Haas 1982).

All nine sampling methods (including the four different plot areas arranged on grids) were applied to each stand. Tests using different start points for the same stand and/or sampling method showed that a thousand samples would produce an average mean stem volume with a coefficient of variation around 0.1%. This amount of variation was considered small enough not to mask any practical differences among methods. A thousand samples were drawn from random start points for each combination of stand and sampling method.

For the four small stands (2, 3, 4, 5), an average of 70 stems was selected in each sample while an average of 200 stems was selected in the large stand (Stand 1). Having similar sample size by method within stand affords easier comparison between methods in terms of measurement effort. Sample stems were selected without replacement.

For each sample, the mean and confidence interval of stem volume were calculated by treating the selected stems as a simple random sample (SRS). That is, the standard error was calculated as:

$$s_{\bar{x}} = \sqrt{\frac{s^2}{n}\left(1-\frac{n}{N}\right)} \quad (1)$$

where s^2 is the sample variance, n is the number of stems sampled and N is the number of stems in the stand.

Confidence intervals for each sample were formed as:

$$\bar{x} \pm t_{0.975,n-1} s_{\bar{x}} \quad (2)$$

where $\bar{x}$ is the estimated mean stem volume and t is a Student's t-test value.

Those samples selected using plot-based methods (quasi-random groups and grid-based) were also analysed as cluster samples using a ratio estimator. Cluster sampling saves costs by measuring a number of stems at each sample point and is more robust in the presence of spatial correlation between stems. However it is less efficient than random sampling and will produce estimates with poorer precision for the same number of measured stems.

Under cluster sampling using a ratio estimator, the mean stem volume is estimated as:

$$\frac{\sum_{i=1}^{n} y_i}{\sum_{i=1}^{n} M_i} \quad (3)$$

Table 2 Simulation Results[1] by Stand

Stand	Sampling Method	Bias (%)[2]	Design Effect	Reliability -SRS (%)	Reliability – Cluster (%)[3]
1	Random	−0.28	1.000	94.6	
1	Stand-boundary	***13.52	0.624	0.1	
1	Quasi-random group	0.08	0.967	94.8	93.8
1	Zigzag line	−0.12	1.019	94.5	
1	Grid (0.01 ha plot)	0.14	1.043	94.3	91.4
1	Grid (0.02 ha plot)	−0.11	0.934	95.2	93.8
1	Grid (0.04 ha plot)	0.11	1.025	94.5	91.0
1	Grid (0.08 ha plot)	*0.23	1.114	94.1	90.8
1	Single- point	−0.28	2.311	81.2	
2	Random	−0.12	1.000	94.2	
2	Stand-boundary	***12.77	0.650	9.0	
2	Quasi-random group	***-0.98	1.161	91.2	83.8
2	Zigzag line	***-0.74	0.844	95.8	
2	Grid (0.01 ha plot)	0.10	0.776	97.7	92.1
2	Grid (0.02 ha plot)	0.24	1.073	95.5	88.3
2	Grid (0.04 ha plot)	0.20	1.100	93.5	89.8
2	Grid (0.08 ha plot)	0.06	1.368	92.1	88.8
2	Single-point	0.12	3.551	63.0	
3	Random	0.06	1.000	94.0	
3	Stand-boundary	***2.83	0.378	95.9	
3	Quasi-random group	**-0.28	1.087	93.9	82.1
3	Zigzag line	**0.26	0.780	96.4	
3	Grid (0.01 ha plot)	−0.09	0.924	94.9	84.9
3	Grid (0.02 ha plot)	0.15	0.767	96.7	85.9
3	Grid (0.04 ha plot)	**0.34	1.651	88.0	66.8
3	Grid (0.08 ha plot)	−0.06	1.713	84.7	65.1
3	Single-point	***-0.47	1.314	89.7	
4	Random	−0.13	1	94.8	
4	Stand-boundary	***11.54	0.487	13.8	
4	Quasi-random group	−0.07	0.922	96.8	89.7
4	Zigzag line	***-0.99	1.070	93.4	
4	Grid (0.01 ha plot)	0.30	1.050	95.2	86.8
4	Grid (0.02 ha plot)	**0.43	1.112	94.7	82.6
4	Grid (0.04 ha plot)	0.10	0.940	96.6	83.2
4	Grid (0.08 ha plot)	−0.23	1.411	90.2	72.0
4	Single-point	***-2.78	1.120	87.9	
5	Random	−0.10	1.000	95.2	
5	Stand-boundary	***12.76	0.644	4.1	
5	Quasi-random group	0.14	1.124	94.4	87.0
5	Zigzag line	***-0.72	1.264	90.1	
5	Grid (0.01 ha plot)	0.21	1.566	89.6	82.7
5	Grid (0.02 ha plot)	0.18	1.176	93.5	86.2

Table 2 Simulation Results[1] by Stand *(Continued)*

5	Grid (0.04 ha plot)	0.30	1.642		91.5	78.3
5	Grid (0.08 ha plot)	−0.15	1.192		93.4	81.4
5	Single-point	***-1.16	1.820		83.2	

[1]Each row in this table is a summary of the thousand samples drawn from the respective stand under the specified sampling method.
[2]Where the Bias is significantly different from zero, the level of significance is indicated by asterisks; *, ** and *** indicate 5, 1 and 0.1% respectively.
[3]Reliability under cluster sampling is calculated only for the plot-based methods.

where y_i is the cluster total volume and M_i is the stem count of the ith cluster. This is identical to the SRS estimate. The variance of the estimate is:

$$\frac{N(N-n)}{M^2 n(n-1)} \sum_{i=1}^{n} (y_i - \bar{x} M_i)^2 \qquad (4)$$

where N and n are the number of clusters in the population and sample respectively and M is the number of stems in the population. This is a function of the variation only among cluster totals and so will be minimised if clusters are similar. A confidence interval was constructed around $\bar{x}$.

Analysis of the samples from each combination of stand and sampling method provided the following statistics:

Bias
The bias is the mean of the thousand sample estimates of mean stem volume, minus the population mean, as a percentage of the mean of the estimates. The significance of the calculated Student's t-test statistic (which tests H_0, i.e. the mean of the sample estimates equals the population mean) was also calculated.

Design effect
To give an indication of the efficiency of a sampling method, the ratio of the variance of the estimate to the variance of the estimate from the random sample was formed. Design effects less than one indicate improvements in efficiency that may be due to the spatial coverage of the sample.

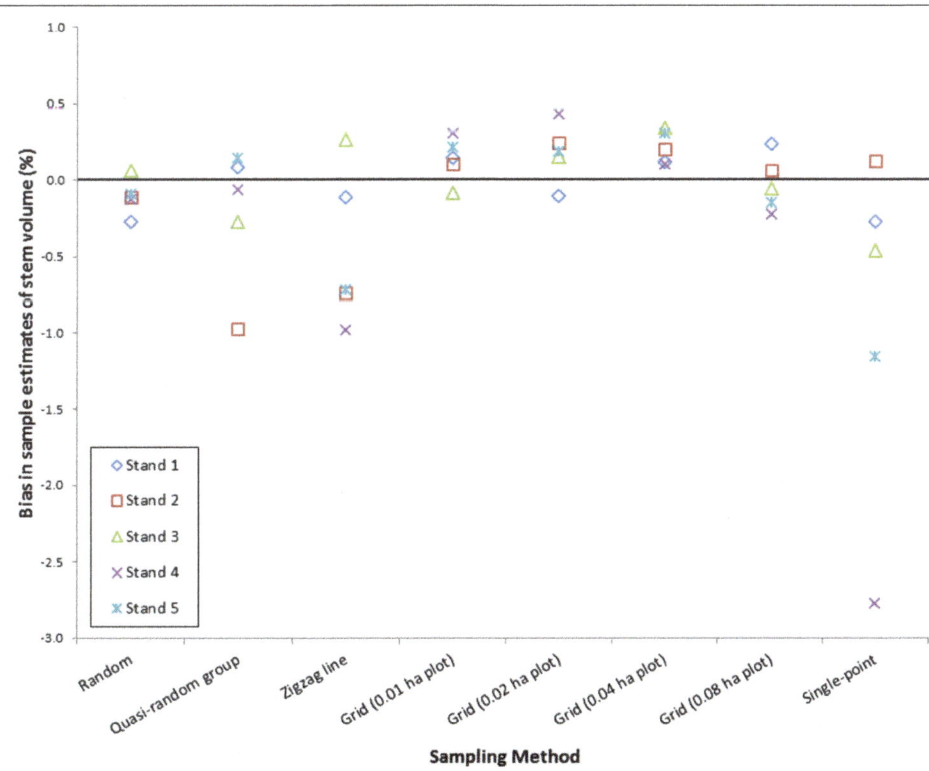

Figure 2 Bias in the estimate of stem volume for eight of the sampling methods studied. Bias in the average sample estimate of mean stem volume, as a percentage of the average estimate, by stand and sampling method. Results from the Stand-boundary sampling method are not shown as the bias was an order of magnitude larger than for most other methods.

Reliability-SRS

This is the percentage of samples that contained the population mean in their individually calculated SRS 95% confidence interval. For one thousand samples, this percentage is expected to lie in the interval from 93.6% to 96.4%.

Reliability-cluster

The same approach was used here as for the SRS Reliability but using the 95% confidence interval from the cluster analysis.

Results and discussion

Results from the simulations for all stands are shown in Table 2. Sampling the stand boundary produced a biased estimate for all five stands tested. The average overestimate for the four older stands (1, 2, 4 & 5) was 12% (range 11.54 – 13.52). Results for the younger stand (Stand 3) showed less bias (2.83%), which is possibly due to recent thinning of this stand that minimised the difference between the boundary and interior stems. The magnitude of this difference is likely to depend on the proximity and age class of adjacent stands as well as the time the edge stems have had to respond to their growing conditions.

The remaining sampling methods showed little bias. Most estimates were within ±0.5% of the population mean (Figure 2), although the Single-point sampling method was clearly less accurate than the others. The Zigzag method produced a significantly biased estimate for four of the five stands. This may be due to a subtle under-sampling of the stand edge, although the bias in all stands was within ± 1%.

Design effect indicates the efficiency of the design relative to a SRS. There appears to be no consistent improvement in efficiency by explicitly including spatial coverage in the sampling design (Table 2). Quasi-random, Zigzag, and Grid 0.01, 0.02, and 0.04 ha groups all showed a range of design effects centred near a value of 1. However, there is indication of a loss in efficiency as plot size exceeds 0.04 ha (Table 2). In the stands used in this study, sample designs with plots larger than 0.04 ha are likely to require an increased sample size to compensate for the drop in efficiency if the same level of precision is required.

Under SRS, grid plots greater than 0.02 ha generated less reliable estimates of the confidence intervals than smaller plots (Figure 3). The most reliable, spatially dispersed method was that for grid-based 0.02 ha plots. This result is similar to the results of Morrison et al.

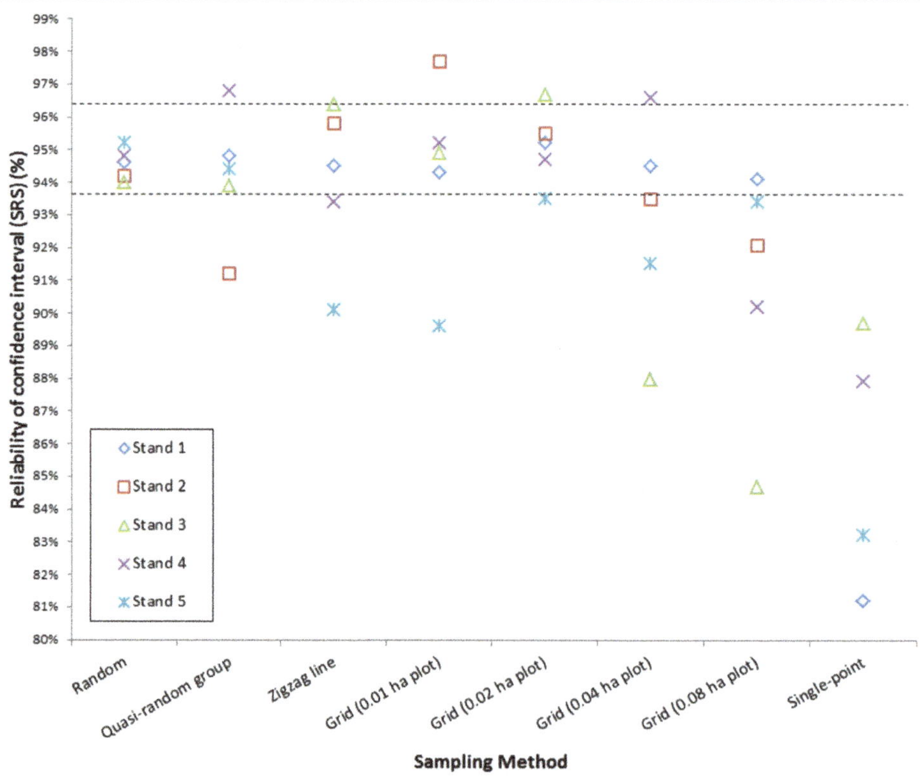

Figure 3 Reliability (SRS) for eight of the sampling methods studied. The percentage of SRS confidence intervals that included the population mean, by stand and sampling method. For 1000 samples, this percentage is expected to lie in the interval from 93.6% to 96.4%. Results from the Stand-boundary sampling method are not shown as the reliability was less than 14% in all but Stand 3.

(2008) who found "grid-based systematic designs were more efficient and practically implemented" than other methods that were compared in a simulation study of sampling rare populations. Theoretical studies (Comas et al. 2011) have also pointed to the worth of using a sampling grid for clustered populations.

The reliability of the confidence intervals of the clustered samples calculated using the ratio estimator was poor (Table 2). All these confidence intervals were overly optimistic with most methods showing reliability values of less than 93%.

Conclusions

As expected, sampling the stand boundary produced an over-estimate of stem volume. Stand-edge stems were about 12% larger than interior stems in stands 1, 2, 4 and 5 (aged 22 to 26 years old).

Most sampling methods produced little bias and good reliability when data were analysed in the form of simple random samples. This result provides the inventory forester with some scope to design sampling procedures that will be practical and operationally efficient while avoiding bias and still producing reliable confidence intervals. Plots 0.02 - 0.04 ha in size, located systematically on a grid with random orientation and origin, gave some of the most unbiased and reliable estimates. However, the Zigzag sampling method may be appropriate for small stands as it shows little bias, good reliability and is likely to be operationally efficient.

Reliable estimates of mean stem volume were obtained using simple random samples even though this approach disregards the clustered nature of plot-based samples. It is possible that the low stand density minimised any spatial correlation among stems in the stands studied, so it would be informative to test this explicitly and repeat the simulations in stands covering a range of stand densities.

Competing interests

The authors declare that they have no competing interests.

Authors' contributions

The main data manipulation, simulation programming, analyses and writing were undertaken by AG. The processing of Stand 1, from ALS data set to a spatially located stem list, was performed by DP who wrote the associated section in the Method. All authors read and approved the final manuscript.

Acknowledgements

Sampling methods were determined in discussions with Lania Holt. Mark Kimberley has provided useful comments and suggestions throughout the study. Funding for some of the initial analyses was provided by Future Forests Research Ltd. Two anonymous reviewers provided comments on an earlier version of the manuscript.

References

Comas C, Mateu J, Delicado P. (2011). On tree intensity estimation for forest inventories: Some statistical issues. *Biometrical Journal, 53*(6), 994–1010.

Conover WJ. (1999). *Practical Nonparametric Statistics*, 3rd ed, John Wiley & Sons, New York.

Culvenor DS. (2002). TIDA: An algorithm for the delineation of tree crowns in high spatial resolution remotely sensed imagery. *Computer & Geosciences, 28*, 33–44.

Ene LT, Næsset E, Gobakken T, Gregoire TG, Ståhl G, Holm S. (2013). A simulation approach for accuracy assessment of two-phase post-stratified estimation in large-area LiDAR biomass surveys. *Remote Sensing of Environment, 133*, 210–224.

Flewelling JW, & Iles K. (2004). Area-Independent Sampling for Total Basal Area. *Forest Science, 50*(4), 512–517.

Gordon AD. (2005). Forest Sampling and Inventory. In M. Colley (Ed.), *Forestry Handbook* (4th ed., Section 6.3; pp. 133–135). Christchurch, New Zealand: New Zealand Institute of Forestry (Inc).

Goulding CJ, & Lawrence ME. (1992). *Assessment Practice for Managed Forests.* (FRI Bulletin No. 171). Wellington, New Zealand: Ministry of Forestry, Forest Research Institute.

Grafström A, & Hedström Ringvall A. (2013). Improving forest field inventories by using remote sensing data in novel sampling designs. *Canadian Journal of Forest Research, 43*, 1015–1022.

Kimberley MO, & Beets PN. (2007). National volume function for estimating total stem volume of *Pinus radiata* stands in New Zealand. *New Zealand Journal of Forestry Science, 37*, 355–371.

Madgwick HAI. (1994). *Pinus radiata - biomass, form and growth*. Rotorua, New Zealand: HAI Madgwick.

Morrison LW, Smith DR, Young CC, Nichols DW. (2008). Evaluating sampling designs by computer simulation: a case study with the Missouri bladderpod. *Population Ecology, 50*(4), 417–425.

Næsset E, Gobakken T, Holmgren J, Hyyppa H, Hyyppa J, Maltamo M, Nilsson M, Olsson H, Persson A, Soderman U. (2004). Laser scanning of forest resources: The Nordic experience. *Scandinavian Journal of Forest Research, 19*(6), 482–499.

Paton VJ. (1988). Predicting the maximum area of pasture covered by radiata pine thinning and pruning slash. In JP Maclaren (Ed.), *Proceedings of the Agroforestry Symposium, Rotorua 24–27 November 1986* (Forest Research Institute Bulletin No. 139). Wellington, New Zealand: Ministry of Forestry, Forest Research Institute.

Pont D, Kimberley MO, Brownlie R, Morgenroth J, Watt MS. (2015). Tree counts from airborne LiDAR. *New Zealand Journal of Forestry, 60*(1), 45-60.

von Schmidt A. (1967). Der rechnerische ausgleich von bestandeshohenkurven. *Forstwissenschaftliches ZentralBlatt, 86*, 370–382.

Schmid-Haas P. (1982). Sampling at the forest edge. In B Ranneby (Ed.), *Statistics in theory and practice: Essays in honour of Bertil Matern* (pp. 263–276). Umea, Sweden: Swedish University of Agricultural Sciences.

Sobol IM. (1967). Distribution of points in a cube and approximate evaluation of integrals. *USSR Computational Mathematics and Mathematical Physics, 7*(4), 86–112.

Strindberg S, & Buckland ST. (2004). Zigzag survey designs in line transect sampling. *Journal of Agricultural, Biological, and Environmental Statistics, 9*(4), 443–461.

Watt MS, Meredith A, Watt P, Gunn A. (2013). Use of LiDAR to estimate stand characteristics for thinning operations in young Douglas-fir plantations. *New Zealand Journal of Forestry Science, 43*: 18.

West PW. (2013). Precision of inventory using different edge overlap methods. *Canadian Journal of Forest Research, 43*, 1081–1083.

11

Within-tree, between-tree, and geospatial variation in estimated *Pinus radiata* bark volume and weight in New Zealand

Glen Murphy[1*] and Dave Cown[2]

Abstract

Background: Many studies have been carried out to quantify the wood properties of radiata pine, but few have explicitly looked at quantifying radiata pine bark. Bark is of increasing interest for many reasons, e.g. energy source, potential source of bioproducts, log handling methods and costs, and phytosanitary methods.

Methods: Over-bark and under-bark diameter measurements recorded from over 1000 discs taken from fixed heights in 150 trees were used to estimate bark volume percentages. The mature trees were from a single seed source and had been planted at 17 sites throughout New Zealand. Bark volume percentages were converted to bark weight percentages using data from 390 trees from the central North Island of New Zealand.

Results and conclusions: This study confirmed earlier research that bark accounts for 12 to 13 % of over-bark volume and 7 to 8 % of over-bark green weight for mature radiata pine boles prior to felling and log handling. It also showed that bark volume percent varied with location in a stem, tree size, and site (mean annual temperature).

Keywords: Green bark density; Temperature; Conversion factors; Over-bark volume percent

Background

The term bark refers to all tissues of a woody stem or root occurring just outside of the vascular cambium, i.e. all tissues that could be stripped away from the woody core. Bark formation is initiated by the process of cell division at the living cambium, which separates the woody stem (xylem on the inside) from the phloem, the food-conducting tissue on the exterior side. Bark is critical to tree survival, serving two very important functions. The outer, mostly dead tissues (outer bark), form a protective barrier between the plant axis and the abiotic (wind, rain, fire, frost, and physical damage) and biotic (insects, fungi, herbivores) environment. Once the tree is felled, however, bark has minimal value and may represent a net financial loss to the forest industry (Marshall et al. 2006). However, increasingly, biomass of traditionally non-commercial components such as broken tops, dead trees, bark, needles, and branches is becoming important for carbon accounting, landscaping products, animal bedding, and substitutes for fossil fuels (Hall 2000; Temesgen et al. 2015). For these purposes, more qualitative and quantitative data are required.

The presence of bark on stems presents a challenge for foresters wishing to estimate the volume of wood contained in stems (Li and Weiskittel 2011), and systems have been developed for predicting stem volume based on bark thickness (BT) measurements at breast height (Gordon 1983; Laasasenaho et al. 2005).

Stem diameter is one of the most obvious commercial indicators in forestry. Bark thickness on a tree varies not only by species (Miles and Smith 2009) but also by the rate of growth, the genetic constitution of each tree, position along the bole (Laasasenaho et al. 2005), and geographic location (Antony et al. 2015). Thus, one BT function with the same set of parameter values cannot be applied to all trees, even for the same species. When under-bark measurements are not available, various approaches have been proposed to estimate and predict diameter under bark (DUB) at a certain stem height. Numerous taper equations (Cao and Pepper 1986; Gordon 1983; Kitikidou et al. 2014) have been published and are commonly used to predict DUB at

* Correspondence: glen.murphy@waiariki.ac.nz
[1]Waiariki Institute of Technology, Rotorua, New Zealand
Full list of author information is available at the end of the article

any given height along the tree bole. The functions commonly incorporate factors such tree age, height in stem, and breast height diameter (DBH).

Miles and Smith (2009) compiled information on the properties of wood and bark for 156 species in the USA for the estimation of biomass. Bark in 24 pine species varied from 8.9 % (*Pinus contorta* Douglas) to 20.4 % (*Pinus jeffreyi* Balf.) with radiata pine (*Pinus radiata* D.Don) reported to have 11.8 % of over-bark volume.

The density and moisture content (MC) of bark and its percentage weight relative to wood are important criteria for biomass calculations and for log handling and transport cost determination. Antony et al. (2015) set out to identify geographical variation in loblolly pine (*Pinus taeda* L.) bark and wood quality and to quantify the responses following silvicultural practices that included planting density, weed control, and fertiliser application. Trees were destructively sampled across the southern USA. Bark and wood properties were measured from discs collected at multiple heights from sampled trees and used to compute the whole-tree bark and wood properties. Significant regional variation was observed for both bark and wood properties. Bark thickness and bark percentage decreased with stem height and were positively correlated with proximity to the ocean (possibly related to temperature), tree age, and DBH. Bark and wood basic density showed an increasing trend from inland to coastal regions and vice versa for bark and wood MC. Effects of silvicultural treatments on the other hand were generally minimal.

Utilisation of bark in New Zealand has been of interest for over 40 years (Harris and Nash 1973). Products currently derived from radiata pine bark include hog fuel, compost, and landscaping products, but potentially, a wider range of options is possible (Ferreira et al. 2015). Bark is comprised of about 65 % extractable chemicals with the remaining insoluble material having a similar composition to that of wood. The extractable chemicals include terpenes, waxes, resin acids, phenols, and polyflavonoids (mainly comprised of tannins). The chemicals that have shown the greatest opportunity for commercial exploitation are the polyflavonoids (Uprichard 1986). The two main areas of interest are in the use of tannins as adhesives and in the antioxidant potential of the lower molecular weight flavonoids (Jorge et al. 2002; Li et al. 2015).

On the assumption that the annual harvest in New Zealand will be around 30 million m^3 by 2020 (MAF 2010), and the proportion of bark around 10–12 % (Webber and Madgwick 1983), up to 3 million m^3 per year of bark could potentially be associated with felled stems. Depending on tree age, season, and the type of harvesting machinery, significant proportions of this will unavoidably end up on the forest floor, at landing sites, at processing plants, and at marine ports. The amount of bark reaching a processing plant or port can vary from 5 % of over-bark volume in spring to 10 % in autumn (Cown 1999).

Radiata pine plantations are intensively managed and generally harvested between 25 and 30 years of age. While numerous wood-quality studies have been carried out, relatively few of them report on bark characteristics because the main focus has been on the wood characteristics of the stem and commercial logs.

Despite its significance, only limited information is available about radiata pine bark properties and bark proportion variation. The quantity and quality of bark produced from plantation-grown radiata pine is important, especially as material to be disposed of or used as an alternative fuel source or bioproduct. Nevertheless, statistics on the quantity of bark produced and used are not widely available. Part of the reason for the paucity of statistics is probably that bark usually has been considered a waste to be disposed of at the lowest possible cost. Data on physical characteristics are important factors related to biomass production and with handling costs of the felled stems through the forest to wood processor supply chain.

Radiata pine wood-quality studies (mostly unpublished) have most commonly involved measuring both diameter over bark (DOB) and DUB, but a few have been specifically aimed at removing bark samples for determining the physical properties of bark independent of the wood. Some of the main results have been the following:

- Standing radiata pine trees had an average of 10 to 18 % bark volume (depending on the assessment method). Bark data were very variable between stems within crops.
- BT was highly influenced by stem age and stem height, although the difference between BT on young and older trees of the same diameter was only of the order of 2–3 mm.
- Bark MCs (based on volume and weight measurements on excised small samples) were strongly affected negatively with tree age and positively with height in stem. In young stems, MC was around 100 % (dry weight basis) at the base of the tree, increasing to around 200 % at the top. Equivalent figures for mature stems were 50 and 100 %.
- The bark basic density (based on volume and weight measurements on excised small samples) varied from 300 kg m^{-3} in thinnings (12–14 years) to around 400 kg m^{-3} in 30-year-old stems.
- Bark properties did not appear to be strongly influenced by geographic location or stem diameter within crops.

In the course of a survey of wood properties in the central North Island region of New Zealand (Cown et al. 1984), extensive measurements were taken from discs of 584 trees of ages between 10 and 50 years to assess stem diameters, wood density, and MC by log position. In the process, bark data were also collected in order to better understand the log weight-to-volume relationship (a common basis for log sale whereby log weight is converted to wood volume under bark (Ellis 1993)). It was concluded that bark volume (based on DOB and DUB data) increased with tree age at all sites and decreased with height in the stem. However, some of the more detailed bark data (e.g. green bark density) was not presented in the published report.

This paper utilises previously unpublished data to quantify the variation in radiata pine bark volume from stems grown at sites ranging from the top of the North Island to the bottom of the South Island of New Zealand. It also uses previously unpublished data to convert percentage volume estimates to percentage green weight estimates.

Methods

Data sources

In the first half of the first decade of the millennium, the New Zealand forest industry funded research to characterise wood quality of the radiata pine resource. Trees from a 1978 genetics gain trial established across 17 sites, stretching from the top of the North Island to the bottom of the South Island of New Zealand, were selected for the wood-quality project (Table 1). The trials were established with known genotypes (three open-pollinated commercial seedlots), sited on former state-owned production forests at 22 sites selected to encompass a broad range of climates. Bark thickness was one of the many attributes measured from 1063 discs obtained from the sample trees. Other attributes included radial trends in green and dry wood density, spiral grain, microfibril angle, heartwood, internal checking, resin, stiffness, and stability. Over time, some variation in silviculture has occurred as a result of changes in forest ownership. Several trials were abandoned due to damage from excessive grazing or severe storms. One seedlot (WN/72/2), produced from the Gwavas seed orchard and based on open-pollinated seed from 25 clones of the 850 series selected from the central North Island, was chosen as the fixed genotype for the wood-quality project. It represented some degree of genetic improvement (GF14) (FRI 1987) and was present in sufficient numbers for sampling at 17 of the sites.

A total of 450 trees were sampled for DBH and outerwood density (increment cores), and a selection was made for a much more intensive study of wood properties. A third of the trees were selected to cover the wood density and diameter range at each site and were felled in 2003/2004 and at ages 25 or 26 years. Diameter over bark and DUB were measured at the butt, 1.4 m, 5 m, and then every 5 m up the stem to an approximate top end diameter of 100 mm. Twenty-five of the 1063 sets of measurements were on discs that had been collected at intermediate heights between the standard 5-m intervals, 17.5 m being the lowest and 38 m being the highest. BT varied from 2 to 50 mm.

Table 1 Location of sites where wood-quality data were collected

Forest	Latitude (° S)	Longitude (° E)	Elevation (m)	No. of trees	Range of diameter breast height (mm)
Aupouri	34.88	173.10	38	10	397–539
Athenree	37.46	145.90	100	10	361–600
Ruatoria	37.73	178.26	140	10	376–670
Kaingaroa	38.45	176.68	560	10	429–614
Kaingaroa	38.54	176.42	220	10	370–642
Mohaka	39.06	176.97	280	10	497–704
Lismore	39.99	175.11	50	6	451–706
Ngaumu	41.05	175.88	250	10	351–615
Rabbit Island	41.26	173.12	7	10	329–575
Golden Downs	41.48	172.89	375	10	404–527
Waimea	41.48	173.07	392	10	335–494
Ashley	43.06	172.47	400	9	439–657
Eyrewell	43.41	172.22	195	10	357–453
Waimate	44.70	170.88	473	10	329–613
Longwood	46.14	167.80	196	6	436–720
Rowallan	46.06	167.66	196	7	430–555
Blackmount	45.74	167.70	350	2	417–429

Expected mean temperature averages from a 30-year-period (temp) for each site were obtained from the New Zealand National Climate Database (NIWA 2005). The nearest weather station was used. Where there were significant differences in elevation between the weather station and the trial site, mean temperature was adjusted by an average atmospheric lapse rate of 0.6 °C per 100 m of elevation change. Mean temperatures ranged between 8.7 °C at the bottom of the South Island and 16.1 °C at the top of the North Island.

Unpublished data associated with a regional wood property survey in the central North Island (Cown et al. 1984) were incorporated to present a fuller picture, specifically green wood density based on whole discs (GWD) and green bark density (GBD), derived using the same methods as the aforementioned genetics trials and Antony et al. (2015). These data were used to calculate green density ratios of bark to wood.

Analysis

Bark volume (BVol%), expressed as a percentage of over-bark volume, was calculated using Eq. 1.

$$\text{BVol\%} = 100 * \left(1 - [\text{DUB}/\text{DOB}]^2\right) \quad (1)$$

The software packages StatGraphics Plus (Version 5) and Microsoft Excel 2010 were used to develop and test models. Correlation analysis showed that BVol% was related to height (−0.60), DOB (0.45), and mean temperature (0.04) in descending order.

A plot of BVol% against height indicated that BVol% decreased non-linearly with height. Height in regression models was, therefore, transformed using a natural log transformation after first adding one to each height to adequately deal with bark volume measurements at the butt (height = 0).

Plots of BVol% against DOB at fixed heights up the stem (e.g. 0 m, 1.4 m, 5 m, etc.) indicated that BVol% was linearly related to DOB.

BVol% data were checked for outliers. One measurement point was deleted because it went against the trend for all other trees where BVol% decreased from the butt to 1.4 m. It also had the largest BVol% and produced a large residual if it was included in any regression models.

Data were randomly split into two sets. Approximately 80 % of the data were used for model construction and approximately 20 % for model validation.

StatGraphics GLM procedures were used to develop regression models. Initial independent variables selected were transformed height, DOB, and mean temperature. Additional independent variables were the interaction of mean temperature and DOB and the interaction of mean temperature with transformed height. Models were compared based on adjusted R^2 and mean absolute error (MAE) values for the constructed models and average residual and MAE values from the validation data set. Residual plots for the validation data set were visually examined to determine if the residuals were heteroscedastic and normally distributed.

The final model selected was of the form

$$\begin{aligned}\text{BVol\%} &= a + b^*\text{Ln}(\text{Ht} + 1) + c^*\text{DOB} \\ &\quad + d^*(\text{Ln}(\text{Ht} + 1) * \text{temp})\end{aligned} \quad (2)$$

The effect of the three sources of variation on BVol% was demonstrated as follows:

- *Within-tree variation.* BVol% was calculated and plotted for various heights for the median tree from all sites, based on DOB at 1.4 m, assuming the overall mean temperature from all sites.
- *Between-tree variation.* Merchantable tree bark volume percent (MTBVol%) was calculated and plotted for three representative trees (the smallest, median, and largest trees from the total data set) for each of the 17 sites. BVol% was assumed to decrease linearly between measurement points.
- *Geospatial variation.* The same calculations used to demonstrate between-tree variation in bark volume were used to demonstrate geospatial variation.

The ratio of GBD to GWD was calculated for up to six 5.5-m log height classes (0–5.5, >5.5–11.0, >11.0–16.5, >16.5–22.0, >22.0–27.5, and >27.5–33.0) and overall for 21- to 30-year-old trees. Since there was no consistent trend in the ratio between low, medium, and high wood density sites within the data collection region, an assumption was made that the ratios would be applicable to stems located at all 17 sites. The parameter MTBVol% can be converted to bark weight percentage (MTBWt%) using Eq. 3. The same approach was used for converting BVol% to bark weight percentage for individual log height classes up a stem.

$$\text{MTBWt\%} = (\text{GBD}/\text{GWD}) * \text{MTBVol\%} \quad (3)$$

Results

The bark volume model

The average bark volume (BVol%) for all sites was 12.6 %. It ranged from 3.4 to 31.3 % for individual discs.

The analysis of variance and summary regression statistics are shown in Table 2 for the final model. The accuracy of the model was highly significant and accounted for 69 % of the variability in bark volume. The MAE for the model was 2.436.

The coefficients for the final model are provided in Table 3. When the model coefficients were applied to the validation data set, the model was unbiased (mean residual = −0.01), the residuals appeared to be homoscedastic and normally distributed, and the MAE was 2.669.

Sources of variation in bark volume

Bark volume decreased exponentially with height in each tree. This trend is illustrated in Fig. 1 and was calculated for the median tree assumed to have grown at the mean temperature from the 17 sites, i.e. 12.4 ° C. In this case, BVol% decreased rapidly from more than 20 % at ground level to 10 % at a 10-m height. It then levelled off to about 8 % for the remaining portion of the stem.

Within a site, MTBVol% was greater for small trees than it was for large trees. This trend is shown in Fig. 2 as calculated for the smallest, median, and largest trees from the combined 17 sites assumed to have grown at all 17 sites. MTBVol% would be expected to be over 7 % greater for the smallest tree than for the largest tree at all sites.

Among sites, there was a 2.1 % difference in calculated MTBVol% for the median tree from all sites with the bark volume being greatest in the warmest climate site (Aupouri) and least in the coldest climate site (Blackmount) (Fig. 2). Assuming that for the same silviculture and rotation age, the median tree for an individual site is likely to be larger for warmer sites than colder sites, then it could be expected that the actual difference between the warmest and coldest sites would be larger than shown.

Conversion to bark weight

The ratio of GBD to GWD for various log height classes is shown in Fig. 3. The increase in this ratio with height is due to three main factors—less fissuring of bark at higher levels, an increase in the actual green density of thinner bark, and an increase in heartwood percentage with height. A whole-tree GBD/GWD ratio of 0.562 was calculated.

For central North Island sites, where the GBD/GWD ratio data were collected, the calculated MTBWt% was 6.8 % for the median tree from the 17 sites. The calculated MTBWt% for the median tree ranged from 6.2 to 7.5 %, being highest for the warmest site.

Table 2 ANOVA and summary statistics for the final regression model

	df	SS	MS	*F*	Significance of *F*
Regression	3	18282	6094.13	616.88	<0.001
Residual	846	8357	9.88		
Total	849	26640			
Adjusted R^2	0.685				
Standard error	3.143				
Observations	850				

Table 3 Coefficients for the final regression model for estimating bark volume (BVol%)

Coefficients	Value	Standard error	*t* stat	*P* value
a	31.2735	0.7850	39.84	<0.001
b	−7.3587	0.3497	−21.04	<0.001
c	−0.0193	0.0013	−14.44	<0.001
d	0.1429	0.0253	5.65	<0.001

Discussion and conclusions

Some wood properties of radiata pine in New Zealand are very well documented, particularly wood density (Cown 1999; Palmer et al. 2013). Information on bark is much sparser.

The data on radiata pine bark volume reported here are in strong agreement with other radiata studies. The bark for the median tree was estimated to be 12.6 % of over-bark volume, similar to earlier estimates of 12 % (Young et al. 1991) and 13 % (Cown et al. 1984) for 50 and 390 mature central North Island radiata pine trees, respectively. It is also similar to the 11.8 % value reported for radiata pine in the USA (Miles and Smith 2009). The amount of bark on a tree was dependent, however, on tree size with the smallest tree estimated to have 7 % more bark than the largest tree, 16.7 and 9.4 % of over-bark volume, respectively.

There was an exponential decrease in bark thickness and volume from the butt upwards in the stem (thickness from about 35 to 5 mm above 30 m; volume from 22 to 8 %). As with the comprehensive southern pine study (Antony et al. 2015), a small decrease in bark volume with mean average temperature was also noted, equivalent to about one quarter of a percent of over-bark volume per degree decrease in mean average temperature.

The data on radiata pine bark weight reported here are also in agreement with other radiata studies. The bark for the median tree was estimated to be 6.8 % of

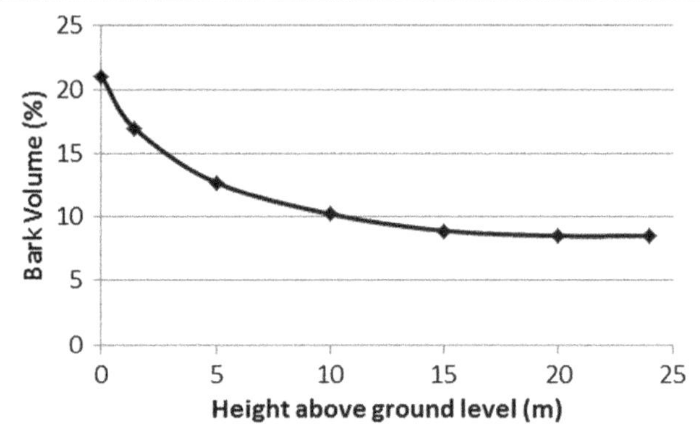

Fig. 1 Variation of calculated radiata pine bark volume (BVol%) with height in the stem for the median tree from all sites

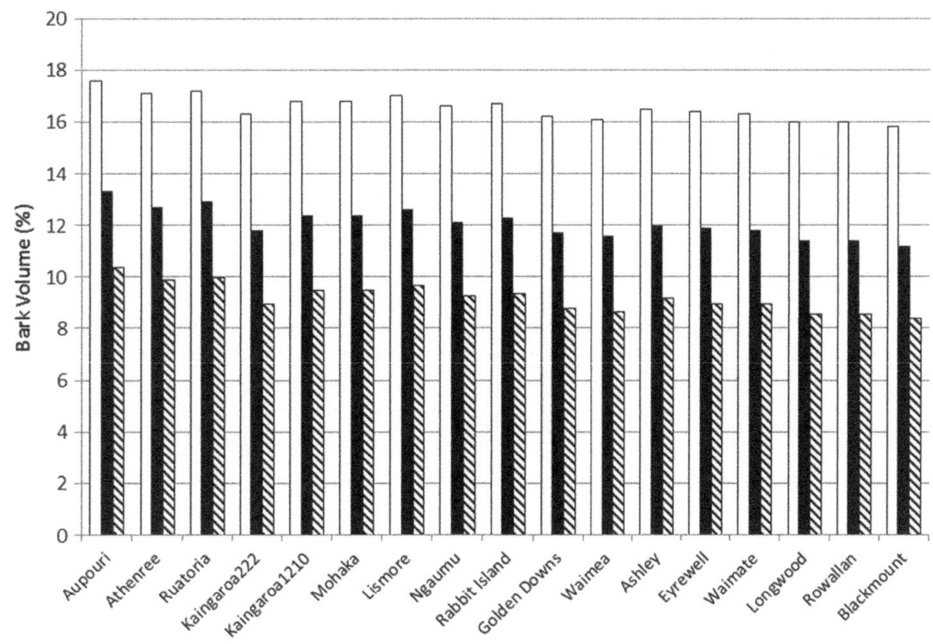

Fig. 2 Variation of calculated merchantable tree bark volume (MTBVol%) with tree size and geospatial location

over-bark weight. This is slightly lower than earlier estimates which ranged from 7 % (Young et al. 1991) to 8.2 % (Cown et al. 1984) on a green weight basis and 8.7 % (Webber and Madgwick 1983) on a dry weight basis for mature central North Island radiata pine tree boles.

The bark weight percentage estimates in this study were dependent on the GBD/GWD ratios. The ratios are likely to be dependent on tree age and tree size. They may also be dependent on season (Gibbs 1958). Chan et al. (2012) found that season exerted no practical effect on whole segment green density for radiata pine. However, it is possible that bark and wood changed differentially in opposite directions as reported by Gibbs (1958) for poplar and willow. Contrary to the work of Chan et al. (2012), Ellis (1993) noted that there is a 4 % difference in weight-volume conversion factors for logs between summer and winter—some of this difference may be due to bark loss during handling.

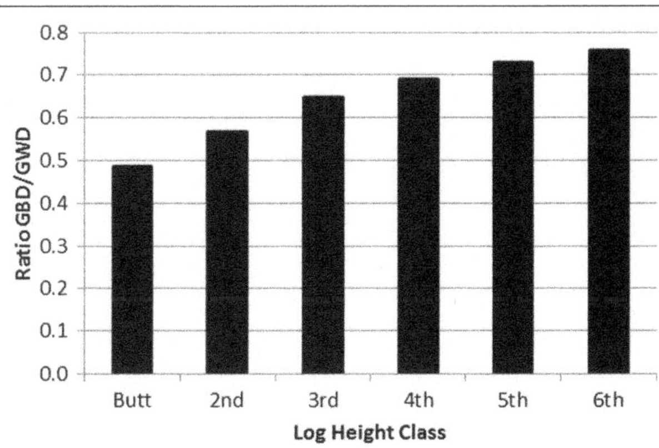

Fig. 3 Variation in the ratio of green bark density (GBD) to green wood density (GWD) with height in the stem for 25- to 26-year-old radiata pine trees from central North Island sites. Log height classes are in increments of 5.5 m: 0–5.5, >5.5–11.0, >11.0–16.5, >16.5–22.0, >22.0–27.5, and >27.5–33.0

It must be noted that the study discs were carefully handled to retain bark, so the results are most relevant to standing trees in the forest. Logs reaching processing sites will inevitably loose a proportion of the bark depending on season and handling systems (Marshall et al. 2006).

This study has confirmed earlier research that bark accounts for 12 to 13 % of over-bark volume and 7 to 8 % of over-bark green weight for mature radiata pine boles. It also shows that bark volume percentage varies with location in a stem, tree size, and site (mean annual temperature). Further work, however, is required to quantify the effects on bark volume percent of tree age and possibly other seed sources. Additionally, further work is required to quantify the effects on green bark to green wood density ratios of tree age, tree size, season, and sites (outside the central North Island) to facilitate the conversion of bark volume percent estimates to bark weight percent estimates.

Competing interests

The authors declare that they have no competing interests.

Authors' contributions

GM was the lead author, carried out the data analyses, and drafted the main manuscript. DC contributed to the collection of the data and the preparation of the manuscript. All authors read and approved the final manuscript.

Acknowledgements

Funding for this study has been provided by a grant from the New Zealand Forest Growers Levy Trust. Data for this study has come from two main sources: WQI Ltd and reworked data from earlier studies by the second author of this manuscript.

Plant licences

The local, national, or international guidelines and legislation relating to plants and the required or appropriate permissions and/or licences for the study have been obtained.

Author details

[1]Waiariki Institute of Technology, Rotorua, New Zealand. [2]New Zealand Forest Research Institute Ltd, Rotorua, New Zealand.

References

Antony, F, Schimleck, LR, Daniels, RF, Clark, A, Borders, BE, Kane, MB, & Burkhart, HE. (2015). Whole-tree bark and wood properties of loblolly pine from intensively managed plantations. *Forest Science,* 61(1), 55–66. doi:10.5849/forsci.12-030.

Cao, Q, & Pepper, W. (1986). Predicting inside bark diameter for shortleaf, loblolly and longleaf pines. *Southern Journal of Applied Forestry, 10,* 220–224.

Chan, JM, Walker, JCF, & Raymond, CA. (2012). Variation in green density and moisture content of radiata pine trees in the Hume region of New South Wales. *Australian Forestry, 5*(1), 31–42.

Cown, DJ. (1999). *New Zealand radiata pine and Douglas fir: suitability for processing. [FRI Bulletin, 216].* Rotorua, New Zealand: New Zealand Forest Research Institute.

Cown, DJ, Love, LG, McConchie, DL, & Colbert, C. (1984). *Wood properties of radiata pine in some forests of the Bay of Plenty/Taupo region. [FRI Bulletin 81].* Rotorua, New Zealand: New Zealand Forest Research Institute.

Ellis, JC. (1993). Measurement of roundwood in New Zealand. *New Zealand Journal of Forestry, 37*(4), 16–20.

Ferreira, JPA, Miranda, I, Gominho, J, & Pereira, H. (2015). Selective fractioning of *Pseudotsuga menziesii* bark and chemical characterization in view of an integrated valorization. *Industrial Crops and Products, 74,* 998–1007.

FRI. (1987). *Which radiata pine seed should you use? What's New in Forest Research, No. 157.* Rotorua, New Zealand: New Zealand Forest Research Institute.

Gibbs, RD. (1958). Patterns in the seasonal water content of trees. In KV Thimann (Ed.), *The physiology of forest trees* (pp. 43–69). New York, USA: Ronald Press Co.

Gordon, AD. (1983). Estimating bark thickness of *Pinus radiata. New Zealand Journal of Forestry Science, 13*(3), 340–345.

Hall, P. (2000). Bioenergy fuel from stem-to-log processing waste using conventional forest harvesting systems. *New Zealand Journal of Forestry Science, 30*(1/2), 108–113.

Harris, JM, & Nash, HT. (1973). Physical properties of bark from radiata pine, Corsican pine and Douglas-fir. In EL Ellis (Ed.), *Bark Utilisation Symposium Proceedings. Paper No. 2. School of Forestry, University of Canterbury, Christchurch, New Zealand.*

Jorge, FC, Neto, CP, Irle, MA, Gil, MH, & De Jesus, JP. (2002). Wood adhesives derived from alkaline extracts of maritime pine bark: preparation, physical characteristics and bonding efficacy. *Holz als Roh - und Werkstoff, 60*(4), 303–310.

Kitikidou, K, Papageorgiou, A, Mileos, E, & Stampoulidis, A. (2014). A bark thickness model for *Pinus halepensis* in Kassandra, Chalkidiki (Northern Greece). *Silva Balcanica, 15*(1), 47–55.

Laasasenaho, J, Melkas, T, & Aldén, S. (2005). Modelling bark thickness of *Picea abies* with taper curves. *Forest Ecology and Management, 206*(1–3), 35–47.

Li, R, & Weiskittel, AR. (2011). Estimating and predicting bark thickness for seven conifer species in the Acadian Region of North America using a mixed-effects modeling approach: comparison of model forms and subsampling strategies. *European Journal of Forest Research, 130*(2), 219–233.

Li, YY, Feng, J, Zhang, XL, & Cui, YY. (2015). Pine bark extracts: nutraceutical, pharmacological, and toxicological evaluation. *Journal of Pharmacology and Experimental Therapeutics, 353*(1), 9–16.

MAF. (2010). *New Zealand wood availability forecasts: 2010–2040.* Wellington, New Zealand: Ministry of Agriculture and Forestry.

Marshall, HD, Murphy, GE, & Lachenbruch, B. (2006). Effects of bark thickness estimates on optimal log merchandising. *Forest Products Journal, 56*(11–12), 87–92.

Miles, PD, & Smith, WB. (2009). Specific gravity and other properties of wood and bark for 156 tree species found in North America. In *[Research Note NRS-38].* Newtown Square, PA, USA: USDA Forest Service Northern Research Station.

NIWA. (2005). *The National Climate Database.* Auckland, New Zealand: National Institute of Water and Atmospheric Research. Online website available at http://cliflo.niwa.co.nz/. Accessed 26 July 2015.

Palmer, DJ, Cown, DJ, Kimberley, MO, & McKinley, RB. (2013). Assessing prediction accuracy in a regression kriging surface of *Pinus radiata* outerwood density across New Zealand. *Forest Ecology and Management, 308,* 9–16.

Temesgen, H, Affleck, D, Poudel, K, Gray, A, & Sessions, J. (2015). A review of the challenges and opportunities in estimating above ground forest biomass using tree-level models. *Scandinavian Journal of Forest Research, 30*(4), 326–335.

Uprichard, JM. (1986). Chemistry of wood and bark. In JA Kininmonth & LJ Whitehouse (Eds.), *Properties and uses of New Zealand radiata pine. Volume one—wood properties.* Rotorua: Ministry of Forestry, Forest Research Institute. Pp. 4–1 to 4–45.

Webber, B, & Madgwick, HAI. (1983). Biomass and nutrient content of a 29-year-old *Pinus radiata* stand. *New Zealand Journal of Forestry Science, 13,* 222–228.

Young, GD, McConchie, DL, & McKinley, RB. (1991). Utilisation of 25-year-old *Pinus radiata. New Zealand Journal of Forestry Science, 21*(2/3), 217–227.

12

Radial variation in selected wood properties of three cypress taxa

Christine L. Todoroki[1], Charlie B. Low[1*], Heather M. McKenzie[2] and Luis D. Gea[3]

Abstract

Background: New Zealand-grown cypresses are typically harvested at 35–40 years of age; however, shorter rotations have been proposed. The objective of this study was to evaluate the feasibility of producing structural products from shorter rotations through examination of wood properties of younger cypress trees. A detailed examination of radial wood property trends was necessary, and the first ring-level models for three cypress taxa were developed because there were no predictive wood property models available.

Methods: Radial trends in wood density, microfibril angle (MFA) and modulus of elasticity (MOE) were examined for 55 trees of three taxa (*Cupressus lusitanica* Mill, *C. macrocarpa* Gordon and *Chamaecyparis nootkatensis* (D.Don) Spach x *Cupressus macrocarpa*). The trees, planted in row-plots, were harvested at age 21 years at which time breast-height increment cores were extracted to determine wood properties. Rings 1 through 18 were examined in detail and used as the basis for developing mixed-effects ring-level models.

Results: Taxon, cambial age, ring width and aspect were highly significant as explanatory variables in wood-density models. Trees with a northern aspect (the warmest side with most sunlight) had greater density than those at the south of the plot. Trees with a southern aspect (least sunlight) had significantly lower MOE, which was about 1–2 GPa below the average of 11–13 GPa at age 18 years. Aspect, however, was not significant in models for MFA. Microfibril angle of the cupressus hybrid declined from about 30° at the pith to 13° near the bark, whereas for *Cupressus lusitanica*, the range was about 36–16° and about 31–16° for *Cupressus macrocarpa*.

Conclusions: The results of this study provide a basis for determining management strategies appropriate to structural timber production from cypress stands. Overall, forecasted ages for production of strength-grade timber were least for the cupressus hybrid. In an application of the models, products of 6 GPa could be obtained from the outer zones of trees of the hybrid from age 22 years and at older ages for the other taxa. Shorter rotations would, therefore, be feasible for the hybrid taxon.

Keywords: Density, Microfibril angle, Modulus of elasticity, Radial variation, Cypress, Mixed-effects models

Background

Cypress timber is known for its stability, attractive grain and natural durability. In comparison with other species, it has a relatively steady wood-density gradient (Brailsford 1999). The timber is commonly used for wall panelling, flooring, architraves, skirting, furniture and interior joinery. It has also been reported as suitable for a wide range of exterior uses, including exterior joinery, weatherboards and boat building, and is an approved building species (Haslett 1986). Timber from New Zealand-grown cypresses is produced from trees that are typically 35–40 years of age when harvested. However, shorter rotations of closer to 20 years have been proposed (Brailsford 1999). The objective of this study was to evaluate the feasibility of producing structural products from shorter rotations through examination of wood properties of younger cypress trees.

Wood properties of particular importance to structural products include wood density, microfibril angle (MFA) and modulus of elasticity (MOE). Wood density is an indicator of strength. Typically, air-dry wood density of New Zealand-grown cypresses ranges from 475 to 495 kg.m^{-3} (Miller and Knowles 1996). Microfibril angle is an indicator of stiffness and is derived from the angle

* Correspondence: Charlie.Low@scionresearch.com
[1]Scion, Private Bag 3020, Rotorua 3046, New Zealand
Full list of author information is available at the end of the article

of cellulose strands in the thickest (S2) layer of the cell wall relative to the long axis of the tracheid (Donaldson 1987). The lower the MFA, the higher the stiffness and the more stable the timber (Yang and Evans 2003, Fang et al. 2006). Together, wood density and MFA determine MOE (Walker and Butterfield 1995, Evans and Ilic 2001, Yang and Evans 2003), a measure of stiffness and an overall indicator of strength. The characteristic bending stiffness of machine-graded pine (MGP) in New Zealand with grades of MGP 6, MGP 8 and MGP 10 is 6.0, 8.0 and 10.0 GPa, respectively, while No. 1 framing and engineering grades (both visually assessed) have minimum requirements of 8.0 and 10.5 GPa, respectively (Gaunt 1998, Standards New Zealand 1993).

The three radial wood properties (density, MFA and MOE), along with ring width, can all be assessed using Silviscan technology (Innventia Ab, Stockholm, Sweden). Following the invention of Silviscan by Evans et al. (1995), numerous studies have used this technology for investigating within-tree wood properties of a range of species including *Eucalyptus globulus* Labill., *E. nitens* H.Deane & Maiden and *E. regnans* F.Muell. (Yang and Evans 2003), *Picea abies* (L.) Karst. (Lundgren 2004, Piispanen et al. 2014), *Pinus taeda* L. (Jordan et al. 2005), *Pseudotsuga menziesii* (Mirb.) Franco (Vikram et al. 2011), *Pinus sylvestris* L. (Auty et al. 2013) and *Pinus radiata* D.Don (Ivković et al. 2013). To our knowledge, there have been no studies investigating within-tree radial variation of any cypress taxa.

In general, radial wood density varies with both age and ring width (Franceschini et al. 2010, Jyske et al., 2008). Ring width, in turn, is influenced by climate (Bouriaud et al., 2004; Wimmer and Downes 2003), aspect (van der Maaten 2012), silvicultural practices (Filipescu et al. 2014) and cambial age. While many different forms of models have been presented for the relationship between ring width and age (or equivalently time), the commonly used Hugershoff (1936) (cit Peschel 1938) curves (Eq. 1) have often been reported as being the most robust (Fang et al. 2010, Bošeľa et al. 2011).

$$y(t) = a \times t^b \times e^{-ct} + d \tag{1}$$

where $y(t)$ is the response variable at time, t, and a, b, c and d are parameters to be estimated.

Microfibril angle also varies with age, being higher nearer the pith and declining with increasing cambial age (McMillin 1973, Zhang et al. 2007). The relationship between MFA and cambial age has been examined in many other studies, with a comprehensive list provided by Donaldson (2008) in a review of MFA measurement, variation and relationships. Relationships with MOE, like those for density, show a general increase with cambial age (e.g. Lasserre et al. 2009, Cortini et al. 2014).

Little research effort has focussed on wood property variation within and among cypress genera. Furthermore, some contradictory results have been found among the few published studies. For example, Watt et al. (2008) studied 4-year-old *Cupressus lusitanica* and found diameter (measured at ground level) to be highly correlated with wood density, but Malimbwi et al. (1992) found density and diameter to be unrelated in a study of 19-year-old *C. lusitanica* trees. McKinley et al. (2000) studied a sample of four cypress taxa (*C. lusitanica* Miller, *C. macrocarpa* Gordon, *Chamaecyparis lawsoniana* (Murray) Parl. and *Chamaecyparis nootkatensis* (D.Don) Spach x *Cupressus macrocarpa*) ranging in age from 13 to 52 years and found wood density was unrelated to tree age. However, wood stiffness of a sample of boards sawn from 21-year-old *Cupressus lusitanica* trees was shown to be strongly influenced by cambial age (Low et al. 2005). The boards (all sawn from butt logs and mechanically tested) were classified as follows: Inner (when sawn from the innermost rings), Outer (when sawn from the outermost rings) and Intermediate (when sawn from intermediary positions). Mean stiffness of the Inner, Intermediate and Outer samples was 4.3, 5.4 and 7.2 GPa, respectively. Given that Low et al. (2005) also found that mean stiffness of *C. lusitanica* boards was less than that of either *C. macrocarpa* or the *Chamaecyparis nootkatensis* x *Cupressus macrocarpa* boards of the same age, a lowered harvesting age of less than 35 years appears plausible.

The objective of this study was to determine whether or not the wood of younger cypress trees is sufficient to meet the requirements of structural products. Firstly, relationships among wood density, MFA and MOE on core samples obtained from trees of three cypress taxa were investigated. The taxa studied were *Cupressus lusitanica*, *C. macrocarpa* and *Chamaecyparis nootkatensis* x *C. macrocarpa*. These are referred to as Lusitanica, Macrocarpa and Leyland, respectively, from here on. Subsequently, mixed-effects models were developed to evaluate cambial age, ring width, distance from pith, edge effect and aspect as suitable candidate explanatory variables. The models were then used to predict wood properties over an extended time horizon, extrapolated beyond the data range and applied to forecast the age at which production of timber of specific dimensions and stiffness ratings would be possible.

Methods

Sample Trees

This study was conducted on a stand of cypress trees growing on pumice soil in the central North Island of New Zealand (location, 38.1° S, 176.3° E, elevation 757 m) where the annual rainfall is about 1350 mm (Chappell 2013). The stand of Leyland, Lusitanica and

Macrocarpa was planted for demonstration purposes on the grounds of the New Zealand Forest Research Institute in 1981 (Low et al. 2005, Fig. 1). There were four Leyland hybrid clones that had been brought out as plants from England by G. W. Sturrock of the Department of Scientific and Industrial Research in the 1950s. The clones were Green Spire, Haggerston Grey, Old Alice and Clone 3. The study trees were raised from cuttings taken from these clones. Lusitanica was grown from seed collected from trees growing in Waipoua forest (location, 35.7° S, 173.6° E; seedlot AK77/23). Macrocarpa was grown from seed collected from a shelterbelt at Bulls nursery (location, 40.1667° S, 175.3833° E; seedlot WN77/7). Vincent and Dunstan (1989) explained seedlot numbers and systems.

A further block of 'fastigiate clone 850.020 *Pinus radiata* Selfs' was also planted within the stand but not used in this study. The trial was set up as a series of row-plots, where all the trees in one row were the same taxon. Rows were numbered 1 on the western side of the block and went through to 27 on the eastern side. Each row contained six trees planted along an approximate North to South axis. Trees within rows were numbered from 1 (southern aspect) to 6 (northern aspect). Intermediary trees (numbered 2 to 5) were categorised as having an 'interior' aspect. Being in the Southern Hemisphere, the north side is the warmest side. The cypress trees on the north side received more light following the harvest of an existing stand to the north of the cypress stand in 1998 when the study trees were about 17 years old. Those northernmost trees were also sheltered from the cold southerly (i.e. Antarctic) winds.

The stand was planted at 1111 stems per hectare and thinned to 550 stems per hectare at an age of about 10 years. Trees were also pruned to between 5 and 8 m depending on tree height, but details of the timing of these pruning lifts is not known. Breast-height (1.4 m) diameters were recorded, and breast-height increment cores were taken from 20 randomly selected trees per taxon in 2002, when the stand was 21 years old. Heights were also measured on some, but not all, trees. After elimination of cores for which ring boundaries could not be estimated, there were a total of 55 samples (20 Leyland, 19 Lusitanica and 16 Macrocarpa) available for use in this study. The location of each tree used in this study, within the row-plot design, is shown in Fig. 2.

Analysis of cores using Silviscan® equipment

The 12 mm cores, taken from the trees, shortly before they were felled in 2002, were sterilised by soaking in a 96 % solution of ethanol then air-dried to a moisture content of 13 %. Due to unforeseen circumstances, there was a delay of 3 years, during which the cores were stored in plastic bags in a freezer, before shipping to Australia for analysis by Silviscan in 2005.

Cores were analysed using Silviscan® III (Evans 2006), an automated suite of instruments that provide quantitative measures of wood density by X-ray densitometry and of microfibril angle by X-ray diffraction. MOE was determined by combining the density and diffraction information as shown in Eq. 2:

$$\mathrm{MOE} = A(I\rho)^B \tag{2}$$

where ρ is density acquired by X-ray diffraction, I is the coefficient of variation of the diffraction profile (and includes scattering from the S2 layer and background scattering from S1 and S3 layers and other cell constituents) and A and B are species-independent calibration constants

Fig. 1 Sample of 21-year-old cypress trees in the demonstration stand

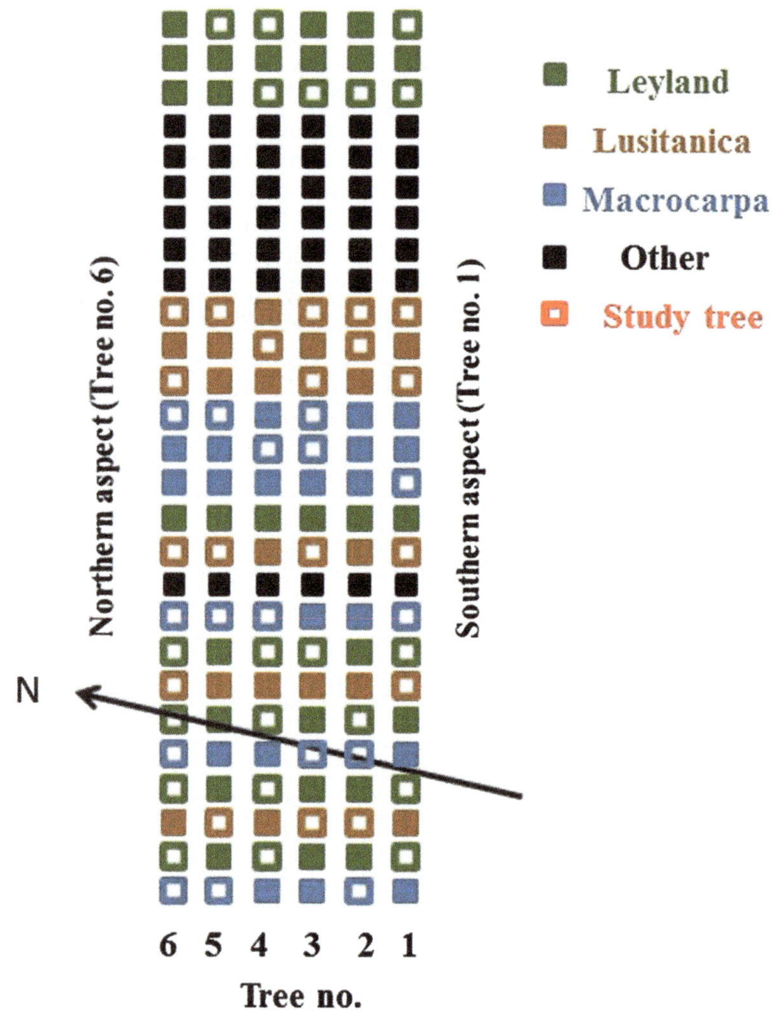

Fig. 2 Layout of the 21-year-old cypress stand. Trees along the northern aspect (tree no. 6 in each row) received more sun and were sheltered from the cold southerly winds to which trees growing along the southern aspect (tree no. 1 in each row) were exposed

that relate to the sonic resonance method used for calibration (Evans 2006).

Density was measured at intervals of 50 μm, while MFA and MOE were estimated at intervals of 5 mm for each core. The positions of the annual ring boundaries for each core were determined from the radial density profile using software written in Interactive Data Language (IDL, proprietary software distributed by Research Systems, Inc. of Boulder, CO, USA now a division of Kodak) (http://www.ittvis.com/ProductServices/IDL.aspx) specifically for the purpose. The boundaries were manually amended where algorithms either failed to identify a boundary or found spurious boundaries. Ring averages were automatically calculated for density, MFA and MOE and the width of each ring recorded when the operator was satisfied with boundary locations. Ring width provided a point estimate of annual diameter growth information.

Less than 50 % of the sample cores had 20 rings, 73 % had 19 or more rings, 84 % had 18 or more rings and 96 % had 17 or more rings. Therefore, at 20 rings, the sample size was small and allowed for the possibility of biased samples. To examine radial patterns across as many rings as possible (while also avoiding small sample sizes) radial properties were examined here on the three cypress taxa from rings 1 to 18 inclusive. Radial properties examined were as follows: ring width, wood density, MFA and MOE.

Statistical analysis

Relationships between radial wood properties (ring width, density, MFA or MOE) and explanatory variables were developed using the linear and nonlinear mixed-effects models package, 'nlme', (Pinheiro et al., 2013) within the R environment (R Core Team, 2013) with a significance level of 0.05 throughout.

Linear models were formulated using the Laird and Ware (1982) model (Eq. 3) and developed following the protocol of Zuur et al. (2009), while nonlinear models followed the Pinheiro (2002) model (Eq. 4).

$$y = \mathbf{X}\boldsymbol{\beta} + \mathbf{Z}\boldsymbol{u} + \boldsymbol{\varepsilon} \tag{3}$$

$$y = f(\mathbf{X}\boldsymbol{\beta} + \mathbf{Z}\boldsymbol{u}) + \boldsymbol{\varepsilon} \tag{4}$$

where $\mathbf{y}$ is the response vector, $\mathbf{X}$ and $\mathbf{Z}$ are matrices of explanatory variables corresponding to fixed and random effects respectively, $\boldsymbol{\beta}$ and $\boldsymbol{u}$ are the corresponding vectors of parameters for the respective fixed and random effects, $\boldsymbol{\varepsilon}$ is a vector of random errors and f is a nonlinear function.

Candidate explanatory variables included (where appropriate) cambial age (i.e. ring number from pith), ring width, distance from pith, taxon, edge-tree and aspect. Interaction terms with taxon, and transformations of explanatory variables (the latter determined through inspection of graphical plots) were also modelled to address the assumption of normality. To address the assumption of homogeneity of variance, heterogeneity was incorporated into the variance structure of each model, using a power function for the ring covariate, and examined through visual plots of residuals versus fitted values. A residual autocorrelation structure, AR(1), was included in the models to address the correlated nature of the consecutive ring measurements. The variance and autocorrelation structures are described in detail by Auty et al. (2013). Random effects were modelled to allow differences due to individual tree. Selection of the most appropriate model structure was based on likelihood ratio tests and the Akaike information criterion (AIC, Akaike 1974).

Model performance was evaluated using mean absolute percentage error (MAPE, a relative measure which expresses errors as a percentage of the actual data), root mean square error (RMSE, which, although highly influenced by extreme values, is a useful measure because it assumes the same units as the response variable) and the coefficient of determination between actual and predicted data, R^2. Predictions were made at the population level, and excluded the estimated random effects, because in practice they would be unknown.

Final models for estimating density, MFA and MOE were presented with both confidence and prediction bands. Confidence intervals contained within the confidence bands indicate the likely location of the true population mean, whereas prediction intervals account for both the uncertainty in the population mean as well as data scatter. For this reason, prediction intervals are always wider than confidence intervals.

At the individual ring level, mean radial wood properties were compared using analysis of variance (ANOVA). A one-way ANOVA was conducted to determine if differences between taxa wood properties were significant. For post-hoc testing, the Tukey honestly significant difference (HSD) test was chosen for conducting the multiple (54 = 3 (taxon) × 18 (rings)) pairwise comparisons. One-way ANOVA followed by the HSD test was also conducted on the stand measurements, again using a significance level of 0.05, to determine if there were differences between mean diameter, height and taper of the three taxa.

Application of models

It is possible to estimate the earliest age at which cypress stands might be expected to meet structural product requirements by running the radial wood property ring-width models, the latter with consideration to tree geometry. For example, consider the production of timber, W mm wide × T mm thick × L m long, and of Y GPa.

To determine the minimum age at which production would be possible:

1. Use the appropriate MOE model (Leyland, Lusitanica or Macrocarpa) to determine the earliest cambial age, A, at which the majority of trees (i.e. lower 95 % model prediction interval) would achieve Y GPa.
2. Use the ring width model to calculate the distance, X, from the pith to A (i.e. cumulative sum of ring widths).
3. Determine the minimum radius, R, at a height of L + 0.1 m (assuming a 0.1 m stump), $R = ((X + T)^2 + (W/2)^2)^{½}$.
4. Estimate the minimum radius at breast height (1.4 m) by accounting for taper, $R = R$ + taper x (L-1.3), where 1.3 is breast height less stump height, and taper is the upper 95 % confidence interval of taper (refer Table 1).
5. Finally, reapply the (cumulative) ring width model to estimate the expected minimum age at which R is attained.

Results

Mean stand characteristics of the three taxa

On average, at age 21 years, trees of the Leyland taxon were the smallest in terms of girth (with a breast-height diameter of 334 mm), Macrocarpa the tallest (21.6 m) and least tapered (17.6 mm.m^{-1}), and Lusitanica the most tapered (20.1 mm.m^{-1}). The Leyland and Lusitanica taxa were, on average, of similar height. Mean stand characteristics of the three taxa and sample sizes by aspect are summarised in Table 1.

Measured ring width

Mean ring width initially increased for the majority of the cypress trees, reaching a peak at a cambial age of 2 to 3 years. After peaking, ring width tended to decrease with cambial age. However, there were some fluctuations and

Table 1 Mean stand characteristics (with 95 % confidence intervals) of the three cypress taxa. Means with the same letter are not significantly different

Taxon	Sample size[a]	Breast-height diameter (mm)	Total height (m)	Taper ($mm.m^{-1}$)
Leyland	20 (4, 11, 5) (13)	334b (317,352)	18.3b (17.6,19.1)	17.8ab (16.3,19.3)
Lusitanica	19 (4,11,4) (19)	387a (357,417)	19.3b (18.6,20.0)	20.1a (18.5,19.2)
Macrocarpa	16 (4, 10, 2) (14)	386a (350,422)	21.6a (19.9,23.3)	17.6b (15.9,19.2)

[a]Number of trees, by aspect (north, intermediary, south) and subsample of trees for which height was measured

spikes, which were particularly pronounced for Lusitanica and Macrocarpa. Aspect appeared to have a small influence on ring width, particularly for Leyland (Fig. 3, top row) and a small thinning response was observed when the trees were aged 12 to 14 years (Fig. 3, bottom row). Subsequent to this period of thinning response, ring width again decreased with cambial age but at a slower rate than prior to the thinning response.

In years 2–8 of growth, mean ring width of Lusitanica was significantly greater than that of Leyland. Differences in mean ring width between Macrocarpa and Leyland were not significant, while differences in mean ring width between Lusitanica and Macrocarpa were significant only for rings 3 ($p = 0.03$) and 5 ($p = 0.04$). After a cambial age of 8 years, mean ring width was not significantly different between taxa.

Measured wood density

Radial trends in mean ring density were unique to each taxon with considerable variation due to individual trees (Fig. 4). The northernmost trees tended to demonstrate higher density than the southernmost trees, particularly those of the Leyland taxon (Fig. 4, top row). For Leyland, wood density declined rapidly from a mean of 570 $kg.m^{-3}$ at ring 1 to a mean of 470 $kg.m^{-3}$ at ring 3. Thereafter, mean ring density of the Leyland taxon was approximately 450 $kg.m^{-3}$, though there was considerable variation (extending to about ±100 $kg.m^{-3}$) due to individual trees. For Lusitanica, mean ring density followed an approximate parabolic form for the first 10 years, with an initial density of 456 $kg.m^{-3}$ and a minimum of 385 $kg.m^{-3}$ at 6 years. After 10 years, mean density was relatively constant (approximately 436 $kg.\ m^{-3}$). For Macrocarpa, mean ring density followed a relatively flat gradient, with a slight trough in the first 3 years and a slight ridge at 11 years.

The Tukey HSD test indicated significant differences among taxon means within the first 13 years of growth. Mean ring density of Lusitanica was significantly less than that of Leyland (until ring 10) and of Macrocarpa (until ring 13). Mean ring density is indicated by grey dots superimposed on the box plots in Fig. 4. Inspection of density-by-ring plots for individual trees suggested that the data points

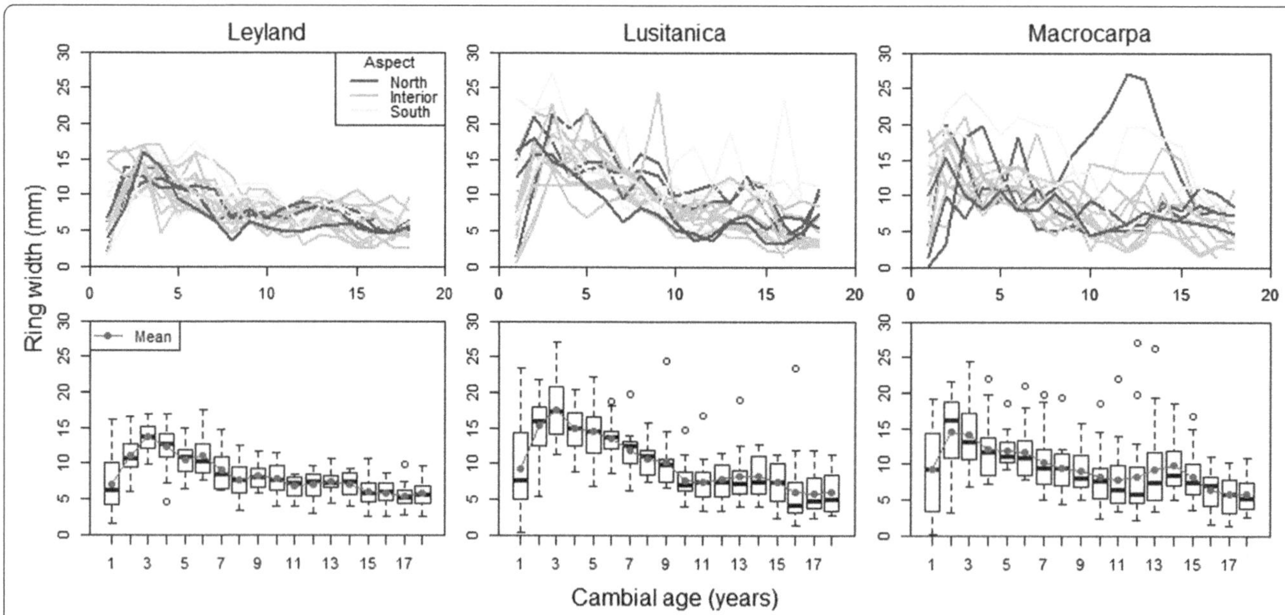

Fig. 3 Radial variation in ring width for three cypress taxa. Outliers, denoted by *black circles* in the bottom row figures, extend beyond 1.5 times the interquartile range. Means, denoted by *grey dots* are connected to illustrate general trends

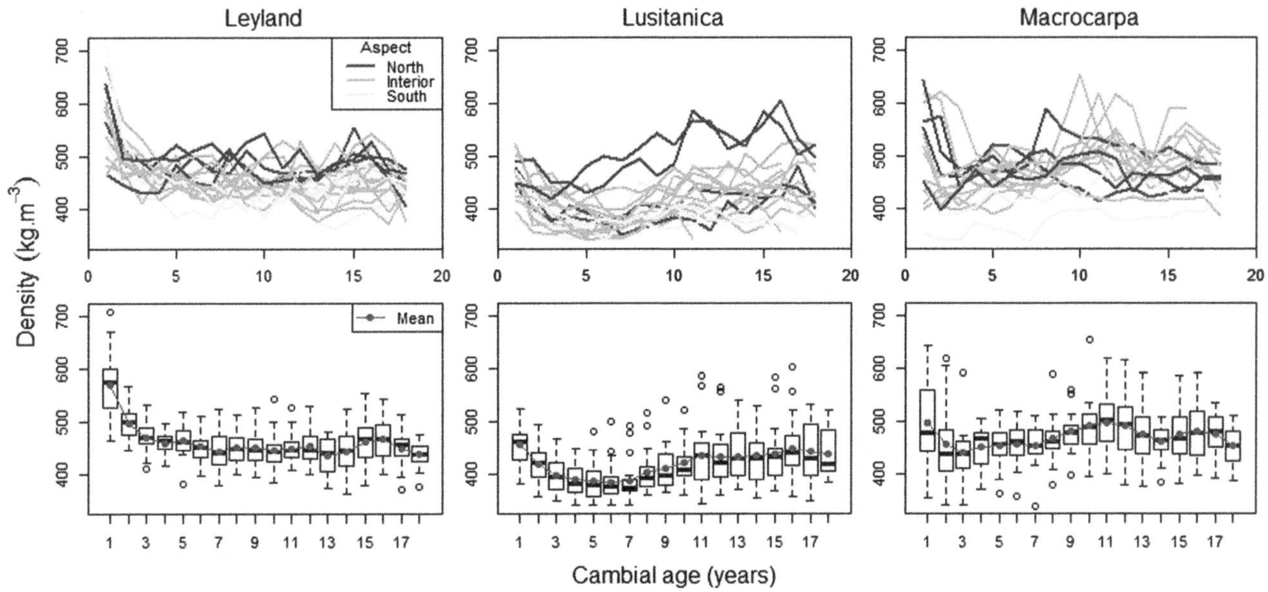

Fig. 4 Radial variation in wood density for three cypress taxa. All denotions are as for Fig. 3

were valid although outliers (which influence mean values) were present (as indicated by the black circles in the figure).

Measured microfibril angle

Mean MFA followed a declining trend with a relatively steep gradient in the early years of growth (up to about age 10 years), before levelling off (Fig. 5). Overall, mean MFA was lowest for Leyland (with an average of 13.9° for the period from 12 to 18 years of age) and highest for Macrocarpa (average 18.4° for the same period). Mean MFA of Lusitanica for the same period was 16.5°. Variation about the means was within ±10° for all trees. Microfibril angle did not appear to be influenced by either aspect or silviculture.

In comparison with Lusitanica, mean MFA of Leyland was significantly lower for all cambial ages up to and including ring 12. In comparison to Macrocarpa, mean MFA of Leyland was not significantly different in the first 4 years of growth, but was significantly lower from

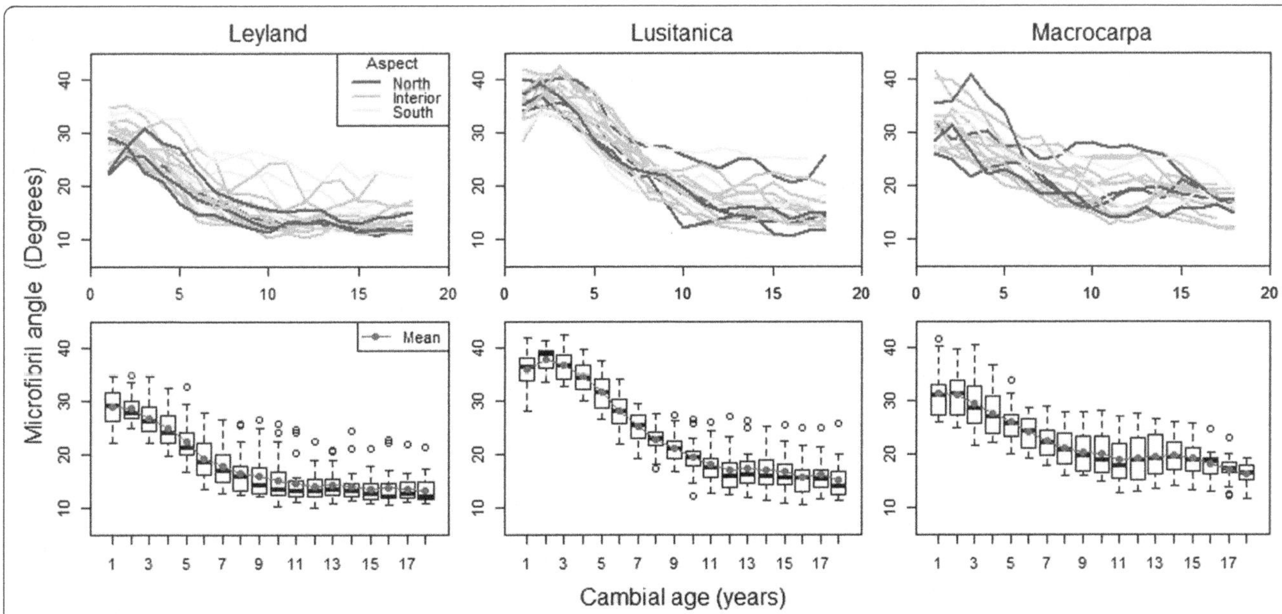

Fig. 5 Radial variation in microfibril angle for three cypress taxa. All denotions are as for Fig. 3

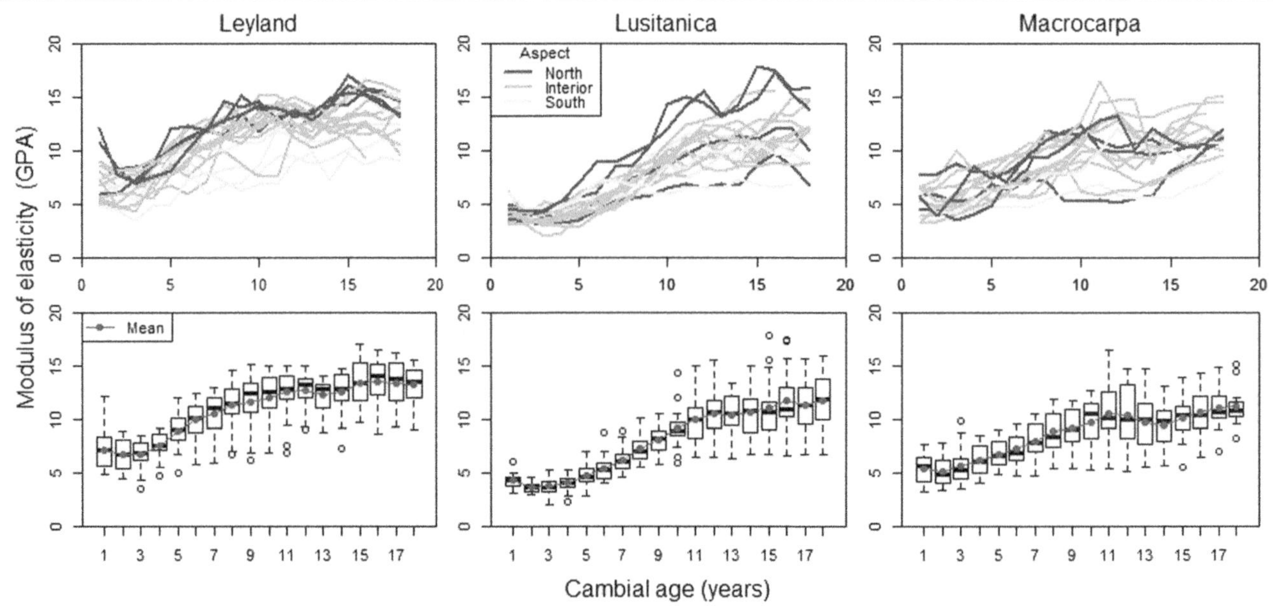

Fig. 6 Radial variation in modulus of elasticity for three cypress taxa. All denotions are as for Fig. 3

age 5 through to 18 years. Mean MFA of Lusitanica was the highest of the three taxa during the early years of growth, being significantly higher than that of Macrocarpa until a cambial age of 7 years (and significantly greater than Leyland until age 12 years).

Measured modulus of elasticity (via Eq. 2)

Mean MOE generally increased with increasing cambial age, though the reverse trend was noted during the initial 2 to 3 years of growth (Fig. 6). The rate of increase in mean MOE was greater in the earlier half of measurements than in the latter half. Mean MOE for the period from age 11 through to 18 years inclusive was 12.9, 10.9 and 10.4 GPa for Leyland, Lusitanica and Macrocarpa, respectively. At age 18 years, MOE averaged 11–13 GPa; however, trees with a southern aspect were about 1–2 GPa lower. The effect of aspect on MOE appeared to be greater for Leyland than the other two taxa (Fig. 6, top row). Effects due to thinning were minimal (Fig. 6, bottom row).

Mean MOE of Leyland was significantly greater than that of Lusitanica until a cambial age of 16 years and,

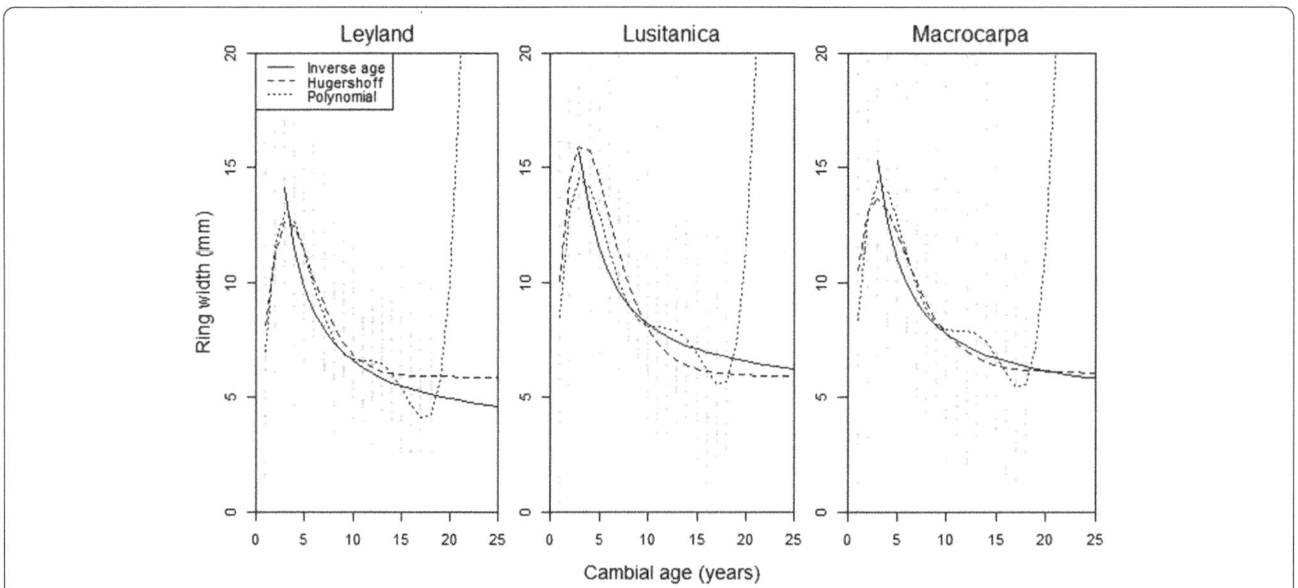

Fig. 7 Models considered for predicting ring width for three cypress taxa

Table 2 Parameters estimates and fit statistics from fitting the linear mixed-effects model to predict mean ring width

Parameter	Value	Std. Error	DF	*t* value	*p* value
(Intercept)	3.57	0.49	802	7.35	0.000
A^{-1}	32.2	2.0	802	16.0	0.000
Lusitanica	1.62	0.56	51	2.86	0.006
Macrocarpa	1.21	0.59	51	2.04	0.046
South	2.28	0.60	501	3.83	0.000

A cambial age

with the exception of ring 11, was significantly greater than that of Macrocarpa throughout the 18-year measurement period. Differences in mean MOE between Lusitanica and Macrocarpa were significant only for the first 8 years of growth during which mean MOE of Lusitanica was significantly less than that of Macrocarpa.

Modelled ring width

Three ring-width models (an inverse transformation of cambial age, Hugershoff curves and a fifth degree polynomial), all augmented by aspect, resulted in similar fit statistics but demonstrated large differences in predictive capability (Fig. 7). The inverse-age model (fitted to rings 3–18 inclusive) demonstrated a gradual decline in ring width with increasing age for predictions beyond ring 18, as would be expected. The nonlinear Hugershoff curves, while providing a better fit in the early years, plateaued early, at about 13 years. The polynomial model provided a reasonable fit to the data until about age 15 years but was impractical for later predictions. Therefore, the final model was a linear mixed-effects model based on the inverse transformation of cambial age, taxon and aspect. Only the southerly aspect was significantly different to the interior aspect. Model parameters are provided in Table 2. Parameter estimates of the autocorrelation and variance structures were 0.591 and −0.142, respectively. Model performance, as judged by the three goodness-of-fit statistics, was reasonably good in view of the data variability demonstrated in Fig. 3. At the population level, MAPE was 32.4 %, RMSE 3.20 mm and R^2 was 0.44.

Modelled wood density

Radial wood density of the three cypress taxa was modelled by extending the model of Franceschini et al. (2010), with additional variables: taxon, aspect, edge-tree and interaction terms. The model of Franceschini et al. (2010) included cambial age, a square-root transformation of cambial age, a transformation of ring width, and the quotient of cambial age and ring width.

Model parameters of the optimal wood density model, which included all explanatory variables proposed by Franceschini et al. (2010) and additional interactions with taxon, are given in Table 3. Parameter estimates of the autocorrelation and variance structures were 0.561 and −0.191, respectively. Edge-tree was not significant as an explanatory variable; however, aspect was highly significant. Trees on the northernmost aspect would be expected to be 44 kg.m^{-3} denser than those on the southernmost aspect.

Table 3 Parameter estimates and fit statistics from fitting the linear mixed-effects model to predict radial density

Parameter	Value	Std. Error	DF	*t* value	*p* value
(Intercept)	503	25	904	20.4	0.000
Lusitanica	−166	31	51	−5.36	0.000
Macrocarpa	−187	33	51	−5.74	0.000
A	12.2	2.7	904	4.46	0.000
$A^{0.5}$	−95.9	15.5	904	−6.17	0.000
A/W	−2.96	1.43	904	−2.07	0.039
$1/(1 + W^{0.5})$	480	36.9	904	13.0	0.000
North	24.3	8.9	51	2.75	0.008
South	−20.2	9.2	51	−2.18	0.034
A: Lusitanica	−7.06	3.90	904	−1.81	0.070*
A: Macrocarpa	−18.1	4.1	904	−4.44	0.000
$A^{0.5}$: Lusitanica	66.3	22.2	904	2.98	0.003
$A^{0.5}$: Macrocarpa	125	23	904	5.36	0.000

A cambial age, *W* ring width
*not significant

Model performance statistics suggested a relatively strong relationship between actual and estimated mean ring density (Table 4). At the population level for the combined sample, MAPE was 6 %, RMSE was 36 kg.m^{-3} and R^2 equalled 0.56. Overall, the population level model explained about 56 % of the variation in mean density.

Modelled regression lines of mean density, using estimated ring widths from the model parameters presented in Table 2, are shown in Fig. 8 together with 95 % confidence and prediction bands. The confidence intervals (shown in the figure by the shaded grey areas) indicate the likely location of the true population mean and have a precision (width of the interval) of about ±16 kg.m^{-3} while precision of the prediction intervals (which include scatter from individual trees) is about ±46 kg.m^{-3}.

Table 4 Performance statistics of the radial density model, by taxon

Taxon	MAPE (%)	RMSE (kg.m^{-3})	R^2
Leyland	5.1	30	0.56
Lusitanica	7.0	38	0.46
Macrocarpa	6.5	41	0.42
All	6.1	36	0.56

MAPE mean absolute percentage error, *RMSE* root mean square error, R^2 coefficient of determination

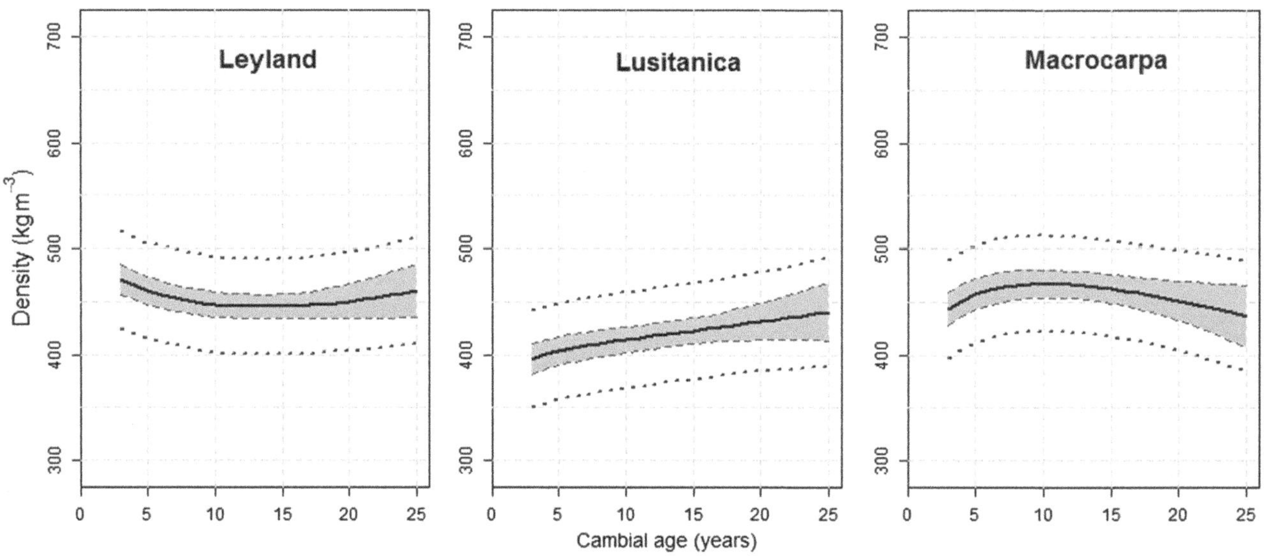

Fig. 8 Prediction of radial density for three cypress taxa. The regression line is shown as a *solid line*, the 95 % confidence bands for the regression line are shown as *dashed lines* with the confidence interval *shaded grey* and the 95 % prediction bands are shown as *dotted lines*

Modelled microfibril angle

A linear mixed-effects polynomial model based on cambial age provided good fit statistics, but due to its symmetric nature, was not reliable for forecasting MFA beyond the sample interval (1–18 years). Improved forecasting was however obtained with an alternative model, based on an exponential equation and similar in form to the MFA model of Cortini et al. (2014). The response variable (MFA) was log-transformed as were two explanatory variables (cambial age and distance from pith), which were transformed following visualisation of the data plots. Taxon was highly significant as an explanatory variable. However, unlike the density model, interaction terms with taxon were not significant, but interaction between ring width and the logarithmic transform of age was highly significant. Aspect and edge-tree were not significant. Parameter estimates of the final linear model are given in Table 5. Parameter estimates for the autocorrelation and variance structures were 0.905 and 0.104, respectively.

Table 5 Parameter estimates and fit statistics from fitting the linear mixed-effects model to predict radial logarithmic microfibril angle, ln(MFA)

Parameter	Value	Std. Error	DF	*t* value	*p* value
(Intercept)	3.50	0.14	799	24.9	0.000
Lusitanica	0.216	0.046	52	4.66	0.000
Macrocarpa	0.146	0.047	52	3.12	0.003
ln(A)	−0.597	0.057	799	−10.4	0.000
W	−0.010	0.004	799	−2.71	0.007
ln(X)	0.136	0.053	799	2.55	0.011
ln(A):W	0.00764	0.00169	799	4.52	0.000

A cambial age, *W* ring width, *X* distance from pith

Performance statistics comparing actual with predicted MFA demonstrated a strong relationship between actual and estimated mean microfibril angle (Table 6). At the population level for all taxa, MAPE was about 16 %, RMSE 3.7°, and R^2 equalled 0.69.

The modelled regression lines of mean microfibril angle are shown in Fig. 9 together with 95 % confidence and prediction bands. Due to the logarithmic transform of the response variable, a bias correction factor (Snowdon 1991) was calculated for each taxon (0.98, 0.99 and 1.03 for Leyland, Lusitanica and Macrocarpa, respectively) and applied when calculating predicted MFA. The confidence intervals shown in the figure have a precision of about ±2.0° while precision of the prediction intervals is about ±3.8°.

Modelled modulus of elasticity

Linear and nonlinear mixed effects models were considered when developing the MOE models. Nonlinear models for each taxon were based on the Mitscherlich equation (Briggs 1925), augmented by aspect. Final

Table 6 Performance statistics of the radial microfibril angle model, by taxon

Taxon	MAPE (%)	RMSE (°)	R^2
Leyland	15.3	3.7	0.64
Lusitanica	16.0	3.7	0.77
Macrocarpa	14.4	3.7	0.63
All	15.6	3.7	0.69

MAPE mean absolute percentage error, *RMSE* root mean square error, R^2 coefficient of determination

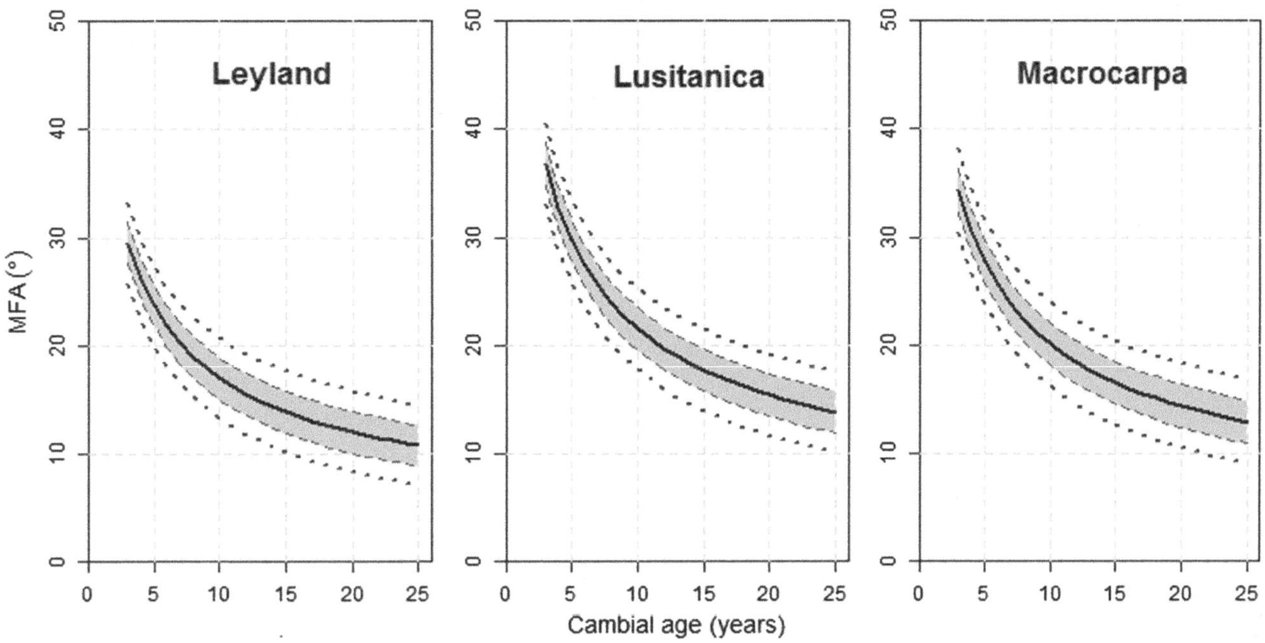

Fig. 9 Prediction of microfibril angle for three cypress taxa. The regression line is shown as a *solid line*, the 95 % confidence bands for the regression line are shown as *dashed lines* with the confidence interval *shaded grey* and the 95 % prediction bands are shown as *dotted lines*

models, with better fit statistics, were of linear form. Explanatory variables comprised taxon, ring width, the logarithmic transform of cambial age and aspect. The southern aspect was significantly different to the intermediary aspect, but the northern aspect was not. An interaction term with ring width was also highly significant for the southern aspect only. Therefore, a new variable that only distinguished the southern aspect was introduced in the model (Table 7). Parameter estimates for the autocorrelation and variance structures were 0.834 and 0.133, respectively. Performance statistics, which suggest a good fit between actual and estimated MOE, are provided in Table 8.

The modelled regression lines of mean modulus of elasticity are shown in Fig. 10 together with 95 % confidence and prediction bands. The confidence intervals have a precision of about ±0.7 GPa, while precision of the prediction intervals is about ±1.4 GPa.

Table 7 Parameter estimates and fit statistics from fitting the linear mixed-effects model to predict radial modulus of elasticity (MOE)

Parameter	Value	Std. Error	DF	*t* value	*p* value
(Intercept)	6.34	0.49	800	13.0	0.000
Lusitanica	−2.54	0.37	51	−6.87	0.000
Macrocarpa	−1.94	0.39	51	−4.98	0.000
Southern	−2.16	0.51	51	−4.24	0.000
ln(A)	2.91	0.17	800	16.8	0.000
W	−0.170	0.014	800	−11.8	0.000
W: Southern	0.102	0.027	800	3.83	0.000

A cambial age, *W* ring width

Forecasts of earliest age for production of timber of specific MOE

The steps outlined under 'Application of Models' were followed to estimate the earliest possible ages for producing 100 × 50 × 2.0 and 150 × 50 × 2.0 timber (width (mm), thickness (mm), length (m)) timber of 6 and 8 GPa. One minor modification to the procedure was required due to the ring-width models being valid for cambial age greater than 2 years. The modification entailed adding the sum of the actual mean ring widths of the first two rings (totalling 18 mm for Leyland and 24 mm for Lusitanica and Macrocarpa) to the cumulative sum obtained from the ring-width models.

With this procedure, the minimum age for producing 100 mm × 50 mm × 2.0 m 6 GPa timber from Leyland stands was estimated to be 22 years whereas that for Lusitanica was 32 years and for Macrocarpa 29 years. At

Table 8 Performance statistics of the radial modulus of elasticity model, by taxon

Taxon	MAPE (%)	RMSE (GPa)	R^2
Leyland	13.3	1.66	0.65
Lusitanica	19.8	1.85	0.74
Macrocarpa	17.2	1.71	0.59
All	16.6	1.74	0.66

MAPE mean absolute percentage error, *RMSE* root mean square error, R^2 coefficient of determination

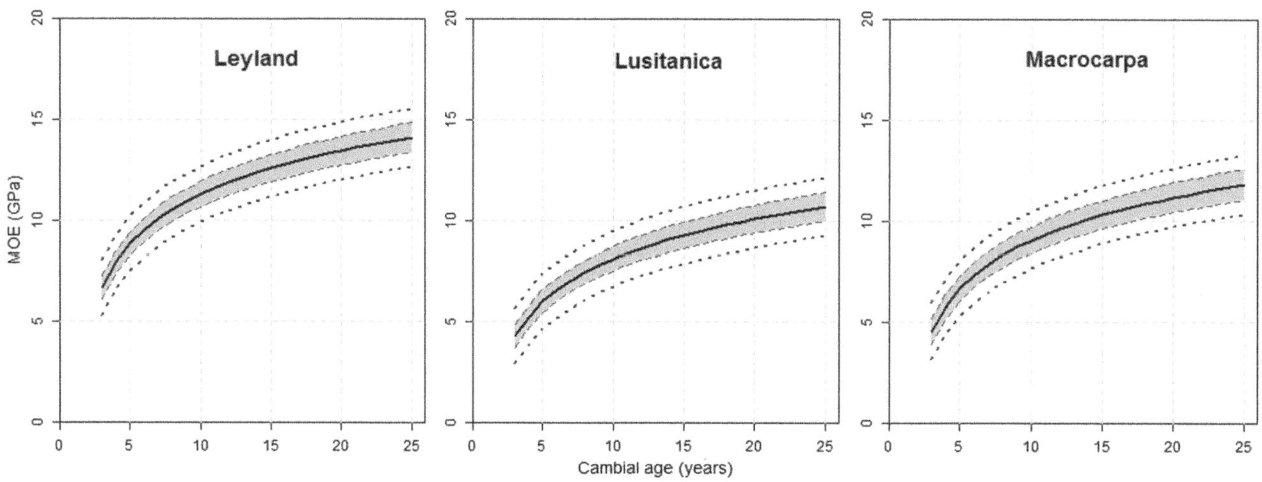

Fig. 10 Prediction of modulus of elasticity for three cypress taxa. The regression line is shown as a *solid line*, the 95 % confidence bands for the regression line are shown as *dashed lines* with the confidence interval *shaded grey* and the 95 % prediction bands are shown as *dotted lines*

25 years, wider timber (150 mm) can be produced from Leyland, and by 30 years, production of 8 GPa timber is forecast to be possible. Age forecasts for 6 and 8 GPa products are summarised in Table 9. Overall, forecasted ages for production of strength-grade timber were greatest for Lusitanica and least for Leyland.

Discussion

All results presented here were based on trees growing in row-plots. Though the row-plot design results in a greater proportion of edge trees than other plot designs, the edge-tree effect was not significant in any of the wood property models. Aspect, on the other hand, was highly significant in models predicting wood density. Trees on the northernmost aspect (warmest aspect with more sunshine in the Southern hemisphere) had greater wood density, amounting to about 44 kg.m^{-3} more than those on the southernmost aspect. As suggested by van der Maaten (2012), these differences may relate to the differing levels of irradiation between aspects. On this basis, it would be expected that a similar plot design in the Northern hemisphere would result in trees with a southern aspect having greater wood density.

Overall, wood density gradients of the three cypress taxa were relatively flat, as suggested by Brailsford (1999). At age 25 years, mean density of the three taxa, using the models whose parameters are given in Table 3, is estimated to be 461, 441 and 437 kg.m^{-3}, for Leyland, Lusitanica and Macrocarpa, respectively. However, with large variation due to individual trees, the 95 % prediction interval suggests that density at age 25 years could be as low as 410, 389 and 385 kg.m^{-3} or as high as 511, 492 and 489 kg.m^{-3} for trees of the three taxa, respectively. As all values are lower than the 540 kg.m^{-3} threshold indicating low-density wood, it can be expected that the timber of these cypress taxa will dent easily when depressed.

Radial patterns of microfibril angle of the three cypress taxa were similar to those of other conifers, with the highest angles occurring in the first five growth rings (McMillin 1973, Zhang et al. 2007), and like MFA of *Pinus radiata* (Moore et al. 2014), the cypress taxa had comparable and low MFA values (Donaldson 1992). Mean MFA of the first five growth rings was 25°, 45° and 29° for Leyland, Lusitanica and Macrocarpa, respectively, while at age 25 years, mean MFA was estimated to be 11°, 14° and 13° for the three taxa, respectively.

Mean modulus of elasticity of Leyland was greater than that of either Lusitanica or Macrocarpa for the majority of the tree rings analysed here (up to and including a cambial age of 18 years). At age 18 years, mean MOE of Leyland was 13 GPa, about 2 GPa greater than

Table 9 Forecasts of the earliest age (years) at which product specifications can be met

Taxon	Product (strength-grade × width)[a]			
	6 GPa × 100 mm	6 GPa × 150 mm	8 GPa × 100 mm	8 GPa × 150 mm
Leyland	22	25	30	32
Lusitanica	32	33	>35	>35
Macrocarpa	29	31	>35	>35

[a]Products in this example forecast are 50 mm thick and 2.0 m long

that of the other two taxa. At age 25 years, predicted mean MOE of the three taxa was 14, 11 and 12 GPa, respectively, while MOE of the lower (95 %) prediction bands was 13, 9 and 10 GPa, respectively. Thus, at age 25 years, it would be expected that products derived from stands of Lusitanica and Macrocarpa would fail to meet the engineering grade criteria of 10.5 GPa and would attain 6 and 8 GPa strength grades only if sufficient growth had occurred.

Forecasts of the age at which specific product density or MFA is attained can be made following the steps outlined in 'Application of Models'. However, while forecasts using the above wood property models address wood quality and form, other degrading features such as knots, warp on drying and internal checking need consideration. For Lusitanica and Macrocarpa, Low et al. (2005) reported problems with warp on drying for the former taxon and problems with internal checking for the latter. However, the Leyland hybrid was reported as having good performance in terms of both appearance and structural products. In general, wood properties can vary for trees growing under different conditions; therefore, if growth rates differ to those presented here, forecasts will need to be revised to match the growing conditions.

Conclusions

Shorter rotations appear to be feasible for Leyland, with the production of 6 GPa timber being possible from the outer log zones from age 22 years. However, benefits of shortened rotations for either Lusitanica or Macrocarpa are questionable. With other studies demonstrating good performance of Leyland timber for both appearances and structural products, this hybrid cypress taxon is a suitable candidate for increased production forestry from exotic species in New Zealand.

Characterisation of radial variation in wood properties is an important step in forecasting not only harvesting ages but also potential product quality. Various product scenarios, of both existing and new products, can be evaluated using methods and models developed here.

Competing interests

The authors declare that they have no competing interests.

Authors' contributions

CLT provided statistical design, analysis and manuscript writing. CBL contributed initial project design, database construction and consolidation and manuscript writing. HMM controlled the logging operation and tree selection process and provided impetus for this study, including its write-up. LDG facilitated the cypress trial and provided input to key concepts culminating in this manuscript. All authors read and approved the final manuscript.

Acknowledgements

The authors would like to thank Mark Miller and Kane Fleet for extracting increment cores; Robert Evans, Sarah King, Sharee Harper, Roxanne Luff and the Silviscan® team for processing the cores; Neena Ranchod-Shaw for estimating ring boundaries from Silviscan® data; and Heidi Dungey for constructive comments on early drafts.

Author details

[1]Scion, Private Bag 3020, Rotorua 3046, New Zealand. [2]Centralis Limited, 3 Ferndale Drive, Waikanae 5036, New Zealand. [3]412 No. 1 Road, RD2, Te Puke 3182, New Zealand.

References

Akaike, H. (1974). A new look at the statistical model identification. *IEEE Transactions on Automatic Control, 19*(6), 716–723. doi:10.1109/TAC.1974.1100705.

Auty, D, Gardiner, BA, Achim, A, Moore, JR, & Cameron, AD. (2013). Models for predicting microfibril angle variation in Scots pine. *Annals of Forest Science, 70*, 209–218.

Bošeľa, M, Kulla, L, & Marušák, R. (2011). Detrending ability of several regression equations in tree-ring research: a case study based on tree-ring data of Norway spruce (*Picea abies* [L.]). *Journal of Forest Science, 57*, 491–499.

Bouriaud, O, Breda, N, Le Moguedec, G, & Nepveu, G. (2004). Modelling variability of wood density in beech as affected by ring age, radial growth and climate. *Trees, 18*, 264–276.

Brailsford, S. (1999). *The cypress growers' handbook*. Christchurch, New Zealand: Brailsford's Forest Management Ltd.

Briggs, GE. (1925). Plant yield and the intensity of external factors—Mitscherlich's 'Wirkungsgesetz'. *Annals of Botany, 39*, 475–502.

Chappell, PR (2013). *The climate and weather of Bay of Plenty*. (3rd edition). [NIWA Science and Technology Series, Number 62]. Wellington, New Zealand: National Institute of Water and Atmospheric Research.

Cortini, F, Groot, A, Tong, Q, & Duchesne, I. (2014). Ring-level models for predicting wood and fibre properties of *Abies balsamea*. *Wood Science Technology, 48*, 1181–1196.

Donaldson, LA. (1987). S3 Lignin concentration in radiata pine tracheids. *Wood Science Technology, 21*, 227–234.

Donaldson, LA. (1992). Within- and between-tree variation in microfibril angle in *Pinus radiata*. *New Zealand Journal of Forestry Science, 22*, 77–86.

Donaldson, L. (2008). Microfibril angle: measurement, variation and relationships—a review. *IAWA Journal, 29*(4), 345–338.

Evans, R, Downes, GM, Menz, DNJ, & Stringer, SL. (1995). Rapid measurement of variation in tracheid transverse dimensions in a radiata pine tree. *Appita Journal, 48*(2), 134–138.

Evans, R, & Ilic, J. (2001). Rapid prediction of wood stiffness from microfibril angle and density. *Forest Products Journal, 51*, 53–57.

Evans, R. (2006). Wood stiffness by X-ray diffractometry. In DD Stokke & LH Groom (Eds.), *Characterization of the cellulosic cell wall. Proceedings of a Workshop Cosponsored by the USDA Forest Service, Southern Research Station; the Society of Wood Science and Technology; and Iowa State University, August 25–27, 2003, Grand Lake, Colorado*. (pp. 138–146). Ames, IA, USA: Blackwell Publishing.

Fang, K, Gou, X, Peters, K, Li, J, & Zhang, F. (2010). Removing biological trends from tree-ring series: testing modified Hugershoff curves. *Tree-Ring Research, 66*, 51–59.

Fang, S, Wenzhong, Y, & Ye, T. (2006). Clonal and within-tree variation in microfibril angle in 10 poplar clones. *New Forests, 31*, 373–383.

Filipescu, CN, Lowell, EC, Koppenaal, R, & Mitchella, AK. (2014). Modeling regional and climatic variation of wood density and ring width in intensively managed Douglas-fir. *Canadian Journal of Forest Research, 44*(3), 220–229. doi:10.1139/cjfr-2013-0275.

Franceschini, T, Bontemps, J-D, Gelhaye, P, Rittie, D, Herve, J-C, Gegout, J-C, & Leban, J-M. (2010). Decreasing trend and fluctuations in the mean ring density of Norway spruce through the twentieth century. *Annals of Forest Science, 67*(816), 1–10. doi:10.1051/forest/2010055.

Gaunt, D. (1998). If you are not winning, change the rules. *Wood Processing Newsletter 23*, (Sawmilling 3–5). Rotorua, New Zealand: Forest Research Institute.

Haslett, AN. (1986). *Properties and utilisation of exotic speciality timbers grown in New Zealand. Part III: Cypresses Chamaecyparis lawsoniana (A. Murr.) Parl. x Cupressocyparis leylandii (Jacks et Dall.) Dall. Cupressus lusitanica Mill. Cupressus macrocarpa Hartw* (FRI Bulletin No. 119). Rotorua, New Zealand: Forest Research Institute.

Hugershoff, R. (1936). *Die mathematischen Hilfsmittel des Kulturingenieurs und Biologen* (Vol. 2). Dresden: Herleitung von gesetzmäßigen Zusammenhängen. als Manuskript veröffentlicht.
Ivković, M, Gapare, W, Wu, H, Espinoza, S, & Rozenberg, P. (2013). Influence of cambial age and climate on ring width and wood density in *Pinus radiata* families. *Annals of Forest Science, 70*, 525–534. doi:10.1007/s13595-013-0290-z.
Jordan, L, Daniels, RF, Clark, A, & He, R. (2005). Multilevel nonlinear mixed-effects models for the modelling of earlywood and latewood microfibril angle. *Forest Science, 51*, 357–371.
Jyske, T, Makinen, H, & Saranpaa, P. (2008). Wood density within Norway spruce stems. *Silva Fennica, 42*, 439–455.
Laird, NM, & Ware, JH. (1982). Random effects models for longitudinal data. *Biometrics, 38*, 963–974. doi:10.2307/2529876.
Lasserre, J-P, Mason, EG, Watt, MS, & Moore, JR. (2009). Influence of initial planting spacing and genotype on microfibril angle, wood density, fibre properties and modulus of elasticity in *Pinus radiata* D. Don corewood. *Forest Ecology and Management, 258*(9), 1924–1931.
Low, CB, McKenzie, HM, Shelbourne, CJA, & Gea, LD. (2005). Sawn timber and wood properties of 21-year-old *Cupressus Lusitanica, C. Macrocarpa*, and *Chamaecyparis Nootkatensis* x *C. Macrocarpa* hybrids. Part 1: sawn timber performance. *New Zealand Journal of Forestry Science, 35*(1), 91–113.
Lundgren, C. (2004). Microfibril angle and density patterns of fertilized and irrigated Norway spruce. *Silva Fennica, 38*(1), 107–117.
McKinley, RB, Shelbourne, CJA, Harris, JM, & Young, GD. (2000). Variation in whole-tree basic wood density for a range of plantation species grown in New Zealand. *New Zealand Journal of Forestry Science, 30*(3), 436–446.
McMillin, CW. (1973). Fibril angle of loblolly pine wood as related to specific gravity, growth rate and distance from the pith. *Wood Science Technology, 7*, 251–255.
Malimbwi, RE, Persson, A, Iddi, S, Chamshama, SAO, & Mwihomeke, ST. (1992). Effects of spacing on yield and some wood properties of *Cupressus lusitanica* at Rongai, Northern Tanzania. *Forest, 65*, 73–82.
Miller, JT, & Knowles, FB. (1996). *Introduced forest trees in New Zealand: recognition, role and seed source. 9. The cypresses* (NZFRI Bulletin No 124(9)). Rotorua, New Zealand: Forest Research Institute.
Moore, J, Cown, DJ, McKinley, RB (2014). Modelling microfibril angle variation in New Zealand-grown radiata pine. *New Zealand Journal of Forestry Science* 44:25. doi:10.1186/s40490-014-0025-4.
Peschel, W. (1938). Die mathematischen methoden zur herleitung der wachstumsgesetze von baum und bestand und die ergebnisse ihrer an wendung. *Tharandter forstliches Jahrbuch, 89*(3/4), 169–247.
Piispanen, R, Heinonen, J, Valkonen, S, Mäkinen, H, Lundqvist, S-O, & Saranpää, P. (2014). Wood density of Norway spruce in uneven-aged stands. *Canadian Journal of Forest Research, 44*, 136–144.
Pinheiro, J. (2002). Model building using covariates in nonlinear mixed-effects models. *Journal de la société française de statistique, 143*, 79–101.
Pinheiro, J, Bates, D, DebRoy, S, Sarkar, D, & the R Development Core Team. (2013). *nlme: Linear and nonlinear mixed effects models.* (R package version 3.1–111).
R Core Team. (2013). *R: a language and environment for statistical computing.* Vienna, Austria: R Foundation for Statistical Computing.
Snowdon, P. (1991). A ratio estimator for bias correction in logarithmic regressions. *Canadian Journal of Forest Research, 21*, 720–724.
Standards New Zealand. (1993). *NZS 3603:1993 timber structures standard.* Wellington: Wright & Carman Ltd.
van der Maaten, E. (2012). Climate sensitivity of radial growth in European beech (*Fagus sylvatica* L.) at different aspects in southwestern Germany. *Trees, 26*, 777–788.
Vikram, V, Cherry, ML, Briggs, D, Cress, DW, Evans, R, & Howe, GT. (2011). Stiffness of Douglas-fir lumber: effects of wood properties and genetics. *Canadian Journal of Forest Research, 41*, 1160–1173.
Vincent, TG, & Dunstan, JS. (1989). *Register of commercial seedlots issued by the New Zealand Forest Service. FRI Bulletin No. 144*. Rotorua, New Zealand: Forest Research Institute.
Walker, JCF, & Butterfield, BG (1995). The importance of microfibril angle for the processing industries. *New Zealand Forestry, 40*(3), 34–40.
Watt, MS, Clinton, PW, Coker, G, Davis, MR, Simcock, R, Parfitt, RL, & Dando, J. (2008). Modelling the influence of environment and stand characteristics on basic density and modulus of elasticity for young *Pinus radiata* and *Cupressus lusitanica. Forest Ecology and Management, 255*(3/4), 1023–1033.
Wimmer, R, & Downes, GM. (2003). Temporal variation of the ring width–wood density relationship in Norway spruce grown under two levels of anthropogenic disturbance. *IAWA Journal, 24*(1), 53–61.
Yang, JL, & Evans, R. (2006). Prediction of MOE of eucalypt wood from microfibril angle and density. *Holz als Roh und Werkstoff, 61*, 449–452.
Zhang, B, Fei, B-H, Yu, Y, & Zhao, R-J. (2007). Microfibril angle variability in Masson pine (*Pinus massoniana* Lamb.) using X-ray diffraction. *Forestry Studies China, 9*, 33–38.
Zuur, AF, Ieno, EN, Walker, NJ, Saveliev, AA, Smith, GM (2009). *Mixed effects models and extensions in ecology with R. Statistics for biology and health.* New York: Springer-Verlag.

13

Effect of logging wounds on diameter growth of beech (*Fagus orientalis* Lipsky) trees following selection cutting in Caspian forests of Iran

Farzam Tavankar[1*], Amireslam Bonyad[2], Enrico Marchi[3], Rachele Venanzi[4] and Rodolfo Picchio[4]

Abstract

Background: Logging damage to residual trees during selection cutting may lead to serious economic losses in terms of both timber quality and diameter growth reduction. In this study, we investigated the effect of logging operations on residual tree damage and the consequence of injuries on diameter growth in an uneven-aged mixed forest dominated by beech (*Fagus orientalis* Lipsky).

Methods: The diameter growth of 56 wounded and 56 unwounded beech trees were compared 12 years after selected logging of other trees had taken place in an Iranian Caspian forest.

Results: Of total logging wounds, 25.0 % (14 wounds) were caused by felling operations and 75.0 % (42 wounds) were caused by winching operations. The reduction in diameter growth of wounded beech trees was only observed in the breast height DBH range below 55 cm. The diameter growth of wounded trees was 8.1 % lower than in unwounded trees. The vertical distance from the ground of wounding was a significant factor in diameter growth. Wounds that were larger than 200 cm^2 in area significantly reduced diameter growth of beech trees. Whether wounds were closed or open wounds had no significant effect on diameter growth but decayed wounds reduced diameter growth by 15.3 %.

Conclusions: This study suggested that intensive logging wounds reduce diameter growth, especially in young beech trees. In addition to the intensity, size and position of wounds and tree age, the reduction of diameter growth was related to the ratio of wound size to stem area.

Keywords: *Fagus orientalis* Lipsky; Felling; Logging damage; Tree growth; Winching

Background

Impact on residual trees and/or to the regeneration of forest after forest harvesting continues to be one of the major concerns of forest managers responsible for implementing selection harvests (Reisinger and Pope 1991; Smith et al. 1994; Nikooy et al. 2010). The reduction of the negative effects of felling and extraction is one of the main goals when achieving sustainable forest management (Sist and Nguyen-Thé 2002). Skinned bark and exposed sapwood, root damage and broken branches are the most common forms of damage occurring during forest logging (Vasiliauskas and Stenlid 2007). When the bark is torn from living trees, physiological changes and cell differentiation occur in the area surrounding the injury and callus starts to develop from the living bark towards the centre of the wound (Neely 1979). Stem wound and root damage are also potential means of entry for organisms causing decay and pitch rings (Shigo 1966; Pechmann 1974; Dimitri 1983). Because of potential decreased tree vigour and increased likelihood of attack by insects or diseases, logging injury to the remaining trees may lead to serious economic losses in terms of timber quality at the final harvest (Kiser 2011), wood losses of damaged trees and tree growth reduction (Ohman 1970; Vasiliauskas 2001).

Several studies on logging damage have highlighted that a number of factors affect the susceptibility to and effect of mechanical injury. These include the following: tree species (Bettinger and Kellogg 1993; Vasiliauskas 2001), silvicultural treatment (Bettinger et al. 1998), logging

* Correspondence: tavankar@aukh.ac.ir
[1]Department of Forestry, Khalkhal Branch, Islamic Azad University, P.O. Box: 56817-31367, Khalkhal, Iran
Full list of author information is available at the end of the article

season (Limbeck-Lilienau 2003), logging methods (Bragg et al. 1994; Spinelli et al. 2010; Marchi et al. 2014), harvesting system (Han and Kellogg 2000), site characteristics (Sist et al. 2003), amount of timber removed during harvest (Fjeld and Granhus 1998; Gullison and Hardner 1993), design of the extraction trails (Gullison and Hardner 1993), and skill of the personnel operating the equipment (Wallentin 2007; Nikooy et al. 2010). The effect of damage on tree growth has been also investigated showing contrasting results. Some studies clearly indicated that growth losses occur in damaged trees, and in some cases, these can seriously offset the gains from harvesting (Vasiliauskas 2001). Shea (1961) recorded a reduction in diameter growth in injured Douglas-fir (*Pseudotsuga menziesii* (Mirb) Franco). Isomäki and Kallio (1974) verified a decrease by 35–40 % in both radial and height growth in Norway spruce (*Picea abies* (L.) H. Karst.) trees with root collar damage following logging and a 15 % reduced growth due to stem damage. Other studies showed stem wounds in Norway spruce may result in growth rate losses ranging from 14 to 25 % (Baader 1956; Vanek 1957). In contrast, Vasiliauskas and Stenlid (2007) recorded an increase in radial growth in injured stems of ash (*Fraxinus excelsior* L.). Finally, a number of investigations failed to reveal any significant effect of stem wounds on the increment of Norway spruce (Vasiliauskas 1994), western hemlock (*Tsuga heterophylla* (Raf.) Srag.) (Shea 1961), oak (*Quercus robur* L.) (Vasiliauskas 1998) and Corsican pine *Pinus nigra* subsp. *salzmannii*var. *corsicana* (Loudon) Hylander (as *Pinus laricio* Poiret) (Picchio et al. 2011). It is difficult, therefore, to make a general conclusion on what effect mechanical stem damage has on tree growth (Vasiliauskas 2001). Also, most of the studies on this topic are quite dated, mainly related to conifer species and, most importantly, approach the effect of damage on tree growth taking into consideration the effect of damage caused by animals rather than logging. Studies on this topic are still limited, especially in broadleaved stands, and the long-term effects of damage are not well known (Mäkinen et al. 2007). In order to improve logging methods, a better knowledge about the long-term impact of forest operations is needed (Whitman et al. 1997; Tavankar and Bonyad 2014).

Beech is the most industrial commercial tree species among more than 80 broadleaved trees and shrubs in the Caspian forests (Marvie-Mohadjer 2006). Pure or mixed oriental beech stands represent the most important forest communities in the natural Caspian forests (Marvie-Mohadjer 2006). These stands constitute approximately 17.6 % of the area, 30 % of standing volume and 23.6 % of stems number in the Iranian Caspian forests (Amiri et al. 2013).

In this study, we investigated the effect of a logging operation on residual tree damage and the consequence of injuries on diameter growth in an uneven-aged mixed forest dominated by beech in the Caspian forest of Iran. In detail, the aims of this study were the following: (i) to assess the long-term effects of logging damage on both diameter growth and wound condition after 12 years since logging and (ii) to analyse the relation of intensity, size and location of injuries with beech diameter growth.

Methods

Study area

This study was conducted in parcel 35 in district 1 of Nav forests, located between 37°38′34′′ to 37°42′21′′ *N* and 48°48′44′′ to 48°52′30′′ E. Elevation in the study area ranged from 950 to 1350 m a.s.l. The mean annual precipitation is approximately 950 mm, and the mean annual temperature is 9.1 °C. The original vegetation of this area is an uneven-aged mixed forest dominated by *Fagus orientalis* Lipsky. The soil type is forest brown, and texture varies between sandy clay loams to clay loam. The silvicultural method applied in this forest is single selection cutting. Logging operations in the area were carried out in December 1999 and January 2000. In total, 216 marked trees (6.75 trees ha^{-1}) were scattered in the parcel area (32 ha), and these have a volume of 688 m^3 (21.5 m^3 ha^{-1}). The diameter at breast height (DBH) of the marked trees ranged from 20 to 135 cm. The marked trees were felled, limbed and topped at 20 cm dib (diameter inside bark) by chain saw. Logs, ranging from 5.2 to 7.8 m in length, were then extracted from the felling site to the roadside landings using a Timber-jack 450 C wheeled skidder. The weight of the skidder was 9.8 t, and its width and length were 3.8 and 6.4 m, respectively.

Data collection and analysis

Immediately after logging (year 2000), all wounded beech trees (56 stems) were identified, numbered and marked. The position of each damaged tree was also identified on a topographical map using the global positioning system. On each damaged tree, the following parameters were recorded after logging: diameter at breast height (DBH, 1.30 m) and diameter at wound height (DWH) (measured by dendrometric calliper in mm), damage severity (i.e. type of damaged tissues: bark, phloem, wood fibres), cause of damage (i.e. felling or extraction), and position and size of wounds. The wound size was determined by measuring the maximum length and width by a ruler (1 mm accuracy) and calculating the ellipse surface area (Picchio et al. 2011). The wound sizes were then classified in three classes: < 100, 101–200 and > 200 cm^2. The position of the wound was determined with a tape measuring the distance between the wound centre and the ground. The position of each wound was recorded in three classes, < 0.3 m, 0.3–1 m and > 1 m (Limbeck-Lilienau 2003). A neighbouring unwounded beech tree with similar characteristics (i.e. DBH, height, vitality, crown class of all the trees—i.e. dominant, codominant subdominant, etc.) was selected for comparison with each wounded tree. After 12 years (i.e. in year 2012), the wounded and unwounded trees were identified in the parcel area, and DBH and condition of wounds were reexamined and classified in three types: closed, open and decayed (Han et al. 2000). The 12-year period diameter growth of damaged and undamaged trees was calculated using Eq. 1 (Clark and Clark 1992):

$$DG = (DBH_2 - DBH_1)/t \quad (1)$$

Where DG is diameter growth (mm × year^{-1}), DBH_1 and DBH_2 are diameter at breast height at the start and end of interval (mm), respectively, and t is time interval between two measurements (years). Reduction of diameter growth was calculated using Eq. 2:

$$RDG = (UDG - WDG)/UDG \quad (2)$$

Where RDG is reduction of diameter growth, UDG is unwounded diameter growth and WDG is wounded diameter growth.

After checking for normality (Kolmogorov–Smirnov test) and homogeneity of variance (Levene test), a paired t test was applied to compare means of diameter growth (DG) in wounded and unwounded trees.

A non-parametric test (Spearman) was applied to test the relationship between the following: DBH vs. wound severity, wound position vs wound severity, wounded area vs. wound severity, wounded area vs. wound position, and wound position vs. DBH.

Regression analysis was applied to test the following relationships between: (i) diameter growth and tree age for both damaged and undamaged trees; (ii) wound size and tree diameter at the time of logging; (iii) diameter growth and wound position; and (iv) reduction of diameter growth and ratio of wound size to stem area. All analyses were performed using SPSS 19 software (IBM, NY, USA).

Results and discussion

Of total logging wounds, 25 % (14 wounds) were caused by the felling operation and 75 % (42 wounds) were caused by the extraction operation (Table 1). Normally, the extraction operation is considered to be one of the major sources of wounds in selection cuttings, but this view is only partially correct. During the logging operations, in particular during the felling phase, some trees marked for retention are accidentally destroyed during the felling of big trees nearby. In these cases, there are no wounds on trees but only dead trees. After 12 years following wound occurrence, 71.4 % of felling wounds were closed, while only 26.2 % of extraction wounds were closed. The number and percentage of open and decayed extraction wounds were higher than felling wounds—38.1 % of extraction wounds were decayed after 12 years. This is a consequence of the correlation between wound position and wound severity. For the extraction wounds, normally the first metre of the stem was damaged and the wounds were large and deep, extending below the bark down into the wood. These are ideal conditions for infection by diseases in particular fungi.

Table 1 Frequency of wounds by cause of wound

Cause of wound	Closed		Open		Decayed	
	(*n*)	(%)	(*n*)	(%)	(*n*)	(%)
Felling	10	71.4	3	21.4	1	7.2
Extraction	11	26.2	15	35.7	16	38.1

Picchio et al. (2012) studied improved winching techniques designed to decrease stand damage in the forests of central Italy. They reported that the use of a snatch block decreased by one-quarter the frequency of wounded trees from 50 to 36 %.

The frequency of wounds for DBH classes by cause of wound is shown in Table 2. From the analysis of all the trees wounded, 4 wounds (7.1 %) were found on trees with DBH < 20 cm, 15 wounds (26.8 %) on trees with DBH ranging from 20 to 40 cm, 24 wounds (42.9 %) on trees with DBH ranging from 41 to 60 cm, and 13 wounds (23.2 %) on trees with DBH > 60 cm. In each DBH class, the frequency of extraction wounds was higher than the frequency of felling wounds. This distribution of wounded trees, with the major frequency in the two central DBH classes (20–40 cm and 41–60 cm) is mainly due to the structure of the uneven-aged forest. In these forests, normally there are small groups of trees coetaneous, with high density and the operation of "thinning" is more intensive in the groups of the two central DBH classes. The probability of damage depends on the felling intensity and the trees density (Picchio et al. 2012). Residually damaged trees were highly concentrated near

Table 2 Frequency of wounds for each DBH class by cause of wound

Cause of wound	DBH (cm)							
	<20		20–40		41–60		>60	
	(*n*)	(%)	(*n*)	(%)	(*n*)	(%)	(*n*)	(%)
Felling	1	25.0	2	13.3	5	20.8	6	46.2
Extraction	3	75.0	13	86.7	19	79.2	7	53.8

the skid trails as showed in other studies (Solgi and Najafi 2007; Ezzati and Najafi 2010).

The results of the Spearman correlation analysis are shown in Table 3. The only statistically significant correlation was between wound severity and wound position ($P < 0.01$). In this case, the correlation was negative because the wounds closer to the ground were more severe. This is due mainly to the skidded logs impacting the standing trees and was rarely due to the impact of the trees with the tractor.

The average annual growth of wounded and unwounded beech trees after 12 years is shown in Table 4. Overall, there was no significant difference ($P > 0.05$) in the diameter increment of wounded trees compared with unwounded trees. In detail, the reduction of diameter growth in wounded beech trees was related only to wound size, severity and condition but only for some specific typologies of wounds characteristics. In order of importance, wounds >200 cm^2 led to greatest reduction of diameter growth (29.3 %; $P < 0.05$), followed by the wound condition, defined as "decayed", (which caused a RDG of 15.3 % ($P < 0.001$)) and lastly the depth of the wound. When the damage has affected the wood fibres, the average RDG was 13.2 % ($P < 0.001$). Also for a Norway spruce forest, width and depth of the logging injury correlated positively with the decrease in growth (Isomäki and Kallio 1974).

Wound size is one of the most important characteristics related to decay (Hunt and Krueger 1962; Han and Kellogg 2000). As found also by Vasiliauskas (2001), most of the resulting wounds occur near the base of the tree and are up to 200 cm^2 in size. Since logging injury usually occurs on the lower parts of trunk, wound decay in a tree affects the most valuable timber (Picchio et al. 2011; Tavankar et al. 2013).

Of all wounds recorded on beech trees, 55.3 % (31 wounds) occurred at a height lower than 0.3 m from the ground, 26.8 % (15 wounds) occurred between 0.3 to 1 m height and 17.9 % (10 wounds) occurred at a height higher than 1 m.

The results of this study indicated that most skidder-logging wounds (55.3 %) occurred in the near ground level of the bole (< 0.3 m) and were associated with a 10.7 % reduction in diameter growth relative to unwounded trees. The location of wounds from ground-based logging operations observed in this study is consistent with results from other published studies (Naghdi et al. 2008; Tavankar et al. 2011). The higher incidence of damage near the base of the tree is important because the butt log is generally the most valuable log in a tree (Bettinger and Kellogg 1993). Han and Kellogg (2000) suggested that artificial tree protection rigging such as rub pads along extraction corridors should be used to prevent damage on stump and stems.

Table 3 Results of Spearman correlation analysis

Variables	N	Correlation coefficient	Sig. (2-tailed)
DBH vs. wound severity	56	0.025	0.853
DBH vs. wound position	56	0.042	0.751
Wound size vs. wound position	56	0.131	0.336
Wound size vs. wound severity	56	−0.018	0.893
Wound position vs. wound severity	56	−0.747	0.000**

**$a = 0.01$.

Twelve years after the trees were wounded, 37.5 % of wounds were completely healed, 32.1 % were still open and 30.4 % were decayed. Trees with decayed wounds showed significant reduced diameter growth in comparison with undamaged ones ($P < 0.01$).

The multiple regression analyses applied to test the relationship between diameter growth and tree DBH (closely related to age) of both wounded and unwounded were statistically significant ($P < 0.001$). The results are shown in Eq. 2 and Fig. 1:

$$DG = 8.27 - (0.037 \times DBH) - (0.415 \times TC) \quad (3)$$

(R^2 adjusted = 0.162; SE = 1.521)

Where DG is the diameter growth of tree (mm $year^{-1}$), and TC is the tree condition (0 for unwounded and 1 for wounded).

Diameter growth of wounded trees was lower than for unwounded trees in small trees (DBH < 55 cm), and the greater the tree diameter the smaller the differences. Thus, growth of wounded and unwounded trees was similar when DBH was greater than 55 cm.

The regression analysis between wound size and tree diameter at the time of logging was also statistically significant ($P < 0.001$), and the results are shown in Eq. 3 and Fig. 2:

$$WS = 72.6 * \ln(DBH) - 128.6 \quad (4)$$

(R^2 adjusted = 0.373; SE = 45.37)

Where WS is the wound size. Only 37.3 % of the variability of the wound size was explained by DBH.

The regression analysis showed diameter growth of wounded trees increased with increasing wound position (height from ground level), (Fig. 3).

The results of the regression analysis between DG of wounded trees and wound position (WP) indicated two different situations depending on wound position. Thus, the relationship was significant for wounds positioned below 0.5 m of height (Fig. 3 and Eq. 4; R^2 adjusted = 0.95; SE = 0.24; F = 579.7, $P < 0.01$), but there was no relationship between DG and wounds positioned over 0.5 m from ground level (R^2 adjusted = 0.25, $F = 6.99$, $P < 0.05$).

Table 4 Mean ± SD of diameter growth of unwounded, wounded and reduction of diameter growth (RDG) of beech trees 12 years after wound occurrence in related to wounds characteristics

Wounds characteristic	*n* paired	Unwounded ($mm.year^{-1}$)	Wounded ($mm.year^{-1}$)	RDG (%)	*t* paired	*P* value
Wound position (m)						
<0.3	31	6.63 ± 1.8	5.92 ± 1.8	10.7	1.48	0.148
0.3–1	15	6.66 ± 0.6	6.06 ± 1.4	9.9	1.64	0.124
>1	10	6.35 ± 2.2	6.45 ± 0.9	−1.6	0.18	0.862
Wound size (cm^2)						
<100	9	5.96 ± 1.1	5.93 ± 0.9	0.5	0.04	0.966
101–200	31	6.25 ± 1.2	6.20 ± 1.0	0.8	0.24	0.812
>201	16	7.36 ± 1.5	5.21 ± 2.1	29.3	2.92*	0.010
Wound severity						
Bark	9	6.74 ± 0.9	6.72 ± 0.9	0.3	0.03	0.972
Phloem	14	6.50 ± 0.9	6.46 ± 1.1	0.6	0.07	0.871
Wood	33	6.60 ± 1.1	5.73 ± 1.4	13.2	4.22**	0.000
Wound condition						
Healed	21	6.72 ± 0.8	6.60 ± 1.0	1.8	1.27	0.217
Still open	18	6.67 ± 1.0	5.88 ± 1.4	11.8	1.94	0.069
Decayed	17	6.54 ± 1.0	5.54 ± 1.5	15.3	5.48**	0.000
All trees	56	6.59 ± 1.65	6.05 ± 1.54	8.1	1.80	0.078

*$a = 0.05$; **$a = 0.01$.

$$DG = 1.09 + 28.46 \times WP \quad (5)$$

Where DG is diameter growth ($mm.year^{-1}$) and WP is wound position (m) from the ground level.

This result suggests that the severity of wounds increased with decreasing height from ground level and/or due to an increased likelihood of pathogen attack in wounds close to ground level.

The regression analysis between the reduction of diameter growth and ratio of wound size to stem area is shown in Fig. 4. The results indicated that the reduction of diameter growth increased with increasing ratio of wound size to stem area above 0.1 ($F = 24.9$, $P < 0.01$).

The results of regression analysis are shown following equation:

$$RDG\% = 491.07(RWS)^2 - 87.81(RWS) + 6.75 \quad (6)$$

(R^2 adjusted = 0.51; SE = 4.62)

Where RDG is the reduction of diameter growth and RWS is the ratio of wound size to stem area.

This study evaluated the effect of logging wounds on diameter growth of standing beech trees in the Caspian forest 12 years after selection logging. Pre-harvest planning and identification of the winching area before logging operation commences can reduce damage to the

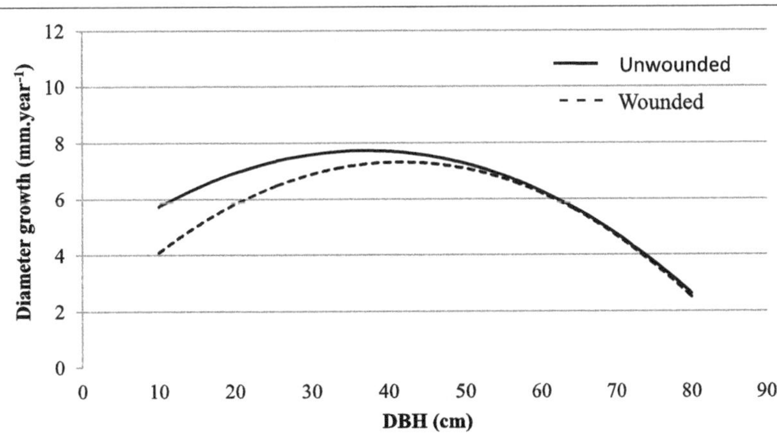

Fig. 1 Relationship between annual diameter growth of wounded and unwounded trees and their DBH in the year 2000

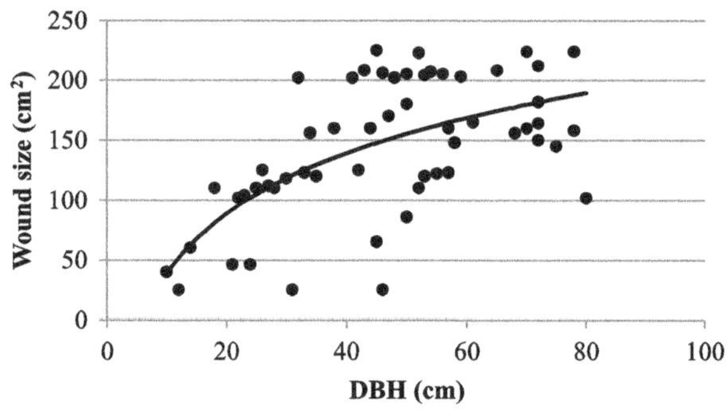

Fig. 2 Relationship between DBH of wounded trees in the year 2000 and wound size

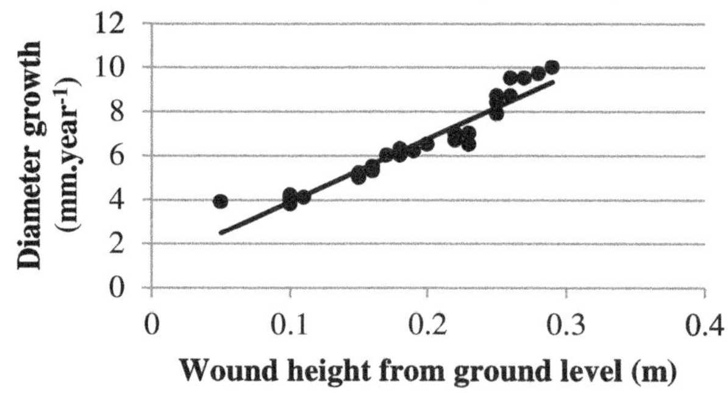

Fig. 4 Relationship between reduction of diameter growth and ratio of wound size to stem area

standing trees in these forests. Also, skid-trail planning before a felling operation can reduce the skidding damage (Naghdi et al. 2008; Majnounian et al. 2009). The skidder and chain saw operators are important factors that can influence productivity and environmental impacts during logging operations. It has been reported that residual stand damage decreases substantially by careful planning and skilful operators (Cline et al. 1991; Pinard et al. 30). Nikooy et al. (2010) reported that skilled operation of a skidder can decrease the level of damage. So training of forest workers can be useful to reducing logging damage to the residual stand. All workers should be made aware of the purpose of the selection cutting and that both minor and major injuries to the residual stands as well as excessive ground disturbance may result in significant volume losses (Davis and Nyland 1991) and natural regeneration of the forest (Picchio et al. 2012).

Conclusions

This study shows that logging wounds cause a substantial reduction (8.1 %) in diameter growth of residual trees left after a selective logging operation. Our results showed that the effect of bole wounds on diameter growth depended on their severity, location and size, and tree age. Young beech trees were more sensitive to logging wounds. Residual stand damage is an unavoidable risk of selection cutting, but the level of damage should be minimised to assure quality timber from future harvests. The results of this study indicated also that tree extraction has a high potential for residual stand damage, and intensive wounds occurred during extraction operations. It is important to minimise damage, both to the number of trees damaged and the extent of damage to any individual tree.

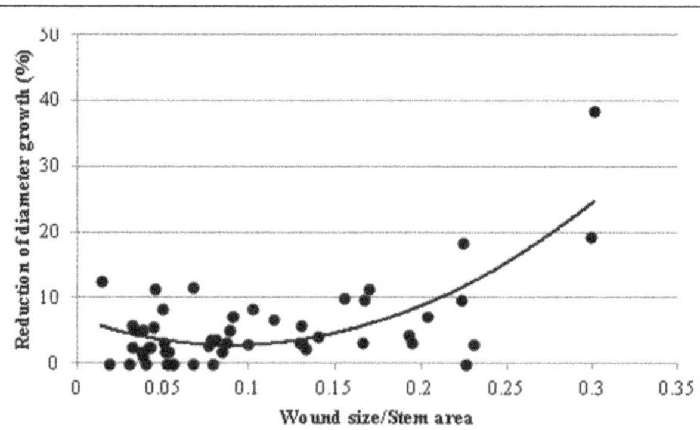

Fig. 3 Relationship between diameter growth and wound position

Competing interests
The authors declare that they have no competing interests.

Authors' contributions
All authors read and approved the final manuscript.

Author details
[1]Department of Forestry, Khalkhal Branch, Islamic Azad University, P.O. Box: 56817-31367, Khalkhal, Iran. [2]Department of Forestry, Faculty of Natural Resources, University of Guilan, P.O. Box: 1144, Somehsara, Iran. [3]Department of Agriculture, Food and Forest Systems, Florence University, Via San Bonaventura, 13 50145 Florence, Italy. [4]Department of Science and Technology for Agriculture, Forest, Nature and Energy (DAFNE), Tuscia University, Via S. Camillo De Lellis, 01100 Viterbo, Italy.

References

Amiri, M, Rahmani, R, Sagheb-Talebi, K, & Habashi, H. (2013). Dynamic and structural characteristics of a natural unlogged oriental beech (*Fagus orientalis* Lipsky) stand during a 5-year period in Shast Kalate forest, northern Iran. *The International Journal of Environmental Resources Research, 1*(2), 107–29.

Baader, G. (1956). Damage by game in Rheinland-Pfalz and the possibilities to minimize it. *Allgemine Forest Jagdztg, 127*, 233–40.

Bettinger, P, & Kellogg, L. (1993). Residual stand damage from cut-to-length thinning of second growth timber in the Cascade Range of western Oregon. *Forest Products Journal, 43*, 59–64.

Bettinger, P, Bettinger, KA, & Boston, K. (1998). Correlation among spatial and non-spatial variables describing a cut-to-length thinning site in the Pacific Northwest, USA. *Forest Ecology and Management, 104*, 139–49. doi:10.1016/s0378-1127(97)00250-8.

Bragg, W, Ostrofsky, W, & Hoffman, B. (1994). Residual tree damage estimates from partial cutting simulation. *Forest Products Journal, 44*, 19–22.

Clark, DA, & Clark, DB. (1992). Life history diversity of canopy and emergent trees in a neotropical rain forest. *Ecological Monographs, 62*, 315–44.

Cline, ML, Hoffman, BF, Cyr, M, & Bragg, W. (1991). Stand damage following whole-tree partial cutting in Northern forests. *Northern Journal of Applied Forestry, 8*, 72–6.

Davis, CJ, Nyland, RD. (1991). Designated skid trails: results after 18 years. *American Society of Agricultural Engineers, ASAE winter meeting, December 17–20, 1991, Chicago Illinois*. [Paper No. 917522].

Dimitri, L. (1983). Wound decay following tree injury in forestry: establishment, significance and possibilities of its prevention. *Forstwissenschaftliches Centralblatt, 102*, 68–79.

Ezzati, S, & Najafi, A. (2010). Long-term impact evaluation of ground-base skidding on residual damaged trees in the Hyrcanian Forest, Iran. *International Journal of Forestry Research, 1*, 1–8. doi:10.1155/2010/183735.

Fjeld, D, & Granhus, A. (1998). Injuries after selection harvesting in multi-storied spruce stands—the influence of operating systems and harvest intensity. *International Journal of Forest Engineering, 9*(2), 33–40. doi:10.1080/08435243.1998.10702716.

Gullison, R, & Hardner, J. (1993). The effects of road design and harvest intensity on forest damage caused by selective logging: empirical results and a simulation model from the Bosque Chimanes, Bolivia. *Forest Ecology and Management, 59*, 1–14. doi:10.1016/0378-1127(93)90067-W.

Han, HS, & Kellogg, LD. (2000). Damage characteristics in Young Douglas-fir stand from commercial thinning with four timber harvesting systems. *Western Journal of Applied Forestry, 15*, 27–33.

Han, HS, Kellogg, LD, Fillip, GM, & Brown, TD. (2000). Scar closure and future timber value losses from thinning damage in western Oregon. *Forest Products Journal, 50*, 36–42.

Hunt, J, & Krueger, KW. (1962). Decay associated with thinning wounds in young-growth western hemlock and Douglas-fir. *Journal of Forestry, 60*, 336–40.

Isomäki, A, & Kallio, T. (1974). Consequences of injury caused by timber harvesting machines on the growth and decay of spruce (*Picea abies* L. Karst.). *Acta Forest Fennica, 136*, 1–25.

Kiser, J. (2011). Histochemical and geometric alterations of sapwood in costal Douglas-fir following mechanical damage during commercial thinning. *Silva Fennica, 45*(4), 729–41.

Limbeck-Lilienau, B. (2003). Residual stand damage caused by mechanized harvesting systems. In T Steinmuller & K Stampfer (Eds.), *Proceedings of high tech forest operations for mountainous terrain. 5–9 Oct. Schlaegl-Austro, 2003*. Vienna: University of Natural Resources and Life Sciences.

Majnounian, B, Jourgholami, M, Zobeiri, M, & Feghhi, J. (2009). Assessment of forest harvesting damage to residual stands and regenerations—a case study of Namkhaneh District in Kheyrud Forest. *Journal of Environmental Sciences, 7*(1), 33–44.

Mäkinen, H, Hallaksela, AM, & Isomäki, A. (2007). Increment and decay in Norway spruce and Scots pine after artificial logging damage. *Canadian Journal of Forest Research, 37*, 2130–41. doi:10.1139/X07-087.

Marchi, E, Picchio, R, Spinelli, R, Verani, S, Venanzi, R, & Certini, G. (2014). Environmental impact assessment of different logging methods in pine forests thinning. *Ecological Engineering, 70*, 429–36. doi:10.1016/j.ecoleng.2014.06.019.

Marvie-Mohadjer, M. (2006). *Silviculture*. Tehran: Tehran University publishing.

Naghdi, R, Rafatnia, N, Bagheri, I, & Hemati, V. (2008). Evaluation of residual damage in felling gaps and extraction routes in single selection method (Siyahkal forest). *Iranian Journal of Forest and Poplar Research, 16*(1), 87–98.

Neely, D. (1970). Healing of wounds on trees. *Journal of the American Society for Horticultural Science, 95*, 536–40.

Nikooy, M, Rashidi, R, & Kocheki, G. (2010). Residual trees injury assessment after selective cutting in broadleaf forest in Shafaroud. *Caspian Journal of Environmental Sciences, 8*(2), 173–9.

Ohman, JH. (1970). Value loss from skidding wounds in sugger maple and yellow birch. *Journal of Forestry, 68*, 226–30.

Pechmann, HV. (1974). The influence of thinning methods on the quality of timber. *Forstarchiv, 45*, 34–8.

Picchio, R, Neri, F, Maesano, M, Savelli, S, Sirna, A, Blasi, S, Baldini, S, & Marchi, E. (2011). Growth effects of thinning damage in a Corsican pine (*Pinus laricio* Poiret) stand in central Italy. *Forest Ecology and Management, 262*, 237–43. doi:10.1016/j.foreco.2011.03.028.

Picchio, R, Magagnotti, N, Sirna, A, & Spinelli, R. (2012). Improved winching technique to reduce logging damage. *Ecological Engineering, 47*, 83–6. doi:10.1016/j.ecoleng.2012.06.037.

Pinard, MA, Putz, FE, Tay, J, & Sullivan, TE. (1995). Creating timber harvesting guidelines for a reduced impact logging project in Malaysia. *Journal of Forestry, 39*, 41–5.

Reisinger, TW, & Pope, PE. (1991). *Impact of timber harvesting on residual trees in a central hardwood forest in Indiana. Proceedings of 8th Central Hardwood Forest Conference, 1991 March 4–6, (pp. 82–91)*. Newtown Square, PA: U.S. Department of Agriculture, Forest Service, Northeastern Forest Experiment Station.

Shea, KR. (1961). *Deterioration resulting from logging injury in Douglas-fir and Western hemlock. [Forest Research Note, 36, 1–5]*. Federal Way, WA, USA: Weyerhaeuser Company.

Shigo, AL. (1966). *Decay and discoloration following logging wounds on northern hardwoods. [USDA Research Paper, 47, 1–43]*. Washington, DC: United States Department of Agriculture.

Sist, P, & Nguyen-Thé, N. (2002). Logging damage and the subsequent dynamics of a dipterocarp forest in East Kalimantan (1990–1996). *Forest Ecology and Management, 165*, 85–103. doi:10.1016/s03781127(01)00549-1.

Sist, P, Sheil, D, Kartawinata, K, & Priyadi, H. (2003). Reduced-impact logging in Indonesian Borneo: some results confirming the need for new silvicultural prescriptions. *Forest Ecology and Management, 179*, 415–27. doi:10.1016/S0378-1127(02)00533-9.

Smith, HC, Miller, GW, & Schuler, TM. (1994). Closure of logging wounds after 10 years. *USDA Tech Bull, 692*, 1–6.

Solgi, A, & Najafi, A. (2007). Investigating of residual tree damage during ground-based skidding. *Pakistan Journal of Biological Sciences, 10*, 1755–8. doi:10.3923/pjbs.2007.1755.1758.

Spinelli, R, Magagnotti, N, & Nati, C. (2010). Benchmarking the impact of traditional small-scale logging systems used in Mediterranean forestry. *Forest Ecology and Management, 260*, 1997–2001. doi:10.1016/j.foreco.2010.08.048.

Tavankar, F, & Bonyad, A. (2014). Long-term effects of logging damages on quality of residual trees in the Asalem Nav forest. *Journal of Environmental Studies, 40*(1), 39–50.

Tavankar, F, Bonyad, A, & Majnounian, B. (2011). Investigation of damages to stand caused by selection cutting using skidding system in the Asalem-Nav forest, Iran. *Journal of Environmental Studies, 37*(3), 89–98.

Tavankar, F, Majnounian, B, & Bonyad, A. (2013). Felling and skidding damage to residual trees following selection cutting in Caspian forests of Iran. *Journal of Forest Science, 59*(5), 196–203.

Vanek, J. (1957). Study on the consequences of bark stripping damage caused by game in forest stands. *Lesnictvi, 3*, 59–78.

Vasiliauskas, R. (1994). Wound healing rate and its influence spread of decay in spruce. *Proceeding of the Lithuanian Forest Research Institute, 34*, 207–12.

Vasiliauskas, R. (1998). Patterns of wounding and decay in stems of *Quercus robur* due to bark peeling. *Scandinavian Journal of Forest Research, 13*, 437–41. doi:10.1080/02827589809383004.

Vasiliauskas, R. (2001). Damage to trees due to forestry operations and its pathological significance in temperate forest: a literature review. *Forestry, 74*, 319–36. doi:10.1093/forestry/74.4.319.

Vasiliauskas, A, & Stenlid, J. (2007). Discoloration following bark stripping wounds on *Fraxinus excelsior. European Journal Forest Pathology, 28*(6), 383–90. doi:10.1111/j.1439-0329.1998.tb01192.x.

Wallentin, C. (2007). *Thinning of Norway spruce. PhD Thesis*. Sueciae, Sweden: Acta Universitatis Agriculture.

Whitman, AA, Brokaw, VLN, & Hagan, JM. (1997). Forest damage caused by selection logging of mahogany (*Swietenia machrophylla*) in northern Belize. *Forest Ecology and Management, 92*, 87–96. doi:10.1016/S0378-1127(96)03941-2.

14

Modelling variation in wood density within and among trees in stands of New Zealand-grown radiata pine

Mark O. Kimberley, Dave J. Cown, Russell B. McKinley, John R. Moore* and Leslie J. Dowling

Abstract

Background: Density is an important wood property due to its correlation with other wood properties such as stiffness and pulp yield, as well as being central to the accounting of carbon sequestration in forests. It is influenced by site, silviculture, and genetics, and models that predict the variation in wood density within and among trees are required by forest managers so that they can develop strategies to achieve certain wood density targets. The aim of the study presented here was to develop a wood density model for radiata pine (*Pinus radiata* D. Don) growing in New Zealand.

Methods: The model was developed using an extensive historical dataset containing wood density values from increment cores and stem discs that were obtained from almost 10,000 trees at over 300 sites. The model consists of two sub-models: (1) a sub-model for predicting the radial variation in breast-height wood density and (2) a sub-model for predicting the distribution of density vertically within the stem.

Results: The radial variation in breast-height wood density was predicted as a function of either ring number or both ring number and ring width, with the latter model better accounting for the effects of stand spacing. Additional model components were also developed in order to convert from annual ring density values to a whole-disc density, predict log density from disc densities, and account for the variation in wood density among individual trees within in a stand. The model can be used to predict the density of discs or logs cut from any position within a tree and can utilise measured outerwood density values to predict the density by log height for a particular stand. It can be used in conjunction with outerwood density to predict wood density distributions by logs for stands of any specified geographic location and management regime and is designed to be able to incorporate genetic adjustments at a later stage.

Conclusions: The analysis has confirmed and quantified much of the previous knowledge on the factors that affect the variation in wood density in radiata pine, particularly the influences of site factors and silviculture. It has also quantified the extent and patterns of variation in wood density within and among trees.

Background

Density is an important wood property that affects the performance of solid wood products, pulp and paper, and some panel products (Kibblewhite 1984; Panshin and de Zeeuw 1980; Xu et al. 2004). It is also an important determinant of carbon sequestration in forests. Radiata pine (*Pinus radiata* D. Don) is the main commercial forestry species in New Zealand and is grown on a wide range of soil types and under differing climatic conditions, particularly with respect to temperature and rainfall. Past research in radiata pine growing in New Zealand established that there are significant regional trends in wood density, related to environmental factors such as mean annual temperature and soil nutrition (Beets et al. 2001; Beets et al. 2007; Cown 1999; Cown et al. 1991; Harris 1965; Palmer et al. 2013). In common with many other conifer species, radiata pine also exhibits a marked radial pattern whereby wood density increases from pith to bark (Cown 1980). In the vertical direction, whole-disc average wood density decreases with height up the stem largely because the

* Correspondence: john.moore@scionresearch.com
Scion, Private Bag 3020, Rotorua, New Zealand

number of rings decreases with height, meaning that proportionately more wood is contained within the lower density inner rings (Cown 1999).

Previous research has also investigated the effects of forest management practices, such as thinning, pruning and rotation age, and genetic differences on wood density (Burdon and Harris 1973; Burdon and Low 1992; Cown 1973, 1974, 1999; Cown and McConchie 1981, 1982; Cown et al. 1992). These studies have reported a positive correlation between wood density and mean annual temperature and a weak negative correlation between growth rate and wood density. Silvicultural treatments have a lesser impact unless stands are suddenly released from intense competition by severe thinning or a nutritional deficiency is corrected by fertiliser application. In these cases, a noticeable drop in density can occur for a few years immediately after treatment (Cown 1973). However, forest management is still expected to have an effect on wood density by varying the proportion of corewood within a tree (Cown 1992).

Forest managers commonly use growth and yield models to predict the impact of different combinations of site, genetics, and silviculture on volume and log product outturn (Weiskittel et al. 2011). There is also a demand to be able to predict the impacts of these factors on wood density as some log grades have a density requirement, often as a proxy for wood stiffness. For example, forest managers want to be able to determine what genetic stock and silvicultural regime (i.e., combination of spacing, thinning, and rotation length) are needed to achieve a particular wood density target for a given site. There are a number of examples of modelling systems that have been developed to link forest growth and timber quality in a range of species (Gardiner et al. 2011; Houllier et al. 1995; Peng and Stewart 2013), but a comprehensive modelling system does not exist for radiata pine wood properties. In New Zealand, an empirical model called STANDQUA was developed in the mid-1990s that enabled within-tree patterns of radiata pine wood density to be estimated from breast-height core samples and the future log-level wood density to be predicted (Tian and Cown 1995; Tian et al. 1995). More recently, a model was developed by Beets et al. (2007) to predict the density of annual growth sheaths in order to estimate carbon sequestration.

Both of these previous radiata pine wood density models were developed using various subsets of the extensive data that have been collected in numerous published and unpublished studies over many years. More comprehensive analyses of these data are required along with the development of models that enable forest managers to predict how different combinations of site, silviculture, and genetics affect wood density. Such analyses and new model development has been undertaken in three phases: (1) the development of a model to predict the spatial variation in breast-height outerwood density (i.e., basic density of the outer five growth rings or outer 50 mm); (2) the development of a model to predict the variation in wood density within a tree and among trees within a stand given a measured or predicted value of breast-height outerwood density; and (3) the incorporation of the effects of genetic gain for wood density into the model.

The results from the first phase of the analysis were presented in Palmer et al. (2013). They developed the concept of a wood density index, which is somewhat analogous to site index in growth and yield analysis as it uses a common base age to account for the known variation in wood density with age. The wood density index is defined as the breast-height outerwood density of 20-year-old radiata pine trees with silvicultural and genetic characteristics typical of stands planted in the 1990s. Across New Zealand, this wood density index was shown to be positively associated with mean annual air temperature, with higher values of wood density found in the warmer northern latitudes (Palmer et al. 2013). In this paper, we present analyses of both the intra-stem variation in wood density and the variation in density among trees within a stand. These analyses were used to develop a model that can be linked to the site-level model developed by Palmer et al. (2013) and also to growth and yield modelling systems (West et al. 2013). This enables investigation of the effects that factors such as rotation age and tree spacing have on within-tree patterns of wood density and the resulting whole-stem average density. Genetic effects can also be included in the modelling system using adjustments to the wood density index. The derivation of these genetic adjustments is described in a separate paper (Kimberley, MO, Moore, JR, & Dungey, HS (in review). Modelling the effects of genetic improvement on radiata pine wood density. New Zealand Journal of Forestry Science).

Methods

Description of datasets used in the analyses

Data from a large number of historic studies were used to develop the within-tree density model. These data were of two main forms: (1) radial profiles of basic density at breast-height and (2) vertical profiles of whole-disc average density. Data on the radial profile of wood density were available from breast-height samples that had been collected from radiata pine stands throughout New Zealand. A total of 336 sites were sampled and approximately 30 trees were typically sampled at each site (Table 1). Data for 250 of these sites were collected in the wood properties survey undertaken by Cown et al. (1991). Wood density was assessed from 5-mm diameter increment cores in five-ring sections from the pith outwards using the Maximum Moisture Content method (Smith 1954). At 219 of these sites, trees were also felled and discs were cut at breast-height (1.4 m above the

Table 1 Summary of the different wood density datasets that were used to develop the various models

Region	Five-ring density samples at breast-height					Disc-level data collected on felled trees					Ring-level densitometry data at breast-height				
	Number of sites	Number of trees	Age (years)			Number of sites	Number of trees	Age (years)			Number of sites	Number of trees	Age (years)		
			Min.	Max.	Mean			Min	Max	Mean			Min.	Max.	Mean
Auckland	50	1592	16	57	32	30	312	14	53	31	4	66	19	33	26
Rotorua	94	2265	10	64	24	117	1264	6	56	25	7	172	17	35	24
Wellington	42	1256	19	53	30	10	63	20	32	26	3	26	25	25	25
Nelson	60	1759	14	56	29	28	220	13	45	27	4	37	25	33	27
Westland	10	234	14	48	23	5	25	28	47	33	–	–	–	–	–
Canterbury	40	1232	14	74	31	8	54	16	30	25	3	29	25	25	25
Southland	40	1181	15	54	30	21	174	14	39	27	4	45	21	25	24
Overall	336	9519	10	74	29	219	2112	6	56	26	25	375	17	35	25

base of the tree) and at 5-m intervals up the stem. Basic density was determined for the whole disc and five-ring groups cut from these discs using gravimetric techniques. X-ray densitometry data were available for 17 sites and eight spacing trials. These data were obtained from either the Silviscan instrument (Evans 2000), located at CSIRO in Melbourne, or Scion's X-ray densitometer (Cown and Clement 1983) and included ring-level values of wood density and ring width.

Total tree height was generally not measured in these studies, which has presented some problems in previous analyses (e.g., Moore et al. 2014; Watt and Zoric 2010). However, it was possible to predict the total height of trees in many studies using the relationship between disc diameter and height up the stem at which the disc was sampled. For each tree, separate quadratic regression models were fitted to predict disc diameter as a function of height up the stem, and total tree height was then predicted by extrapolating these models to a diameter of zero. To ensure this procedure provided reliable predictions of tree height, the models were only fitted using discs cut at greater than 5-m height, as above this height, stem taper was found to be adequately approximated by a quadratic model. Furthermore, predictions were only made when the diameter of the uppermost disc was less than 200 mm and where the uppermost disc was sampled from at least 70 % of the predicted total tree height. This procedure made it possible to estimate total tree height at 150 of the 219 sites where disc samples had been collected at multiple heights up the stem. Relative height of each disc sample was then calculated as disc height divided by predicted total tree height.

Model development

Various component models were developed from this extensive database as described below. In most cases, a nonlinear mixed modelling approach was used with models fitted using the NLMIXED procedure in Version 9.3 of the SAS software (SAS Institute Inc. 2011).

Breast-height density by ring number

The radial variation in breast-height density was modelled as a function of ring number from pith (cambial age). The data used to develop this model consisted of breast-height density data measured in five-ring groups from 336 sites throughout New Zealand. Prior to fitting the model, values of wood density were averaged across all trees at each site for each five-ring group. The fitted model had the following form:

$$D_{BH} = a + fL + (b - a + L)\left(1 - e^{-c\text{Ring}}\right)^{d} \quad (1)$$

where D_{BH} is the basic density of a five-ring group at breast-height, Ring is the midpoint ring number from the pith for this five-ring group, and a, b, c, d, and f are model parameters estimated from the data. Because wood density varies considerably between sites (Palmer et al. 2013), a normally distributed site-level random term with a mean of zero, L, was included in the model to account for local site effects. When the model is applied to predict wood density for a particular site it is calibrated using this local parameter L. This parameter can be obtained using either the wood density index map (Palmer et al. 2013) or a wood density measurement from a sample of trees of known age (see Appendix 1 for details).

Breast-height density by ring number and ring width

An alternative breast-height density model using ring width as its principal independent variable was fitted to densitometry data averaged by ring at each of 25 sites. Data from eight of these sites were obtained from spacing trials, with each trial consisting of several replicate plots covering a range of post-thinning stand densities. Several different model forms from the published literature (e.g.,

Gardiner et al. 2011) were examined, with the following final model form selected on the basis of goodness of fit to the data:

$$D_{BH} = a + L + b(1 + cL)ln\text{Rwidth}-de^{-(f\text{Ring})} \quad (2)$$

where D_{BH} is the basic density of a given annual ring (Ring) at breast-height, Rwidth is the width of the annual ring (mm), a, b, c, d, and f are model parameters estimated from the data, and L is a normally distributed site-level random term with mean zero.

Disc density by height in the stem

By plotting whole-disc average density against relative height, it was apparent that this property exhibited a sigmoidal pattern with relative height, well-approximated by a cubic function. Therefore, a model with the following form was fitted to the data:

$$D_{disc} = (a + L) + (b + fL)(H-0.4) + c(H-0.4)^2 + d(H-0.4)^3 \quad (3)$$

where D_{disc} is the whole-disc average density, H is the relative height of the disc up the stem, and a, b, c, and d are model parameters estimated from the data. As with the radial density models, this model is calibrated for a particular site using a local parameter L.

Conversion between outerwood basic density and whole-disc average density

When linking the breast-height radial density models (given by either Eqs. (1) or (2)) with the model for predicting the variation in whole-disc average density with height up the stem (Eq. (3)), it is necessary to convert from a measure of the density of the outer five rings (outerwood basic density) to whole-disc average density at breast-height. The ratio of disc density to outerwood density would be expected to be close to one both at very young stand ages (e.g., at about age 6 years when the disc will consist of only about five annual rings) and very old stand ages (because the pith-to-bark density profile reaches an asymptotic value at a cambial age of approximately 20 years). This was modelled using the following nonlinear function:

$$\frac{D_{disc}}{D_{ow}} = 1-\frac{1}{a + b(\text{age}-6) + c(\text{age}-6)^2} \quad (4)$$

where D_{disc} is the whole-disc average density, D_{ow} is the outerwood basic density, age is the stand age (years), and a, b, and c are model parameters.

Variation in outerwood density among individual trees in a stand

In order to quantify the tree-to-tree variation in outerwood basic density, we used data from 82 stands which were sampled using two or more cores per tree. With these data, it was possible to separate out the small-scale variability of cores within a tree from the general variation between trees using variance component analysis.

The first step in this analysis was to examine the form of the statistical distribution of outerwood density between trees. Secondly, the existence of any potential relationships between easily predicted tree characteristics and wood density were examined. More specifically, the potential relationship between DBH and outerwood density was investigated by calculating correlation coefficients between these two variables for each stand. Following this, between- and within-tree variance components were estimated for each stand. Stands were classified on the basis of mean basic density into the following classes: 300–350, 351–400, 401–450, 451–500, and 501–550 kg m^{-3}, and the variance components were averaged across stands in each class. Variance components were expressed as coefficients of variation (i.e., standard deviation as a percentage of the mean).

Application of the model

The wood density models developed in this paper were designed to be coupled to growth and yield models, so that the effects of site, silviculture, and genetics on wood density can be predicted. The implementation of these models within Forecaster, which is the most widely used growth and yield prediction system for radiata pine in New Zealand (West et al. 2013), is described in detail in Appendix 1 and Appendix 2. To illustrate the utility of the models developed in this study, we used them to predict within- and among-tree variation in wood density for different radiata pine stands growing on four sites: the northern part of the North Island (Northland); the central North Island (CNI); the upper part of the South Island (Nelson); and the southern part of the South Island (Southland). Values of wood density index, site index (mean height of the 100 largest trees per hectare based on diameter at breast-height at age 20 years) and 300 index (a volume productivity index for radiata pine defined as the mean annual volume increment of a stand containing 300 stems ha^{-1} at age 30 years (Kimberley et al. 2005)) were obtained from the relevant geospatial layers (Palmer et al. 2013; Watt et al. 2010) and are given in Table 2. At each site, three different silvicultural regimes were simulated. These consisted of two different initial planting densities (500 and 1000 stems ha^{-1}). Stands were then thinned when the trees reached a mean height of 10 m. A constant thinning ratio of 2.5:1 was used resulting in residual stand densities after

Table 2 Characteristics of the four sites selected to demonstrate the utility of the models

Site	Site index (m)	300 index ($m^3\ ha^{-1}\ yr^{-1}$)	Wood density index ($kg\ m^{-3}$)
Northland	27.4	17.5	450
Central North Island	32.9	27.7	398
Nelson	28.1	24.9	402
Southland	25.4	30.2	382

thinning of 200 and 400 stems ha^{-1}, respectively. In addition, an un-thinned stand established at an initial density of 1000 stems ha^{-1} was also simulated. For each stand, whole-log average density was predicted for up to eight logs from each stem (each log had a nominal length of 5 m) and for rotation lengths of 20, 30, and 40 years.

Results

Modelling radial variation in wood density at breast-height

Basic density at breast-height showed a monotonic increase with increasing ring number from the pith, with the rate of increase diminishing after approximately 20 rings from the pith (Fig. 1). The model given by Eq. (1) explained 93 % of the variation in breast-height density for the dataset containing five-ring average values from 336 sites. This variation could be partitioned into variation between sites (which explained 45 % of the variation) and the pith-to-bark trend described by the model (which explained a further 48 % of the variation). Parameter estimates for the model are given in Table 3, and a comparison of model predictions against the five-ring average density values are shown in Fig. 1.

Although the model given by Eq. (1) performed well in general, it did not explicitly account for the effects of stand stocking on wood density. For example, in one of the spacing trials in the dataset located in Tarawera Forest (near Rotorua in the central North Island), basic density at ring ten was approximately 400 kg m^{-3} at a stand density of 200 stems ha^{-1} but almost 500 kg m^{-3} at a stand density of 2000 stems ha^{-1} (Fig. 2). There was a strong negative linear relationship between the natural

Table 3 Parameter estimates and associated standard errors (s.e.) and tests of significance for the models given by Eqs. (1)–(4)

Parameter	Estimate	s.e.	*t* value	*p* value
Eq. (1)				
a	339.9	2.5	135.96	<0.0001
b	459.2	2.5	183.68	<0.0001
c	0.1447	0.0092	15.73	<0.0001
d	2.44	0.29	8.42	<0.0001
f	0.657	0.088	7.47	<0.0001
Eq. (2)				
a	579.8	17.7	32.76	<0.0001
b	−58.61	3.59	−16.33	<0.0001
c	0.002763	0.000224	12.33	<0.0001
d	80.17	4.73	16.95	<0.0001
f	0.1845	0.0260	7.10	<0.0001
Eq. (3)				
a	390.3	2.2	177.41	<0.0001
b	−42.30	2.67	−15.84	<0.0001
c	31.93	4.07	7.85	<0.0001
d	−293.0	16.6	−17.65	<0.0001
f	−0.5181	0.0307	−16.88	<0.0001
Eq. (4)				
a	44.0	13.3	3.31	0.0013
b	−2.34	1.10	2.13	0.036
c	0.0467	0.0218	2.14	0.035

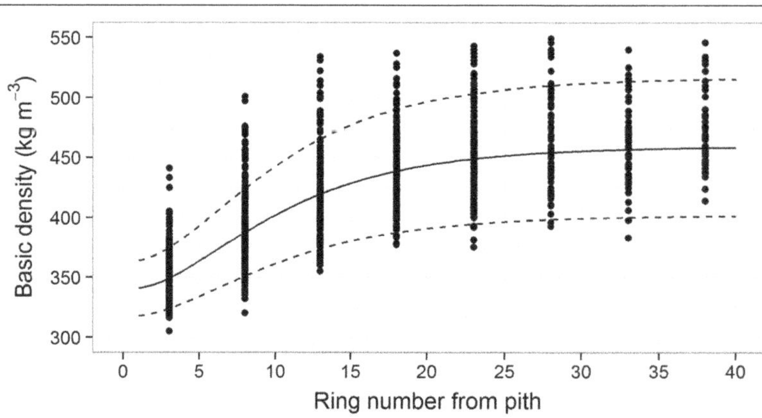

Fig. 1 Pith-to-bark radial trend in breast-height wood density. *Black dots* correspond to mean densities by five-ring group at each site. The *black lines* show the density predicted by model 1 for an average site (*middle solid line*, $L = 0$) and for sites at the 5th and 95th percentiles (*lower and upper dashed lines*, $L = -34.6$ and $L = +34.6$, respectively)

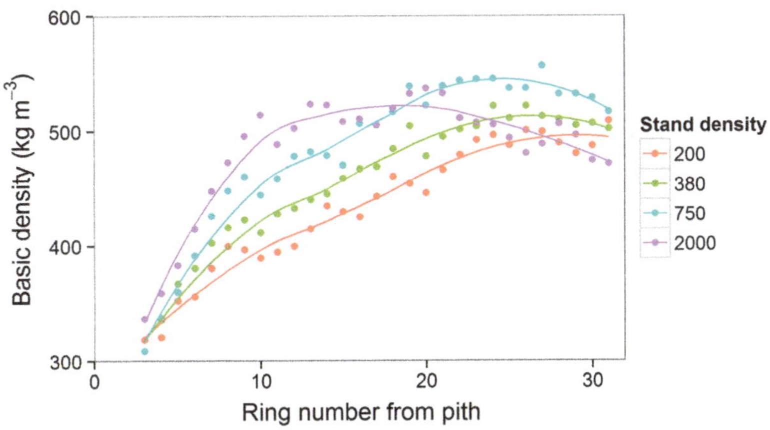

Fig. 2 Relationship between breast-height wood density and ring number from the pith for four post-thinning levels of stand density. Data are from a thinning trial in Tarawera Forest, Rotorua

logarithm of ring width and basic density in this trial, with ring width seemingly able to explain the differences in basic density between the spacing treatments (Fig. 3). Across the 25 sites for which data on ring width and density were available, there was also a strong negative relationship between the natural logarithm of ring width and basic density (Fig. 4). However, ring width did not explain site differences in basic density. For this dataset, the site effects explained 63 % of the variation in basic density, while the model given by Eq. (1) explained 88.7 % of the variation. A model using the natural logarithm of ring width in combination with site (i.e., using a local parameter L) performed better than Eq. (1), explaining 91.5 % of the variation in basic density. However, the model given by Eq. (2) (which included both ring width and ring number as independent variables) explained 93.1 % of the variation in ring-level basic density. Parameter estimates for this model are given in Table 3.

Modelling vertical variation in whole-disc average density along a tree stem

The data used to model longitudinal variation in whole-disc average density consisted of measurements of disc diameter and density made at approximately 5-m intervals in trees from stands at 219 sites throughout New Zealand. By plotting disc density against relative height, it was apparent that density has a sigmoidal pattern with relative height well-approximated by a cubic function (Fig. 5) and was modelled using Eq. (3). In this model, 45.4 % of the total variation in density between all discs in the dataset was due to differences among sites (i.e., the L parameter) and the model explained 73.2 % of the total variation in whole-disc average density. This implied that the vertical pattern predicted by the model explained an additional 27.8 % of the variation. Equation (3), which was fitted using absolute height as the independent variable, explained 71.0 % of the variation, indicating that relative height was a slightly better predictor

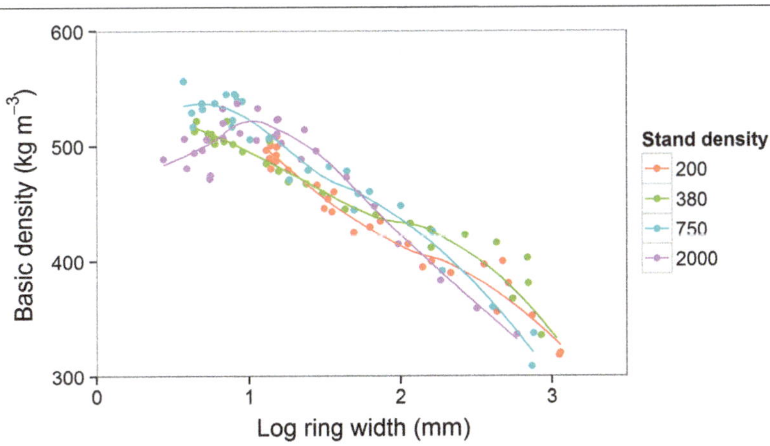

Fig. 3 Relationship between breast-height wood density and ring width (natural log base e) for four post-thinning levels of stand density. Data are from a thinning trial in Tarawera Forest, Rotorua

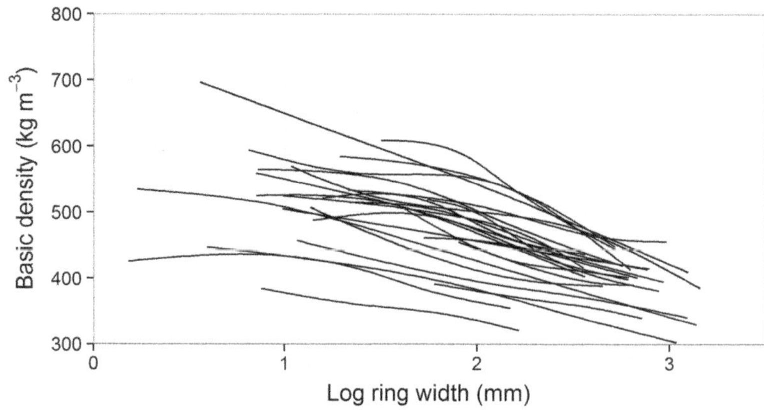

Fig. 4 Relationship between breast-height wood density and ring width from 25 sites across New Zealand. *Each line* represents a cubic spline curve fitted to data from each site

of whole-disc average density than absolute height. Furthermore, it was obvious when examining plots of density versus absolute and relative height (data not shown) that the vertical trend of whole-disc average density was more consistent between sites when using relative height as the independent variable. Parameter estimates for the model given by Eq. (3) are given in Table 3, and the model is shown plotted against the data in Fig. 5.

Converting outerwood density to whole-disc average density

The ratio of whole-disc average density to outerwood density at breast-height was found to be approximately constant with stand age, averaging 0.942 with standard error 0.0032 (Fig. 6). A simple conversion to whole-disc average density could, therefore, be achieved by multiplying the outerwood density by this mean value. However, a slightly better result was obtained using Eq. (4) with parameter estimates given in Table 3.

Variation in outerwood density among individual trees within a stand

The distribution of outerwood density among trees within a stand was approximately normal. For example, the distribution of breast-height outerwood density is shown in Fig. 7 for stands grouped into low and high density classes. Overall, the variation among trees expressed as a coefficient of variation averaged 6.46 % and did not vary with mean density. In contrast, within-tree variation (i.e., variation between multiple breast-height outerwood cores taken from the same tree) increased with mean density, or equivalently, with age (Fig. 8). This implies that as trees become larger and older, the variation in outerwood density at different locations around the stem increases. However, more importantly, the variation in mean density does not increase when expressed as a percentage of the mean density. There was almost no relationship between tree diameter and wood density within a stand. The average of the correlation coefficients calculated for each stand

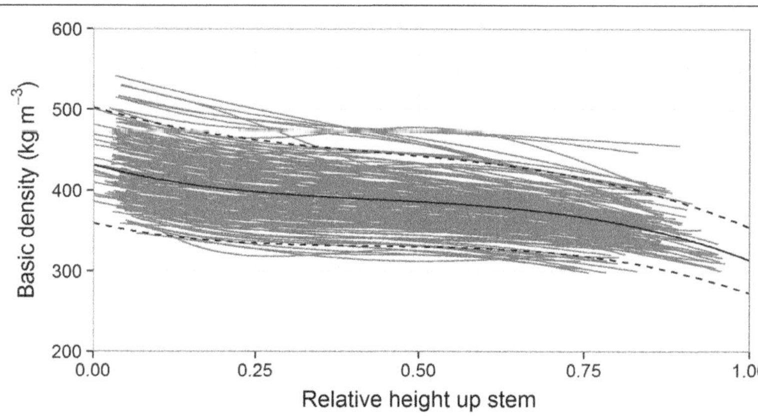

Fig. 5 Variation in whole-disc density with relative height up the stem. The *grey lines* are smoothing curves fitted to the disc densities from each of 150 sites. The *black lines* show disc density predicted by model 3 for an average site (*middle line*, $L = 0$) and for sites at the 5th and 95th percentiles (*lower and upper dashed lines*, $L = -59.3$ and $L = +59.3$, respectively)

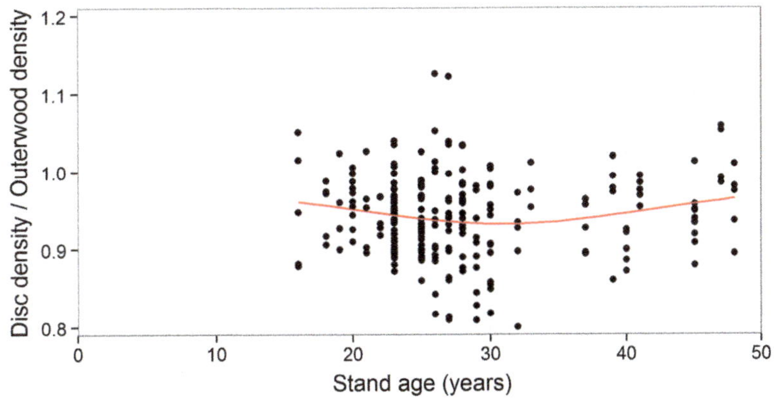

Fig. 6 Variation in the ratio between disc density and density of the outer five rings at breast-height with stand age. The red line shows Eq. (4)

was −0.07, and the individual correlation coefficients were symmetrically distributed about this mean.

Application of the model

The values of log average density predicted for the four sites followed the expected pattern given the differences in wood density index (Fig. 9). For the stand growing on a site in Northland, whole-log average wood density was 410 kg m^{-3} at age 30 years. Conversely, whole-log average wood density for the stand growing at a site in Southland was 325 kg m^{-3} at the same age. The effect of silvicultural regime was not as large as the effect of site, although the stands thinned to 200 stems ha^{-1} typically had a whole-log average wood density that was 18–30 kg m^{-3} lower than stands thinned to 400 stems ha^{-1}. The difference in wood density between un-thinned stands and stands thinned to 400 stems ha^{-1} was much less, typically 5–10 kg m^{-3}. Whole-log average density decreased with increasing log height up the stem. At a stand age of 30 years, it was approximately 20 kg m^{-3} lower in the fifth log than in the butt log at the Southland site and 50 kg m^{-3} lower at the Northland site. For an equivalent log height, there was an increase in log average density of 5–8 kg m^{-3} for each 10-year increase in rotation length.

Discussion

The models described in this paper represent a comprehensive analysis of the data collected for radiata pine over the past 50 years. While the general within- and among-stem trends in radiata pine wood density are generally well-known (Cown 1973, 1974; Cown and McConchie 1981, 1982; Cown et al. 2002; Palmer et al. 2013), this is the first time that a complete model linking tree growth and wood density has been developed for the species. This modelling system will allow the relative effects of environmental, silvicultural, and genetic effects on radiata pine wood density to be accurately quantified. The breast-height radial density model linked to the height-within-stem density model can predict densities of logs cut from any position within a stem and at any age. When linked to a growth modelling system capable of predicting ring widths at any height, it is also possible to use these models to estimate density by annual ring

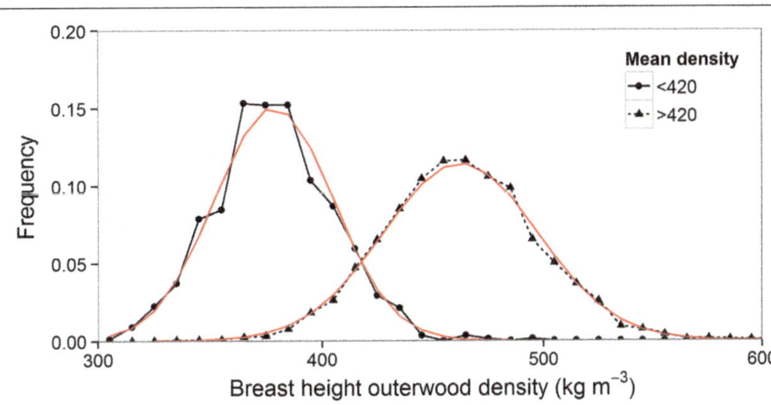

Fig. 7 Distribution of outerwood density in low (<420 kg m^{-3}) and high density (>420 kg m^{-3}) stands. For illustrative purposes, densities were subtracted from stand means and plotted around the mean for each density class. *Red lines* show fitted normal distributions

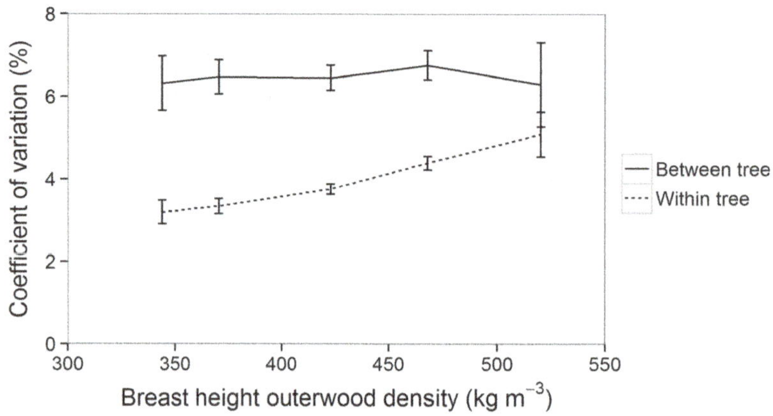

Fig. 8 Variation between and within tree (expressed as coefficients of variation) for stands in five mean density classes. *Error bars* show standard errors of each coefficient of variation

and height within stem. This can be achieved by estimating the disc densities annually, and using the predicted ring width at any height, to derive the density of each ring. Finally, the distribution of individual logs cut from different stems in a stand is modelled on the assumption that they are normally distributed with a fixed coefficient of variation. Model implementation details are explained in Appendix 1 and Appendix 2.

The purpose of the analysis reported here was to investigate the variation in wood density within and among trees within a stand. However, analysis of the breast-height five-ring and the whole-disc datasets showed that approximately 45 % of the variation in these datasets was due to differences between sites. These site effects are incorporated into the model through a local parameter, which can be estimated from density samples collected from trees or based on the location of the stand, from the national wood density index map described by Palmer et al. (2013). Differences in wood density among sites have been shown to be due to factors such as temperature, rainfall, and soil fertility (Beets et al. 2001; Beets et al. 2007; Cown et al. 1991; Harris 1965; Palmer et al. 2013). While it is conceivable that the effect of these factors on wood density might be explained by ring width as they are known to also affect productivity (Watt et al. 2010), our analysis showed that

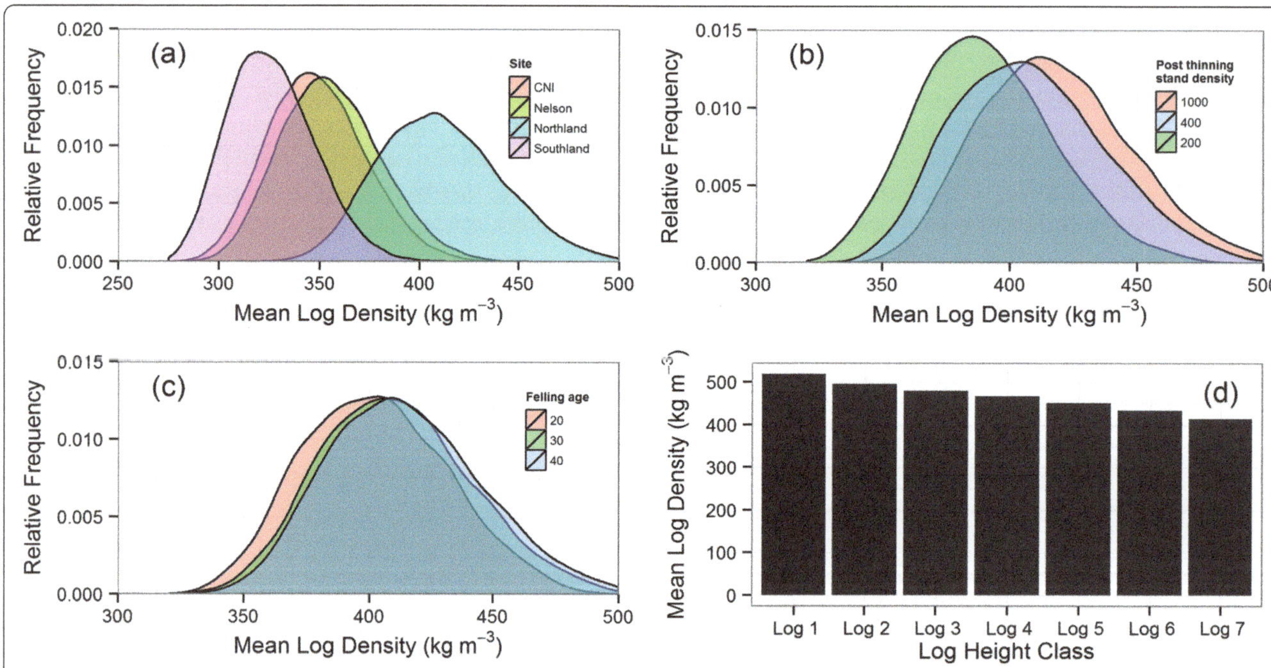

Fig. 9 Predicted values of whole-log average density **a** for stands grown at four contrasting sites, **b** under different silvicultural regimes (combinations of initial and post-thinning stand density), **c** for three different rotation lengths, and **d** at different heights up the stem (5-m height classes)

even after accounting for ring width, there was still a large amount of unexplained site-level variation (Fig. 4). Therefore, the wood density model given by Eq. (2) that included ring width as an independent variable also requires a local parameter to account for site differences.

The negative relationship between ring width and wood density, after accounting for ring number, is consistent with findings in studies on other species (e.g., Gardiner et al. 2011; Jozsa and Middleton 1994; Jyske et al. 2008; Kantavichai et al. 2010; Zhang 1998). This apparent relationship between wood density and ring width has been the subject of some controversy in the scientific literature (e.g., Sutton and Harris 1974; Wimmer and Downes 2003) where, in some cases, the negative correlation between ring number and ring width has not been accounted for. Our results showed that ring width can be used to explain the effect of growth differences on wood density among stands growing at the same site but not differences among sites. In fact, the negative relationship between ring width and wood density would suggest that lower density wood would be found on those sites where radial growth rate (all other factors being equal) is higher. However, for the same silvicultural regime both wood density and ring width are higher on warmer northern sites than on cooler southern sites (Palmer et al. 2013; Watt et al. 2010). A previous study by Cown and Ball (2001) showed that latewood percentage was the main factor associated with regional variation in wood density. More detailed investigation is required in order to better understand the relationship between ring width and wood density, particularly how within-ring density components (i.e., earlywood density, latewood density, and latewood percentage) are affected by ring width. Studies in Norway spruce (*Picea abies* Karst.) have shown that the partial correlation between ring width and density, after controlling for latewood percentage, is much lower than the simple correlation (Jyske et al. 2008; Wimmer and Downes 2003). Furthermore, the correlation between ring width and wood density fluctuates between years depending on climatic and silvicultural factors (Wimmer and Downes 2003).

This level of variation in wood density among trees within a stand confirms the results from previous studies (Cown 1999; Cown et al. 1991). Overall, there was no evidence of a correlation between wood density and diameter at breast-height. However, in a few of the stands there was a weak negative correlation between these traits. A closer examination of the few stands that showed a significant association between DBH and density indicated that this occurred only in older, highly stocked stands, containing small suppressed trees. Such trees tended to be of higher than average density producing a weak negative correlation between density and DBH in these stands. It is important to note that this result applies to individual trees within a stand, not to stands growing at different stand densities at the same site, where mean wood density is related to mean ring width. While this study found that there was generally no phenotypic correlation between wood density and diameter at breast-height, most studies in radiata pine have found that there is a negative genetic correlation between these two traits (Wu et al. 2008). The wood density model is able to assist tree breeders who want to be able to predict the implications of early selections on later age wood properties and whole-log density. At present, tree breeders rely on knowledge of age-age correlations to assist with this (Kumar and Lee 2002; Li and Wu 2005).

The radial and vertical trends in wood density were broadly similar to those observed in other studies (Burdon et al. 2004; Cown et al. 1991). Previous research has suggested that the gradients of wood properties found in radiata pine are underpinned by differences in the rate of cambial cell division, differences in the rate and duration of tracheid wall thickening, and differences in gene expression (Cato et al. 2006). In the radial direction, wood density increased rapidly with increasing ring number from the pith before reaching a quasi-asymptotic value such that little change in density occurred after approximately ring 20. This typical radial pattern has frequently been used to determine the extent of the corewood zone, which is defined as the region in which the radial rate of change in wood properties is greatest (Lachenbruch et al. 2011; Zobel and Sprague 1998). In radiata pine, corewood has typically been defined as the innermost ten growth rings from the pith (Burdon et al. 2004; Cown 1992), although a threshold density of 400 kg m^{-3} has also been proposed (Cown 1992). The results from the current study show that wood density is still increasing rapidly with cambial age at ten rings from the pith, suggesting that the arbitrary ten ring definition may be inadequate. However, other properties, such as microfibril angle, also need to be considered when determining the age of transition from corewood to outerwood (Mansfield et al. 2009). The models developed in this study can be used to further investigate the extent of the corewood zone, particularly if defined based on a wood density threshold, and how this is affected by different factors, such as environment and silviculture.

Conclusions

An extensive database of radiata pine wood density data collected over 50 years of research has enabled models to be generated to predict within-tree, within-stand, and among-stand variation in wood density. These models can be linked to growth and yield models so that forest managers can predict the impacts of factors such as site, silvicultural regime, and genetics on wood density. The research here has confirmed and quantified much of the previous knowledge on the factors that affect the variation in wood density in radiata pine.

Appendix 1

Implementation of models within a growth and yield prediction system

The density models use local parameters to calibrate them to particular sites or stands. For the breast-height radial density models, this parameter is calculated from the density at a specified ring number. This can be obtained either from a measurement of density (e.g., from outerwood density cores taken from sample trees), or more commonly, using the map showing national variation in wood density index developed by Palmer et al. (2013). For the breast-height radial density model given by Eq. (1), the local parameter is calculated by solving Eq. (1) for L, such that

$$L = \frac{D_{\mathrm{BH}} - a - (b-a)\left(1 - e^{-c\mathrm{Ring}}\right)^{d}}{f + \left(1 - e^{-c\mathrm{Ring}}\right)^{d}} \qquad \text{(A1)}$$

The wood density index is defined as breast-height outerwood density (density of the outermost five rings) for a 20-year-old stand. The mean ring number for an outerwood sample taken from breast-height in a 20-year-old stand is calculated using the following equation:

$$\mathrm{Ring} = 18 - \left(7.8 - 0.329SI + 0.00388SI^{2}\right) \qquad \text{(A2)}$$

where SI is the site index (mean top height (m) of a stand at a reference age of 20 years). The bracketed term in this equation is the number of years for a stand growing on a site with a given site index to achieve a mean height of 1.4 m. It was derived by running the 300 index growth model (Scion, unpublished data) for a range of site indices and using a quadratic function to model the relationship between site index and the age at which mean height reached 1.4 m.

For the breast-height radial density model given by Eq. (2), the local parameter is calculated by solving Eq. (2) for L such that

$$L = \frac{D_{\mathrm{BH}} - a - b\ln\mathrm{Rwidth} + de^{-f\mathrm{Ring}}}{1 + c\ln\mathrm{Rwidth}} \qquad \text{(A3)}$$

Equation (A3) requires a measurement of both wood density and ring width at a known ring number. These could be obtained from outerwood cores taken from a sample of trees, in which case the mean ring number and ring width of the density sample would be used. However, if the national wood density index map is used to calibrate the model, it is necessary to use a ring width that is compatible with the site selected on this map. Because the model given by Eq. (2) is intended to be used in conjunction with the 300 index growth model, the 300 index (Kimberley et al. 2005) and site index must be either specified by the user or estimated from a growth measurement. In either case, *Ring* is estimated using Eq. (A2), while ring width is calculated using the following equation:

$$\mathrm{Rwidth} = 11.29 + 0.263I_{300} - 0.340SI + 0.543SI^{2} - 0.00807I_{300}SI \qquad \text{(A4)}$$

where I_{300} is the 300 index (m^3 ha^{-1} yr^{-1}). This equation was developed by predicting mean ring width of the five outermost growth rings at breast-height for a 20-year-old stand using the 300 index growth model. The model was run for sites where the 300 index ranged from 15 to 40 m^3 ha^{-1} year^{-1} and site index ranged from 20 to 40 m. The silvicultural regime that these stands were grown under was assumed to be typical of those contributing to the database used to develop the wood density index geospatial layer (i.e., an initial planting density of 833 stems ha^{-1}, pruned to 6 m and thinned to a residual stand density of 400 stems ha^{-1} at age 9 years).

Once the local parameter has been estimated, the radial profile of basic density at breast-height can be predicted using the models given by either Eq. (1) or Eq. (2). When the model given by Eq. (2) is used in conjunction with a growth model, the width of each annual ring is predicted by the growth model, and these can be used to estimate the whole-disc density at breast-height as an area-weighted average of the ring-level values. When the model given by Eq. (1) is used, breast-height disc density is estimated using Eq. (4). The model given by Eq. (3) is then used to predict whole-disc average density at any position along the stem. This requires a local parameter, which is calculated by solving Eq. (3) for L, such that

$$L = \frac{\left(D_{\mathrm{disc,BH}} - a - bh - ch^{2} - dh^{3}\right)}{1 + fh} \qquad \text{(A5)}$$

where $D_{\mathrm{disc,BH}}$ is the breast-height disc density, $h = \frac{1.4}{H_{\mathrm{total}} - 0.4}$, and H_{total} is total tree height (m).

Appendix 2

Derivation of a method to calculate whole-log density from disc density

The simplest approach would be to take the arithmetic average of the density of the disc at each end of the log. However, as logs are usually tapered, slightly more weight should be given to the density of the disc at the large end of the log, compared with the disc from the small end. If both disc diameter and density vary linearly with distance along a log (a reasonable assumption over the length of a typical log), the average density of the log is

$$D_{\log} = WD_{\mathrm{disc,L}} + (1-W)D_{\mathrm{disc,S}} \tag{A6}$$

where: $D_{\mathrm{disc,L}}$ and $D_{\mathrm{disc,S}}$ are the whole-disc average density values at the large and small ends of the log, respectively, and W is the weighting factor. This weighting factor is a function of the ratio of small- to large-end diameter of the log (*SED/LED*):

$$W = \frac{\frac{(\mathrm{SED/LED})^2}{4} + \frac{\mathrm{SED/LED}}{2} + \frac{3}{4}}{(\mathrm{SED/LED})^2 + \mathrm{SED/LED} + 1} \tag{A7}$$

This result can be established using basic calculus. Without loss of generality, assume that log length is 1 and large-end diameter (LED) is 1. Let $R = \mathrm{SED/LED}$ and let the whole-disc densities of each end of log be $D_{\mathrm{disc,L}}$ and $D_{\mathrm{disc,S}}$. The density at any position x along the log is $D_{\mathrm{disc,L}} - x(D_{\mathrm{disc,L}} - D_{\mathrm{disc,S}})$ and the diameter of the log at position x is $1 - (1 - R)x$. The density of the log is the integral over the length of the log of density multiplied by the square of the diameter, divided by the integral of the diameter, i.e.,

$$\int_0^1 \left(D_{\mathrm{disc,L}} - x\left(D_{\mathrm{disc,L}} - D_{\mathrm{disc,S}}\right)\right)(1-(1-R)x)^2 dx \Big/ \int_0^1 (1-(1-R)x)^2 dx \tag{A8}$$

The solution to this integral (steps not shown) is:

$$\left((R^2 + 2R + 3)D_{\mathrm{disc,L}} - (3R^2 + 2R + 1)D_{\mathrm{dics,S}}\right) \Big/ \left(4R^2 + 4R + 4\right) = WD_{\mathrm{disc,L}} + (1-W)D_{\mathrm{disc,S}} \tag{A9}$$

Using the above result, log density will be obtained in the growth model implementation of the density model using the predicted diameter of the stem at each end of the log. In a stand-alone version of the density model, the weight W could be calculated by making reasonable assumptions on the values of log diameter, length, and taper. As an example, in a 5-m log with 400-mm SED and 6-mm/m taper, the value of W calculated using Eq. (A7) is 0.512. It can be seen from this example that a simple average of the two log end densities (i.e., assuming $W = 0.5$) would generally be adequate.

Competing interests

All authors declare that they have no competing interests.

Authors' contributions

MOK developed the modelling approach, undertook the data analysis, and contributed to writing the manuscript. DJC and RBMcK assembled the wood density data and contributed to the interpretation of the results and writing of the manuscript. JRM contributed to writing the manuscript, and LJD undertook the forecaster analysis. All authors read and approved the final version of the manuscript.

Acknowledgements

Funding for the analysis of these data was provided by Future Forests Research Ltd. We would like to thank numerous current and former colleagues who collected density samples over many years.

References

Beets, PN, Gilchrist, K, & Jeffreys, MP. (2001). Wood density of radiata pine: effect of nitrogen supply. *Forest Ecology and Management, 145*, 173–80.

Beets, PN, Kimberley, MO, & McKinley, RB. (2007). Predicting wood density of *Pinus radiata* annual growth increments. *New Zealand Journal of Forestry Science, 37*(2), 241–66.

Burdon, RD, & Harris, JM. (1973). Wood density in radiata pine clones on four different sites. *New Zealand Journal of Forestry Science, 3*(3), 286–303.

Burdon, RD, & Low, CB. (1992). Genetic survey of *Pinus radiata*. 6. Wood properties: variation, heritabilities, and interrelationships with other traits. *New Zealand Journal of Forestry Science, 22*, 228–45.

Burdon, RD, Kibblewhite, RP, Walker, JCF, Megraw, RA, Evans, R, & Cown, DJ. (2004). Juvenile versus mature wood: a new concept, orthogonal to corewood versus outerwood, with special reference to *Pinus radiata* and *P. taeda*. *Forest Science, 50*(4), 399–415.

Cato, S, McMillan, L, Donaldson, L, Richardson, T, Echt, C, & Gardner, R. (2006). Wood formation from the base to the crown in *Pinus radiata*: gradients of tracheid wall thickness, wood density, radial growth rate and gene expression. *Plant Molecular Biology, 60*(4), 565–81. doi:10.1007/s11103-005-5022-9.

Cown, DJ. (1973). Effects of severe thinning and pruning treatments on the intrinsic wood properties of young radiata pine. *New Zealand Journal of Forestry Science, 3*, 379–89.

Cown, DJ. (1974). Comparison of the effects of two thinning regimes on some wood properties of radiata pine. *New Zealand Journal of Forestry Science, 4*, 540–51.

Cown, DJ. (1980). Radiata pine: wood age and wood property concepts. *New Zealand Journal of Forestry Science, 10*, 504–7.

Cown, DJ. (1992). Corewood (juvenile wood) in *Pinus radiata*—should we be concerned? *New Zealand Journal of Forestry Science, 22*(1), 87–95.

Cown, DJ. (1999). *New Zealand pine and Douglas-fir: suitability for processing. (FRI Bulletin 216)*. Rotorua: New Zealand Forest Research Institute.

Cown, DJ, & Ball, RD. (2001). Wood densitometry of 10 *Pinus radiata* families at seven contrasting sites: influence of tree age, site, and genotype. *New Zealand Journal of Forestry Science, 31*(1), 88–100.

Cown, DJ, & Clement, BC. (1983). A wood densitometer using direct scanning with X-rays. *Wood Science and Technology, 17*(2), 91–9.

Cown, DJ, & McConchie, DL. (1981). Effects of thinning and fertiliser application on wood properties of *Pinus radiata*. *New Zealand Journal of Forestry Science, 11*, 79–91.

Cown, DJ, & McConchie, DL. (1982). Rotation age and silvicultural effects on wood properties of four stands of *Pinus radiata*. *New Zealand Journal of Forestry Science, 12*(1), 71–85.

Cown, DJ, McConchie, DL, & Young, GD. (1991). *Radiata pine wood properties survey. (FRI Bulletin 50, revised edition)*. Rotorua, New Zealand: Ministry of Forestry, Forest Research Institute.

Cown, DJ, Young, GD, & Burdon, RD. (1992). Variation in wood characteristics of 20-year-old half-sib families of *Pinus radiata*. *New Zealand Journal of Forestry Science, 22*(1), 63–76.

Cown, DJ, McKinley, RB, & Ball, RD. (2002). Wood density variation in 10 mature *Pinus radiata* clones. *New Zealand Journal of Forestry Science, 32*(1), 48–69.

Evans R. (2000). *SilviScan and its future in wood quality assessment*. Melbourne, Australia: 54th APPITA Annual General Conference 2000, (pp. 271–4). APPITA.

Gardiner, B, Leban, J-M, Auty, D, & Simpson, H. (2011). Models for predicting wood density of British-grown Sitka spruce. *Forestry, 84*(2), 119–32. doi:10.1093/forestry/cpq050.

Harris, JM. (1965). A survey of the wood density, tracheid length and latewood characteristics of radiata pine grown in New Zealand. (FRI Technical Paper 47). Rotorua: New Zealand Forest Service, Forest Research Institute.

Houllier, F, Leban, J-M, & Colin, F. (1995). Linking growth modelling to timber quality assessment for Norway spruce. *Forest Ecology and Management, 74*, 91–102.

Jozsa, LA, & Middleton, GR. (1994). *A discussion of wood quality attributes and their practical applications. Special publication no. SP-34)*. Vancouver: Forintek Canada Corp.

Jyske, T, Mäkinen, H, & Saranpää, P. (2008). Wood density within Norway spruce stems. *Silva Fennica, 42*(3), 439–55.

Kantavichai, R, Briggs, D, & Turnblom, E. (2010). Modeling effects of soil, climate, and silviculture on growth ring specific gravity of Douglas-fir on a drought-prone site in Western Washington. *Forest Ecology and Management, 259*(6), 1085–92. doi:10.1016/j.foreco.2009.12.017.

Kibblewhite, RP. (1984). Radiata pine wood and kraft pulp quality relationships. *Appita, 37*(9), 741–7.

Kimberley, M, West, G, Dean, M, & Knowles, L. (2005). The 300 index—a volume productivity index for radiata pine. *New Zealand Journal of Forestry, 50*(2), 13–8.

Kumar, S, & Lee, J. (2002). Age-age correlations and early selection for end-of-rotation wood density in radiata pine. *Forest Genetics, 9*(4), 323–30.

Lachenbruch, B, Moore, JR, & Evans, R. (2011). Radial variation in wood structure and function in woody plants, and hypotheses for its occurrence. In FC Meinzer, B Lachenbruch, & TE Dawson (Eds.), *Size- and age-related changes in tree structure and function* (pp. 121–64). Berlin: Springer-Verlag.

Li, L, & Wu, HX. (2005). Efficiency of early selection for rotation-aged growth and wood density traits in *Pinus radiata*. *Canadian Journal of Forest Research, 35*(8), 2019–29. doi:10.1139/x05-134.

Mansfield, SD, Parish, R, Di Lucca, CM, Goudie, J, Kang, KY, & Ott, P. (2009). Revisiting the transition between juvenile and mature wood: a comparison of fibre length, microfibril angle and relative wood density in lodgepole pine. *Holzforschung, 63*(4), 449–56.

Moore, JR, Cown, DJ, & McKinley, RB. (2014). Modelling microfibril angle variation in New Zealand-grown radiata pine. *New Zealand Journal of Forestry Science, 44*:25. doi:10.1186/s40490-014-0025-4.

Palmer, DJ, Kimberley, MO, Cown, DJ, & McKinley, RB. (2013). Assessing prediction accuracy in a regression kriging surface of *Pinus radiata* outerwood density across New Zealand. *Forest Ecology and Management, 308*, 9–16.

Panshin, AJ, & de Zeeuw, C. (1980). *Textbook of wood technology* (4th ed.). New York: McGraw-Hill.

Peng, M, & Stewart, JD. (2013). Development, validation, and application of a model of intra- and inter-tree variability of wood density for lodgepole pine in western Canada. *Canadian Journal of Forest Research, 43*(12), 1172–80. doi:10.1139/cjfr-2013-0208.

SAS Institute Inc. (2011). *Base SAS® 9.3 Procedures guide*. Cary, NC, USA: SAS Institute Inc.

Smith, DM. (1954). *Maximum moisture content method for determining specific gravity of small wood samples*. Madison, WI, USA: United States Department of Agriculture, Forest Service, Forest Products Laboratory. Report 2014.

Sutton, WRJ, & Harris, JM. (1974). Effect of heavy thinning on wood density in radiata pine. *New Zealand Journal of Forestry Science, 4*, 112–5.

Tian, X, & Cown, DJ. (1995). Modelling of wood properties in New Zealand. In K Klitscher, DJ Cown, & L Donaldson (Eds.), *Wood quality workshop 95. FRI bulletin 201* (pp. 72–81). Rotorua, New Zealand: Forest Research Institute.

Tian, X, Cown, DJ, & McConchie, DL. (1995). Modelling of radiata pine wood properties. Part 2: wood density. *New Zealand Journal of Forestry Science, 25*(2), 214–30.

Watt, MS, & Zoric, B. (2010). Development of a model describing modulus of elasticity across environmental and stand density gradients in plantation-grown *Pinus radiata* within New Zealand. *Canadian Journal of Forest Research, 40*(8), 1558–66.

Watt, MS, Palmer, DJ, Kimberley, MO, Hock, K, Payn, TW, & Lowe, DJ. (2010). Development of models to predict *Pinus radiata* productivity throughout New Zealand. *Canadian Journal of Forest Research, 40*(3), 488–99.

Weiskittel, AR, Hann, DW, Kershaw, JA, Jr, & Vanclay, JK. (2011). *Forest growth and yield modeling*. Chichester, UK: John Wiley & Sons, Ltd.

West, GG, Moore, JR, Shula, RG, Harrington, JJ, Snook, J, Gordon, JA, & Riordan, MP. (2013). Forest management DSS development in New Zealand. In J Tucek, R Smrecek, A Majlingova, & J Garcia-Gonzalo (Eds.), *Implementation of DSS tools into the forestry practice* (pp. 153–163). Slovakia.: Technical University of Zvolen.

Wimmer, R, & Downes, GM. (2003). Temporal variation of the ring width-wood density relationship in Norway spruce grown under two levels of anthropogenic disturbance. *IAWA Journal, 24*(1), 53–61.

Wu, HX, Ivković, M, Gapare, WJ, Matheson, AC, Baltunis, BS, Powell, MB, & McRae, TA. (2008). Breeding for wood quality and profit in *Pinus radiata*: a review of genetic parameter estimates and implications for breeding and deployment. *New Zealand Journal of Forestry Science, 38*(1), 56–87.

Xu, P, Donaldson, L, Walker, J, Evans, R, & Downes, G. (2004). Effects of density and microfibril orientation on the vertical variation of low-stiffness wood in radiata pine butt logs. *Holzforschung, 58*(6), 673–7.

Zhang, SY. (1998). Effect of age on the variation, correlations and inheritance of selected wood characteristics in black spruce (*Picea mariana*). *Wood Science and Technology, 32*, 197–204.

Zobel, BJ, & Sprague, JR. (1998). *Juvenile wood in forest trees*. Berlin: Springer.

15

Early root development of field-grown poplar: effects of planting material and genotype

Grant B. Douglas[1*], Ian R. McIvor[2] and Catherine M. Lloyd-West[1]

Abstract

Background: Poplar trees (*Populus* spp.) are used widely for soil conservation. A key advantage is their ability to establish from unrooted stem sections of varying dimensions, ranging from small cuttings to large poles. This study determined root length and biomass of young trees from three different-sized stem sections and quantified clonal variation.

Methods: Two concurrent field trials were conducted: trial 1 compared root attributes of trees from cuttings, stakes, and poles of a single poplar clone, while trial 2 compared those of trees from cuttings of six poplar clones. Excavations of entire trees were conducted in autumn for three (trial 1) or two (trial 2) years after planting.

Results: Total root mass averaged over 3 years was in the order poles (364 g) > stakes (70 g) > cuttings (17 g), and total root length was in the order poles (73 m) > stakes (21 m) > cuttings (7 m). Maximum lateral root extension was approximately 2.6 m from poles, 1.7 m from stakes, and 0.8 m from cuttings. Clonal variation in trees from cuttings was found for both mean total root mass (10.4–45.9 g) and total root length (3.5–11.8 m). In both trials, root mass and length increased, decreased, or were unchanged with increasing 0.5-m increments of the distance from stem and soil depth, depending on year, planting material, root diameter, and their interaction.

Conclusions: Early root development from poles was greater than from cuttings, with development from stakes being intermediate. Different poplar clones exhibited large variation in root biomass development within 2 years of planting. The results provide an understanding of the differences in early root development of poplar planting materials and clones used for soil conservation and other purposes and guidance on appropriate tree spacings of different planting materials to achieve root interlock.

Keywords: Pasture-tree systems, Erosion control, Root systems, Root diameter, Tree spacing

Background

Soil erosion in diverse ecosystems and landscapes modified by human activity is a significant global problem (Toy et al. 2002; Blanco and Lal 2008; Liu et al. 2011). Methods to control various erosion types and processes comprise built structures (Gebrernichael et al. 2005; Posthumus and De Graaff 2005; Yang et al. 2009; Wang et al. 2012) or the use of biological solutions, principally the establishment of a live vegetation cover (Zuazo and Pleguezuelo 2008; Stokes et al. 2014). The choice of approach depends on factors such as the extent and severity of the erosion problem, initial and ongoing costs of implementation, anticipated maintenance requirements, and expected useful life of works. Vegetation reduces erosion through above- and below-ground mechanical and hydrological effects that are governed by factors including species (e.g. growth form, morphology, root depth, and distribution), age, and spatial and temporal configurations (Bischetti et al. 2005; de Baets et al. 2009; Stokes et al. 2009; Genet et al. 2010). Vegetation used in erosion control programmes ranges from indigenous or introduced herbaceous grasses, legumes, and herbs through to woody vegetation comprising assorted shrub and tree species (Blanco and Lal 2008; Evette et al. 2009; Kuzovkina and Volk 2009; Wu et al. 2010).

* Correspondence: grant.douglas@agresearch.co.nz
[1]AgResearch, Private Bag 11008, Palmerston North, New Zealand
Full list of author information is available at the end of the article

Species and hybrids of the genus *Populus* (poplar) are important examples of trees used in erosion control and remediation programmes in many regions of the world (Wu et al. 1994; Lammeranner et al. 2005; Licht and Isebrands 2005; Reisner et al. 2007; Blanco and Lal 2008). The advantages of using poplar for erosion control include rapid growth, high evapotranspiration rates, extensive lateral root systems, abundant fine root production, and rapid tree establishment from vegetative material. There continue to be many above-ground attributes of poplars assessed, including in Europe (Pulkkinen et al. 2013; Toillon et al. 2013; Verlinden et al. 2013), China (Fang et al. 2013), and North America (Hart et al. 2013; Kaczmarek et al. 2013), mainly with respect to the rapidly developing use of short-rotation coppice systems for biofuels and other products. In contrast, fewer below-ground investigations have been conducted recently (Benomar et al. 2013; Berhongaray et al. 2013; Hajek et al. 2014; Phillips et al. 2014), and this imbalance needs addressing because root characteristics and their spatial and temporal changes are particularly important in determining efficacy for soil stabilisation (Stokes et al. 2009). Furthermore, in view of the many natural and bred poplar genotypes available (Eckenwalder 1996), variation in root characteristics between species/hybrids of poplar in the field has received scant attention (Al Afas et al. 2008; Benomar et al. 2013; Berhongaray et al. 2013; Phillips et al. 2014). Consequently, there is limited knowledge of the genetic variation in root attributes of populations of poplar which hinders identifying the most appropriate clones for stabilising soil.

Poplars have been planted in significant numbers in New Zealand since the 1960s to reduce the extent and severity of mass-movement erosion processes such as shallow landslides, earthflows, and gully erosion (Thompson and Luckman 1993). Plantings are particularly widespread on hill country supporting a pastoral cover for livestock (predominantly sheep and beef cattle) grazing enterprises, where established trees at densities of less than 100 stems per hectare (sph) provide effective slope stabilisation and enable understorey pasture production (Wilkinson 1999; Guevara-Escobar et al. 2007; Benavides et al. 2009). For example, the area of shallow landsliding was reduced by 50–80 % by trees of *Populus* spp. and species of other genera at 70 sph or greater (Hicks 1995), and trees of mostly *Populus* spp. at 32–65 sph reduced the occurrence of shallow landslides by an average of 95 % (Douglas et al. 2013). Annual pasture production beneath widely spaced poplars is reduced by 20–50 % depending mainly on tree spacing and tree size/age (Douglas et al. 2006; Wall et al. 2006; Guevara-Escobar et al. 2007).

In pasture-poplar systems, the growth and distribution of 'Veronese' poplar (*Populus deltoides* Marshall × *nigra* L.) roots on erodible slopes have been measured for entire trees aged 5–11.5 years (McIvor et al. 2008; McIvor et al. 2009), but there appear to be scant data for other clones, regardless of age. Nine months after establishing poplar clones from various planting materials (PMs) on flat, cultivated terrace soils, the mean root depth of Veronese was almost twice that of *P. deltoides* × *yunnanensis* clone 'Kawa' (Phillips et al. 2014). There were no significant differences between the clones in below-ground biomass, maximum lateral root diameter, and total root length. In a greenhouse study of up to 10 weeks, genetic variation was detected for number of root nodes, root mass, and root length among ten poplar genotypes established from 0.2-m-long stem sections (McIvor et al. 2014). Additionally, root development was reduced significantly with increasing soil bulk density, which has implications for likely field performance of the genotypes in different substrates.

Poplars can be established in the presence of grazing livestock by planting stem sections protected with a plastic sleeve (Wilkinson 1999), rather than using nursery-grown seedlings and ceasing grazing within at least the first few years following transplanting. In New Zealand, unrooted stems 3 m long, 50–70 mm diameter, and aged 2 years, henceforth referred to as poles, are typically planted vertically in grazed pastures. Smaller and younger vegetative material such as 1-m-long stems of 20–30 mm diameter and aged 1 year, henceforth referred to as stakes, may be planted where livestock are excluded (Wilkinson 1999). Stakes also have potential use in non-farm applications such as protecting roadside and railway embankments. Similarly, shorter and thinner stems, e.g. 0.4 m long and 15–20 mm diameter, henceforth referred to as cuttings, are used in agricultural and non-agricultural situations. The effect of the size of PM on early tree development, particularly in relation to root length and biomass, is poorly understood (Sidhu and Dhillon 2007; DesRochers and Tremblay 2009). It is hypothesised that over the first few years of tree establishment, poles will have significantly greater rates of above- and below-ground biomass development than stakes and cuttings.

The first objective of this study was to determine the effect of PM (poles, stakes, and cuttings) on above- and below-ground biomass development of Veronese poplar within the first 3 years after planting in the field. The second objective aimed to quantify the variation in root development of selected experimental and released poplar clones in the first 2 years after planting as cuttings. The results from two concurrent trials are reported.

Methods

Study sites

Two trials were conducted in a cultivated block on flat land at a commercial plant propagation nursery (175°39′ E, 40°21′ S), approximately 10 km from Palmerston North in the south-west of the North Island. A flat site was deemed likely to maximise tree growth and facilitate whole-tree excavation. The ground was cultivated 2 months before planting when soil water levels were high, so it did not develop a fine tilth. Data obtained from the National Institute of Water and Atmospheric Research (NIWA) showed that the site has a temperate climate, with mean monthly air temperature over the period 1981–2010 ranging from 8.6 °C in July (winter) to 18.3 °C in February (summer), and mean annual rainfall of 917 mm (NIWA 2013). The soil was a Manawatu fine sandy loam (Fluvial Recent soil, Dystric Fluventic Eutrochrept derived from greywacke alluvium) (Hewitt 1998). Soil samples from 0 < 25, 25 < 50, and 50–75 cm depth at the site were analysed for pH (1:2.1 *v*/*v* water slurry), Olsen phosphate (Olsen extraction), and sulphate-sulphur content (potassium phosphate extraction).

Trial 1: planting material comparison

Plant material and experimental design

Hybrid poplar clone Veronese was established from vegetative PM of three dimensions and ages: poles, stakes, and cuttings, as described previously. Veronese has been available commercially for 15+ years (Wilkinson 1999) and is used widely in erosion control programmes throughout New Zealand. Experience has found that the unrooted stems of Veronese are able to establish roots readily in a range of field and glasshouse conditions. Cuttings and stakes were removed from the middle of vertical stems of parent trees, and poles were obtained from lower-middle parts of stems. All PMs were dormant, obtained from a local nursery, and soaked in a water bath for 1–3 days before planting.

In August 2009, three units of each PM were planted in each of five randomised complete blocks that provided five replicates for destructive sampling in each of three consecutive years. Each PM unit was treated as an experimental unit, and in total, there were 45 experimental units in the trial. The volume (V) of each of the three PMs was estimated using the formula $V = \pi d^2 L 4^{-1}$ where d = diameter of PM and L = length of PM. A single value was calculated for each of the three PMs assuming all poles were 3 m long and 60 mm diameter (midpoint of 50–70 mm range), all stakes were 1 m long and 25 mm diameter (20–30 mm), and all cuttings were 0.4 m long and 17.5 mm diameter (15–20 mm). Units were at 2-m spacings within two adjacent rows 3 m apart, with two or three buffer rows of stakes planted either side to reduce any border effects. The PMs were planted vertically at depths of 0.3 m for cuttings, 0.5 m for stakes, and 0.8 m for poles to facilitate anchorage of the developing trees in the substrate. Soil surrounding the PM was compacted to reduce the likelihood of air pockets (van Kraayenoord et al. 1986). Weeds, principally grasses (mostly *Holcus lanatus* L.), but also legumes (e.g. *Lotus uliginosus* Schkuhr) and other broadleaf species (e.g. *Ranunculus repens* L.), were controlled by hand-weeding and application of herbicide (glyphosate at 1.8 kg a.i. ha^{-1}). For example, in the first year, weeds were sprayed once after planting and once in summer.

Measurements

Above- and below-ground attributes were measured in autumn 2010 (year (Y)1), 2011 (Y2), and 2012 (Y3). Five replicates of establishing trees from stakes and poles were measured each year whereas the death of several trees established from planted cuttings resulted in the measurement of four replicates of this PM in Y1 and Y2 and three replicates in Y3. Above-ground attributes measured were as follows: longest shoot length, henceforth referred to as shoot length (m, measured in Y1–Y3); diameter of the base of the longest shoot, henceforth referred to as shoot diameter (mm, Y1 and Y3); distance from ground level to the base of the longest shoot, henceforth referred to as shoot position (m, Y1–Y3); root collar diameter (mm, Y1); and shoot mass (g), comprising leaf and stem (Y1–Y3). All leaves and stems were severed from the PM units, and their total mass determined by oven-drying at 70 °C for 24 h. Mass of the destructively sampled PMs (g) was determined in Y1–Y3 by drying at 70 °C to a constant weight (up to 4 days).

Whole-tree root excavations were conducted using handheld implements within a series of 0.5-m radial widths radiating outwards from the establishing trees and at 0.5-m soil-depth increments (Fig. 1). Width and depth each comprised three classes of 0 < 0.5, 0.5 < 1.0, and ≥1.0 m. Roots extending beyond 1.0 m laterally or vertically were followed to their end point, often characterised by a venation of very thin roots. Roots were removed from the original PMs and classified into diameters of <1 (fibrous), 1 < 2, 2 < 5, 5 < 10, 10 < 20, and ≥20 mm. The total length of roots in each diameter class, except fibrous, was measured, and all material was oven-dried (70 °C to a constant weight) to determine root mass.

Trial 2: clonal variation comparison

Plant material and experimental design

Cuttings from six different poplar clones ('San Rosa', 'Fraser', 'Veronese', 'Kawa', 'Geyles' and 'PN471') from various poplar species/hybrids (Table 1) (Van Kraayenoord and Hathaway 1986; MAF 2011) were planted in a randomised design with five replicate cuttings of each clone in each of 2 years. Spacings between cuttings were 2 m within rows and 3 m between rows and, as in trial 1, were planted in August 2009. Experimental units were as defined previously, and the total number of units in the experiment was 60 (6 clones × 5 reps × 2 years).

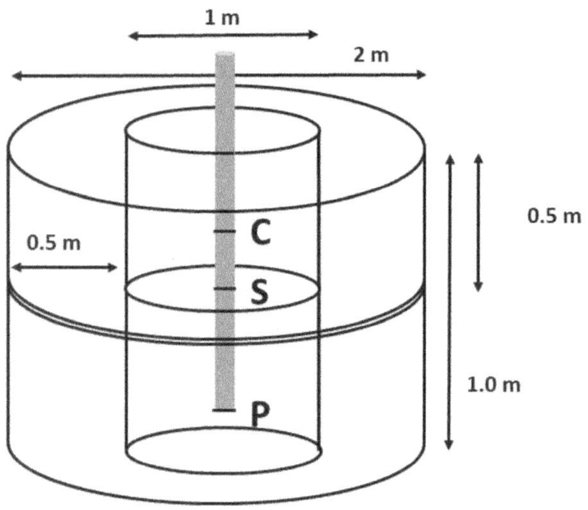

Fig. 1 Arrangement of sampling depths and widths around poplar trees established from cuttings (*C*), stakes (*S*), and poles (*P*). *Horizontal bars* on central axis represent depth of planting of cuttings (*C*, 0.3 m), stakes (*S*, 0.5 m) and poles (*P*, 0.8 m). Planting materials have different diameters (not shown). Excavations exceeded 1.0-m radius and depth when roots extended beyond shown dimensions

Planting method and depth, and plot management, were as in trial 1.

Measurements

Above-ground measurements were conducted in autumn 2010 (Y1) and 2011 (Y2) for shoot length (m), shoot diameter (mm, Y1 only), and shoot mass (g), as in trial 1. Mass of cuttings (g) was determined as previously. The number of replicates measured per clone varied in Y1 (three to five) and Y2 (one to four) because of inadvertent tree deaths occurring after planting.

Below-ground attributes were measured using the same protocol used in trial 1, including soil width and depth increments of $0 < 0.5$, $0.5 < 1.0$, and ≥ 1.0 m. Roots were classified into diameter classes of <2, $2 < 5$, $5 < 10$, and ≥ 10 mm. The total soil volume excavated for each clone was relatively low because of the small root systems developed over the initial 2 years of establishment compared to those for stakes and poles in trial 1.

Statistical analyses of data

All analyses were conducted using GenStat© 16th edition (VSN International 2014). Data for above-ground attributes were analysed using mixed effects models with a restricted maximum likelihood (REML) approach. In trial 1, data for root collar diameter, shoot mass, and PM mass were log-transformed to achieve variance homogeneity. Similarly in trial 2, shoot mass data were log-transformed. In trial 1 for above-ground attributes measured in each of 3 years, sources of variation in the model were block (4 degrees of freedom (df)), year (2 df), PM (2 df), year × PM interaction (4 df), and residual (28 df), giving a total of 40 df. The total df in models was reduced for attributes measured in one (13 df) or two (26 df) years. In trial 2, key sources of variation in the model were year (1 df), clone (5 df), and year × clone interaction (5 df). Significant terms in the fitted models were examined further using Fisher's protected least significant differences (LSDs) at the 5 % level. For transformed data, back-transformed predicted means and approximate standard errors are presented.

Root data in both trials were analysed using a two-step method, similar to that used previously (Douglas et al. 2010), comprising (1) analysis of the proportion of width × depth × year combinations (cells) with roots and (2) analysis of data for cells where roots were present. This method was used because of the large number of zero values in the datasets. Logistic regression was used to analyse presence/absence data for step 1. Data for step 2 were log-transformed and analysed using REML, and back-transformed predicted means and approximate standard errors are presented.

In trial 1, two separate analyses of root data (A, B) were conducted for step 1. Analysis A was for all PMs and

Table 1 Clones of *Populus* spp. established from cuttings

Clone	Description
Populus deltoides Marshall × *P. ciliata* (Wall. Ex Royle) clone San Rosa	New Zealand (NZ)-bred vigorous clone, best suited to drier regions because of rust susceptibility; copes well with wind.
Populus deltoides Marshall × *P. nigra* L. clone Fraser	NZ-bred, very narrow tree with a light open canopy that casts minimal shade; light stems are prone to breakage at windy sites.
Populus deltoides Marshall × *P. nigra* L. clone Veronese	European-bred, straight-stemmed, narrow-crowned tree that has good drought and wind tolerance.
Populus deltoides Marshall × *P. yunnanensis* (Dode) clone Kawa	NZ-bred, narrow-crowned tree, resistant to browsing by possums; well suited to moist slopes but not to strong winds.
Populus maximowiczii (A. Henry) × *P. nigra* L. clone Geyles	NZ-bred, straight-stemmed tree with a narrow crown, high rust resistance; grows well on moist sites.
Populus trichocarpa (Torrey and Gray) clone PN471	European-selected, American balsam poplar, narrow-crowned tree for slope stabilisation and gully plantings; not recommended for planting on windy and exposed sites.

comprised data for 3 years, two depths ($0 < 0.5$, $0.5 < 1.0$ m) and two widths ($0 < 0.5$, $0.5 < 1.0$ m). The data were restricted to a maximum of $0.5 < 1.0$ m width because trees from cuttings did not produce roots beyond 1.0 m within the study time frame. Terms in the model comprised year, depth, width, PM, and depth × width, depth × year, and depth × PM interactions, with a regression df of 11 and total df of 35. Analysis B was for pole data only and was conducted because the roots of trees from poles extended frequently beyond 1 m. This enabled three ($0 < 0.5$, $0.5 < 1.0$, ≥ 1.0 m) rather than two widths to be included in the analysis to obtain a greater understanding of variation between widths. The model used comprised terms of depth, width, year, and depth × year interaction with a regression df of 7 and total df of 89. For step 2, total values for root attributes were calculated in two ways. Firstly, data were summed over all diameter classes, soil depths, and widths to calculate total root mass (TRM) and total root length (TRL) per tree. Secondly, data were summed over all diameter (D) classes only, providing attributes of TRMD and TRLD for all other factor combinations (width and depth). As an example of models used, TRM and TRL data were analysed using a model with terms block (4 df), year (2 df), PM (2 df), PM × year interaction (4 df), and residual (28 df). There was considerable data imbalance from the absence of roots, for example, in the larger diameter classes for trees from all PMs in Y1 and from many trees from cuttings in Y2 and Y3. The widest distribution of diameter classes (<2 to ≥ 20 mm) occurred for roots from trees from poles in Y3. For these data, root mass and root length in each diameter class were analysed separately to determine the effect of depth ($0 < 0.5$, ≥ 0.5 m) and width ($0 < 0.5$, $0.5 < 1.0$, ≥ 1.0 m). The model comprised block (4 df), depth (1 df), width (2 df), depth × width interaction (2 df), and residual (17 df).

In trial 2, root presence/absence was analysed (step 1). Roots were absent or detected rarely in the larger diameter classes ($5 < 10$, ≥ 10 mm) and width zones ($0.5 < 1.0$, ≥ 1.0 m), and therefore, diameter was reduced to three classes for root mass (<2, $2 < 5$, ≥ 5 mm) and root length ($1 < 2$, $2 < 5$, ≥ 5 mm), and width from the trees was reduced to two classes ($0 < 0.5$, ≥ 0.5 m). Classification by depth was discarded because only 3 of 25 trees in Y1 and 4 of 23 trees in Y2 had root development exceeding 0.5-m soil depth. A ratio of TRL to TRM was calculated.

Results

Soil properties

Soil had a pH of 6.0 at all depths: $0 < 25$, $25 < 50$, and 50–75 cm and corresponding values for Olsen phosphate of 16, 11, and 4 μg mL^{-1}, respectively. Sulphate-sulphur levels were 1 ppm or less at the three soil depths.

Trial 1: planting material comparison

Above-ground attributes

There was a large variation in stem volume between PMs—approximately 8490 cm^3 for poles, 490 cm^3 for stakes, and 95 cm^3 for cuttings. Tree growth over the first 3 years of establishment varied markedly between PMs with the largest growth achieved from poles. Mean shoot length over 3 years was in the order poles (3.4 m) > stakes (2.9 m) > cuttings (1.7 m), and differences among PMs increased over time (PM × year interaction $P < 0.001$; Fig. 2a). In Y3, shoot length from cuttings averaged 2.3 m which was less than half that of the longest shoots from stakes (4.8 m) and poles (5.6 m).

Shoot diameter averaged over years was in the order poles (41 mm) > stakes (34 mm) > cuttings (20 mm)

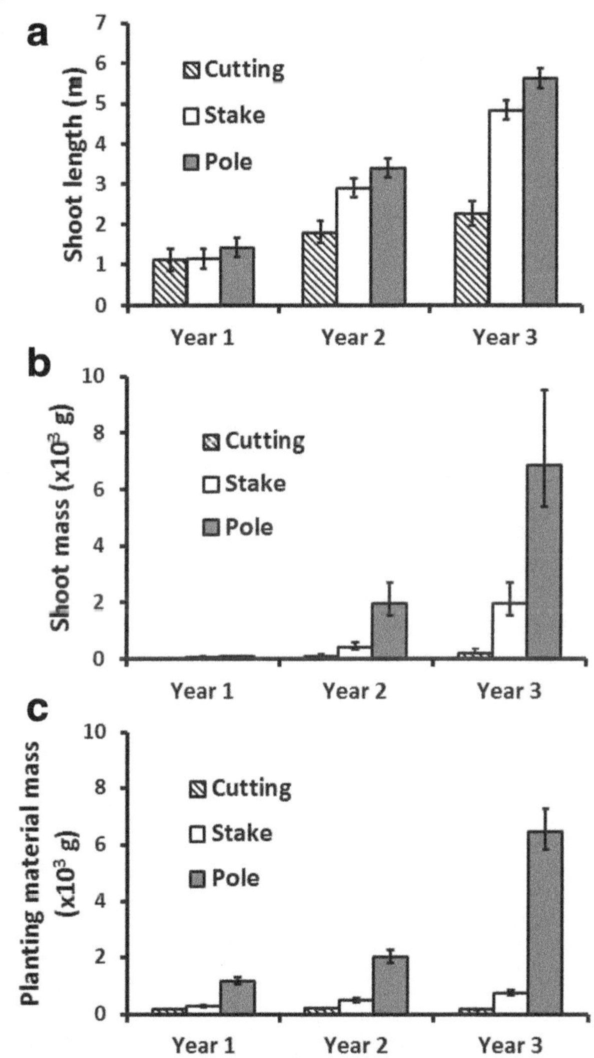

Fig. 2 Above-ground attributes of **a** shoot length, **b** shoot mass, and **c** planting material mass of Veronese poplar established from three planting materials. *Values* are means and *capped vertical lines* are standard errors

($P < 0.001$) and increased from an average across PMs of 15 mm in Y1 to 56 mm in Y3 ($P < 0.001$). Shoot diameter of PMs was similar in Y1 (range 13–18 mm) whereas, in Y3, the diameter of poles (66 mm) was 16 % greater than from stakes (57 mm) and 113 % greater than from cuttings (31 mm). The longest shoots from poles originated at an average height above the ground of 1.77 m which was significantly greater than those from cuttings (0.05 m above ground) and stakes (0.34 m). There were more and thicker shoots from PMs in the order poles > stakes > cuttings. In Y1, root collar diameter averaged 58 mm for poles, 43 mm for stakes, and 35 mm for cuttings, which all differed from each other ($P < 0.001$).

Shoot mass of trees from the three PMs in Y1 ranged from 30 to 70 g whereas, in Y2 and Y3, shoot mass was in the order poles > stakes > cuttings (PM × year interaction $P < 0.01$; Fig. 2b). Stakes and poles produced significant increases in shoot mass over the 3 years. Similar trends occurred over time for the total mass of pole PM, with a marked increase occurring between Y2 and Y3 (Fig. 2c). In contrast, between Y1 and Y3, there was only a small increase in mass of stakes and no significant change in cutting mass.

Root attributes

Root presence increments decreased with increasing width from stem and soil depth ($P < 0.001$). Roots were present within the top 0.5 m of soil in 72 % of cells for cuttings, 90 % for stakes, and all cells for poles whereas, at 0.5 < 1.0 m depth, root presence was 33 % for cuttings, 47 % for stakes, and 60 % for poles (PM × depth interaction $P = 0.014$). A decrease in root presence at 0 < 0.5 m depth with increasing width (100 % at 0 < 0.5 m width, 78 % at 0.5 < 1.0 m) was smaller than at 0.5 < 1.0 m depth (66 vs. 29 %) (depth × width interaction $P = 0.024$). In the analysis of pole data, root presence decreased with increasing width ($P < 0.001$), being 87, 77, and 50 % for 0 < 0.5, 0.5 < 1.0, and ≥1.0 m, respectively, and this trend was not influenced by other factors. There were large increases in root presence over time at 0.5 < 1.0 m depth compared with at 0 < 0.5 m depth (depth × year interaction) in the analyses comprising data for all PMs ($P = 0.001$) and poles only ($P = 0.049$). For example, root presence of poles at 0.5 < 1.0 m depth increased from 27 % in Y1 to 87 % in Y3, whereas, at 0 < 0.5 m depth, roots were found in 73 % of cells in Y1 and all cells in Y2 and Y3.

Each year, roots from cuttings did not extend beyond 0.5-m width, except for one instance where a root extended between 0.5 and 1.0 m, in contrast to those from poles where extension was >1.0 m, but no intersection between roots from different trees was observed. An exception was the intersection of the roots of two trees from poles in Y3. Root extension from stakes was intermediate between that from cuttings and poles, with roots found at both 0 < 0.5 m and 0.5 < 1.0 m widths in Y1 and Y2 and at ≥1.0 m in Y3. Roots extended between 0.5- and 1.0-m soil depth at all widths from poles in Y2 and Y3, and Y3 roots were also found at ≥1.0-m depth within 0.5 m of poles. No roots were found at this depth for the other PMs. The greatest radial distance attained by the individual roots of PMs was about 0.8, 1.7, and 2.6 m for those from cuttings, stakes, and poles, respectively. A root 4.6 m long was found in Y3 from a pole, but it was not in a straight line. Roots <5 mm diameter were produced by all PMs each year whereas roots 5 < 10 mm diameter were only found in Y2 and Y3. Roots ≥10 mm diameter were only excavated from poles in Y2 and Y3 and from stakes in Y3.

The value of TRM averaged over 3 years was in the order poles (364 g) > stakes (70 g) > cuttings (17 g) ($P < 0.001$). There was much greater variation between years for TRM of trees from poles than from cuttings and stakes (PM × year interaction $P < 0.05$; Fig. 3a). In Y3, the TRM of poles was 8-fold greater than stakes and 80-fold greater than cuttings.

The value of TRMD decreased dramatically with increasing width from trees and with increasing soil depth, and a greater increase in TRMD occurred at 0 < 0.5 m depth over time than at ≥0.5 m (depth × year interaction $P = 0.033$; Fig. 4a). In Y1 and Y3, TRMD at the shallower

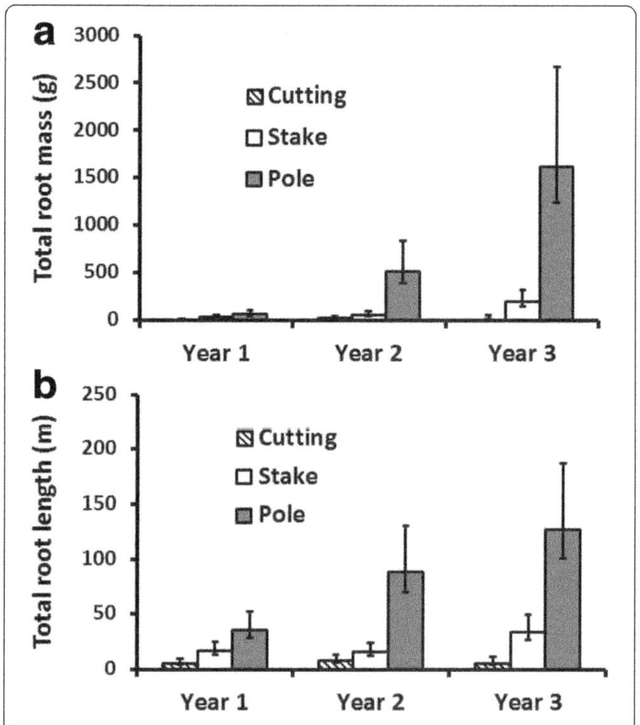

Fig. 3 Total root mass (**a**) and length (**b**) of Veronese poplar established from three planting materials. *Values* are means and *capped vertical lines* are standard errors

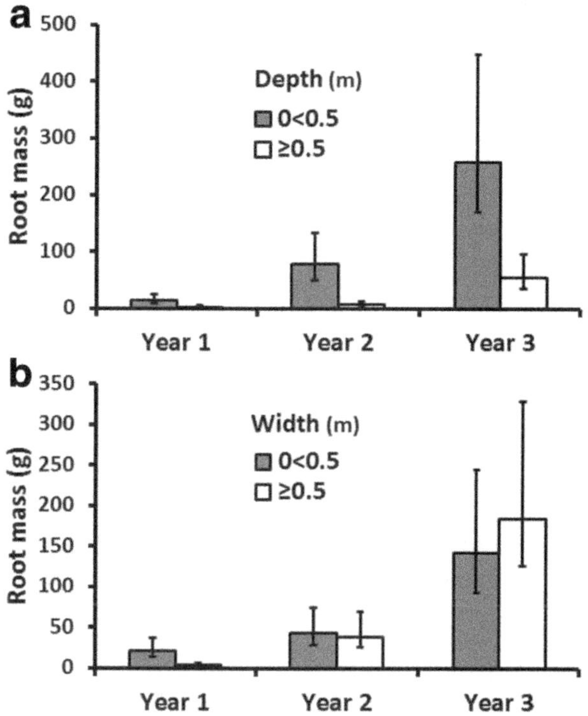

Fig. 4 Root mass of Veronese poplar with **a** soil depth and **b** width over time. *Values* are means and *capped vertical lines* are standard errors

depth was 5-fold greater than at ≥0.5-m depth, whereas, in Y2, it was 11-fold greater. TRMD in Y1 within 0.5 m of trees was significantly greater than beyond 0.5 m in contrast to other years, where no difference was found between widths (width × year interaction $P < 0.001$; Fig. 4b).

For trees established from poles, in Y3, mass of roots of <2, 2 < 5, and 5 < 10 mm diameter was 3–6-fold greater in 0 < 0.5 m soil depth than deeper in the profile (Table 2). Within 0.5 m of trees, mass of the thinnest roots was 3-fold greater than at 0.5 < 1.0 and ≥1.0 m widths, which were not significantly different from each other (Table 2). Trends for depth and width were inconsistent for these smaller-diameter classes (depth × width interaction $P < 0.05$).

The relative differences in TRL between trees established from the three PMs were less than those found for TRM but still followed a similar order of poles (73 m) > stakes (21 m) > cuttings (7 m) ($P < 0.001$), and there was also less variation between years (Fig. 3b). The values of TRLD decreased with increasing distance from trees and increasing soil depth (not presented), but responses varied with year. Over time, the difference between root lengths at 0 < 0.5 m and ≥0.5 m depths increased ($P = 0.017$; Fig. 5a). Root length at 0 < 0.5 m width exceeded that beyond 0.5 m from trees in Y1, whereas, in subsequent years, length did not vary significantly between widths (width × year interaction $P < 0.001$; Fig. 5b).

The length of roots of trees established from poles of <2, 2 < 5, and 5 < 10 mm diameter was 4–6-fold greater in 0 < 0.5 m soil depth than deeper in the profile ($P < 0.001$; Table 2). Within 0.5 m of trees, the length of 1 < 2 mm roots was over 2-fold greater than at 0.5 < 1.0 m and ≥1.0 m widths. For 5 < 10 mm roots, length did not vary significantly between widths at 0 < 0.5 m depth whereas, at ≥0.5-m depth, length beyond 1.0 m from trees (0.8 m) was significantly greater than at 0 < 0.5 m width (0.6 m) (depth × width interaction $P = 0.044$).

The partitioning of biomass between the root and shoot changed over time for the three sizes of PM, with the root as a proportion of total biomass (shoot + root, excluding PM mass) in trees from poles reducing from

Table 2 Mean root mass and length of Veronese poplar trees 3 years after establishment from poles

	Root mass (g)					Root length (m)				
	Root diameter (mm)					Root diameter (mm)				
Factor	<2	2 < 5	5 < 10	10 < 20	≥20	1 < 2	2 < 5	5 < 10	10 < 20	≥20
Soil depth (m)										
0 < 0.5	11.1^a (2.0)	29.8^a (6.4)	96.1^a (19.3)	114.1 (32.6)	120.4 (67.3)	7.3^a (1.6)	10.3^a (1.8)	6.7^a (1.4)	2.6 (0.6)	2.2^a (0.4)
≥0.5	2.0^b (0.4)	10.1^b (2.4)	28.4^b (6.4)	66.2 (25.4)	11.8 (9.1)	1.2^b (0.3)	2.6^b (0.5)	1.7^b (0.4)	1.0 (0.3)	1.1^b (0.2)
P value	*<0.001*	*0.004*	*<0.001*	0.556	0.087	*<0.001*	*<0.001*	*<0.001*	0.102	*0.040*
Width (m)										
0 < 0.5	10.7^a (2.4)	27.0 (7.1)	77.3 (19.1)	199.5 (74.7)	93.2 (54.7)	5.8^a (1.6)	8.8 (1.9)	4.2 (1.1)	3.3 (0.9)	2.6 (0.6)
0.5 < 1.0	3.5^b (0.8)	13.9 (3.9)	37.2 (9.7)	45.9 (18.1)	27.4 (21.5)	2.4^b (0.7)	4.2 (1.0)	3.4 (0.9)	1.2 (0.4)	1.4 (0.4)
≥1.0	3.7^b (0.9)	17.9 (5.3)	70.0 (19.3)	88.7 (40.0)	103.9 (180.1)	2.6^b (0.8)	5.2 (1.3)	4.0 (1.1)	2.1 (0.7)	1.1 (0.3)
P value	*0.002*	0.190	0.184	0.084	0.689	*0.047*	0.054	0.794	0.141	0.067
Depth × width										
P value	*0.029*	*0.049*	*0.010*	0.250	0.261	0.072	0.145	*0.044*	0.187	0.067

Within factors within columns, means with different letters differ significantly at $P = 0.05$. *P* values that were significant ($P < 0.05$) are shown in italics. Figures in brackets are standard errors

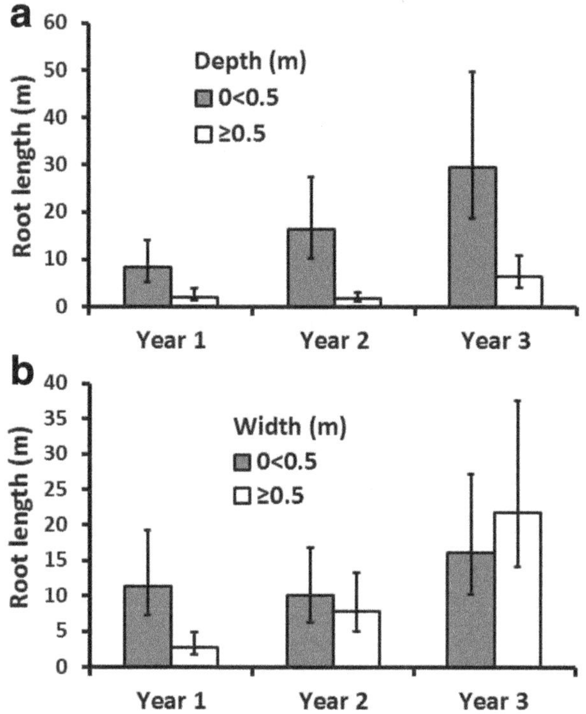

Fig. 5 Root length of Veronese poplar with **a** soil depth and **b** width over time. *Values* are means and *capped vertical lines* are standard errors

48 % in Y1 to 21 % in Y2 and 19 % in Y3. Corresponding values for stakes were 39, 13, and 9 % and for cuttings were 25, 19, and 9 %, respectively.

Trial 2: clonal variation comparison

Above-ground attributes

Clones varied in shoot length, shoot mass, and cutting mass (Table 3) but not in shoot diameter ($P = 0.078$). Mean shoot length of Kawa was greater than for San Rosa, Fraser, Veronese, and PN471, and it had greater shoot mass than all other clones, except for the mass of Geyles, which did not vary significantly from that of Kawa. The cutting mass of Veronese was twice that of San Rosa, and it was also significantly greater than for PN471 and Geyles (Table 3). Across clones, shoot length increased ($P < 0.001$) 77 % between Y1 (1.19 m) and Y2 (2.09 m), shoot mass increased 4.8-fold between Y1 (29.1 g) and Y2 (168.0 g), and cutting mass increased only slightly over the same period (133 to 144 g). Shoot diameter of clones averaged 13 (s.e. = 0.7) mm.

Root attributes

The presence of roots varied between clones, years, and widths from trees. The root presence of all clones was more than 80 % in both years except for San Rosa and PN471 in their first year (50 %) (clone × year interaction $P = 0.043$), where they only produced roots within 0.5 m of trees. In Y1, all roots of clones were <5 mm diameter except for those of San Rosa and Geyles where roots ≥5 mm were also found.

Mean TRM per tree varied between clones ($P = 0.001$) and ranged from 10.4 g for PN471 to 45.9 g for Geyles (Table 3). Clones Kawa and Geyles had similar TRM that was 2.5–4.5-fold greater than for the other clones, all of which did not vary significantly from each other. Across all clones, mean TRM in Y2 (43.5 g) was almost 5-fold greater than in Y1 (9.2 g) ($P < 0.001$).

Clones varied in mass of roots <2 and 2 < 5 mm diameter, but there were no significant differences between them for ≥5-mm roots (Table 4). The root mass of Kawa and Geyles exceeded that of most other clones. For all root diameters, mass of roots beyond 0.5 m was at least 75 % less than within 0.5 m of trees (Table 4). There was a trend of increasing root mass between Y1 and Y2, with mass of 2 < 5 mm roots increasing 5-fold ($P < 0.001$). The mass of this root class beyond 0.5-m width was 12 % of that at 0 < 0.5 m in Y1, increasing to 48 % in Y2 (width × year interaction $P < 0.001$).

The TRL of Kawa and Geyles was greater than for Fraser and PN471, and the mean for San Rosa was not significantly different from that for any other clone (Table 3). Clone Kawa had approximately 3-fold greater TRL than PN471. The length of 2 < 5 mm diameter roots of Geyles was greater than for Fraser and PN471 but was not significantly different from that for San Rosa, Veronese, and Kawa (Table 4). There was no

Table 3 Mean above-ground attributes and total root mass and length of clones of poplar established from cuttings

Clone	Shoot length(m)	Shoot mass(g)	Cutting mass(g)	Total root mass(g)	Total root length(m)
P. deltoides × *P. ciliata* San Rosa	1.52^{bc} (0.34)	48.8^{b} (23.6)	93.6^{c} (26.9)	15.2^{b} (5.0)	6.4^{abc} (1.7)
P. deltoides × *P. nigra* Fraser	1.47^{bc} (0.24)	58.7^{b} (19.5)	141.4^{abc} (19.5)	13.7^{b} (4.4)	4.1^{bc} (1.1)
P. deltoides × *P. nigra* Veronese	1.51^{bc} (0.21)	60.1^{b} (16.6)	190.0^{a} (16.5)	16.4^{b} (5.3)	7.3^{ab} (2.0)
P. deltoides × *P. yunnanensis* Kawa	2.30^{a} (0.17)	156.2^{a} (27.1)	164.7^{ab} (13.3)	39.1^{a} (12.7)	11.8^{a} (3.2)
P. maximowiczii × *P. nigra* Geyles	1.84^{ab} (0.20)	85.9^{ab} (22.4)	108.6^{c} (15.6)	45.9^{a} (14.9)	10.3^{a} (2.8)
P. trichocarpa PN471	1.20^{c} (0.22)	51.0^{b} (15.3)	132.1^{bc} (17.8)	10.4^{b} (3.4)	3.5^{c} (0.9)
P value	*0.004*	*0.021*	*0.009*	*0.001*	*0.006*

Means are averaged over 2 years. Means for shoot mass and root attributes are back-transformed from logarithms. Within columns, means with different letters differ significantly at $P = 0.05$. *P* values that were significant ($P < 0.05$) are shown in italics. Figures in brackets are standard errors

Table 4 Mean root mass and length of six clones of poplar established from cuttings

	Root mass (g)			Root length (m)		
	Root diameter (mm)			Root diameter (mm)		
Factor	<2	2 < 5	≥5	1 < 2	2 < 5	≥5
Clone						
P. deltoides × *P. ciliata* San Rosa	1.7^{ab} (0.5)	2.6^{bc} (0.9)	2.0 (1.3)	1.5 (0.4)	1.2^{abc} (0.3)	0.2 (0.1)
P. deltoides × *P. nigra* Fraser	1.0^{b} (0.3)	2.8^{bc} (0.9)	4.5 (3.0)	1.2 (0.3)	1.1^{bc} (0.3)	0.3 (0.2)
P. deltoides × *P. nigra* Veronese	1.1^{b} (0.3)	3.8^{b} (1.2)	2.0 (1.3)	1.3 (0.4)	1.5^{abc} (0.4)	0.2 (0.1)
P. deltoides × *P. yunnanensis* Kawa	2.7^{a} (0.8)	5.9^{ab} (1.9)	12.9 (8.6)	2.1 (0.6)	2.1^{ab} (0.6)	0.8 (0.6)
P. maximowiczii × *P. nigra* Geyles	2.5^{a} (0.8)	9.4^{a} (3.0)	8.6 (5.7)	1.6 (0.4)	2.8^{a} (0.8)	0.6 (0.4)
P. trichocarpa PN471	0.8^{b} (0.2)	1.5^{c} (0.5)	6.7 (4.5)	1.1 (0.3)	0.7^{c} (0.2)	0.5 (0.4)
P value	*0.027*	*0.005*	0.118	0.333	*0.013*	0.322
Width (m)						
0 < 0.5	4.2^{a} (0.7)	7.2^{a} (1.2)	9.8^{a} (3.6)	3.2^{a} (0.5)	2.1^{a} (0.3)	0.7^{a} (0.3)
≥0.5	0.5^{b} (0.1)	1.8^{b} (0.3)	2.4^{b} (0.9)	0.7^{b} (0.1)	1.0^{b} (0.1)	0.2^{b} (0.1)
P value	*<0.001*	*<0.001*	*<0.001*	*<0.001*	*0.002*	*0.006*
Year after planting						
1	1.2 (0.2)	1.5^{b} (0.3)	2.9 (1.3)	1.2 (0.2)	0.9^{b} (0.1)	0.3 (0.1)
2	1.9 (0.4)	8.9^{a} (1.7)	8.2 (3.7)	1.7 (0.3)	2.2^{a} (0.4)	0.7 (0.3)
P value	0.085	*<0.001*	0.137	0.081	*<0.001*	0.158

Within factors within columns, means with different letters differ significantly at $P = 0.05$. *P* values that were significant ($P < 0.05$) are shown in italics. Figures in brackets are standard errors

significant variation between clones for the length of the other root diameter classes. For each diameter, the length decreased 52–78 % with increasing width from trees. For all clones and in both years, roots ≥5 mm diameter made negligible contribution to mean root length. The ratio of TRL to TRM was 0.22 for Geyles, 0.30 for Kawa and Fraser, 0.34 for PN471, 0.42 for San Rosa, and 0.45 for Veronese.

Discussion

The ability to establish trees of *Populus* spp. rapidly and easily using unrooted stems is a major reason for their widespread use for stabilising erodible pastoral hill country (Wilkinson 1999). This study showed the significant benefits in early establishment from planting poles compared with shorter and thinner material in terms of above-ground (shoot length, shoot diameter, shoot mass) and below-ground (root length, root mass, lateral and vertical root distribution) attributes. The different responses were predominantly due to the large variation in stem volume between PMs, ranging from 95 cm^3 for cuttings to 8490 cm^3 for poles, which suggested significant differences in stored energy reserves available to initiate growth under suitable conditions. The differences in surface area between the PMs would have influenced the number of root initials and potential root production. Variation in planting depth of the PMs would have ensured root initiation to varying depths and subsequent access to water. These factors favour larger PM over smaller such as cuttings and aid survival. Poles would have also been able to hold more water than the two other PMs, and proportionally less of their surface area was cut and exposed to air (DesRochers and Thomas 2003), which would have slowed water loss and facilitated growth. However, larger PMs will also be at greater risk of desiccation because they produce more foliage rapidly after planting, when roots are limited (DesRochers and Tremblay 2009).

In a range of species, the carbohydrate content used for initiating and supporting adventitious root growth and shoot meristem development is related directly to the dimensions of the PM (diameter and length), although it is affected by many other factors including environment (e.g. temperature, soil moisture, soil texture), physiology (e.g. phytohormonal balances, particularly of auxin and cytokinins), condition of the parent trees from which the PM was obtained, and sampling position within parent trees (Zalesny Jr et al. 2003; Leakey 2004; da Costa et al. 2013; Zhao et al. 2014). In this study, the sampling of specific parts of parent trees to obtain the different PMs suggests that stem units of each PM had similar physiological states and propensities for root initiation and development.

Comparisons between the PMs were confounded because of factors including different depths of burial and variation in the ratio of above- to below-ground stem length. These were considered minor issues because the study aimed to compare PMs established using best practice techniques. For example, depth of pole planting in this study aligned with recommendations of planting at depths of 0.6–0.8 m or even up to one third of their length depending on site conditions such as dryness and exposure to wind (Hathaway 1986; National Poplar and Willow Users Group 2007).

The development of fewer and thinner shoots from the smaller PMs was probably because of relatively low stored energy reserves and fewer epicormic buds (Meier et al. 2012) along the shorter stem sections. Greater shoot initiation and growth from the upper compared with lower parts of all PMs may have been in response to slightly elevated irradiation higher up the stems; variation in bark thickness along the above-ground stem sections, particularly for the longer poles; and differential distribution of soluble sugars and carbohydrate reserves (Morisset et al. 2012). Field observations in Y3 indicated that the significantly greater shoot mass produced by trees from poles than from cuttings and stakes was likely because of more and larger shoots and the greater total photosynthetic leaf area. No carbohydrates were assayed, but the contribution to growth of reserves and current photosynthate (Tromp 1983; Magel et al. 2000) in trees from poles was likely considerably greater than in those from the other PMs.

The lower variation among PMs and among years within each PM for root length compared with root mass was considered due to the predominance of roots <5 mm contributing to the length in all treatments and the investment in root diameter of trees established from poles in Y3. The ≥10-mm roots contributed 55 % to root mass but only 10 % to root length. Proliferation of fine roots is essential for enhancing water and nutrient uptake, but there is variation within and among seasons in their longevity and mass depending on soil factors (e.g. water content), tree species, age, and root diameter (Block et al. 2006). Conducting root excavations at the end of the growing season each year, as was done in this study, would have minimised any confounding effects when comparing responses between years. The increased partitioning of biomass to shoots and reduced allocation to roots with increasing age, found in trial 1, has been reported for young hybrid poplars grown from cuttings from plantation stock in North America (Wullschleger et al. 2005). Such results have implications for growth and survival of young trees in dry environments.

Root extension is important in aiding tree survival and growth. The larger PM produced greater root extension and accessed a greater soil volume for water and nutrients than smaller PMs over the same time period. The results reflected the greater resources available in the larger PM. Greater abundance of roots in shallow soil layers than deeper in the soil profile is a common occurrence because of factors such as enhanced soil organic matter content and higher soil nutrient status (e.g. trends for Olsen phosphate in this study) (Block et al. 2006; Johnston et al. 2009; McIvor et al. 2009). The spacings used were appropriate to allow development of young root systems unaffected by those of neighbouring trees because, for all trees except for two from poles in Y3, no intersection between roots from different trees was observed.

The variation among clones grown from cuttings in trial 2 supported findings for field-grown hybrid poplar clones evaluated elsewhere (Pregitzer et al. 1990; Dickmann et al. 1996; Wullschleger et al. 2005; Phillips et al. 2014). Clones Kawa and Geyles had relatively high above- and below-ground growth. Kawa is one of the main poplar clones grown in New Zealand for soil stabilisation and other purposes (National Poplar and Willow Users Group 2007), and this study demonstrated its ability to rapidly establish an extensive root system. Clone Geyles would also seem worthy of further appraisal for root proliferation in soil and was released for commercialisation in 2011 (McIvor et al. 2011). The role of cutting mass in contributing to the shoot and root growth of Kawa and Geyles was uncertain because Geyles had 30 % lower cutting mass than Kawa whereas the two clones did not differ significantly with respect to shoot length, shoot mass, and root attributes. Veronese is grown for a range of applications (National Poplar and Willow Users Group 2007), but this study found that its early growth from cuttings was less than Kawa for shoot length and shoot mass and less than Kawa and Geyles for total root mass. The results suggest that where cuttings are used in applications, clones such as Kawa and Geyles could be used as alternatives to plantings of Veronese or be included in mixtures with Veronese. Phillips et al. (2014) found that 9-month-old Veronese and Kawa trees established from 0.5-m-long stem sections were similar for above- and below-ground attributes, except for maximum root depth where Veronese was about 40 % greater than Kawa (1.0 vs. 0.7 m). They did not recommend a particular clone for planting. Based on its overall low above- and below-ground growth, PN471 could not be recommended for providing rapid and extensive early root development when grown from cuttings and could be difficult to establish from cuttings in regions where summer rainfall is low and irrigation is not available.

The development of relatively thick roots (≥5 mm diameter) in Y1, as shown by trees of San Rosa and Geyles, could be advantageous for storing reserves for maintenance and growth, enhancing anchorage of the developing tree to the soil, and providing a foundation for further development and position within the soil of thinner roots essential for water and nutrient uptake

(Stokes et al. 2009). The differences between all clones in TRL:TRM ratio indicated clonal variation in root development and morphology. Higher TRL:TRM ratio for clones such as Veronese and San Rosa suggested that they had more numerous roots, particularly fine roots, than clones such as Geyles and Kawa. It is possible that the focus of some clones is root extension into new nutrient sources, whereas other clones tend to exploit smaller nutrient sources but to a greater degree. This has possible implications for risk from drought in the first few establishment years.

The death of trees established from cuttings could have been due to several factors. Cuttings had potentially fewer reserves to initiate growth above and below ground. They also had shorter stem lengths buried than stakes and poles, presenting lower surface areas for potential root development and extraction of soil water and nutrients. Inspection of cuttings and the other PMs within 3 months of planting found that all were alive and therefore that post-planting factors were responsible for any tree mortality. There was no evidence of pest or disease attack, the area of the trials was not grazed by livestock, and browsing damage by small mammals was not observed.

Although soil pH at the trial sites was within the range of 5.5 to 7.0 recommended for poplar-nursery production (van Kraayenoord et al. 1986), levels of Olsen phosphate were low, particularly at 50–75-cm soil depth, and levels of sulphate-sulphur were very low. Raising the levels of phosphate and sulphur through strategic fertiliser application may have accentuated the differences found between the PMs in root and shoot responses (van den Driessche 1999; Guillemette and DesRochers 2008), particularly by favouring trees from poles with their larger and more extensive root systems and consequent exploration of greater soil volumes.

Poles are more costly to produce than cuttings and stakes and are heavier and bulkier to transport. However, these disadvantages are more than compensated for in field applications with their potentially higher survival and lower maintenance requirements. Also, the planting of poles (with sleeve protection) is the only practicable option for establishment in pastoral areas continuing to be grazed by livestock (Wilkinson 1999).

In a comparison of poplar PMs at Gisborne on the east coast of the North Island, Phillips et al. (2014) found that 9 months after planting, the above- and below-ground growth of trees from poles exceeded that from stakes and cuttings, as found presently. Both studies also found no significant differences in root length between clones Veronese and Kawa. However, growth at Gisborne was considerably greater than found in this study with, for example, root collar diameter of Veronese PMs at Gisborne (77–104 mm) being almost 2-fold greater than in Y1 (35–58 mm). At Gisborne, TRL ranged from 155 m for trees established from stakes to 255 m for trees established from poles, compared with less than 40 m for trees from any PM after one growing season in this study. At Gisborne, roots extended 3 to 5 m from the stem after 9 months, whereas, currently, maximum radial extension was 2.6 m. Finally, the lowest estimates of TRM of Veronese at Gisborne ranged from 1.50 to 3.05×10^3 g, which were considerably greater than those found in Y1. By Y3, pole TRM was at the lower end of the earlier estimates (Phillips et al. 2014). The much greater tree growth at Gisborne than Palmerston North may have been because of higher soil nutrient status, higher average air temperatures (long-term values of 12 to 23 °C at Gisborne; 8.6 to 18.3 °C at Palmerston North), irrigation during summer (Gisborne only), and loose structured soil at Gisborne compared with more compacted upper soil layers at Palmerston North, which would have facilitated root extension.

The rapid interlocking of roots of neighbouring trees on man-made or natural slopes is an important attribute to minimise the risk of mass movement erosion of the ground between adjacent trees (Reubens et al. 2007; Stokes et al. 2009). The lower lateral root extension from stakes and particularly from cuttings, compared with that from poles, indicated that these smaller PMs need to be planted at much closer spacings than poles to achieve, within a given time period, a similar amount of root interlocking in a defined volume of shallow soil. The results indicated that for roots of trees established from each PM to touch those of neighbours by Y3, spacings would need to be about 5.2 m for poles (370 sph), 3.4 m for stakes (865 sph), and 1.6 m for cuttings (3900 sph). Spacings of established poplar trees on erodible pastoral hill country are often 10–15 m (100–45 sph), which are effective in reducing erosion processes such as shallow landslides (Cairns et al. 2001; Douglas et al. 2013). These spacings are achieved by planting at closer spacings with thinning later or the more common (but higher risk) approach of planting at final spacings and relying on near 100 % survival to achieve desired root coverage of the slope. There would also need to be negligible mass movement erosion in the meantime. Another important element in slope stabilisation by woody vegetation is vertical root development and its potential penetration and anchoring into underlying bedrock (Reubens et al. 2007; Stokes et al. 2009). In trial 1, the early soil conservation advantage of poles was shown further by the development in Y3 of vertical roots from the bottom or near bottom of poles that extended deeper than 1 m within 0.5-m radius of poles. The ability of these roots to penetrate stable strata would depend on factors such as their depth, pedological structure, and the presence of cracks in the strata.

Conclusions

Poles of Veronese hybrid poplar produced trees that had considerably greater above- and below-ground growth and development than those from cuttings within the first 3 years after planting. Trees from stakes were intermediate between poles and cuttings in their growth and development. After 3 years, maximum lateral root extension was about 0.8, 1.7, and 2.6 m for trees grown from cuttings, stakes, and poles, respectively. This indicated that large differences in spacing would be required to achieve similar levels of contact between roots of neighbouring trees established from the different PMs over the same period of time. Survival of trees from cuttings was <100 % whereas all trees from stakes and poles survived. The proportion of biomass allocated to roots decreased with age for trees from the three PMs. Among six different clones from cuttings, there was a 4.4-fold variation in TRM and a 3.4-fold variation in TRL. Hybrid poplar clones Kawa and Geyles produced the greatest amount of above- and below-ground biomass. This study indicates that there are poplar clones that root more readily and produce greater root mass in the first year than Veronese; they should be considered for greater commercial production than occurs presently.

The findings on early root development of poplar add to the very limited information available. Different-sized stem sections are used in a range of applications, and it is important to know how they influence root development spatially and temporally. This impacts directly on decisions regarding initial spacing of stem sections to achieve interlock of roots within a specific time frame. Practitioners will have to accommodate the delay in time of root interlock of trees established from stakes, and particularly cuttings, in their applications. The data for Veronese complement those for trees aged 5+ years collected for trees established from poles on slopes and will enable an improved understanding of root development over time. Results are for clones used widely in New Zealand, such as Veronese and Kawa, and therefore have immediate applicability.

Abbreviations

LSD: least significant difference; PM: planting material; REML: restricted maximum likelihood approach; sph: stems per hectare; TRL: total root length; TRLD: total root length diameter; TRM: total root mass; TRMD: Total root mass diameter; Y1: year 1; Y2: year 2; Y3: year 3.

Competing interests

The authors declare that they have no competing interests.

Authors' contributions

GD participated in the design of the study, data acquisition, and drafted the manuscript. IM conceived of the study, participated in its design, coordination and data acquisition, and helped to draft the manuscript. CL-W conducted the statistical analyses and helped draft the section on analyses. All authors read and approved the final manuscript.

Acknowledgements

Funding from New Zealand Foundation for Research, Science and Technology's 'Sustainable Land Use Research Initiative' (SLURI) programme (Contract No. C02X0813, Objective 2); large teams of post-graduate students from Massey University, Palmerston North, for assisting with tree excavations and sample processing; Drs. Mike Dodd and Andrew Wall, AgResearch, and two anonymous reviewers for constructive comments and helpful suggestions.

Author details

[1]AgResearch, Private Bag 11008, Palmerston North, New Zealand. [2]Plant and Food Research, Private Bag 11600, Palmerston North, New Zealand.

References

Al Afas, N., Marron, N., Zavalloni, C., & Ceulemans, R. (2008). Growth and production of a short-rotation coppice culture of poplar—IV: Fine root characteristics of five poplar clones. *Biomass and Bioenergy, 32*(6), 494–502.

Benavides, R., Douglas, G. B., & Osoro, K. (2009). Silvopastoralism in New Zealand: Review of effects of evergreen and deciduous trees on pasture dynamics. *Agroforestry Systems, 76*, 327–350.

Benomar, L., DesRochers, A., & Larocque, G. (2013). Comparing growth and fine root distribution in monocultures and mixed plantations of hybrid poplar and spruce. *Journal of Forestry Research, 24*(2), 247–254.

Berhongaray, G., Janssens, I. A., King, J. S., & Ceulemans, R. (2013). Fine root biomass and turnover of two fast-growing poplar genotypes in a short-rotation coppice culture. *Plant and Soil, 373*(1–2), 269–283.

Bischetti, G., Chiaradia, E., Simonato, T., Speziali, B., Vitali, B., Vullo, P., et al. (2005). Root strength and root area ratio of forest species in Lombardy (Northern Italy). *Plant and Soil, 278*, 11–22.

Blanco, H., & Lal, R. (2008). *Principles of soil conservation and management.* Dordrecht, Netherlands: Springer Science+Business Media B.V.

Block, R., Van Rees, K., & Knight, J. (2006). A review of fine root dynamics in *Populus* plantations. *Agroforestry Systems, 67*(1), 73–84.

Cairns, I., Handyside, B., Harris, M., Lambrechtsen, N., & Ngapo, N. (2001). *Soil conservation technical handbook* (New Zealand Ministry for the Environment Series; no. 404). Wellington: New Zealand Association of Resource Management.

da Costa, C. T., de Almeida, M. R., Ruedell, C. M., Schwambach, J., Maraschin, F. S., & Fett-Neto, A. G. (2013). When stress and development go hand in hand: main hormonal controls of adventitious rooting in cuttings. *Frontiers in Plant Science, 4*, 1–19.

de Baets, S., Poesen, J., Reubens, B., Muys, B., de Baerdemaeker, J., & Meersmans, J. (2009). Methodological framework to select plant species for controlling rill and gully erosion: Application to a Mediterranean ecosystem. *Earth Surface Processes and Landforms, 34*(10), 1374–1392.

DesRochers, A., & Thomas, B. (2003). A comparison of pre-planting treatments on hardwood cuttings of four hybrid poplar clones. *New Forests, 26*(1), 17–32.

DesRochers, A., & Tremblay, F. (2009). The effect of root and shoot pruning on early growth of hybrid poplars. *Forest Ecology and Management, 258*(9), 2062–2067.

Dickmann, D. I., Nguyen, P. V., & Pregitzer, K. S. (1996). Effects of irrigation and coppicing on above-ground growth, physiology, and fine-root dynamics of two field-grown hybrid poplar clones. *Forest Ecology and Management, 80*(1–3), 163.

Douglas, G. B., Walcroft, A. S., Hurst, S. E., Potter, J. F., Foote, A. G., Fung, L. E., et al. (2006). Interactions between widely spaced young poplars (*Populus* spp.) and introduced pasture mixtures. *Agroforestry Systems, 66*, 165–178.

Douglas, G. B., McIvor, I. R., Potter, J. F., & Foote, L. G. (2010). Root distribution of poplar at varying densities on pastoral hill country. *Plant and Soil, 333*, 147–161.

Douglas, G. B., McIvor, I. R., Manderson, A. K., Koolaard, J. P., Todd, M., Braaksma, S., et al. (2013). Reducing shallow landslide occurrence in pastoral hill country using wide-spaced trees. *Land Degradation and Development, 24*, 103–114.

Eckenwalder, J. E. (1996). Systematics and evolution of *Populus*. In R. F. Stettler, H. D. Bradshaw Jr., P. E. Heilman, & T. M. Hinckey (Eds.), *Biology of Populus and its implications for management and conservation* (pp. 7–32). Ottawa: NRC Research Press.

Evette, A., Labonne, S., Rey, F., Liebault, F., Jancke, O., & Girel, J. (2009). History of bioengineering techniques for erosion control in rivers in Western Europe. *Environmental Management, 43*(6), 972–984.

Fang, S., Zhai, X., Wan, J., & Tang, L. (2013). Clonal variation in growth, chemistry and calorific value of new poplar hybrids at nursery stage. *Biomass and Bioenergy, 54*, 303–311.

Gebremichael, D., Nyssen, J., Poesen, J., Deckers, J., Haile, M., Govers, G., et al. (2005). Effectiveness of stone bunds in controlling soil erosion on cropland in the Tigray Highlands, northern Ethiopia. *Soil Use and Management, 21*(3), 287–297.

Genet, M., Stokes, A., Fourcaud, T., & Norris, J. E. (2010). The influence of plant diversity on slope stability in a moist evergreen deciduous forest. *Ecological Engineering, 36*(3), 265–275.

Guevara-Escobar, A., Kemp, P. D., Mackay, A. D., & Hodgson, J. (2007). Pasture production and composition under poplar in a hill environment in New Zealand. *Agroforestry Systems, 69*, 199–213.

Guillemette, T., & DesRochers, A. (2008). Early growth and nutrition of hybrid poplars fertilized at planting in the boreal forest of western Quebec. *Forest Ecology and Management, 255*(7), 2981–2989.

Hajek, P., Hertel, D., & Leuschner, C. (2014). Root order- and root age-dependent response of two poplar species to belowground competition. *Plant and Soil, 377*(1–2), 337–355.

Hart, J., de Araujo, F., Thomas, B., & Mansfield, S. (2013). Wood quality and growth characterization across intra- and inter-specific hybrid aspen clones. *Forests, 4*(4), 786–807.

Hathaway, R. L. (1986). Plant establishment. In C. W. S. Van Kraayenoord & R. L. Hathaway (Eds.), *Water and soil miscellaneous publication no. 93* (pp. 21–37). Wellington, New Zealand: Water and Soil Directorate, Ministry of Works and Development.

Hewitt, A. E. (1998). *New Zealand soil classification.* (2nd ed., Landcare Research Science Series No. 1.). Lincoln, Canterbury, New Zealand: Manaaki Whenua Press.

Hicks, D. L. (1995). *Control of soil erosion on farmland: a summary of erosion's impact on New Zealand agriculture, and farm management practices which counteract it (MAF Policy Technical Paper 95/4).* Wellington, New Zealand: MAF Policy, Ministry of Agriculture.

Johnston, A., Poulton, P., & Coleman, K. (2009). Soil organic matter: its importance in sustainable agriculture and carbon dioxide fluxes. *Advances in Agronomy, 101*, 1–57.

Kaczmarek, D. J., Coyle, D. R., & Coleman, M. D. (2013). Survival and growth of a range of *Populus* clones in central South Carolina USA through age ten: Do early assessments reflect longer-term survival and growth trends? *Biomass and Bioenergy, 49*, 260–272.

Kuzovkina, Y. A., & Volk, T. A. (2009). The characterization of willow (*Salix* L.) varieties for use in ecological engineering applications: Co-ordination of structure, function and autecology. *Ecological Engineering, 35*, 1178–1189.

Lammeranner, W., Rauch, H., & Laaha, G. (2005). Implementation and monitoring of soil bioengineering measures at a landslide in the middle mountains of Nepal. *Plant and Soil, 278*(1–2), 159–170.

Leakey, R. R. B. (2004). Physiology of vegetative reproduction. In J Burley, J Evans, & J Youngquist (Eds.), *Encyclopaedia of Forest Sciences* (pp. 1655–1668). Oxford, UK: Elsevier Academic Press.

Licht, L. A., & Isebrands, J. G. (2005). Linking phytoremediated pollutant removal to biomass economic opportunities. *Biomass and Bioenergy, 28*(2), 203–218.

Liu, X., Zhang, S., Zhang, X., Ding, G., & Cruse, R. M. (2011). Soil erosion control practices in Northeast China: A mini-review. *Soil and Tillage Research, 117*, 44–48.

MAF (2011). *Poplars for the farm. Specially bred clones for New Zealand conditions.* Poplar [Brochure No. 5]. http://www.poplarandwillow.org.nz/documents/brochure-5-poplars-for-the-farm.pdf. Accessed 27 June 2014.

Magel, E., Einig, W., & Hampp, R. (2000). Carbohydrates in trees. In G. Anil Kumar & K. Narinder (Eds.), *Developments in crop science* (Vol. 26, pp. 317–336). Amsterdam, Oxford: Elsevier.

McIvor, I. R., Douglas, G. B., Hurst, S. E., Hussain, Z., & Foote, A. G. (2008). Structural root growth of young Veronese poplars on erodible slopes in the southern North Island, New Zealand. *Agroforestry Systems, 72*, 75–86.

McIvor, I. R., Douglas, G. B., & Benavides, R. (2009). Coarse root growth of Veronese poplar trees varies with position on an erodible slope in New Zealand. *Agroforestry Systems, 76*, 251–264.

McIvor, I. R., Hedderley, D. I., Hurst, S. E., & Fung, L. E. (2011). Survival and growth to age 8 of four *Populus maximowiczii* × *P. nigra* clones in field trials on pastoral hill slopes in six climatic zones of New Zealand. *New Zealand Journal of Forestry Science, 41*, 151–163.

McIvor, I. R., Sloan, S., & Pigem, L. R. (2014). Genetic and environmental influences on root development in cuttings of selected *Salix* and *Populus* clones – a greenhouse experiment. *Plant and Soil, 377*(1–2), 25–42.

Meier, A. R., Saunders, M. R., & Michler, C. H. (2012). Epicormic buds in trees: a review of bud establishment, development and dormancy release. *Tree Physiology, 32*(5), 565–584.

Morisset, J. B., Mothe, F., Bock, J., Bréda, N., & Colin, F. (2012). Epicormic ontogeny in *Quercus petraea* constrains the highly plausible control of epicormic sprouting by water and carbohydrates. *Annals of Botany, 109*(2), 365–377.

NIWA. (2013). *Climate data and activities.* http://www.niwa.co.nz/education-and-training/schools/resources/climate. Accessed 21 May 2014.

Phillips, C., Marden, M., & Suzanne, L. (2014). Observations of root growth of young poplar and willow planting types. *New Zealand Journal of Forestry Science, 44*, 15.

National Poplar and Willow Users Group. (2007). *Growing poplar and willow trees on farms. Guidelines for establishing and managing poplar and willow trees on farms.* Palmerston North, New Zealand: National Poplar and Willow Users Group.

Posthumus, H., & De Graaff, J. (2005). Cost-benefit analysis of bench terraces, a case study in Peru. *Land Degradation & Development, 16*(1), 1–11.

Pregitzer, K. S., Dickmann, D. I., Hendrick, R., & Nguyen, P. V. (1990). Whole-tree carbon and nitrogen partitioning in young hybrid poplars. *Tree Physiology, 7*, 79–93.

Pulkkinen, P., Vaario, L.-M., Koivuranta, L., & Stenvall, N. (2013). Elevated temperature effects on germination and early growth of European aspen (*Populus tremula*), hybrid aspen (*P. tremula* × *P. tremuloides*) and their F2-hybrids. *European Journal of Forest Research, 132*(5–6), 791–800.

Reisner, Y., de Filippi, R., Herzog, F., & Palma, J. (2007). Target regions for silvoarable agroforestry in Europe. *Ecological Engineering, 29*(4), 401–418.

Reubens, B., Poesen, J., Danjon, F., Geudens, G., & Muys, B. (2007). The role of fine and coarse roots in shallow slope stability and soil erosion control with a focus on root system architecture: A review. *Trees - Structure and Function, 21*(4), 385–402.

Sidhu, D. S., & Dhillon, G. P. S. (2007). Field performance of ten clones and two sizes of planting stock of *Populus deltoides* on the Indo-gangetic plains of India. *New Forests, 34*(2), 115–122.

Stokes, A., Atger, C., Bengough, A., Fourcaud, T., & Sidle, R. (2009). Desirable plant root traits for protecting natural and engineered slopes against landslides. *Plant and Soil, 324*, 1–30.

Stokes, A., Douglas, G., Fourcaud, T., Giadrossich, F., Gillies, C., Hubble, T., et al. (2014). Ecological mitigation of hillslope instability: Ten key issues facing researchers and practitioners. *Plant and Soil, 377*(1–2), 1–23.

Thompson, R. C., & Luckman, P. G. (1993). Performance of biological erosion control in New Zealand soft rock hill terrain. *Agroforestry Systems, 21*, 191–211.

Toillon, J., Fichot, R., Dallé, E., Berthelot, A., Brignolas, F., & Marron, N. (2013). Planting density affects growth and water-use efficiency depending on site in *Populus deltoides* × *P. nigra. Forest Ecology and Management, 304*, 345–354.

Toy, T. J., Foster, G. R., & Renard, K. G. (2002). *Soil erosion: Processes, prediction, measurement and control.* New York: John Wiley & Sons Inc.

Tromp, J. (1983). Nutrient reserves in roots of fruit trees, in particular carbohydrates and nitrogen. *Plant and Soil, 71*(1–3), 401–413.

van den Driessche, R. (1999). First-year growth response of four *Populus trichocarpa* × *Populus deltoides* clones to fertilizer placement and level. *Canadian Journal of Forest Research, 29*(5), 554–562.

Van Kraayenoord, C. W. S., & Hathaway, R. L. (Eds.). (1986). *Plant materials handbook for soil conservation. Volume 2, introduced plants.* (Water and Soil Miscellaneous Publication No. 94). Wellington, New Zealand: Water and Soil Directorate, Ministry of Works and Development.

van Kraayenoord, C. W. S., Wilkinson, A. G., & Hathaway, R. L. (1986). Nursery production of soil conservation plants. In CWS Van Kraayenoord, & RL Hathaway (Eds.), *Plant materials handbook for soil conservation. Volume 1, Principles and practices.* (pp. 149–160, Water and Soil Miscellaneous Publication No. 93.). Wellington, New Zealand: Water and Soil Directorate, Ministry of Works and Development.

Verlinden, M. S., Broeckx, L. S., Van den Bulcke, J., Van Acker, J., & Ceulemans, R. (2013). Comparative study of biomass determinants of 12 poplar (*Populus*) genotypes in a high-density short-rotation culture. *Forest Ecology and Management, 307*, 101–111.

VSN International. (2014). *GenStat for Windows* (17th ed.). United Kingdom: Hemel Hempstead. Retrieved from: http://www.vsni.co.uk/software/genstat/.

Wall, A. J., Kemp, P. D., & MacKay, A. D. (2006). Predicting pasture production under poplars using canopy closure images. *Proceedings of the New Zealand Grassland Association, 68*, 325–330.

Wang, Z., Qi, L., & Wang, X. (2012). A prototype experiment of debris flow control with energy dissipation structures. *Natural Hazards, 60*(3), 971–989.

Wilkinson, A. G. (1999). Poplars and willows for soil erosion control in New Zealand. *Biomass and Bioenergy, 16*, 263–274.

Wu, Q. X., Liu, X. D., & Zhao, H. Y. (1994). Soil water characteristics in mountain poplar stand and its benefits to soil and water conservation in loess hilly region. *Journal of Environmental Sciences, 6*(3), 347–354.

Wu, J. Y., Huang, D., Teng, W. J., & Sardo, V. I. (2010). Grass hedges to reduce overland flow and soil erosion. *Agronomy for Sustainable Development, 30*(2), 481–485.

Wullschleger, S. D., Yin, T. M., DiFazio, S. P., Tschaplinski, T. J., Gunter, L. E., Davis, M. F., et al. (2005). Phenotypic variation in growth and biomass distribution for two advanced-generation pedigrees of hybrid poplar. *Canadian Journal of Forest Research, 35*(8), 1779–1789.

Yang, Q., Zhao, Z., Chow, T. L., Rees, H. W., Bourque, C. P. A., & Meng, F.-R. (2009). Using GIS and a digital elevation model to assess the effectiveness of variable grade flow diversion terraces in reducing soil erosion in northwestern New Brunswick, Canada. *Hydrological Processes, 23*(23), 3271–3280.

Zalesny, R. S., Jr., Hall, R. B., Bauer, E. O., & Riemenschneider, D. E. (2003). Shoot position affects root initiation and growth of dormant unrooted cuttings of *Populus. Silvae Genetica, 52*(5–6), 273–279.

Zhao, X., Zheng, H., Li, S., Yang, C., Jiang, J., & Liu, G. (2014). The rooting of poplar cuttings: A review. *New Forests, 45*(1), 21–34.

Zuazo, V. H. D., & Pleguezuelo, C. R. R. (2008). Soil-erosion and runoff prevention by plant covers. A review. *Agronomy for Sustainable Development, 28*(1), 65–86.

Development of *Pinus radiata* suspension cultures from xylogenic callus

Anna R. Caird[1*], Gerty J. H. P. Gielen[1], Peter Dare[1], Christine Devillard[2], Christian Walter[2] and Daniel Gapes[1]

Abstract

Background: The ability to grow xylogenic *Pinus radiata* D.Don in a liquid medium rather than on a solid one would produce a more homogeneous culture, and this in turn would improve cell and gene studies. We report the development of a liquid culture system for two xylogenic *P. radiata* cell lines and compare the subsequent formation of tracheary elements induced on the conventional solid media.

Findings: The cell viability (fluorescein diacetate staining) in liquid cultures and subsequent tracheary element (TE) differentiation was as high as, or higher than, that observed with conventional callus cultures on solid media. The growth of cells in liquid culture was confirmed by comparing organic carbon consumption and dry weight increase. Conditions for optimal growth were determined by measuring substrate consumption and cell dry weight with two different cell lines, flask volumes, and starting inoculum densities. Changes to flask volume and cell line were observed to modify substrate carbon consumption within the cell culture, whilst having no significant impact on overall cellular yield. Inoculum density and cell line were the most significant factors affecting the percentage of TE produced.

Conclusion: Overall, these preliminary findings confirm that *P. radiata* xylogenic cells were able to be grown in liquid cultures and did produce TE when induced on solid medium. Therefore, liquid culture has the potential to replace the current standard solid medium system for xylogenic culture of *P. radiata*.

Keywords: Xylogenic, *Pinus radiata*, Suspension culture, Tracheary element

Introduction

A unique in vitro technique developed at Scion (Wagner et al. 2013) induces *Pinus radiata* D.Don xylogenic callus to differentiate, forming cells with lignified secondary cell walls similar to tracheary elements (TE) (Möller et al. 2003). This system enables candidate genes to be tested in vitro on solid media in order to achieve further genetic gains in *P. radiata* production. However, the heterogeneous characteristics of cells grown on solid media introduce difficulty when testing for functional genes. Callus cultures on solid media have differences in spatial location of cells relative to the base media, other cells, and the air-callus interface. Liquid-phase media-based systems have many potential advantages over solid systems, in particular because plant cells cultured in suspension are relatively homogenous. Therefore, their biochemistry and modifications to their biochemistry can be studied easily (Blee et al. 2001). These attributes make high-throughput assessments possible.

Work with *P. radiata* cells in liquid culture has been reported (Ishii and Teasdale 1997; Campbell et al. 1992; Teasdale and Richards 1991), but there are no published results on cell suspension cultures of xylogenic lines from *P. radiata* (Devillard and Walter 2014).

Liquid culture has been used successfully for tracheary element differentiation in a number of other species, including *Zinnia elegans* L. (Fukuda and Komamine 1980), *Arabidopsis thaliana* (L.) Heynh. (Oda et al. 2005), and *Phyllostachys nigra* (Lodd. ex Lindl.) Munro (Ogita et al. 2012). Induction of TE directly within the liquid phase has been achieved in the *Z. elegans* and *A. thaliana* systems, but converting the entire *P. radiata* xylogenic system to a liquid phase has some technological issues to be overcome (such as

* Correspondence: anna.caird@scionresearch.com
[1]Scion (The Trading Name for the New Zealand Forest Research Institute Ltd.), Private Bag 3020, Sala Street, Rotorua, New Zealand
Full list of author information is available at the end of the article

the need for activated charcoal during the induction phase) that are not examined in this study. The aims of this study were (i) to determine whether or not primary *P. radiata* xylogenic cells could be cultured in liquid medium and, if so, (ii) test the efficacy of cultured cells in inducing secondary cell walls thereby demonstrating the potential for enhanced biotechnological development of these cell lines. The following hypothesis was tested: that liquid media can be as effective as callus culture in growing primary cell cultures of *P. radiata* amenable to subsequent induction of TE on solid medium.

Material and methods

Plant/callus starting material

Callus cultures were initiated from xylem strips of *P. radiata* as described by Möller et al. (2003). Two xylogenic cultures were used, Xy8 and Xy14. These cell lines are able to form xylogenic (Xy) cells and were labelled numerically from when they were first isolated. The callus cultures were maintained in the dark at 21.5 ± 0.5 °C on standard solid P6-SHv medium (Möller et al. 2003). These cultures were sub-cultured every 14 days as described in Möller et al. (2003).

Culture on solid medium Calli were grown either on standard solid P6-SHv medium (which contains 30 g L^{-1} of sucrose) or P6-SHvSu6, which contained an increased level of sucrose (60 g L^{-1}). Cells were collected every second day for 14 days.

Culture in liquid medium Fourteen-day-old tissue grown on solid P6-SHv medium as described above was collected, pooled then added to liquid P6-SHv medium containing sucrose (60 g L^{-1}), and mixed thoroughly using a magnetic stirrer. Aliquots (10 or 100 mL) were then transferred to Erlenmeyer flasks (25- or 250-mL flasks). Flasks were wrapped in aluminium foil to exclude light and kept on an orbital shaker (throw 50 mm) at 180 rpm, at 23.7 ± 1 °C. Flasks were moved randomly around the shaker throughout the experiment to minimise effects of layout-related variables. Cells were collected every second day for 18 days.

Cell viability

Callus from two plates with two calli per plate were sampled. Four samples from each of three locations were taken. These cells were stained with fluorescein diacetate (FDA) to determine the percentage of live cells (see the 'Analytical methods' section). Aliquots (1 mL) of cells grown in liquid medium were also sampled and stained with FDA to determine the percentage of live cells.

Effects of cell line and culture environment

A separate experiment was conducted with each cell line independently. Cell line was included as a factor in the subsequent analysis, which contained the following factors: (i) cell line—Xy8 or Xy14; (ii) starting tissue concentration (50 or 100 g L^{-1}) in liquid P6-SHvSu6; and (iii) working volume, 10 mL (in a 25-mL flask) or 100 mL (in a 250-mL flask). Three flasks were replicated for each of the treatments and sampled at day 15. Day 15 was chosen for sampling based on cellular yield plateau seen in previous experiments (data not shown). In addition, a full set of replicate flasks containing the above suspensions were cultured for the 18 days, without sub-sampling for cell viability. Aliquots of suspensions were sub-cultured on to solid induction medium to form secondary cell walls after 10 days (see the 'Analytical methods' section).

Solid plates were also inoculated with the starting suspension and cultured in the dark at 21.5 ± 0.5 °C on standard solid P6-SHv medium. Sub-samples were collected every second day for up to 18 days and stained with FDA (see the 'Analytical methods' section). Organic carbon and dry weight were also measured (as described in the 'Analytical methods' section).

Analytical methods

Total and dissolved organic carbon (TOC and DOC, respectively), pH, and total suspended solid dry weight were determined according to the Standard Methods of the American Public Health Association (APHA) (Clesceri et al. 1998).

Fresh and dry cell weights were obtained by filtering each cell suspension under vacuum using a pre-weighed glass fibre filter then rinsing it with deionised H_2O before weighing. The cells were placed in an oven at approximately 100 °C for at least 24 h and then reweighed.

Filtrates were analysed for dissolved organic carbon and total organic carbon using an Elementar HiTOC machine (Elementar GmbH, Hanau, Germany) that was operated according to the manufacturer's recommendations. The method and machine complied with method 5310 B Standard Methods for the Examination of Water and Wastewater (Clesceri et al. 1998).

Cell viability determination (solid and liquid media) Fluorescein diacetate (FDA) was used to visually assess cell viability (Widholm 1972). Cell viability in liquid culture was assessed by adding FDA solution directly to a sub-sample of the liquid culture before counting the number of fluorescing cells. Fluorescing cells were deemed alive due to enzyme activity causing fluorescence. Cell viability on solid media was assessed by excising a small sample of callus and suspending it in distilled water to a final concentration of FDA of approximately 0.1 % (Razdan

1993). All samples were viewed under UV radiation using a Zeiss Axiovert inverted microscope (filter set 09, 450–490-nm excitation and 515-nm emission, Carl Zeiss, Jena, Germany) and the number of fluorescing cells recorded as a proportion of the total number of cells (two samples taken with a minimum of four fields of view counted).

Formation of tracheary elements Cells were collected from liquid media and sub-cultured, as during maintenance, but on to solid induction medium (Möller et al. 2006; Möller et al. 2003). Solid induction medium contained 5 g L^{-1} activated charcoal (Duchefa Biochemie B.V., The Netherlands). Cells were cultured using a 16-h photoperiod under cool-white fluorescent light (TLD58 W/33 cool-white fluorescent tubes; Philips, Thailand) with a photon flux density (PFD) of 90 μmol m^2 s^1 at 23.3 ± 2 °C, for 10 days. After this time, calli were collected and suspended in distilled water then viewed under bright field and polarised light using a Zeiss Axiovert inverted microscope (Carl Zeiss, Jena, Germany). Tracheary elements with bi-refringent secondary wall thickenings were counted as a proportion of the total number of cells (two samples taken with a minimum of four fields of view counted).

Where factorial designs were utilised, statistical analysis was conducted using SAS statistical software (SAS Institute Inc 2011).

Findings

Cell viability

Aim (i) of this study was to determine whether or not *P. radiata* xylogenic cells could be cultured in liquid medium. Liquid cell cultures were indeed successful and maintained similar cell viability to cells of a similar age taken from callus culture on solid medium (Fig. 1).

Factorial experiment for effects of cell line and liquid culture environment

Cell line Xy8 consumed significantly more substrate carbon than cell line Xy14. Also, in most cases, substrate carbon consumption was larger in 250-mL flasks than in 25-mL flasks suggesting that working volume is important (Fig. 2).

Irrespective of shake flask mass transfer implications, the elevated substrate consumption did not generally lead to any enhancement of cellular yield or TE formation.

Aim (ii) of this study was to test the efficacy of cultured cells in inducing secondary cell walls thereby demonstrating the potential for enhanced biotechnological development of these cell lines. Inoculum density and cell line were the most significant factors affecting the percentage of TE produced (Fig. 3). Cell line Xy8 produced significantly greater proportion of TE than cell line Xy14. It also produced more TE from the liquid primary cultures than in the solid callus indicating that cells previously cultured in liquid medium were capable of inducing secondary cell walls.

The significance testing is summarised in Table 1. Details of the full statistical analysis are provided in Additional file 1.

Lower inoculum level increased the percentage of TE, which might be expected since lower inoculum level results in a lower oxygen demand and thus leads to a higher relative oxygen level in the liquid phase.

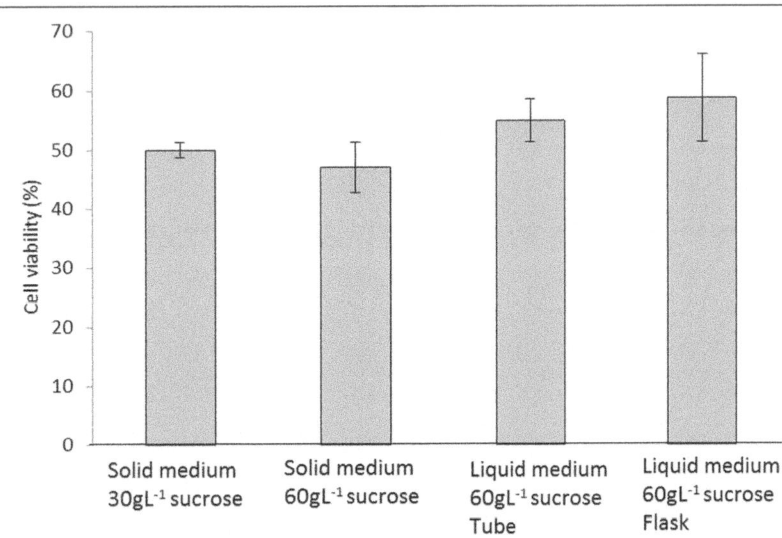

Fig. 1 Cell viability for various solid media and liquid cultures averaged over days 10, 12, 14, 16, and 18. *Error bars* indicate standard errors

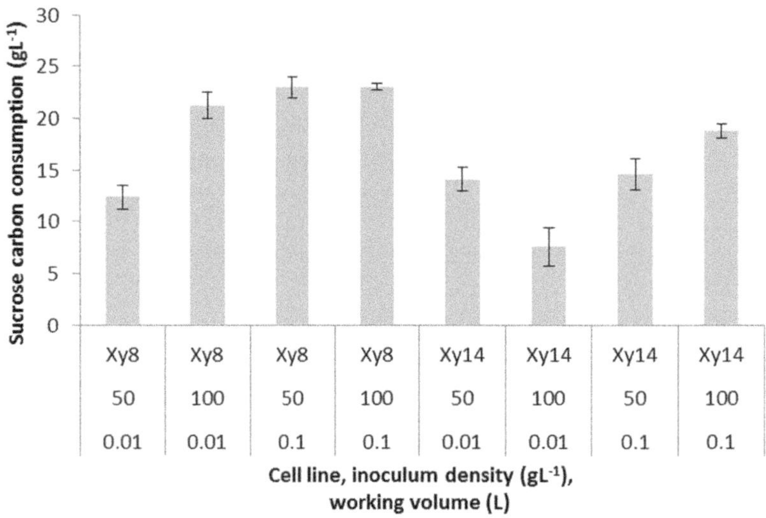

Fig. 2 Substrate carbon consumption in factorial experiment. Presentation of *x* axis datasets; cell line (*top*), inoculum density (g L^{-1}, *middle*), and working volume (L, *bottom*). *Error bars* represent standard error of replicate data ($n = 6$)

The cells originate and had been maintained in aerobic conditions; therefore, oxygen level is likely to be an important factor. Inoculum density is also known to have an effect on enzymes involved in the phenylpropanoid pathway (Hahlbrock and Wellmann 1973). The phenylpropanoid pathway is involved in the formation of lignin in TE (Douglas 1996). Therefore, further optimisation of inoculum density is crucial.

Tested as one of the factors, cell line was found to be significant in both the growth and viability of the liquid cultures. Interestingly, the cellular yields between the two lines were not significantly different, despite significant differences in growth rate (data not shown) and substrate consumption. For optimisation, a greater number of cell lines would need to be tested, and an efficient cell-line screening protocol developed. Further investigation on this aspect of culturing is also warranted because of the limite knowledge on programmed cell death and autolysis in the TE of conifers.

Conclusion

Liquid cultures of *P. radiata* xylogenic cells maintained good viability and TE differentiation rates compared with the traditional callus culturing technique. Overall, these findings support our hypothesis that liquid media

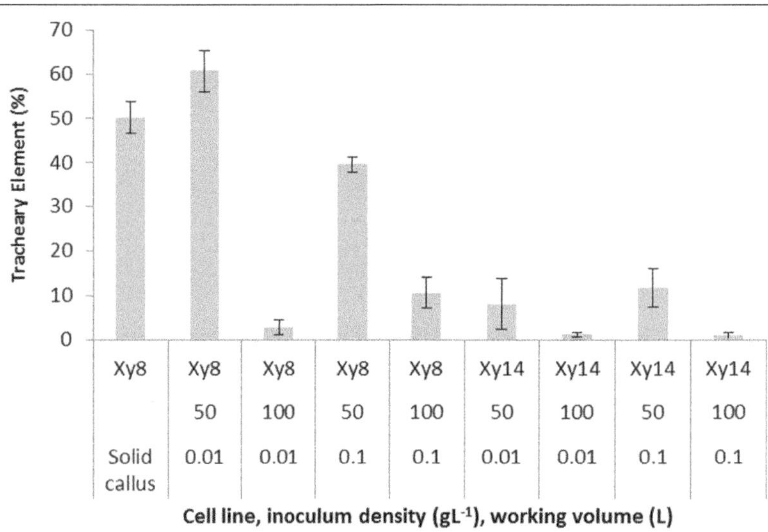

Fig. 3 Tracheary element percentage from factorial experiment. Presentation of *x* axis datasets; cell line (*top*), inoculum density (g L^{-1}, *middle*), and working volume (L, *bottom*). *Error bars* represent standard error of replicate data ($n = 6$)

Table 1 Significance of factors on effects from factorial flask experiment

Parameter	Experimental factor[a]						
	Inoculum density (I)	Working volume (W)	Cell line (C)	IxW interaction	IxC interaction	WxC interaction	IWC interaction
Substrate carbon consumption	++	++	++			+	+
Dry weight increase		++	++			+	+
Cellular yield	+						
% live cells	++			+	+		
% TE	++		++		++		++

[a]Type III sums of squares Pr > F; + significant, between 0.05 and 0.005; ++ highly significant <0.005; empty cells = not significant, >0.051

can be as effective as callus culture in growing primary cell cultures of *P. radiata* that are amenable to subsequent induction of TE. Whilst more development is required to optimise the system, the current work demonstrates that a liquid culture has the potential to replace the current standard solid medium system.

Abbreviations
TE: tracheary elements; Xy: xylogenic.

Competing interests
The authors declare that they have no competing interests.

Authors' contributions
AC was primary author and undertook tissue culture and data collection. GG contributed to the conception and technical design of the study and provided critical revisions of the manuscript. PD undertook carbon measurements and provided technical advice. CD Provided technical advice and revisions of the manuscript. CW contributed to technical design of the study, and revisions of the manuscript. DG contributed to the conception and technical design of the study and prepared manuscript. All authors read and approved the final manuscript.

Acknowledgements
The authors would like to thank Louise Flagstad and Christina Maher for technical assistance and Mark Kimberley for statistical analysis.

Author details
[1]Scion (The Trading Name for the New Zealand Forest Research Institute Ltd.), Private Bag 3020, Sala Street, Rotorua, New Zealand. [2]Formerly of Scion, Private Bag 3020, Sala Street, Rotorua, New Zealand.

References
Blee, K. A., Wheatley, E. R., Bonham, V. A., Mitchell, G. P., Robertson, D., Slabas, A. R., et al. (2001). Proteomic analysis reveals a novel set of cell wall proteins in a transformed tobacco cell culture that synthesises secondary walls as determined by biochemical and morphological parameters. *Planta, 212*(3), 404–415.

Campbell, M. A., Kinlaw, C. S., & Neale, N. B. (1992). Expression of luciferase and β-glucuronidase in *Pinus radiata* suspension cells using electroporation and particle bombardment. *Canadian Journal of Forest Research, 22*(12), 2014–2018. doi:10.1139/x92-265.

Clesceri, L. S., Greenberg, A. E., & Eaton, A. D. (1998). *Standard methods for the examination of water and wastewater* (20th ed.). Washington, DC: American Public Health Association.

Devillard, C., & Walter, C. (2014). Formation of plant tracheary elements in vitro – a review. *New Zealand Journal of Forestry Science, 44*, 22. doi:10.1186/s40490-014-0022-7.

Douglas, C. J. (1996). Phenylpropanoid metabolism and lignin biosynthesis: from weeds to trees. *Trends in Plant Science, 1*(6), 171–178.

Fukuda, H., & Komamine, A. (1980). Establishment of an experimental system for the study of tracheary element differentiation from single cells isolated from the mesophyll of *Zinnia elegans*. *Plant Physiology, 65*, 57–60.

Hahlbrock, K., & Wellmann, E. (1973). Light-independent induction of enzymes related to phenylpropanoid metabolism in cell suspension cultures from parsley. *Biochimica et Biophysica Acta (BBA) - General Subjects, 304*(3), 702.

Ishii, K., & Teasdale, R. D. (1997). Effects of xylooligosaccharides on suspension-cultured cells and protoplasts of *Pinus radiata*. *Plant Cell, Tissue and Organ Culture, 49*(3), 189–193.

Möller, R., Ball, R. D., Henderson, A. R., Modzel, G., & Find, J. (2006). Effect of light and activated charcoal on tracheary element differentiation in callus cultures of Pinus radiata D. Don. *Plant Cell, Tissue and Organ Culture, 85*(2), 161–171. doi:10.1007/s11240-005-9065-z.

Möller, R., McDonald, A. G., Walter, C., & Harris, P. J. (2003). Cell differentiation, secondary cell-wall formation and transformation of callus tissue of *Pinus radiata* D. Don. *Planta, 217*(5), 736–747. doi:10.1007/s00425-003-1053-0.

Oda, Y., Mimura, T., & Hasezawa, S. (2005). Regulation of secondary cell wall development by cortical microtubules during tracheary element differentiation in arabidopsis cell suspensions. *Plant Physiology, 137*(3), 1027–1036.

Ogita, S., Nomura, T., Kishimoto, T., & Kato, Y. (2012). A novel xylogenic suspension culture model for exploring lignification in *Phyllostachys* bamboo. *Plant Methods, 8*, 1. doi:10.1186/1746-4811-8-40.

Razdan, M. K. (1993). *An introduction to plant tissue culture*. Andover: Intercept Limited.

SAS Institute Inc. (2011). *SAS* (93rd ed.). United States of America: SAS Institute Inc.

Teasdale, R. D., & Richards, D. K. (1991). Study of a factor produced by suspension-cultured *Pinus radiata* cells which enhances cell growth at low inoculum densities. *Plant Cell, Tissue and Organ Culture, 26*(1), 53–59.

Wagner, A., Tobimatsu, Y., Goeminne, G., Phillips, L., Flint, H., Steward, D., et al. (2013). Suppression of CCR impacts metabolite profile and cell wall composition in *Pinus radiata* tracheary elements. *Plant Molecular Biology, 81*(1–2), 105–117.

Widholm, J. M. (1972). The use of fluorescein diacetate and phenosafranine for determining viability of cultured plant cells. *Stain Technology, 47*(4), 189–194.

17

Needle fungi in young Tasmanian *Pinus radiata* plantations in relation to elevation and rainfall

Istiana Prihatini[1,2], Morag Glen[1*], Timothy J. Wardlaw[3], David A. Ratkowsky[1] and Caroline L. Mohammed[1]

Abstract

Background: Needle fungi in conifers have been extensively studied to explore their diversity, but environmental factors influencing the composition of fungal communities in *Pinus radiata* D.Don needles have received little attention. This study was conducted to examine the influence of the environment as defined by rainfall, elevation and temperature on the composition of fungal communities in pine needles at an age prior to that at which spring needle cast (SNC) is generally observed. Elucidating the entire fungal community in the needles is a first step towards understanding the cause of the disease.

Methods: Needle samples were collected from 5-year-old *P. radiata* trees, their age predating the onset of SNC, from 12 plantations in Tasmania. Interpolated data for the climate variables, including seasonal components for rainfall and temperature, were obtained from an enhanced climate data bank. Pooled needle samples were examined for the fungi they contained using DNA sequencing of cloned polymerase chain reaction (PCR) products. Clones were grouped into operational taxonomic units (OTUs) and identified to their lowest possible taxonomic level by comparison with reference isolates and public DNA databases.

Results: DNA sequencing revealed that needle fungal communities differed greatly, depending upon the total annual rainfall and needle age. Needle fungi that have been previously associated with pathogenic behaviour, such as *Cyclaneusma minus*, *Dothistroma septosporum*, *Lophodermium pinastri*, *Strasseria geniculata* and *Sydowia polyspora*, were all found in the needles in this study. Also, of 70 ascomycetous OTUs that were discriminated by their DNA sequences, 21 were identified as belonging to the Teratosphaeriaceae.

Conclusions: An overview of the endophytic fungi present in the needles of *P. radiata* has been obtained from trees whose age precedes the onset of SNC. All of the previously known pathogenic species for needle cast diseases were present, in addition to a range of unidentified OTUs belonging to the Teratosphaeriaceae.

Keywords: Needle fungi, *Pinus radiata*, Rainfall, Elevation, Direct PCR

Background

As described in previous studies (Ganley & Newcombe 2006; Koukol et al. 2012; Terhonen et al. 2011), the *Pinus* phyllosphere (i.e. the environment for microorganisms on and below the needle surface) is a rich and varied microbial community. Its composition is strongly influenced by both genotypic and environmental factors (Botes et al. 1997; Van Maanen et al. 2000; Wang & Guo 2007). These factors can affect the complex interactions between microbial organisms, which in turn influence their interactions with their host plant (Giauque & Hawkes 2013). Abiotic or environmental factors influencing the composition of fungal communities associated with *Pinus radiata* D.Don needles have received little attention.

Studies have shown that endophyte communities in *Pinus sylvestris* L. needles can vary according to geographic location (Gourbière & Debouzie 2003) and elevation (Gourbière et al. 2001). Communities in *Picea mariana* (Mill.) Britton, Sterns & Poggenb. needles have

* Correspondence: Morag.Glen@utas.edu.au
[1]Tasmanian Institute of Agriculture, University of Tasmania, Private Bag 98, Hobart, Tasmania 7001, Australia
Full list of author information is available at the end of the article

been reported to vary with latitude (Sokolski et al. 2007). A study by Hata and Futai (1996) showed that the endophytic mycota associated with 45 pine species or varieties was similar for those pines belonging to the same taxonomic group. That study also concluded that endophytic colonisation might reflect the degree of adaptation of the host to local biotic and abiotic factors. However, considerable controversy exists about whether an endophyte, usually an organism such as a bacterium or a fungus, can become a pathogen. Some authors have suggested that endophytes may play a role in protecting trees from pathogens (Arnold et al. 2003; Bradshaw 2004; Gao et al. 2010), but others maintain that a commensal or mutualist fungal species can be triggered under certain environmental conditions to act as a virulent parasite (Krabel et al. 2013). The definition of an endophyte differs greatly, but in a comprehensive chapter on endophytic fungi, Stone et al. (2004) wrote that, in general, the term endophytic fungi applies to 'fungi capable of symptomless occupation of apparently healthy plant tissue'. Broadly, this means that endophytes are viewed as colonising living plant tissues without causing them any immediately obvious negative effect. Nevertheless, others, such as Brown et al. (1998), who elucidated the endophytic fungal communities of the *Musa acuminata* Colla species (banana) complex and found 24 taxa of which almost all were ascomycetes, raised the question 'Can endophytes of *Musa* spp. become pathogens?' They noted the strong correlation between lists of pathogenic taxa of *Musa* spp. and frequently isolated tropical endophyte genera, but in the context of their study, they determined that it was not possible to conclude if banana pathogens isolated as species of endophytes in healthy leaves were the same pathogenic strains. As the fungi may be non-pathogenic strains, they concluded that experiments following Koch's postulates were required to confirm any pathogenicity. Similarly, Larran et al. (2002) and Douanla-Meli et al. (2013) found that their lists of fungal species in the healthy leaves of soybean and citrus, respectively, were predominantly or almost exclusively made up of common plant pathogens, but each study concluded that further investigations were necessary to understand the precise role of these endophytes.

Spring needle cast (SNC) in Tasmania appears in pine plantations at canopy closure (age 6–7 years) and is manifested by the rapid browning of 1-year-old needles followed by the heavy casting of needles. Spring needle cast is not considered to be a classic needle blight disease caused by a primary fungal pathogen (Podger & Wardlaw 1990); instead, it is thought to be due to a suite of endophytic fungi that are triggered into secondary pathogenic activity by an environmental stress. In their study of 53 stands selected for climatic analysis, Podger and Wardlaw (1990) found that 23 of the sites were visually assessed as having SNC, and that these stands were restricted to a mean annual rainfall range of 1200–2000 mm and mean annual temperature range of 8–11.5 °C. Podger and Wardlaw (1990) putatively associated *Cyclaneusma minus*, *Lophodermium pinastri* and *Strasseria geniculata* with the disease in Tasmania, but there were many areas with severe SNC where Podger and Wardlaw (1990) found no fungal fruiting bodies of these species on attached needles, dead or alive. However, these three species of fungi were ubiquitous on fallen needles at sites both affected and unaffected by SNC. More recently, fungal communities were surveyed in a *P. radiata* family trial established in 1999 in Oonah, Tasmania. The trial was assessed at 8 years of age in 2007 using DNA extraction from needles, polymerase chain reaction (PCR) with fungal-specific primers and sequencing to detect the presence of fungal species (Prihatini et al. 2015a). Trees with contrasting levels of SNC disease severity were found to have significantly different needle fungal communities. The four fungi, *L. pinastri*, *C. minus*, *S. geniculata* and *Dothistroma septosporum*, previously reported as common pathogens or endophytes in *P. radiata* in Tasmania implicated in causing SNC (Podger 1984; Podger and Wardlaw 1990) were also identified in the Prihatini et al. (2015a) study. Of these, only *L. pinastri* was correlated with high levels of disease. Several species of Teratosphaeriaceae were also detected. Teratosphaeriaceae sp. 23 was detected in almost all of the needle samples and was presumed to be endophytic as it was associated with needles from healthy trees as well as those affected by SNC. In contrast, Teratosphaeriaceae sp. 3 was detected most frequently in needles from severely diseased trees and not at all in needles from trees with <5 % disease. An overview of the fungal species detected in these two studies (Prihatini et al. 2015b) highlighted the discrepancy between fungal associates of *P. radiata* detected by PCR and sequencing and those detected by isolation.

The present study was conducted in parallel with Prihatini et al. (2015a, b) to explore the types and diversity of needle fungi present in 5-year-old *P. radiata* in Tasmania, an age at which all trees have green, non-symptomatic needles, and to examine the influence of climate and landform as defined by rainfall, elevation and temperature on the composition of fungal communities. The motivation for this was to obtain a catalogue of fungal species from healthy tissues that could point towards the fungi that caused, or might be implicated in, SNC. This follows the approach of Petrini (1986) and Carroll (1988), who advocated systematic sampling of healthy plant tissue for internal fungi rather than the usual practice of the time of examining diseased or necrotic plant organs.

Methods

Sampling sites and environmental data

Twelve *P. radiata* plantations were chosen in Tasmania for the collection of needle samples from 5-year-old trees (Fig. 1). At each site, five 20-tree plots had been established for a previous study (unpublished), and needle samples were obtained from those plots. Interpolated estimates (data drill) of various aspects of rainfall and temperature data were downloaded from SILO (an enhanced climate data bank; Queensland Government 2014). These included average annual mean temperature, average annual minimum and maximum temperature, average winter minimum and maximum temperature, average annual rainfall and total monthly rainfall in winter and in summer. Rainfall was used both as a categorical predictor variable in a permutational analysis of variance and as a continuous variable in a permutational regression analysis (see the Statistical Analysis section below). For the analysis of variance, there were three replications (sampling sites), two classes for elevation (above and below 320 m) and three classes for mean annual rainfall: dry sites (<800 mm); intermediate sites (800–1100 mm) and 'wet' sites (>1100 mm) (Table 1).

Needle samples

Four composite needle samples were obtained from each site, corresponding to needles of four different age classes (1-year-old, 2-year-old, 3-year-old and fallen needles), as described in Prihatini et al (2015b), to reduce the number of cloning reactions without losing information about the effect of needle age on fungal communities. Needles were collected between December 2006 and April 2007. Five trees were randomly selected from each of the five 20-tree plots, and a single fascicle of each class was taken from the middle section of each tree. Fallen needles were collected from beneath each tree. Fascicles in the same class were pooled for each site, placed into paper bags and stored at 10 °C for 1–2 days before drying at 42 °C.

DNA extraction

Approximately 1 cm was cut from the middle part of each needle, placed into a 1.5-mL microcentrifuge tube and stored at −80 °C prior to grinding with a mortar and pestle Prihatini et al. (2015b). DNA extractions were performed using the procedure described in Glen et al. (2002).

Polymerase chain reaction and electrophoresis

To amplify the ITS region, PCR was carried out using primers ITS4A (Larena et al. 1999) and ITS5 (White et al. 1990) with the conditions as described in Prihatini et al. (2015b). Amplification products that were clearly visible on an agarose gel stained with ethidium bromide were cloned directly after purification with an UltraClean® PCR Clean-up Kit (MO BIO Laboratories, USA). Samples with insufficient product after the first round of PCR were diluted 1 in 5 and re-amplified by nested PCR using the primers ITS1 and ITS4 (White et al. 1990). Nested PCR products were purified and cloned as above. All 48 samples obtained by standard PCR or nested PCR were cloned.

Cloning of amplified DNA, screening and DNA sequencing

The PCR and nested PCR products were cloned using pGEM®-T Vector Systems (Promega) according to the technical manual provided. Clones were screened by PCR-restriction fragment length polymorphisms (PCR-RFLP), and three or four clones from each RFLP group from each needle sample were sequenced. Chromatograms were viewed in ChromasPro version 1.34 software and edited to remove poor quality sequences at each end; the sequence was then saved in FASTA format.

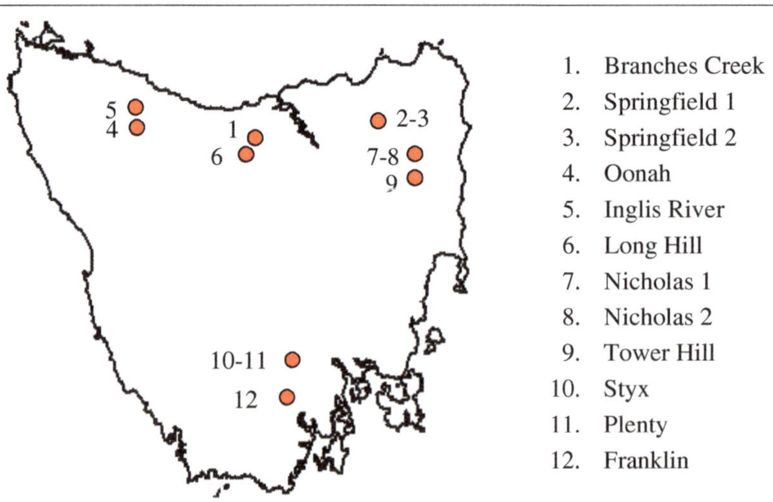

Fig. 1 Location of pine plantations in Tasmania from which materials were collected (latitude, longitude and other information of these sites are presented in Table 1)

Table 1 Location (latitude and longitude), elevation and annual rainfall for each site

Site	Coupe	Latitude-longitude[a]		Elevation (m)[b]	Rainfall (mm)[d]		Average minimum temperature (°C)	
		North	East		Annual[c]	Monthly (winter)[d]	Annual[d]	Winter[d]
Branches Creek	BC029A	41.2653	146.6629	131 (L)	744 (D)	87.7	8.6	5.2
Springfield 1	SF121F	41.2118	147.6107	311 (L)	785(D)	89.4	8.1	4.8
Springfield 2	SF121B	41.2114	147.6261	294 (L)	785 (D)	89.4	8.1	4.8
Long Hill	LH106A	41.3426	146.4901	120 (L)	988 (M)	131.8	6.9	3.7
Franklin	FN014H	43.0573	146.8787	293 (L)	1123 (W)	115.2	5.3	2.6
Inglis River	IR035D	41.1087	145.5974	111 (L)	1353 (W)	170.8	7.0	4.2
Styx	SX038Z	42.7770	146.8269	539 (H)	714 (D)	74.1	5.6	2.1
Tower Hill	TH135A	41.5277	147.9119	512 (H)	716 (D)	66.9	6.0	2.3
Nicholas 1	NI162B	41.4511	147.9764	338 (H)	915 (M)	82.9	5.7	2.2
Nicholas 2	NI166A	41.4719	147.9829	324 (H)	915 (M)	82.9	5.7	2.2
Plenty	PL020X	42.8659	146.8929	427 (H)	876 (M)	85.2	4.7	2.0
Oonah	OO072B	41.2299	145.6152	454 (H)	1439 (W)	177.5	6.9	4.1

[a]Average from Global Positioning System data
[b]Sites with elevation below 320 m above sea level are categorised as low (L), and sites with elevation at or above 320 m are categorised as high (H)
[c]Average of annual rainfall for the years 2000–2008 (samples were collected in 2008). Sites with mean annual rainfall below 800 mm are categorised as dry (D), sites with mean annual rainfall above 1100 mm are categorised as wet (W), and sites with mean annual rainfall between these extremes are categorised as moderate (M)
[d]Additional climate data obtained from SILO (Queensland Government 2014): total rainfall per winter month (TotRnW); average annual minimum temperature (AvTmin); and average winter minimum temperature (AvTminW)

Sequencing of PCR products was performed by Macrogen Inc. (Seoul, Korea).

Fungal DNA identification

A search of public DNA databases (BLAST, Altschul et al. 1990) retrieved sequences of high similarity. Sequences were grouped according to BLAST search results and aligned with sequences of high similarity using ClustalW (Larkin et al. 2007). If single nucleotide polymorphisms occurred between clones from the same operational taxonomic unit (OTU), chromatograms were rechecked to confirm these. Isolates or clones with less than 2 % sequence variation were grouped into OTUs. The OTUs were identified to the lowest possible taxonomic level based on sequence similarity to known fungi from public DNA databases and reference cultures (Prihatini et al. 2015b). Phylogenetic analysis helped to refine the discrimination and identification of OTUs. Several sequences were selected from the best matches retrieved by BLAST searches of public databases for each OTU and included in phylogenetic analysis for that OTU. One or two sequences from more distantly related taxa were also included as outgroups. All sequences for each OTU were aligned using the ClustalW program in BioEdit version 7.0.9.0 (Hall 1999) prior to phylogenetic analysis using DNAml from the PHYLIP package (Felsenstein 1989). Trees were drawn by TreeView software (Page 1996) and edited using MEGA4 software (Tamura et al. 2007).

Statistical analysis

All statistical analyses to study the community of needle fungi present were conducted using PRIMER v6 (Clarke & Gorley 2006), with its add-on PERMANOVA+ v1.0.2 (Anderson et al. 2008). The basic data matrix consisted of the presence or absence of 70 ascomycetous OTUs discriminated from the phylogenetic analysis on the 34 pooled samples that yielded clones. A Bray-Curtis resemblance (similarity) matrix constructed from all possible pairs of samples was the input for each of the routines used. Both permutational multivariate analysis of variance (PERMANOVA) and canonical analysis of principal coordinates (CAP) were used to test separately whether the factors needle age, elevation (low vs. high) and total annual rainfall (dry, intermediate, wet) produced significantly different fungal assemblages. The advantage of CAP is that it is specifically designed to find canonical axes that maximise differences amongst pre-defined groups in the data cloud, and it also tests individual samples for the correctness or otherwise of their group membership by cross-validation, employing a leave-one-out procedure. The third routine used, DistLM, modelled whether the climate variables elevation, rainfall and temperature could explain a significant proportion of the variation in the data cloud. Total annual and monthly rainfall data were available along with average annual and seasonal data for maximum, minimum and average temperatures. There were high correlations amongst them so principal component analysis (PCA) was first applied to the raw data, which was standardised to have unit variance. Subsequently, the DistLM modelling was conducted on the first three principal components, which accounted for almost all of the total variation, and the Bayesian selection criterion BIC was used as the 'stopping rule'. All three routines of the PERMANOVA+ program that were used here employed 9999 permutations.

Results

Identification of fungi amplified from *P. radiata* needles

A total of 504 clones were sequenced but only 439 clones gave legible sequence results, allowing OTU assignment and identification. Overall, 407 ascomycetous sequences were identified. Sequences were grouped into OTUs with over 98 % sequence similarity. After phylogenetic analysis, 70 ascomycetous OTUs were discriminated (Table 2), of which 36 were detected from more than one sample. Seventeen species of hitherto unknown members of the Teratosphaeriaceae were identified as part of the *P. radiata* marker-aided selection trial (Prihatini et al. 2015a) run in parallel with the current study, and nine of these were also detected in the current study.

Fungal prevalence across sites

Clones for DNA identification were obtained for only 34 out of the 48 pooled samples. However, all 12 plantation sites were represented by at least one pooled sample that gave rise to clones. Each OTU observed in this study (Table 2) was scored as present or absent in each pooled sample. The ascomycetous OTUs present in at least three of the 34 pooled samples are listed in Table 3, except for *S. geniculata* and *D. septosporum*, which were less frequent but are also included in Table 3 because of their putative previous involvement as causal agents of needle cast disease. Forty OTUs were identified only from a single site, but seven of these were detected in more than one cloning sample at that site. The remaining 30 OTUs were present in more than one site, but only six OTUs occurred at six or more of the sites. Several of the commonly found OTUs were well-known pine associates belonging to the genera *Lophodermium* and *Cyclaneusma*, but their distribution varied amongst the 12 plantation sites, and none was found at all 12 sites. The two most dominant OTUs were Teratosphaeriaceae sp. 26, found in 15 pooled samples at seven sites, and *L. pinastri*, found in 14 pooled samples and at all but one of the 12 sites. *C. minus* 'simile' and *C. minus* 'verum' were each identified at half of the sites.

Fungal communities

Effect of climate

Permutational multivariate analysis of variance (PERMANOVA) produced a significant separation ($P = 0.0217$) of the fungal communities present at the three levels of annual rainfall (<800 mm, dry; 800–1200 mm, intermediate; >1200 mm, wet). Use of CAP provided further evidence of the importance of rainfall, strongly separating the three groups ($P = 0.0048$; see Fig. 2). Neither PERMANOVA nor CAP analyses of fungal communities provided evidence for an association between communities and elevation ($P = 0.182$ and 0.217, respectively). Similarly, using DistLM, no evidence was found of a correlation between fungal communities and temperature ($P = 0.167$ using a composite score from PCA), whereas there was evidence of an association with rainfall ($P = 0.040$ using a composite score from PCA), thereby supporting the results from PERMANOVA and CAP that rainfall is a significant environmental variable influencing fungal community assemblages.

Effect of needle age

Significant differences in fungal assemblages as a function of needle age were also obtained using permutational multivariate analysis of variance. From PERMANOVA, pairwise differences were indicated between 1-year-old needles and fallen needles ($P = 0.0128$) and between 2-year-old needles and fallen needles ($P = 0.0022$), with marginal evidence of an assemblage difference between 3-year-old needles and fallen needles ($P = 0.0584$). Using CAP was an alternative method of testing whether or not the multivariate means are different for the four different age classes. Results were unequivocal in indicating a significant difference between age classes ($P = 0.0181$ (trace test) or 0.006 (test based on the first squared canonical correlation)), with eight out of the 11 fallen needle samples being correctly classified, i.e. placed in the correct needle class by the cross-validation procedure of CAP. However, no 1-year-old needle sample was correctly classified; only two of the eight 2-year-old needle samples and only two of the eight 3-year-old needle samples were correctly classified. This indicates that fallen needles have a different fungal assemblage to those of the three age classes, with their OTUs being more distinct than those in the needles retained by the trees. The widespread species *L. pinastri* was present in 14 of the cloning samples, nine of which were in fallen needles. The sites at which *L. pinastri* was not present in the fallen needle sample were Tower Hill, Styx and Branches Creek, the three driest sites.

Discussion

The main purpose of the study was to produce an inventory of the fungi present in the needles of healthy trees at an age that precedes the onset of SNC disease. This was achieved with the identification of 70 distinct OTUs. A number of these species were also found as part of the *P. radiata* marker-aided selection trial (Prihatini et al. 2015a) run in parallel with the current study, including some members of the Teratosphaeriaceae. However, there is some doubt about the taxonomic placement of

Table 2 List of all OTUs observed in this study with the number of cloning samples in which each OTU was present and a GenBank sequence accession representative of that OTU

Class; order; family	OTUs	No. of cloning samples	GenBank accession
Dothideomycetes; Botryosphaeriales			
Incertae sedis	Botryosphaeriales sp. 3	1	KM216350
Dothideomycetes; Capnodiales			
Davidiellaceae	*Davidiella* sp.	8	KM216336
	Davidiellaceae sp. 1	3	KM216333
	Davidiellaceae sp. 2	1	KM216347
Incertae sedis	Capnodiales sp. 2	1	KJ406757
	Capnodiales sp. 3	1	KM216329
	Phaeotheca fissurella	3	KM216349
Mycosphaerellaceae	*Dothistroma septosporum*	2	KM216330
	Mycosphaerellaceae sp. 1	3	KJ406801
	Phaeothecoidea sp. 1	4	KJ406802
	Phaeothecoidea sp. 2	1	KJ406792
	Pseudocercospora sp. 1	1	KJ406795
	Ramularia stellenboschensis	1	KJ406791
Teratosphaeriaceae	*Teratosphaeria* aff. *associata*	2	KJ406767
	Teratosphaeria aff. *capensis*	1	KJ406771
	Teratosphaeria aff. *parva*	1	KJ406775
	Teratosphaeria aff. *suttonii*	1	KJ406763
	Teratosphaeriaceae sp. 3	3	KJ406765
	Teratosphaeriaceae sp. 4	1	KM216332
	Teratosphaeriaceae sp. 5	2	KJ406766
	Teratosphaeriaceae sp. 6	2	KJ406768
	Teratosphaeriaceae sp. 7	1	KJ406760
	Teratosphaeriaceae sp. 8	2	KJ406759
	Teratosphaeriaceae sp. 9	1	KJ406770
	Teratosphaeriaceae sp. 10	3	KJ406764
	Teratosphaeriaceae sp. 14	4	KJ406758
	Teratosphaeriaceae sp. 16	1	KJ406786
	Teratosphaeriaceae sp. 17	1	KJ406785
	Teratosphaeriaceae sp. 18	2	KJ406787
	Teratosphaeriaceae sp. 19	1	KJ406788
	Teratosphaeriaceae sp. 21	2	KJ406769
	Teratosphaeriaceae sp. 23	9	KJ406784
	Teratosphaeriaceae sp. 24	1	KM216331
	Teratosphaeriaceae sp. 26	15	KJ406761
Dothideomycetes; Dothideales			
Dothioraceae	*Aureobasidium pullulans*	2	KJ406825
	Sydowia polyspora	10	KJ407006
Incertae sedis	Dothideales sp. 1	1	KJ406831
Dothideomycetes; Incertae sedis			
	Dothideomycetes sp. 2	1	KJ406840

Table 2 List of all OTUs observed in this study with the number of cloning samples in which each OTU was present and a GenBank sequence accession representative of that OTU *(Continued)*

Dothideomycetes; Pleosporales			
Didymellaceae	*Leptosphaerulina* sp.	1	KJ406837
Incertae sedis	Pleosporales sp. 2	2	KJ406836
Montagnulaceae	*Paraphaeosphaeria michotii*	3	KJ406849
Phaeosphaeriaceae	*Phaeosphaeria* sp. 3	1	KJ406845
Sporormiaceae	*Sporormiella* sp. 3	1	KJ406867
	Sporormiella sp. 7	2	KJ406863
Eurotiomycetes; Chaetothyriales			
Herpotrichiellaceae	*Exophiala eucalyptorum*	2	KJ406880
Incertae sedis	Chaetothyriales sp. 1	3	KJ406877
	Chaetothyriales sp. 2	4	KJ406875
	Chaetothyriales sp. 3	1	KJ406884
Eurotiomycetes; Eurotiales			
Trichocomaceae	*Penicillium* sp. 1	1	KJ406888
	Penicillium sp. 2	4	KJ406887
	Penicillium sp. 3	1	KJ406890
	Penicillium sp. 4	1	KJ406894
Leotiomycetes; Helotiales			
Helotiaceae	*Varicosporium* aff. *elodeae*	1	KM216334
Incertae sedis	Helotiales sp. 4	1	KM216351
Sclerotiniaceae	*Torrendiella eucalypti*	2	KM216355
Leotiomycetes; Incertae sedis			
Incertae sedis	*Cyclaneusma minus* 'simile'	8	KJ406925
	Cyclaneusma minus 'verum'	9	KJ406907
	Leotiomycetes sp. 1	3	KJ406966
	Meliniomyces sp.	1	KM216335
Leotiomycetes; Phacidiales			
Phacidiaceae	*Phacidium laceratum*	2	KJ406930
Leotiomycetes; Rhytismatales			
Incertae sedis	*Fulvoflamma* sp.	1	KM216341
Rhytismataceae	*Lophodermium* aff. *conigenum*	4	KJ406948
	Lophodermium pinastri	14	KJ406957
Sordariomycetes; Sordariales			
Incertae sedis	Sordariales sp. 1	1	KJ406980
Lasiosphaeriaceae	*Bagadiella* sp. 1	1	KM216344
Sordariomycetes; Xylariales			
Amphisphaeriaceae	*Pestalotiopsis* sp. 2	1	KJ406983
Ascomycota; Incertae sedis			
	Ascomycota sp. 1	3	KM216353
	Ascomycota sp. 3	1	KM216354
	Ascomycota sp. 8	2	KJ406964
	Strasseria geniculata	2	KM216342

Table 3 The major OTUs present in 12 young *P. radiata* plantations in Tasmania listed in decreasing order of prevalence in terms of number of pooled samples

	Site (no. of pooled samples)											
	BC (2)	SP1 (3)	SP2 (4)	ST (3)	TH (4)	LH (2)	N1 (3)	N2 (3)	PL (3)	FR (4)	IR (1)	ON (2)
Environmental conditions[a]	DL	DL	DL	DH	DH	ML	MH	MH	MH	WL	WL	WH
OTUs[b]:												
Teratosphaeriaceae sp. 26	0	0	2	2	3	2	2	3	0	0	1	0
Lophodermium pinastri †	1	2	2	0	1	1	1	1	1	2	1	1
Sydowia polyspora	0	1	1	0	2	1	1	0	0	2	1	1
Cyclaneusma minus 'verum' †	0	0	1	0	3	2	1	1	0	0	1	0
Teratosphaeriaceae sp. 23	0	3	3	1	0	0	0	1	0	1	0	0
Cyclaneusma minus 'simile' †	1	0	2	0	2	0	1	0	0	1	1	0
Davidiella sp.	0	1	0	1	2	1	1	0	1	1	0	0
Chaetothyriales sp. 2	0	1	1	0	0	1	0	0	1	0	0	0
Lophodermium aff. *conigenum* †	0	0	0	0	0	0	0	1	0	2	1	0
Penicillium sp. 2	0	1	0	0	0	0	0	0	2	1	0	0
Phaeothecoidea sp. 1	0	0	0	0	0	1	1	0	2	0	0	0
Teratosphaeriaceae sp. 14	0	3	0	0	0	0	0	0	0	1	0	0
Ascomycota sp. 1	0	1	0	0	0	0	2	0	0	0	0	0
Chaetothyriales sp. 1	0	0	1	0	0	0	0	0	0	0	1	1
Davidiellaceae sp. 1	0	1	0	0	2	0	0	0	0	0	0	0
Leotiomycetes sp. 1	0	0	0	0	0	0	0	1	0	2	0	0
Mycosphaerellaceae sp. 1	0	0	1	0	0	1	0	0	0	1	0	0
Paraphaeosphaeria michotii	0	1	0	0	0	0	0	0	0	1	0	1
Phaeotheca fissurella	0	0	2	0	0	0	0	0	1	0	0	0
Teratosphaeriaceae sp. 3	0	0	0	0	0	1	0	0	0	0	1	1
Teratosphaeriaceae sp. 10	0	1	1	0	0	0	0	0	0	0	0	1
Dothistroma septosporum †	0	0	2	0	0	0	0	0	0	0	0	0
Strasseria geniculata †	0	0	0	0	1	0	1	0	0	0	0	0

BC Branches Creek, *SP1* Springfield 1, *SP2* Springfield 2, *ST* Styx, *TH* Tower Hill, *LH* Long Hill, *N1* Nicholas 1, *N2* Nicholas 2, *PL* Plenty, *FR* Franklin, *IR* Inglis River, *ON* Oonah

[a]Environmental conditions were categorised as dry (D), intermediate (M) or wet (W) based upon annual rainfall and low (L) or high (H) elevation

[b]Entries in columns 2 to 13 indicate the number of pooled DNA samples in which the OTU was found at a particular site. OTUs marked with † are fungal species that have been previously found to be associated with SNC or other needle cast diseases

some of the OTUs listed in Table 2. For example, *Mycosphaerella* was originally thought to be monophyletic but has more recently been shown to be polyphyletic (Crous et al. 2007). Also, genera such as *Davidiella* and *Teratosphaeria* were split off from the genus and have formed the basis of new families. Despite this, the separation between *Teratosphaeria* and *Mycosphaerella* is not always clear, and many phylogenetically distinct taxa still remain in '*Mycosphaerella*', '*Teratosphaeria*' or in one of the associated asexual genera, with their correct generic disposition requiring clarification of their taxonomy and DNA phylogeny (Crous et al. 2009). Obtaining living cultures of previously described species, coupled with advances in molecular techniques, is assisting in the resolution of these generic complexes. With this in mind, some of the 21 OTUs of the present study listed in Table 2 as a 'Teratosphaeriaceae sp.' may prove to belong to another family.

The needle fungal communities varied significantly from site to site in the 12 different plantations of 5-year-old *P. radiata* that were investigated. Other studies also indicate similar significant variation in the needle fungal communities of another *Pinus* species (*P. sylvestris*) across different sites in the Massif Central in France, colonised by the same or similar needle cast fungi active in Tasmania, i.e. *L. pinastri* and *C. minus* (Gourbière et al. 2001; Gourbière and Debouzie 2003). Results from Prihatini et al. 2015a also indicated that needle fungal communities may be highly variable amongst trees at a single site. A key part of the original experimental design involved conducting a needle-health survey in the 12 sites sometime during the following 3 years, when the

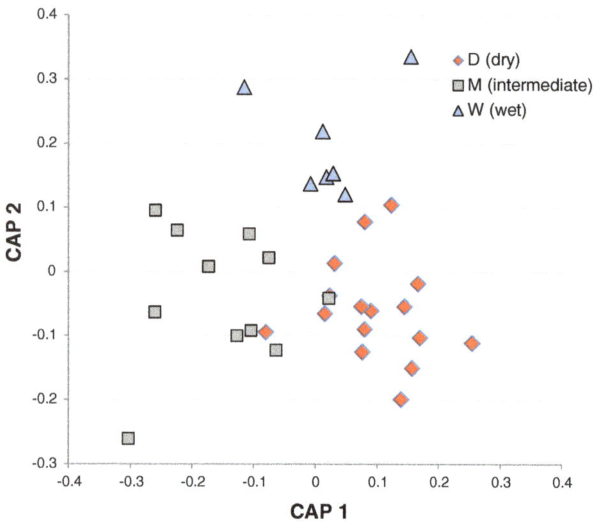

Fig. 2 CAP analysis of fungal communities in 34 pooled DNA samples from three rainfall categories (dry, intermediate, wet) of *P. radiata* plantations in Tasmania

trees would be susceptible to SNC. Unfortunately, changing circumstances (particularly changes in ownership and management of the plots) prevented this from taking place. Therefore, no definitive conclusions can be drawn from the present study regarding the influence of needle fungal communities on subsequent development of SNC.

Of the two main variables reflecting climate, i.e. temperature and rainfall, only the latter influenced the composition of needle fungal communities in these 5-year-old trees. Podger and Wardlaw (1990) predicted that an annual rainfall range of 1200–2000 mm was a climatic characteristic of the disease distribution of SNC. Unpublished reports from Forestry Tasmania records suggest that the two wettest sites studied here (Inglis River and Oonah) subsequently suffered from SNC but further research is required to confirm these observations.

Conclusions

Many of the fungal species previously linked to needle-cast diseases were present in needles of young, healthy *Pinus radiata*. This indicates that environmental or host factors may have a significant role in the development of SNC. A range of previously unknown Teratosphaeriaceae species were detected in young pine needles, some of these appear to be almost ubiquitous endophytes.

Competing interests

The authors declare that they have no competing interests.

Author's contributions

IP carried out the sampling and laboratory work, DNA sequence analyses, some statistical analyses and drafted the manuscript, MG supervised the laboratory work, DNA analyses and revised the manuscript, TJW selected the study sites, provided input to the design of the experiments and feedback on early drafts of the manuscript, DAR provided statistical advice, carried out the main statistical analyses, aided in their interpretation and assisted in revision of the manuscript, CLM conceived of the study, supervised some of the laboratory work and critically revised the manuscript. All authors read and approved the final manuscript.

Acknowledgements

This study was part of a project funded by the Australian Research Council, Forestry Tasmania, Taswood Growers, Norske-Skog, Forest NSW and Hosking Ltd. New Zealand. We thank David Page for his help with the sampling. Istiana Prihatini is the recipient of a John Allwright Fellowship, ACIAR.

Author details

[1]Tasmanian Institute of Agriculture, University of Tasmania, Private Bag 98, Hobart, Tasmania 7001, Australia. [2]Centre for Forest Biotechnology and Tree Improvement, Jl. Palagan Tentara Pelajar Km 15, Purwobinangun Pakem, Yogyakarta 55582, Indonesia. [3]Forestry Tasmania, GPO Box 207, Hobart, Tasmania 7001, Australia.

References

Altschul, SF, Gish, W, Miller, W, Myers, EW, & Lipman, DJ. (1990). Basic local alignment search tool. *Journal of Molecular Biology, 215*, 403–410.

Anderson, MJ, Gorley, RN, & Clarke, KR (2008). *PERMANOVA+ for PRIMER: guide to software and statistical methods* (p. 214). Plymouth, UK: PRIMER-E and Auckland, New Zealand: University of Auckland.

Arnold, AE, Mejía, LC, Kyllo, D, Rojas, EI, Maynard, Z, Robbins, N, Herre, EA. (2003). Fungal endophytes limit pathogen damage in a tropical tree. *Proceedings of the National Academy of Sciences of the United States of America, 100*, 15649–15654.

Botes, WM, Swart, WJ, & Crous, PW. (1997). Distribution of fungal endophytes in needles of *Pinus radiata. South African Journal of Science, 93*,v-vi. ISSN 00382353. Abstracts of the annual meeting of the Southern African Society for Plant Pathology, Badplaas.

Bradshaw, RE. (2004). *Dothistroma* (red-band) needle blight of pines and the dothistromin toxin: a review. *Forest Pathology, 34*, 163–185.

Brown, KB, Hyde, KD, & Guest, DI. (1998). Preliminary studies on endophytic fungal communities of *Musa acuminata* species complex in Hong Kong and Australia. *Fungal Diversity, 1*, 27–51.

Carroll, G. (1988). Fungal endophytes in stems and leaves: from latent pathogen to mutualistic symbiont. *Ecology, 69*, 2–9.

Clarke, KR, & Gorley, RN (2006). *PRIMER v6: user manual/tutorial* (p. 192). Plymouth, UK: PRIMER-E.

Crous, PW, Braun, U, & Groenewald, JZ. (2007). *Mycosphaerella* is polyphyletic. *Studies in Mycology, 58*, 1–32.

Crous, PW, Summerell, BA, Carnegie, AJ, Wingfield, MJ, Hunter, GC, Burgess, TI, Andjic, V, Barber, PA, & Groenewald, JZ. (2009). Unravelling *Mycosphaerella*: do you believe in genera? *Persoonia, 23*, 99–118.

Douanla-Meli, C, Langer, E, & Talontsi Mouafo, F. (2013). Fungal endophyte diversity and community patterns in healthy and yellowing leaves of *Citrus limon. Fungal Ecology, 6*, 212–222.

Felsenstein, J. (1989). PHYLIP—phylogeny inference package (version 3.2). *Cladistics, 5*, 164–166.

Ganley, RJ, & Newcombe, G. (2006). Fungal endophytes in seeds and needles of *Pinus monticola. Mycological Research, 110*, 318–327.

Gao, FK, Dai, CC, & Liu, XZ. (2010). Mechanisms of fungal endophytes in plant protection against pathogens. *African Journal of Microbiology Research, 4*, 1346–1351.

Giauque, H, & Hawkes, CV. (2013). Climate affects symbiotic fungal endophyte diversity and performance. *American Journal of Botany, 100*, 1435–1444.

Glen, M, Tommerup, IC, Bougher, NL, & O'Brien, PA. (2002). Are Sebacinaceae common and widespread ectomycorrhizal associates of *Eucalyptus* species in Australian forests? *Mycorrhiza, 12*, 243–247.

Gourbière, F, & Debouzie, D. (2003). Local variations in microfungal populations on *Pinus sylvestris* needles. *Mycological Research, 107*, 1221–1230.

Gourbière, F, van Maanen, A, & Debouzie, D. (2001). Associations between three fungi on pine needles and their variation along a climatic gradient. *Mycological Research, 105*, 1101–1109.

Hall, TA. (1999). BioEdit: a user-friendly biological sequence alignment editor and analysis program for Windows 95/98/NT. *Nucleic Acids Symposium Series, 41*, 95–98.
Hata, K, & Futai, K. (1996). Variation in fungal endophyte populations in needles of the genus *Pinus*. *Canadian Journal of Botany, 74*, 103–114.
Koukol, O, Kolařík, M, Kolářová, Z, & Baldrian, P. (2012). Diversity of foliar endophytes in wind-fallen *Picea abies* trees. *Fungal Diversity, 54*, 69–77.
Krabel, D, Morgenstern, K, & Herzog, S. (2013). Endophytes in changing environments—do we need new concepts in forest management? *iForest – Biogeosciences and Forestry, 6*, 109–112.
Larena, I, Salazar, O, González, V, Julián, MC, & Rubio, V. (1999). Design of a primer for ribosomal DNA internal transcribed spacer with enhanced specificity for ascomycetes. *Journal of Biotechnology, 75*, 187–194.
Larkin, MA, Blackshields, G, Brown, NP, Chenna, R, McGettigan, PA, McWilliam, H, Valentin, F, Wallace, IM, Wilm, A, Lopez, R, Thompson, JD, Gibson, TJ & Higgins, DG. (2007). ClustalW and ClustalX version 2. *Bioinformatics, 23*, 2947–2948.
Larran, S, Rollán, C, Ángeles, HB, Alippi, HE, & Urrutia, MI. (2002). Endophytic fungi in healthy soybean leaves. *Investigacion Agraria Produccion y Proteccion Vegetales, 17*(1), 173–178.
Page, RDM. (1996). TreeView: an application to display phylogenetic trees on personal computers. *Computer Applications in the Biosciences, 12*, 357–358.
Petrini, O (1986). Taxonomy of endophytic fungi of aerial plant tissues. In NJ Fokkema & J van den Heuvel (Eds.), *Microbiology of the phyllosphere* (pp. 175–187). Cambridge: Cambridge University Press.
Podger, FD. (1984). *Dothistroma* in Tasmania. *Australasian Plant Pathology, 13, 65.*
Podger, FD, & Wardlaw, TJ. (1990). Spring needle-cast of *Pinus radiata* in Tasmania: I. Symptoms, distribution, and association with *Cyclaneusma minus. New Zealand Journal of Forest Science, 20*, 184–205.
Prihatini, I, Glen, M, Wardlaw, TJ, & Mohammed, CL (2015a). *Lophodermium pinastri* and an unknown species of Teratosphaeriaceae are associated with needle cast in a *Pinus radiata* selection trial. *Forest Pathology 45*, 281-289. doi:10.1111/efp.12169.
Prihatini, I, Glen, M, Wardlaw, TJ, & Mohammed, CL. (2015b). Diversity and identification of fungi associated with needles of *Pinus radiata* in Tasmania. Southern Forests (in press) http://dx.doi.org/10.2989/20702620.2015.1092345.
Queensland Government (2014). SILO climate data. Department of Science, Information Technology, Innovation and the Arts, The State of Queensland. http://www.longpaddock.qld.gov.au/. Accessed 2 January 2014.
Sokolski, S, Bernier-Cardou, M, Piché, Y, & Bérubé, JA. (2007). Black spruce (*Picea marinara*) foliage hosts numerous and potentially endemic fungal endophytes. *Canadian Journal of Forest Research, 37*, 1737–1747.
Stone, JK, Polishook, JD, & White, JF, Jr (2004). Endophytic fungi. In GM Mueller, GF Bills, & MS Foster (Eds.), *Biodiversity of fungi: inventory and monitoring methods* (pp. 241–270). Burlington: Elsevier Academic Press.
Tamura, K, Dudley, J, Nei, M, & Kumar, S. (2007). MEGA4: Molecular Evolutionary Genetics Analysis (MEGA) software version 4.0. *Molecular Biology and Evolution, 24*, 1596–1599.
Terhonen, E, Marco, T, Sun, H, Jalkanen, R, Kasanen, R, Vuorinen, M, & Asiegbu, F. (2011). The effect of latitude, season and needle-age on the mycota of Scots pine (*Pinus sylvestris*) in Finland. *Silva Fennica, 45*, 301–317.
Van Maanen, A, Debouzie, D, & Gourbière, F. (2000). Distribution of three fungi colonising fallen *Pinus sylvestris* needles along altitudinal transects. *Mycological Research, 104*, 1133–1138.
Wang, Y, & Guo, LD. (2007). A comparative study of endophytic fungi in needles, bark, and xylem of *Pinus tabulaeformis*. *Canadian Journal of Botany, 85*, 911–917.
White, TJ, Bruns, T, Lee, S, & Taylor, J (1990). Amplification and direct sequencing of fungal ribosomal RNA genes for phylogenetics. In MA Innis, DH Gelfand, JJ Sninsky, & TJ White (Eds.), *PCR protocols: a guide to methods and applications* (pp. 315–322). London: Academic Press.

18

New Zealand school children's perceptions of local forests and the Montréal Process Criteria and Indicators: comparing local and international value systems

Karen M. Bayne[1*], Barbara K. Höck[1], Harley R. Spence[2], Kirsten A. Crawford[2], Tim W. Payn[1] and Tim D. Barnard[1]

Abstract

Background: A review of the international Montréal Process Criteria and Indicators (MP C&I) sustainable forests reporting framework required member countries to benchmark the Indicators against locally held forest values. Children constitute an important subgroup of the local stakeholders in forests: they may respond to forested environments in different ways and may hold values which are unique to their age group. This study reports on the values school children attributed to their local forests, across three New Zealand regions, and details the agreement of these values to the Montréal Process C&I, a values framework developed by international forestry experts.

Methods: Workshops were held with local school children at three locations neighbouring a forested area (Tapawera, Nelson, South Island; Hunua and Kaiata, Auckland, North Island; and Lynmore, Rotorua, Bay of Plenty, North Island). Workshops were conducted using a similar facilitated structure, tailored to the age group of the children. The workshops used both interactive drawings and dialogue exercises to elicit the features of importance and cultural values the students held for their local forests. Following the workshops, these values were mapped to seven broad categories, and analysed for agreement with the MP C&I.

Results: Children aged 9–15 years valued forest access and experiences created by visits to their local forest. The children demonstrated an active relationship with their local forest environment, being aware of its landmarks and facilities. Deeper awareness extended across only quite a small geographical area of their forest. They had personal tacit knowledge of the usefulness of the forested environment, as opposed to abstract, generalised ideas, but only indirect allusion was made to legal, institutional and economic topics. Living organisms were by far the most frequently mentioned forest feature. Landscape amenity, also attachment to and identity with their local forest appear to be important to school children. Although the values held by children living close to forests in New Zealand differed across the three regions, they all fell within the purview of the Montréal Process C&I. Conversely, Criteria 5 (Carbon) and 7 (Legal and institutional) were only tenuously held values.
(Continued on next page)

* Correspondence: karen.bayne@scionresearch.com
[1]New Zealand Forest Research Institute Ltd (Scion), University of Canterbury, Ilam, Christchurch, PO Box 29237, Forestry Road, Fendalton, Christchurch 8540, New Zealand
Full list of author information is available at the end of the article

(Continued from previous page)

Conclusions: School children's views and understanding of forests in New Zealand offer an insight into the integration of local communities with forests and forestry, forestry activities and forest amenities. Taking children's values into consideration has broadened our understanding of the comprehensiveness of the Montréal Process C&I. Results from our workshops provided input for New Zealand's contribution to a revision of the Montréal Process C&I because delegates could cross-reference the proposed alterations to the Indicators against the nationally derived values, which included those of school children.

Keywords: Children; Forest; Values; Montréal Process; New Zealand; Sustainable forest management

Background

Today's society is becoming increasingly aware of the importance of environmental quality, global and local risk, and protection of human rights (Dunlap and Jorgenson 2012; Farrow 2015; Kotler et al. 2011). Internationally, sustainable forestry is required to adhere to stakeholder values, particularly those of the local community; identifying and understanding values for a forest environment is considered a necessary part of forest management (Lavallee 2010). People are prepared to challenge practices they consider pose an environmental risk (Graffy and Booth 2008), and such challenges can have important ramifications. For example, the ability of the forest sector to operate and/or to access local and international markets could be reduced, or the standing of a country could be diminished in international arenas. In order to mitigate or nullify this possibility, governments must be able to demonstrate to their constituents that forestry practices reflect community values.

Criteria and indicators (C&I) as guiding principles for sustainable forests have been developed internationally (for example the Montréal Process C&I described below); however, these represent one values system—that of international experts. The values system that represents community values is best represented by locals' experiences with, and conations regarding, their local forest environment. Here we compare and contrast the international Montréal Process C&I for reporting on the sustainability of national temperate and boreal forests against the local experiences and values of school-aged children in New Zealand to determine whether the Criteria and Indicators are sufficiently comprehensive to encompass the values of children as stakeholders of forests, and explore how extensive and comprehensive the children's values are for their local forests.

The Montréal Process

Set up as the Working Group on Criteria and Indicators for the Conservation and Sustainable Management of Temperate and Boreal Forests, the Montréal Process (MP) is an example of an international framework of C&I for monitoring the sustainability of forests arising from societal concerns about the future of forests (Montréal Process Working Group 2015). It was formed in Geneva, Switzerland, in June 1994 to develop and implement an internationally agreed approach for monitoring the sustainability of member country's forests. In 1995, the twelve member countries endorsed a set of criteria and related indicators to be used by policymakers. Member countries publish five yearly reports on the state and progress of their forests towards sustainability across all the indicators.

The Montréal Process Criteria and Indicators (MP C&I) have been developed through several reviews (see Additional file 1), which required agreement to be reached among all member countries. The Technical Advisory Committee (TAC) meetings that approved the MP review process were in Buenos Aires, Argentina in November 2007, where revised indicators for criteria 1–6 were signed off, and on Jeju Island, Korea, in June 2009, where revised indicators for criterion 7 were also agreed. The resulting set of indicators has been used since 2010 for national reporting.

At the national policy level, the principle that demonstration of sustainability should reflect stakeholder values is considered to be valid (Wyatt et al. 2011). The 2004 review required each country to consult its stakeholders in order to "lead... to improved co-operation and harmonisation between domestic stakeholders and international processes" (Minutes of the MP Technical Advisory Committee Meeting, June 7–12, 2004, in Iguazú, Argentina). Surveying the values held by their respective residents would allow the 12 country Technical Advisory Committee members to discover inadequacies, lack of representation or even inappropriateness in the coverage of MP C&Is for all aspects of sustainable forest management. Conversely, it can be used to inform residents about the social, economic and environmental importance of forests and to engage local communities in the Montréal Process review and stimulate discussion on what sustainable forestry means to their respective countries.

New Zealand forest environment

The case study site for this research was based on New Zealand forests. There are some unique aspects to forests and forestry in this country affecting responsibilities for the forests and influencing how the public interact with forests. Forests in New Zealand are largely divided into two types: native untended forest conservation areas (23 % of the land cover; 6.2 million ha), which afford ecosystem services such as biodiversity and recreational values (Smaill et al. 2014) along with picturesque tourism landscapes (Swaffield and Fairweather 2003), and production planted forests (7 % of the land cover; ~1.8 million ha) that are managed predominantly for commercial benefit through timber harvest (Yao et al. 2013). These production forests are not public parks, but allow public access out of good will though this may be restricted to certain areas, or at certain times such as during harvesting. While most consist of an exotic planted resource, their location and history can vary and, except for a few cases (e.g. Bottle Lake Forest; Whakarewarewa), are usually quite distant from residential areas.

In contrast to publically owned conservation areas managed by the Department of Conservation, production forests in New Zealand are almost completely in private ownership. Historically, some planted forests have been accessible to the public for example for hunting and fishing, but only those in close proximity to residential areas have any history of being visited regularly by the public (Turner et al. 2011). Hence most public visits to forested areas are within the national conservation estate. Children therefore do not visit nearby forests alone or with friends for play purposes, but may visit with their families or on school visits, particularly as limits to recreational access in private forests have increased in response to safety concerns during forest operations and permits for access are often required. Despite such restrictions, production forests can be desirable destinations for some people, even remote forests, as there are more roads and tracks through them than for native conservation forests (Burgin and Hardiman 2012).

Ownership differences are also reflected in different requirements and objectives for management to incorporate community forest values (Smaill et al. 2014). Forests in the conservation estate are managed for conservation values including public values for this land. Privately owned planted forests, on the other hand, address public values more indirectly, as part of achieving a social "license to operate" (Hock et al. 2009) and to mitigate challenges to forest operations.

Despite their differences, both forest types contribute to people's experience of forest environments and to the sustainability of forests within New Zealand. Children are likely to be unaware of such distinguishing features as management objectives.

Incorporating stakeholder values

New Zealand, as co-signatory to the Montréal Process, set about to: (i) identify the values that New Zealanders hold for their local forests; and (ii) categorise these values in order to compare them with the international values outlined in MP C&I.[1]

Children as stakeholders in forestry

A number of studies have demonstrated that children and adults use and value the natural world in a different way (e.g. Balling and Falck 1982; Barraza 1999; Blizard and Schuster 2004; Kaplan and Kaplan 1989; Lovell and O'Brien 2009; Rydberg and Falck 2000; Schneekloth 1989; Snaddon et al. 2008; Strommen 1995), which may also hold for forested environments. However, the available literature does not identify specific attributes of a forest that are important from a child's perspective or how their values of forested landscapes differ from that of adult counterparts (Fjørtoft and Sageie 2000; Gundersen and Frivold 2008; Lovell and O'Brien 2009; Rydberg and Falck 2000).

Preferences for forested or urban landscape areas as a play environment for children are well documented (e.g. Blizard and Schuster 2004; Fjørtoft and Sageie 2000; Francis 1988; Francis and Lorenzo 2006). However, children's concepts and values associated with forest environments have received less attention. Studies undertaken with children about their understanding of types of forest environments (e.g. rainforest), those in which animals and plants may be present (Rydberg and Falck 2000; Snaddon et al. 2008; Strommen 1995), where the value of forests to the environment (Barraza and Pineda 2003; Greaves and Stanisstreet 1993; Lovell and O'Brien 2009) and the economy (Barraza and Pineda 2003), all focus on perceptions about generic forests, rather than knowledge of and experiences gained from interactions with local forests.

Understanding children's perceptions about forests is important because childhood interaction with nature through play activity has a profound impact on physical and mental wellbeing (Fjørtoft 2001; Fjørtoft and Sageie 2000; King and Church 2013; Rydberg and Falck 2000) and leads to latent awareness of the importance of environmental stewardship (Chipeniuk 1998). Hence it is valuable to know how forests in a country are contributing to this.

Snaddon et al. (2008) considered that childhood experience of the natural world is critical for the development of adult environmental stewardship. Visits to natural environments can help children to understand nature and natural processes. For example, early personal contact, often from foraging in forests for plants and wild animals, leads to increased competence in identification of biota (Rydberg and Falck 2000). Schneekloth (1989) noted that "an individual's relationship with nature can be formed and/or reinforced either from beneficial experiences with nature, such as walks in the woods, or from negative experiences..." p. 16. According to Blizard and Schuster (2004),

"Children who grow up with limited exposure to diverse natural environments may develop into adults who are unconcerned about the loss of biodiversity" p. 57. Understanding children's exposure to forested environments was thus considered helpful for understanding the childhood basis they were developing for the forest environment, on which their sense of attachment and care for the forest environment in adulthood would grow. This extends beyond the forest itself to include the potential for increased eventual understanding of the importance of forest ecology to the local community (Dwyer et al. 1991; Fjørtoft 2001).

There is little doubt that today's children are spending less time in natural environments than many of their forebears (Francis and Lorenzo 2006) and that this could affect future attitudes to forestry practice. Those aged younger than 15 years represent 20 % of the national population (Statistics New Zealand [StatsNZ] 2014) who may be experiencing the forest environments very differently to the current policymakers. New Zealand children's daily exposure to production forestry is usually, if at all, as a distant landscape view, or a timber crop on a logging truck. Native forests are usually experienced as a weekend or holiday activity as are the production forests located near to a residential area.

In this study we sought to identify aspects of a local forested environment that different groups of school children consider to be important. We provide a school-age perspective on New Zealand forest values that contributed to discussions about the relevance of MP C&I to forest sustainability in New Zealand, and conversely information being gathered for the MP review process on any inadequacies of the indicators.

Methods

We conducted a series of workshops throughout New Zealand between 2004 and 2009 to engage forest user groups, management agencies and school children. The research method for this project was based on the Canadian Council of Forest Ministers Criteria and Indicators Framework Review: Focus Group Research (Compas 2002). This paper reports on the workshops held for school children[2] considered to be important stakeholders in New Zealand's forests. Parallel results from workshops conducted with adult stakeholders as part of the wider project will be reported elsewhere.

Participant selection

We selected six case study areas to represent different geographical locations, different levels of proximity to urban environments, differences in forest type and scale, different degrees of involvement in the forestry industry and different community characteristics. Six schools from within three of these areas (Tapawera, Nelson, South Island; Hunua and Kaiata, Auckland, North Island; and Lynmore, Rotorua, Bay of Plenty, North Island) were selected for participation in facilitated workshops (Fig. 1 and Table 1). The six school sites all had a landscape view of the local forest from their property, although these three regions had very different types of forest.

The Decile Rating of participating schools ranged from 4 to 9.[3] Each school was largely made up of European as well as Māori students (New Zealand aboriginal people) with a smaller percentage of Asian and other ethnicities.

Students participating in the workshops were aged between 9 and 15 years old. The research approach was approved through the Scion Ethics Committee. Following ethical practices, we contacted school principals to seek permission and discussed objectives with teachers prior to meeting the children. Teaching staff assisted with our management of the workshops, which were held during ordinary class time.

Workshop structure and facilitation

All workshops followed the same format, designed to be appropriate for a variety of population groups differing in location, background, age and experience. They were structured to achieve robust engagement with participants in a workshop setting.

We used a participatory, interactive workshop approach following a variety of internationally recognised best-practice procedures. The workshop method was tailored to suit a classroom environment and used age-appropriate terminology.

All the workshops were conducted by the same two trained facilitators. One facilitator led the workshop, and a second observed the process, recorded dialogue and comments, and made observations on behaviour. This second person played a passive role during group discussions but encouraged children to participate when working in small groups.

The process involved a formal welcome followed by an explanation of the purpose of the workshop and two exercises:

Exercise 1: drawings and representations of forests

This exercise focussed participant children on significant features of forests and on the activities that take place in local forests. It was designed to capture information in visual form for later analysis. Small groups (3–5 participants) were asked to draw a picture of their local forest on an A1 sheet of paper and to include in it the following:

- Activities that they associated with forests, including those that they valued most in their local forests; and
- Important features that they associated with the forest.

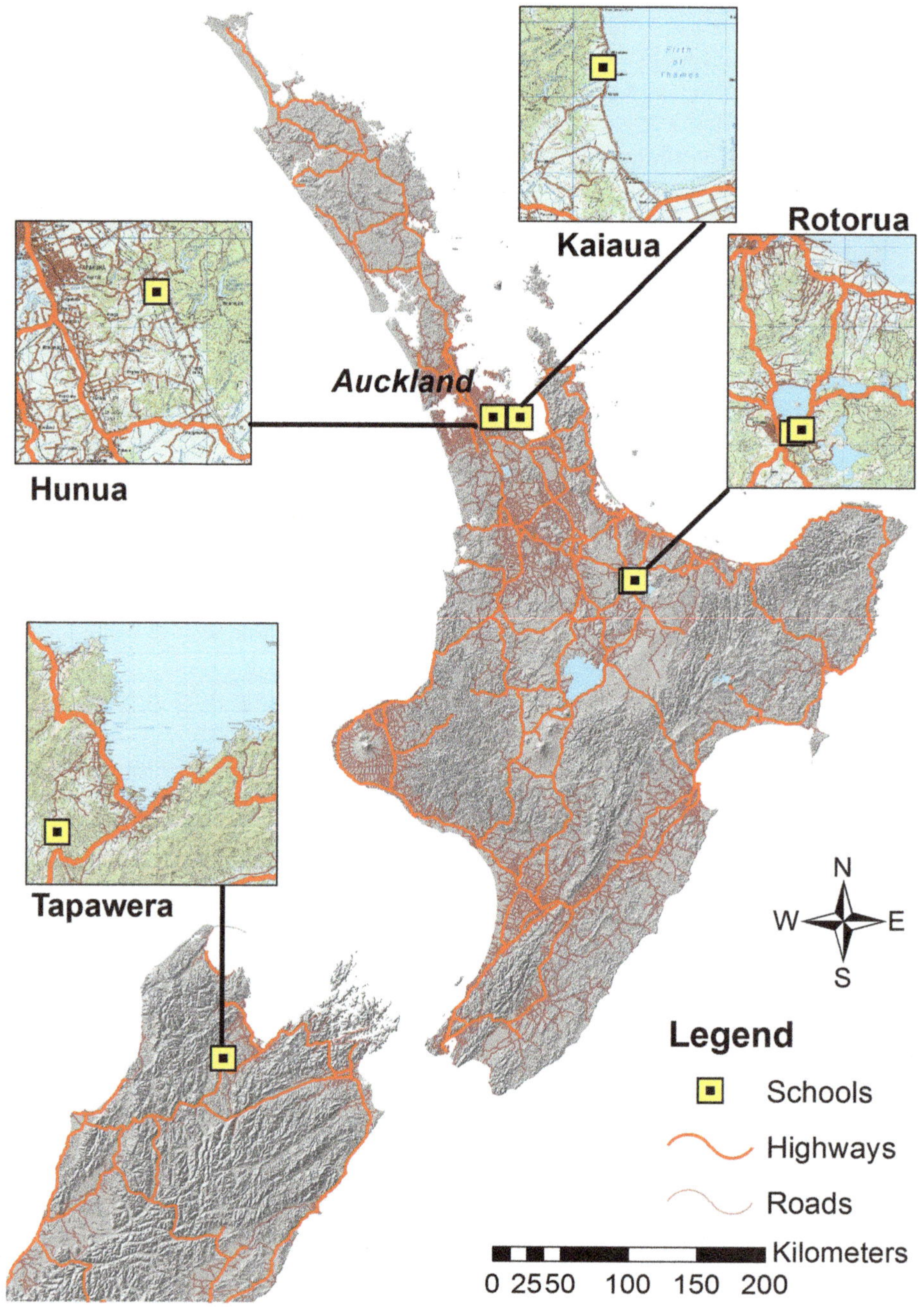

Fig. 1 Regions in New Zealand where workshops with school children were held

Participants were then invited to explain their drawings (e.g. Fig. 2) to the whole workshop group. Points of clarification were sought where drawings were unclear, and discussion was encouraged.

Exercise 2: identification and grouping of valued features
Children working in small groups were asked to write on pieces of paper the things that they thought were most important about their local forest. Teachers and facilitators assisted younger students with this writing task. Each piece of paper had one feature or value of importance and was placed on a wall. When all the groups had run out of contributions, the facilitator worked with the whole workshop group to achieve collective understanding by (a) grouping together values that were similar and (b) giving each collection of values a theme name. The students were then invited to discuss the themes and to make sure that all their thoughts had been captured.

Table 1 Workshop participants

School	Number of workshops held	Number of participants
Nelson region		
Tapawera Area School, South Richmond	3	23
Year 0–13 students from the horticultural and farming area surrounding the Tapawera township		18 26
Auckland region		
Hunua School	3	23
A full primary school (year 1–8) serving a small rural community situated in the Hunua Ranges		17
Kaiaua School		21
A small rural primary school (year 1–8) located in the coastal settlement of Kaiaua on the Firth of Thames		
Rotorua region		
Lynmore Primary School	3	27
A large year 1–8 suburban school situated in Rotorua		
Mokoia Intermediate School		One class (~25)
A year 7–8 school located in the eastern suburbs of Rotorua		
Rotorua Lakes High School		32
A coeducational secondary school (year 9–13) in the eastern suburbs of Rotorua		

Post-workshop processing

After the workshop, information from written words, drawings and dialogue notes were tabulated. If a drawing was not clear, or a child had not told the wider group what the picture depicted, the contribution was ignored.

Two tables were drawn up for each workshop. In the first, each valued feature, drawn or mentioned, was assigned to one of the four following categories:

Fig. 2 An example of children's drawings produced in Exercise 1

- Key species or landmarks (animal names, plant types, waterways and local landmarks etc.);
- Other natural features (clouds, hills, rocks, ponds etc.);
- Infrastructure (manmade features such as tracks and roads, bridges, signs and machinery); and
- Activities (what was happening or what the children enjoyed doing in the forest).

The second table listed all features and themes named in Exercise 2. These features were also combined in a word cloud using Wordle™ software (Feinberg 2013) which gave visual prominence to the most frequently mentioned items. Using all the features mentioned in the workshops, a word cloud was prepared for each of the three regions.

The intent was for features and themes to be aligned to the seven Montréal Process Criteria (Additional file 1). Given the youth of the participants, discussions were tailored to a more general level than these criteria and so for the analysis, their values were mapped into seven broad categories:

- *Biodiversity*—a range of diversity of species living in a forest ecosystem
- *Human needs*—forest values for human needs; for example environmental services, wood products, work, income, and food
- *Forest stewardship*—people's responsibility of actions towards forest protection and management for different reasons
- *Social wellbeing*—for example identity, sensory factors, peacefulness
- *Environmental context*—soil, water and carbon
- *Social benefits—recreation*—particularly having fun and enjoyment in the forest
- *Economic and institutional issues*—employment and maintenance of the forest

An overview of this process is depicted in Fig. 3.

The broad categories were then related back to the seven MP Criteria. Presence or absence of the features associated with each of the 2007 MP C&Is was noted. This formed the input to the MP review. In addition, a post-MP C&I review check was undertaken to ensure the children's values matched the indicators to confirm that the relevant values were (still) being addressed.

Results

In total, some 200 children took part in the nine workshops. Of these, 50 were aged 13–15 and the rest aged between 9 and 12 years. A similar number of boys and girls were present in each workshop.

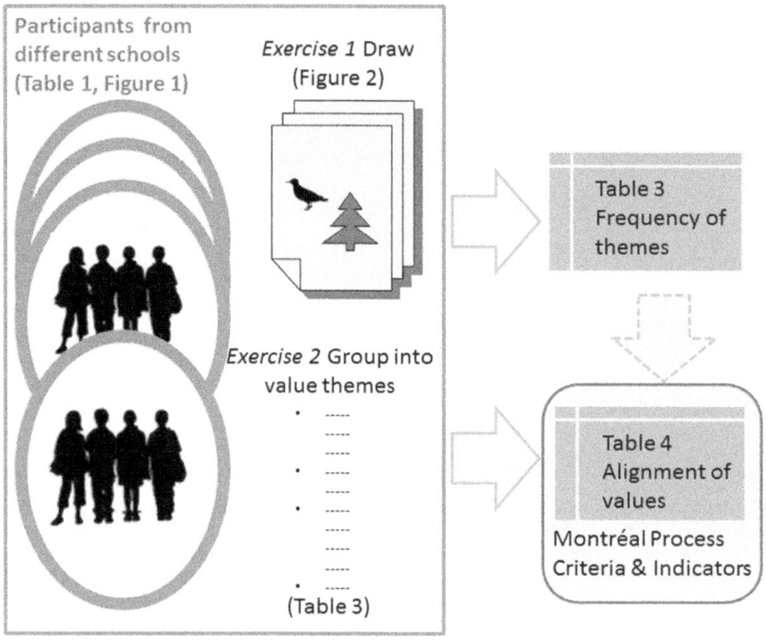

Fig. 3 Outline of methodology used in school children's workshops

Data management

An example of the table listing all features and themes named in Exercise 2 is shown in Table 2; this table shows theme names developed in one of the schools.

* Numbers in brackets denote the number of times a particular feature was mentioned. Drawings from Exercise 1 and the naming of features in Exercise 2 indicated that the children were aware of a wide range of natural components, landscape, infrastructure, industry workings and activities associated with their local forests.

Themes

Each of the workshops reported a slightly different range of forest features considered to be important by participants. This may reflect the different forest cover present, and also the regional difference in forest-based employment. Grouping of the named themes into broad categories (Table 3) showed that those most frequently identified were related to nature, plants and animals (five of the six schools); water and recreational activities (three schools each).

Valued features drawn and named in Exercises 1 and 2

A. *Biodiversity*—the range of species living in the forest
At every workshop, native birds and animals, agricultural herds, fish, a number of plant species (exotic and native) and/or introduced pests were identified. Some children referred to fungi and insects, others to the importance of nests, bees, and flowering plant species. Drawings were very detailed, showing animals *doing things* (e.g. a possum eating a leaf).

B. *Human needs*—e.g. environmental services, wood products, work, income, food
Children talked about wood for houses and furniture, indicating awareness that forests are a source of physical human requirements. They identified infrastructure present in the forest and described its purpose. Recreational activities, hunting and fishing were also discussed. There were many drawings and comments about hunting (often pigs) and getting food from the forest.

C. *Forest stewardship*—responsible action relating to forest protection and management
Children at Hunua spoke of the importance of having people in the forest to help keep the environment clean (i.e. Rangers). People were seen to be important for the forest because they plant trees, take care of the bush (forest) and get rid of pests.

D. *Social wellbeing*—e.g. identity, sensory factors, peacefulness
Drawings from Exercise 2 conveyed both emotional and sensory experience. Sounds of water and other forest components were written on the pictures, indicating that stimulation and sensory observation were associated with experience of the forest environment. Students at Tapawera told stories about being in the forest, indicating that their local forest was a normal part of their everyday experience.

Table 2 Theme names suggested for sets of valued features

Theme name	Feature*
Native plants	Native plants
Animals	Animals, wildlife, pest control
Creatures	Birds, pigs, animals (2), insects, fungus, humans
Nature	Birds, insects, animals (2), mosses and plants, nature
Nature and the world	The nature, dirt and soil so trees can grow, grasses for animals, trees (3), carbon dioxide, poison berries and good berries, plants to keep cold away [medicines from plants]
Clean water	Clean water, fresh drinkable water
Water	Water, rivers and lakes, rain
Needs	Logging, the trees supply our everyday needs... paper, houses, oxygen, gives us wood for furniture, houses etc., trees turn carbon dioxide into oxygen
Looks good [forest] instead of towers	Looks good [forest] instead of towers
Fresh air	Fresh air
Smell	We like the fresh air—smell
Forests are protected	Forests are protected
Work	Overseas income, good income, creates employment, logging (2), jobs, timber, roads so people can get to the forest
Logging	Jobs, timber, roads so people can get to the forest, logging (2)
4 × 4s	4 × 4s [NB sparked discussion on whether these were important, half of the class thought they were because they helped people get into the forest, half of the class did not think they were important]
Recreation	Camping, good to go exploring, walking tracks and huts, recreation
Outdoor activities	Tramping, people can visit, outdoor activities
Hunting (2)	We like hunting pigs, hunting (2)

*The numbers in brackets indicate how often this particular feature was mentioned

Children spoke about what they or family members did in the forest, suggesting that it was normal practice to enter the forest. Some children talked about what they themselves had seen while in the forest, and described particular locations that they had visited.

E. *Environmental context*—soil, water, carbon
Some children talked about how trees turn carbon dioxide into oxygen for us to breathe, indicating awareness that forests provide environmental services.

F. *Social benefits*—e.g. *recreation*—particularly fun and enjoyment
Drawings and stories depicted many activities carried out in the forest, e.g. a person going hunting, a fisher catching a fish. Comments included mountain biking and walking. Paths, tracks, bridges and signposts were commonly shown features reflecting use of recreational facilities. There were stories about hunting with parents and repeated accounts of their parents' hunting experiences.

G. *Economic and institutional issues*—employment and forest maintenance
Some children talked about a parent's employment in the forest. Tapawera students were well-versed in the terminology of the logging industry, naming heavy machinery correctly and understanding the growing and logging of trees and the importance of associated income to the region. Their drawings showed logging trucks, haulers and other machinery in great detail. Some children spoke about the importance of signage and track infrastructure. Older children were also aware of pest control activities and the impacts of pests, particularly pigs and possums.

Table 3 Frequency of workshop participant proposals for theme names relating to valued forest features.

Category	Themes	Number of schools identifying theme topics (maximum = 6)
A. Biodiversity	Native plants	5
	Animals	
	Creatures	
	Nature	5
	Nature and the world	
B. Human needs	Clean water	3
	Needs	2
C. Forest stewardship	Forests are protected	1
D. Social wellbeing	Looks good	3
	Fresh air	2
	Smell	2
	Recreation	3
	Outdoor activities	3
E. Environmental context	Water	3
F. Social benefits—recreation	Hunting	1
G. Economic and institutional issues	4 × 4s	1
	Work	1
	Logging	1

Additional values

Some groups included a wide landscape view in their picture, depicting mountains. This indicated appreciation of the forest as a component of the landscape. The following example from one of the Hunua workshops suggests that a varied landscape is important:

Facilitator: "What do you mean by 'if there weren't any forests New Zealand would be bare'?..."

"...well New Zealand would look bare...there would just be grass..."
"...there would be no trees..."
Facilitator: "Do you mean to look at?...so the forest is important to look at?"
"...yeah, it would be bare otherwise..."

Individual statements made concerning landscape attributes included the vastness of the forest, and the vista that is created by forestry in contrast to urban or agricultural land use.

The facilitators noted a strong sense of pride and local identity associated with nearby forests.

Regional differences

Although Wordle™ assessment confirmed that trees and animals (including pests) were considered by most children to be important forest features, the choice of words and images conveyed different perceptions of local forests in the three regions investigated.

Nelson—The local forests in the Nelson region are predominantly of exotic radiata pine (*Pinus radiata* D.Don) grown for production purposes. Drawings in Exercise 1 were more detailed than those from other regions and depicted more activity. At the Primary School level, Tapawera children drew attention to a greater range of values than Hunua and Kaiaua (Auckland) or Lynmore (Rotorua) children (Fig. 4a–c). These included employment and production activities (particularly logging) and forest biodiversity (birds, insects, fungi, soil, animals and trees).

Auckland—Workshops in this region elicited a strong theme of nature conservation. This may reflect the Regional Park status of the forest in the selected area, and the native forest cover present. Drawings were orientated towards nature and activity. They included a range of flora and fauna: trees, flowers, birds, fish, eels etc. Activities included camping, swimming and fishing. Native trees were a prominent feature along with birds and water. Drawings included rivers and views of the Hunua Ranges. Water (particularly waterfalls) was depicted, and the locally iconic Hunua Falls were often placed in the middle of the picture. A wide range of activities and features was included. Although children at Kaiaua drew very detailed pictures outlining very diverse activities (logging, planting trees, tramping), they depicted a small range of values. When asked what was most important about the forest, they cited trees and animals, humans and food.

Rotorua—Fewer values overall were educed from participants in this region. The children were more aware of tracks and signage. The value of recreation and particularly the peaceful nature of the forest was prominent. Forestry plantations surrounding Rotorua are mostly of mixed exotic species. Despite some similarity with other regions in the broad range of activities and depth of knowledge, the values here fell into the three broad categories of physical ecology, recreation, and emotive connection. Rotorua was the only region in which school children did not mention production forestry verbally, although drawings included pine trees, logging trucks and piles of logs. This is interesting in view of the fact that Rotorua is one of the major forestry centres of New Zealand.

Age differences

Values were largely similar across the school-age groups; however, older children were able to provide greater detail, displaying greater social awareness and depicting more political images. Older children were more politically aware than 9–12-year-olds, mentioning specific provocative issues such as illegal cannabis plantations, and use of 1080 poison to control pests. Older school children in Nelson were able to relay tales of hunting in detail, while younger children relayed camping and tramping stories.

The school children at the Hunua workshops talked a lot about why people have to manage forests. However, where adults (forest users and management agencies) in this region placed important value on historic features of the local Hunua ranges, the school children did not. Children put a lot of detail on their drawings, often preferring to draw than write words. Their drawings showed people and animals doing specific things, for example, animals eating leaves, cows in streams, camp fires burning. Adults in the wider workshops often would draw a tree as a symbol for all trees; in this way, they were much more generic and considered it sufficient to speak to a diagram than an elaborate drawing.

Agreement with values of the MP Criteria

Elements corresponding to most of the Montréal Process Criteria and their Indicators were found among the workshop responses (Table 4).

Criterion 1—Children were aware of and appreciated the presence of many different living organisms in the forest.
Criterion 2—Children recognised the ability of the forest to provide wood and to enable clean water and carbon dioxide–oxygen cycling.
Criterion 3—Children were aware of the health benefits of the forest, although features mentioned relate to *human* health rather than the health of the forest ecosystem.
Criterion 4—Children showed limited understanding of the broader ecological value of soil conservation and

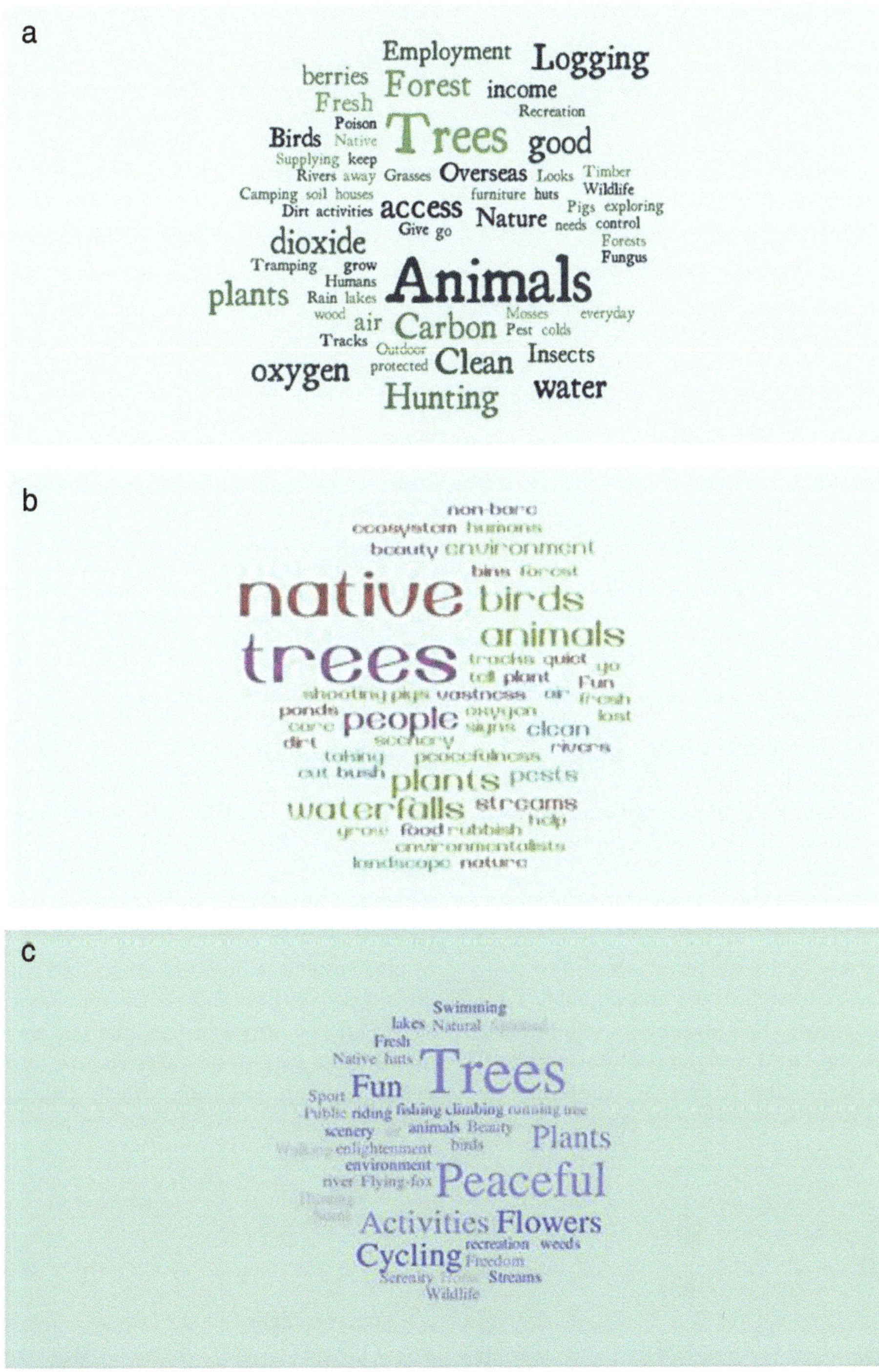

Fig. 4 **a** Tapawera children's values. **b** Hunua children's values. **c** Rotorua children's value

water quality, although they knew that soil and water are needed for trees to grow.

Criterion 5—There was little reference to global carbon cycles besides knowledge that trees utilise carbon dioxide and produce oxygen.

Criterion 6—The children appreciated recreation and scenery associated with forests, particularly those encompassing waterways. The sense of pride and local identity associated with nearby forests was not considered to be aligned with Criterion 6 but rather with a place attachment value extraneous to MP C&Is.

Criterion 7—Children appreciated the role of people in maintaining the productive capacity of the forest and provision of infrastructure for human use.

Table 4 Alignment of childrens' values with the Montréal Process Criteria

Montréal Process Criteria	Valued forest features drawn or named
Criterion 1: Conservation of biological diversity	Creatures (including native and exotic species), wildlife, native plants, trees, flowers, fungi, mosses, poison berries and good berries, weeds, plants to keep colds away [medicinal plants], animals, native animals, native birds, birds, humans, insects; forest ecosystem, nature and the world/nature/life
Criterion 2: Maintenance of productive capacity of forest ecosystems	Grasses for animals, fresh air, forest ecosystem, O_2, food, nature
Criterion 3: Maintenance of forest ecosystem health and vitality	The trees help us survive, trees with fresh air, clean environment, nature
Criterion 4: Conservation and maintenance of soil and water resources	Dirt and soil so trees can grow, clean water, fresh drinkable water, water, rivers, streams, ponds, lakes, waterfalls, rain
Criterion 5: Maintenance of forest contribution to global carbon cycles	Carbon dioxide, nature and the world, O_2
Criterion 6: Maintenance and enhancement of long-term multiple socio-economic benefits to meet the needs of societies	Plants to keep colds away [medicinal plants], fresh air—we like the smell, trees supply our everyday needs (wood, houses, furniture), recreation—full range of activities presented, we like all the activities there are to do, a public place for recreational activity, walking tracks and huts, people can visit, hunting, it is ours, it is the Hunua, fun, beauty, peacefulness and quiet, peaceful scenery, serenity, spiritual enlightenment, scent of the forest, fresh air, freedom
Criterion 7: Legal, institutional and economic framework for forest conservation and sustainable management	Forests are protected, overseas income, roads so everyone can get to the forest, tracks help us not get lost, signs to tell you where to go, shooting pests, people plants trees, people take care of the bush, people get rid of pests, people cut down trees, humans, environmentalists, rubbish bins, logging, good income, creates employment

Agreement with MP indicators

Most of the forest features identified by children in the workshops could be related to one or more of the seven MP Criteria. Some of the MP Indicators were found to be more relevant to the children's perceptions than others (Fig. 5).

Living organisms were by far the most frequently mentioned forest features. The children showed awareness of diversity of species associated with forests, both flora and fauna (Indicator 1.1.a: Ecosystem types and arguably 1.1.b: Protected ecosystem types). This awareness extended to biosecurity threats to the forest and the need for pest control for species such as possums (Indicator 3.a: Biotic processes and agents).

Visiting and experiencing the forest were considered to be important (Indicator 6.4.a: Recreation and tourism areas; 6.4.b Recreation and tourism visits). Visits were implied through depiction of many forest-based activities. These were often culturally based, e.g. picnics, hunting (Indicator 6.5.a: Cultural protection). Forests evidently had high relevance and importance for these children (Indicator 6.5.b: Importance of forests) although the sense of pride and local identity associated with nearby forests is considered by the authors to represent place attachment rather than cultural or spiritual connection. The children identified a variety of infrastructure associated with their visits, e.g. car parks and tracks (Indicator 6.2.a: Capital investment).

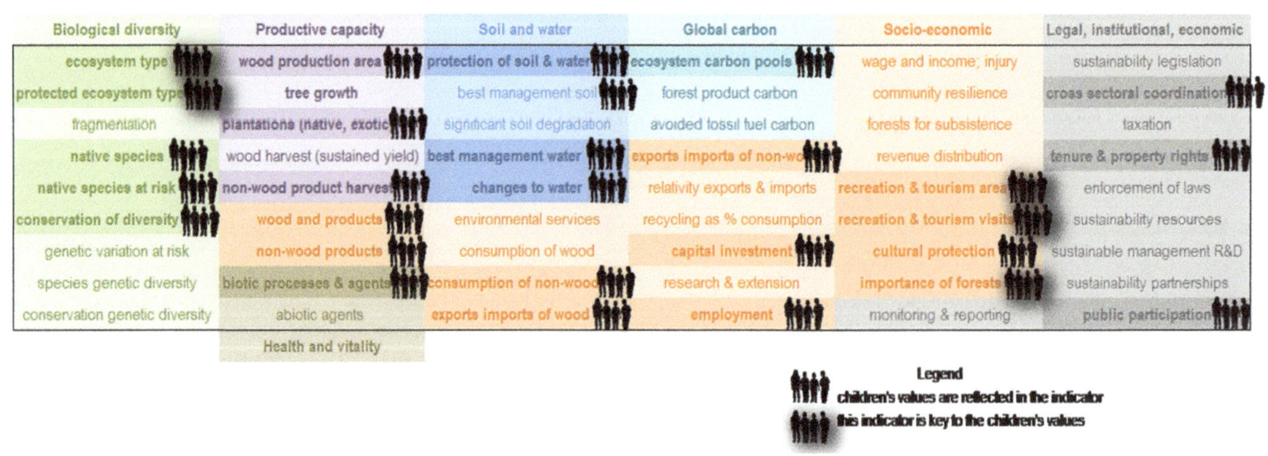

Fig. 5 MP indicators most relevant to the children's values (relevant ones are written in *bold* with slightly darker background)

Non-wood forest products were commonly mentioned or implied, e.g. food from hunting and fishing, and Christmas trees (Indicator 2.e: Non-wood product harvest, and 6.1.b: Non-wood products). These were mentioned at least as frequently as wood products.

While school children were aware of the range of activities and of biodiversity present within a forest, there was little connection with Criterion 5: Global carbon cycles. Some of the children knew that trees produce oxygen (an indirect link with Indicator 5.a; Ecosystem carbon pools and fluxes) but there was limited understanding about soil conservation or water quality (Criterion 4). They appreciated recreation and scenery associated with waterways in forests (possible connection with Indicator 4.3.b Changes to water).

Criterion 7 and its Indicators appeared to have little relevance. Only indirect allusion was made to legal, institutional and economic topics. Other industrial sectors were indirectly acknowledged through reference to sheep and grass (Indicator 7.1.b: Cross sectorial coordination). Mention of fencing could be indicative of an emergent consciousness of tenure and property rights (Indicator 7.3.a).

Discussion

Forest features considered by children to be important

Drawings in the present investigation pertained largely to nature and recreation. The frequency of named values confirmed the perception that trees and animals were the most important forest features. This agrees with the findings of Snaddon et al. (2008) who reported that children perceive forests primarily as either a habitat for animals or a wilderness, but contrasts with those of Strommen (1995) who analysed drawings produced by First Grade children in the USA when asked to draw a rainforest and describe what lived there. He found that although trees and animals predominated, understanding about which animals inhabited specific forest environments (e.g. jungle or woodland) tended to be confused. He concluded that "Children's ideas about forests and their inhabitants appear to be very global and imprecise" and that they are "aware of only stereotypical [habitat] extremes" p. 697. Barraza (1999), from a study of 7–9-year-old children's drawings, concluded that their perceptions of the environment were relatively pessimistic. They tended to draw rural settings, were aware of environmental problems and showed concern about the future. This negativity differs from our generally positive results.

In order to position our findings on children's awareness of forest structure within the literature, the results of several studies are first discussed. Snaddon et al. (2008) suggested that children lack understanding about forest ecosystem functionality, particularly the role of insects and soil. Other investigations (e.g. Rydberg and Falck 2000; Blizard and Schuster 2004; Lovell and O'Brien 2009) indicate that although children are aware of forest features they do not understand forest structure or the function and importance of forest ecosystem services. In a study of Mexican secondary school students participating in forest management educational programmes, Barraza and Pineda (2003) found that "Young people perceived forests as extraneous to them and were unaware of their biological, ecological, and economic value" p. 10. They "lacked knowledge about forest issues and about the consequences of the loss of their forests" p. 16. Responses indicated that perceptions were not affected by proximity to forest management activities but were positively influenced by forest protection programmes in which they were able to participate. Greaves and Stanisstreet (1993) investigated children's understanding of rainforests in an open interview with secondary pupils. They reported appreciation of the effect of forests on climate change, but also misconceptions about the role of oxygen in the process. Children emphasised the importance of rainforest habitat for endangered animal species, but they were less aware of the presence of endangered plants and other organisms.

In contrast to these reports, our workshops involving children living near to forests in New Zealand have revealed awareness of forest structure, the presence of many types of living organisms in forests and the activities and environmental services associated with forests. The children demonstrated an active relationship with their local forest environment. In particular, they were aware of its landmarks and facilities. There was clear evidence that they had personal tacit knowledge of the usefulness of the forested environment, as opposed to abstract, generalised ideas. They did not regard the forest as extraneous to their lives. There were comments relating to the economic importance of forests, and to opportunities for employment and recreation taken up by their families. Proximity to the local forest appeared to influence their understanding and their identity with its associated activities and landscape vistas.

Results of each case study region reflect the local forest of that area. For example, in Tapawera, school children drew lots of forest machinery and logging trucks, depicting the growing importance of forestry to the local economy. The Rotorua school children most associated with the nearby Whakarewarewa forest, which has large recreational areas set aside while areas where forestry activities are occurring are often inaccessible to the public for safety reasons. This may account for the emphasis on recreational activity, and lack of production forestry as an important value. Given the long-term association of forestry in the Rotorua region, it may also be that this is no longer recognised as anything special, in comparison with Tapawera where a rise in forestry activities occurred in the years preceding this study. Results from Hunua reflect the Regional Park status of the area, through drawings and values of nature conservation and recreation, and the importance of the local Hunua Falls. Drawings were of varied activities, and this was relayed when asked what they enjoyed most about forests:

"Bike riding"
"Motor biking over a bridge through puddles"
"We like how much nature we have in our forests"
"We like streams and waterfalls"
"We like to hear the birds singing"

These results indicate that school children were drawing what they know of their *local* forest, and not depicting values relating to forests in general. The results also show an awareness of not only what is occurring in their local forest, but the importance of the forested environment to the local region. For example, Tapawera children displayed greatest awareness of forest ecosystem services (fresh air and clean water). From the literature, we know that for children the level of experience of a forested area is linked to awareness of ecosystem processes and the production of more detailed drawings in workshop investigations (Strommen 1995), hence the Tapawera children's awareness may reflect a greater level of exposure to their local forest.

Knowledge of the ways in which environmental values are interlinked was considered by our facilitators to be undeveloped. In common with the results of other studies, we noted a number of misconceptions about the environmental and economic value of forests, particularly the importance of forests to the wider community.

Considerations for the future

The workshops held with the children contributed to the knowledge base on New Zealand forest stakeholder values and were also used to direct New Zealand's research on the environmental aspects of sustainable forest management (Hock 2013). Children rely on experiences and knowledge transfer regarding forestry practices from their peers, parents and teachers, but the information must be well informed to avoid bias from unsubstantiated viewpoints (e.g. Greer 2014).

Future forests will be shaped by the ways in which they are valued and used by society. People are beginning to recall the influence that forests have on the quality of life. There is a consciousness among children as well as adults that forests are a place for play, for relaxation and escape as well as a source of economic and environmental benefits. Results from our workshops suggest that children value forests for a range of reasons that can be categorised as ecological, environmental, social, emotional, productive or economic. The school children identified values that, while sometimes expressed in different terminology, were similar to adult participant responses of the wider research programme. Adults that were subsequently interviewed in Nelson spoke of "conservation", "biodiversity" and "economic value" while their school children spoke of "creatures", "work" and "income" (K Crawford pers comm.). These differences align with the literature which indicates that adults perceive landscapes as forms; children as functions (Fjørtoft and Sageie 2000) for which they often give a very sensory depiction of what they see, smell, and hear or how exciting they find the activity.

While children participating in our study did regard their local forest as places for recreation, we were not able to determine if their behaviours emulated Western societies' current tendency for children to interact less with nature and to spend less time outdoors (Francis and Lorenzo 2006). Certainly, the results supported the concept that when natural settings were made available for play they were valued and enjoyed. It would be interesting to study if all areas would have similar responses to those in the example whereby children became angry and upset when a forest area adjacent to a schoolyard was harvested by the owner, and the play areas they had created within it were subsequently destroyed in the process (Blizard and Schuster 2004).

The opportunity for being unrestrained and active was also highly valued by the children in our study. The literature explains that forest environments offer privacy (secret, non-adult spaces) and provide opportunities for imaginative and adventurous recreation (Blizard and Schuster 2004; Francis 1988). From a child's perspective, one major feature of an accessible forested area is the opportunity for modifying surroundings through play and recreation (Blizard and Schuster 2004; Francis 1988; Rydberg and Falck 2000; Strommen 1995). Children participating in our study appreciated their local forest as a place for physical recreation, particularly walking, biking, hunting and fishing. Some owners of planted forests in New Zealand have responded to the desire for recreation, even catering for thrill-seeking adventure sports (Cloke and Perkins 1998) and setting aside dedicated areas for mountain biking. There is evidence of activities associated with mountain biking (e.g. within Whakarewarewa) as well as off-road 4 × 4 activities modifying the planted forest environment to suit, through the establishment of dedicated trails (Burgin and Hardiman 2012; King and Church 2013), creating a modicum of "common good" under a private business paradigm that more commonly tends to exclude the public.

The responses were also considered in regard to place attachment and landscape preference; these may be influenced by familiarity, or else developed by learning to belong to the land through foraging and other childhood experiences (Chipeniuk 1998). Māori have a strong affiliation to their land or *whenua*, grounded in *whakapapa* as the creation stories relate to landscapes that link with their genealogy. While a place attachment value relates to MP Indicator 6.5.a, it is not clear if this is spiritually based or culturally learned, and a limitation of the study is the lack of any distinction between local Māori school children and other ethnicities in terms of local forest values. People care more for environments considered to be "uniquely theirs" (Chipeniuk 1998); this attachment is usually with indigenous landscapes, rather than to flora and fauna of the region (Chipeniuk 1998). In our study, Nelson and Rotorua children drew attention to features associated with exotic forests in their vicinity (e.g. mountain biking, redwoods, productive economy) whereas Hunua children had a strong feeling for the surrounding indigenous/native bush. This reflects an affinity for the familiar, rather than for provenance.

We argue that consideration of the perceptions of very young people during decision-making process about forest management procedure could have a positive effect on their development and on the values they hold into the future. Involvement of young persons in activities specific to the forest environment could also enhance their understanding of ecosystem processes and how these benefit the wider community. For more knowledge on quite how children prefer and value forests, however, other methods may illuminate more detail than was possible in this study, for example observing children during their outdoor play or by experimental photo studies.

Input into the Montréal Process

Our results contribute to fulfilling New Zealand's obligations for reviewing and incorporating stakeholder values into the MP framework. The children's workshops, together with results from workshops with adult stakeholders (unpublished data), achieved the country's obligations to understand how the within-country values aligned with the MP C&I, and whether stakeholder values were adequately incorporated within the MP framework (Ministry of Agriculture and Forestry [MAF], 2009). Montréal Process Indicators that were not raised by the children may represent gaps in their knowledge, and this provides a guideline for extending their knowledge base.

The findings provided input into the Technical Advisory Committee (TAC)'s negotiations during the review of the MP C&I ensuring the inclusion of the country's views (Fig. 6). The Technical Advisory Committee members, who included one of the authors (Tim Barnard), were able to cross-reference proposed alterations to the MP indicators against within-country values, so ensuring adequate representation of the indicators and the comprehensiveness of MP.

Limitations

This study reports on the difference in values between local (in this case school childrens') experience of forests as compared to those of international experts through the MP C&I set. As such, our methods focussed on eliciting from children via participatory workshops how they perceived their local forest and tabulated drawings and dialogue to match against the seven broad criteria of the Montréal Process. While this approach sufficed for the purpose of this study, alternative methods, such as facilitator-led classroom dialogues to provide greater depth around children's stories of personal experiences of the forest and what they did there. Alternatively, observations or short surveys (most likely at the weekend or public holiday) of families with children who were seen experiencing the forest environment may have provided a greater body of evidence for childhood forest interactions but were outside the scope of this study.

Discussing each of the seven criteria in turn with the children during the workshop to elicit the values the forests afforded for each criterion would have provided direct evidence to compare and contrast value agreement with the established seven C&I. However, such an adult's perspective of important forest values (i.e. the seven broad criteria) would have imposed a top–down framework onto the discussions. Instead, by not directly discussing the criteria, our indirect route afforded a bottom–up approach of eliciting the values that the children held themselves, rather than only in the context of the values important to international experts.

Information obtained from the workshops was predominantly qualitative. Participating schools were selected from specific case study areas and were not intended to be representative of children throughout the country. The information collected does however provide some insight into the way that very young New Zealanders regard forestry as a whole, and their local forests in particular. Our sample of children aged 9–15 years was enthusiastic about sharing experiences of local forests and demonstrated good understanding and appreciation of many forest features. This does not imply that all New Zealand children have a similar level of understanding—those from urban regions or areas without readily accessible forests are likely to have different perceptions.

It is important to note that the children were likely to omit mention of things they disliked (Tamoutseli and Polyzou 2010) and also that the level of detail in a picture is not always commensurate with actual experience. Group dynamics also probably influenced the results to

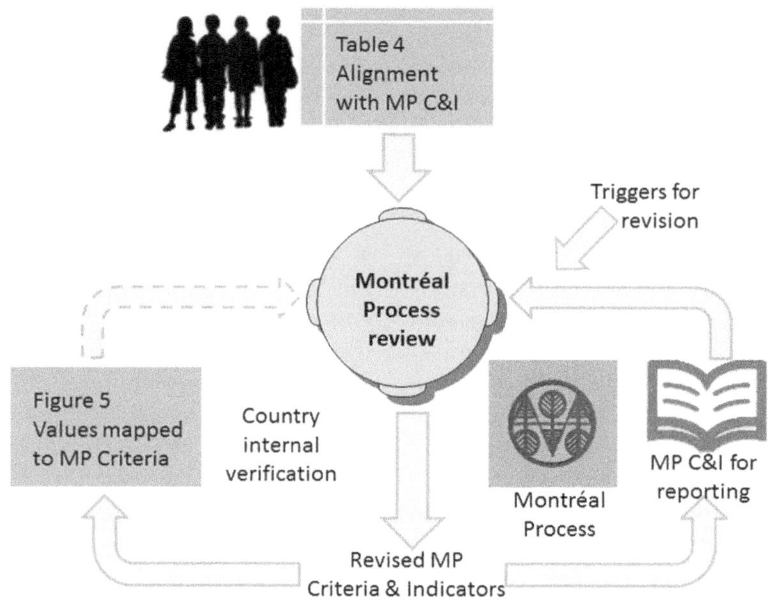

Fig. 6 Incorporating New Zealand workshop results into the MP review

some extent. The children talked loudly and excitedly and individual thoughts were often elaborated by others in the group. Although the facilitators endeavoured to be consistent in their approach, there was some inevitable variability in the workshop experience. For example, input from the classroom teacher sometimes changed the outcome of a discussion.

Conclusions

The mapping of children's values onto the MP C&I provides an opportunity for exploring how the C&I framework could be used to inform and maintain an education curriculum on forests around the seven criteria. Values held by children aged 9–15 years living close to forests in New Zealand are broadly aligned with most of the MP Criteria, although connection with Nos. 5 and 7 is somewhat tenuous.

School children in our study demonstrated a broad knowledge about activities in their local forests and about forest components such as different living organisms. They were aware of the environmental services, employment, wellbeing and recreational opportunities that are associated with forests. It is clear that, while similar forest values are important to both school children and adults within a region, children experience and express these differently to older members of the public and appear to hold identity with a smaller geographical forested area.

The participants in our study held a wide range of values for their local forest, and while they may not express their concepts and ideas in the same terminology as adults, nor use the same rationale, the expressed values encapsulate almost all those values present within the Montréal Process Criteria. Landscape amenity, also attachment to, and identity with their local forest appear to be important to school children.

Forest features valued by children in this study are regionally distinct and relate to the particular forested environment with which they are familiar. Children are known to gain a greater appreciation for conservation policies and develop a sense of care for the forested environment when they are familiar with natural settings.

Conservation awareness is critical to the acceptance of sustainable forest management policy, and there is a need to ensure that school children are exposed to environments in which forest management can be seen in action and appreciated.

School children's views and understanding of forests in New Zealand offer an insight into the integration of local communities with forest components, forestry activity and forest amenities. These views and attachment are likely to be preserved and maintained throughout their lives.

The iterative development of the Montréal Process C&I provided a best-practice monitoring approach for the sustainability of forests. Taking local values into consideration, including children's values, during the iterations enhanced the New Zealand technical experts' knowledge base when representing the country in the international discussions on the developments and revisions to the MP indicators.

Policymakers should be aware of the ways in which children, as a vitally important component of society, relate to forests. Children are not only the inheritors of managed forests but, given the opportunity, are current major recreational users of the resource.

Endnotes

[1]These aims are outlined in an unpublished report: Tim Barnard, Harley Spence and Kirsten Crawford. (Feb 2006). *New Zealand Montréal Process Review: Forest Values in New Zealand.* Contract report for Ministry of Agriculture & Forestry. Rotorua, New Zealand: ENSIS

[2]This research article draws from the following unpublished reports and datasets:

- Tim Barnard, Harley Spence and Kirsten Crawford (2007). *If you go down to the woods today...Young peoples' forest values in New Zealand.* Technical Report for the New Zealand Montréal Process Review. Rotorua, New Zealand: Scion and Coastline Consultants.
- Tim Barnard, Harley Spence and Kirsten Crawford (Dec 2007). *New Zealand Montréal Process Review: Forest Values in New Zealand - Young Persons Technical Report. Forest Values in New Zealand. Report Volume Two: Young Persons Values.* Rotorua, New Zealand: ENSIS and Coastline Consultants.
- Tim Barnard, Harley Spence and Kirsten Crawford. (Feb 2006). *New Zealand Montréal Process Review: Forest Values in New Zealand.* Contract report for Ministry of Agriculture & Forestry. Rotorua, New Zealand: ENSIS
- Tim Barnard, Harley Spence and Kirsten Crawford. (Jul 2005). *Forest Values in New Zealand and their alignment with the Montréal Process Criteria and Indicators. Forest Values in New Zealand: Engaging in the Montréal Process Review.* Contract report for Ministry of Agriculture & Forestry. Rotorua, New Zealand: ENSIS.
- Tim Barnard, Harley Spence and Kirsten Crawford. (July 2010). *Adding Value to New Zealand Forests.* Report no. ES008 for Future Forests Research.
- Tim Barnard, Harley Spence and Kirsten Crawford. (July 2010). *Giving Regional Context to International Sustainable Forestry Indicators.* Report for Future Forests Research.

[3]The Ministry of Education uses a decile rating (ranking) system for school funding purposes. Each decile contains approximately 10% of schools. Schools in decile 1 have the highest proportion of students from low socio-economic backgrounds. Schools in decile 10 have the lowest proportion of these students. (Education Review Office [ERO] 2015).

Abbreviations

1080: sodium fluoroacetate (a metabolic pesticide used in New Zealand forests); 4 × 4: Four-wheel drive vehicle; C&I: Criteria and Indicators; CO_2: carbon dioxide; MP C&I: Montréal Process Criteria and Indicators; MP: Montréal Process; O_2: oxygen; TAC: Technical Advisory Committee.

Competing interests

The authors declare that they have no competing interests.

Authors' contributions

TB conceived of the study and designed the workshops that were facilitated by HS and KC. The original unpublished reports were authored by KC with analytical assistance from HS and TB. KB reviewed the literature and with BH led the drafting of the manuscript, analysis for the discussion section and submission process. BH presented results from the wider study to industry. TP liaised with Alan Reid and MAF staff members and with TB assisted in the incorporation of the results into MP C&I reporting at TAC meetings. All authors have read and approved the manuscript.

Authors' information

Karen Bayne is a social scientist with specialist knowledge in marrying consumer requirements with the needs of the market. Her research involves understanding consumer behaviour in terms of forest products and human values of forest environments. Barbara Hock specialises in environmental spatial analysis and modelling. Her current research involves visualisations for sustainable forest management. Harley Spence has long-term experience in the resource management sector. He has particular expertise in working with communities and stakeholder engagement processes. Harley has worked in both the private and public sectors and is currently the Marine and Coastal Area Manager for the Office of Treaty Settlements. Kirsten Crawford has long-term experience in the resource management sector with a particular interest in partnerships between public and industry, as well as education. She is a skilled analyst and project manager and currently manages the Southern Seabird Solutions Trust, an alliance between the New Zealand Government and fishing interests. Dr. Tim Payn comes from a background in soil science and supports global efforts around sustainable forest management. Tim currently chairs the 12-country Montréal Process Technical Advisory Committee which steers policy aimed at defining and measuring sustainable practice. Tim Barnard is a social scientist specialising in policy analysis and environmental planning. He represents New Zealand on the Technical Advisory Committee of the Montréal Process.

Acknowledgements

The authors are indebted to the teachers and pupils at participating schools for their time and enthusiasm. Alan Reid (Ministry for Primary Industries) is acknowledged for discussions on the international context and implications of the research, and Ruth Gadgil (Scion) for assistance with earlier drafts of this paper. We thank the two anonymous reviewers for useful comments, which have improved the quality of the manuscript.

Author details

[1]New Zealand Forest Research Institute Ltd (Scion), University of Canterbury, Ilam, Christchurch, PO Box 29237, Forestry Road, Fendalton, Christchurch 8540, New Zealand. [2]Coastline Consultants, Rotorua 3010, New Zealand.

References

Balling, JD, & Falck, JH. (1982). Development of visual preference for natural environments. *Environmental Behaviour, 1*, 5–28.

Barraza, L. (1999). Children's drawings about the environment. *Environmental Education Research, 5*(1), 49–67.

Barraza, L, & Pineda, J. (2003). How young people see forests in Mexico: A comparison of two rural communities. *Unasylva, 54*(213), 10–17.

Blizard, C, & Schuster, R. (2004). *They all cared about the forest: Elementary school children's experiences of the loss of a wooded play space at a private school in upstate New York.* Paper presented at the 2004 Northern Recreation Research Symposium, Bolton Landing, New York.

Burgin, S, & Hardiman, N. (2012). Extreme sports in natural areas: Looming disaster or a catalyst for a paradigm shift in land use planning? *Journal of Environmental Planning and Management, 55*(7), 921–940.

Chipeniuk, R. (1998). Childhood foraging as regional culture: Some implications for conservation policy. *Environmental Conservation, 25*(3), 198–207.

Cloke, P, & Perkins, H. (1998). "Cracking the canyon with the awesome foursome": representations of adventure tourism in New Zealand. *Environment and Planning D: Society and Space, 16*, 185–218.

Compas. (2002). Focus Group Research. In CCFM (Ed.), *Canadian Council of Forest Ministers Criteria and Indicators Framework Review*. Ottawa, Canada: Canadian Council of Forest Ministers .

Dunlap, RE, & Jorgenson, AK. (2012). Environmental problems. In *The Wiley-Blackwell Encyclopedia of Globalization*. Hoboken: Wiley-Blackwell.

Dwyer, JF, Schroeder, HW, & Gobster, PH. (1991). The significance of urban trees and forests: toward a deeper understanding of values. *Journal of Arboriculture, 17*(10), 276–284.

Education Review Office [ERO]. (2015). *FAQs 5. What is a decile rating of ranking?* Retrieved 30 September 2015 from http://www.ero.govt.nz/Help/FAQs

Farrow, G. (2015). *The rise of the conscious consumer*. New Zealand: New Zealand Herald.

Feinberg, J. (2013). *Wordle*. Retrieved from http:// www.wordle.net

Fjørtoft, I. (2001). The natural environment as a playground for children: The impact of outdoor play activities in pre-primary school children. *Early Childhood Education Journal, 29*(2), 111–117.

Fjørtoft, I, & Sageie, J. (2000). The natural environment as a playground for children. Landscape description and analyses of a natural playscape. *Landscape and Urban Planning, 48*(1–2), 83–97.

Francis, M. (1988). Negotiating between children and adult design values in open space projects. *Design Studies, 9*(2), 67–75.

Francis, M, & Lorenzo, R. (2006). Children and city design: proactive process and the 'renewal' of childhood. In C Spencer & M Blades (Eds.), *Children and their environments: learning, using and designing spaces* (pp. 217–236). Cambridge: Cambridge University Press.

Graffy, EA, & Booth, NL. (2008). Linking environmental risk assessment and communication: An experiment in co-evolving scientific and social knowledge. *International Journal of Global Environmental Issues, 8*(1–2), 132–146.

Greaves, E, & Stanisstreet, M. (1993). Children's ideas about rainforests. *Journal of Biological Education (Society of Biology), 27*(3), 189–195.

Greer, G. (2014). Our hunger for wood is destroying forests. *The Telegraph*. Retrieved from http://www.telegraph.co.uk/earth/environment/forests/10863451/Germaine-Greer-Our-hunger-for-wood-is-destroying-forests.html

Gundersen, VS, & Frivold, LH. (2008). Public preferences for forest structures: A review of quantitative surveys from Finland, Norway and Sweden. *Urban Forestry and Urban Greening, 7*(4), 241–258.

Hock, BK. (2013). Value-driven sustainable forest management in New Zealand. *International Journal of Environmental Sustainability, 8*(4), 71–85.

Hock, BK, Payn, T, Clinton, P, & Turner, J. (2009). Towards green markets for New Zealand plantations. *New Zealand Journal of Forestry, 54*(1), 9–18.

Kaplan, R, & Kaplan, S. (1989). *The experience of nature. A psychological perspective*. New York: Cambridge University Press.

King, K, & Church, A. (2013). 'We don't enjoy nature like that: Youth identity and lifestyle in the countryside. *Journal of Rural Studies, 31*, 67–76.

Kotler, P, Kartajaya, H, & Setiawan, I. (2011). *Marketing 3.0: from products to customers to the human spirit*. Hoboken, NJ: Wiley.

Lavallee, LF. (2010). *Human values associated with forests. Forests and Forestry in the Americas: An Encyclopedia* (29 Oct 2010th ed.). New York: Routledge Taylor & Francis Group.

Lovell, R, & O'Brien, L. (2009). Wood you believe it? Children and young person's perceptions of climate change and the role of trees, woods and forests. In *Research Summary* (p. 2). Farnham, UK: Forest Research - Social and Economic Research Group.

MAF. (2009). *Sustainable management of New Zealand's forests - The 2008 New Zealand country report on the Montreal Process criteria and indicators. Report by Ministry of Agriculture and Forestry*. Wellington: Policy Publications.

Payn, T. (2015). Criteria and indicators for the conservation and sustainable management of temperate and boreal forests (5, September 2015th ed.). Montréal: Montréal Process Working Group. [http://www.montrealprocess.org/documents/publications/techreports/MontrealProcessSeptember2015.pdf]

Rydberg, D, & Falck, J. (2000). Urban forestry in Sweden from a silvicultural perspective: A review. *Landscape and Urban Planning, 47*(1–2), 1–18.

Schneekloth, L. (1989). "Where did you go?" "The forest." "What did you see?" " Nothing. *Children's Environments Quarterly, 6*(1), 14–17.

Smaill, S, Bayne, K, Coker, G, Paul, T, & Clinton, P. (2014). The right tree for the job? Perceptions of species suitability for the provision of ecosystem services. *Environmental Management, 53*(4), 783–799.

Snaddon, JL, Turner, EC, & Foster, WA. (2008). Children's perceptions of rainforest biodiversity: Which animals have the lion's share of environmental awareness? *PLoS One, 3*(7):e2579.

StatsNZ. (2014). *New Zealand in Profile 2014. Report from Statistics New Zealand*. Wellington: New Zealand Government.

Strommen, E. (1995). Lions and tigers and bears, oh my! Children's conceptions of forests and their inhabitants. *Journal of Research in Science Teaching, 32*(7), 683–698.

Swaffield, S, & Fairweather, J. (2003). *Contemporary public attitudes to landscape*. Paper presented at the *Reclaiming our Heritage, Environmental Defence Society/New Zealand Institute of landscape Architects conference, Bruce Mason Centre, Takapuna, Auckland, New Zealand*.

Tamoutseli, N, & Polyzou, E. (2010). Using drawings to assess children's perceptions of schoolyard environment: A case study of a primary school in Drama Greece. In P Dondon & O Martin (Eds.). *Latest Trends on Engineering Education* (pp. 54–57). Sofia, Bulgaria : WSEAS Press. http://www.wseas.us/books/2010/Corfu/EDUCATION.pdf

Turner, J, Dhakal, B, Yao, R, Barnard, T, & Maunder, C. (2011). Non-timber values from planted forests: recreation in Whakarewarewa forest. *New Zealand Journal of Forestry, 55*(4), 24–31.

Wyatt, S, Rousseau, MH, Nadeau, S, Thiffault, N, & Guay, L. (2011). Social concerns, risk and the acceptability of forest vegetation management alternatives: Insights for managers. *Forestry Chronicle, 87*(2), 274–289.

Yao, R, Barry, L, Wakelin, S, Harrison, D, Magnard, L-A, & Payn, T. (2013). Planted forests. In JR Dymond (Ed.), *Ecosystem services in New Zealand – conditions and trends*. Lincoln, NZ: Manaaki Whenua Press.

Susceptibility of New Zealand flora to *Phytophthora kernoviae* and its seasonal variability in the field

Judy Frances Gardner*†, Margaret Anne Dick† and Martin Karl-Friedrich Bader

Abstract

Background: The oomycete *Phytophthora kernoviae* is known from the United Kingdom and the Republic of Ireland, where it is considered to be a recent invader, from Chile where it was only discovered in 2014, and New Zealand where records date back to 1953. As there is little information in New Zealand linking *P. kernoviae* with plant disease, it may have been present for much longer and may be indigenous. Seasonal activity of *P. kernoviae* in a site known to have infested soil was tested by isolation from soil and foliage of existing shrubs and the use of indicator plants. In greenhouse studies, the susceptibility of a range of indigenous plants to *P. kernoviae* was tested via stem and foliar inoculation.

Methods: Soil, litter and understorey vegetation samples were collected for isolation of *P. kernoviae* at monthly intervals for a year. Plants of *Rhododendron catawbiense*, which is known to be susceptible to foliar and shoot infection, were placed in the stand as indicator species. In laboratory and greenhouse studies, stem and foliar inoculations of a selection of arborescent plants representing major groups within the New Zealand flora were carried out and compared with three exotic plants of known susceptibility.

Results: *Phytophthora kernoviae* was not isolated from foliage of understorey plants at the study site, but it was recovered from soil and litter from April to November (autumn through spring) inclusive. Little disease developed on the *Rhododendron catawbiense* indicator plants.
All of the exotic, and most of the indigenous, species developed a lesion in response to stem inoculation. Stem lesions were more developed on exotic species than on indigenous hosts. In contrast, few species formed foliar symptoms. *Phytophthora kernoviae* was recovered from asymptomatic tissue, stems and foliage, of a number of species.

Conclusions: Based on the results of the inoculations and the lack of historical records of disease of indigenous plants associated with *P. kernoviae*, this oomycete does not appear to be a damaging pathogen of New Zealand's indigenous flora. Although presence in the soil or litter was demonstrated at the study site, little disease developed on the indicator plants suggesting that at least part of the New Zealand population is of low pathogenicity to *R. catawbiense*.

Keywords: *Phytophthora kernoviae*; Plant disease; Forestry; New Zealand flora; Susceptibility

* Correspondence: judy.gardner@scionresearch.com
†Equal contributors
New Zealand Forest Research Institute (Scion), Private Bag 3020, Rotorua 3046, New Zealand

Background

Phytophthora kernoviae Brasier, Beales and S.A. Kirk is an oomycete pathogen that infects a range of plant genera (Brasier et al. 2005; Beales et al. 2006; Beales et al. 2009) in the United Kingdom and the Republic of Ireland causing foliar necrosis and shoot dieback on both tree and shrub hosts, and bleeding cankers on some tree hosts. Response to infection ranges from host mortality to minor foliage spotting depending on the host species affected (Denman et al. 2006b). Inoculum in the form of sporangia and zoospores is produced on infected foliage but generally not on bleeding cankers of the trees which are thereby designated "dead-end hosts" (e.g. Davidson et al. 2005). Denman et al. (2009b) also demonstrated that *P. kernoviae* could infect the foliage of some plants without symptom expression.

More recently, *P. kernoviae* has been found to naturally infect plants from 20 genera representing 13 families (Anon 2012; Dick et al. 2014) in New Zealand and other parts of the world. Host plants are predominately angiosperms although three conifers, each from a different family, are also recorded hosts (Anon 2012; Dick et al. 2014). Additional host species have been shown to be susceptible with artificial inoculations (Anon 2012, Fichtner et al. 2012).

The existence of *P. kernoviae* in New Zealand was first recognised early in 2006 (Ramsfield et al. 2009) although it had been recorded earlier as an undescribed species (McAlonan 1970). In 2006, it was known as a pathogen only of the introduced cherimoya (*Annona cherimola* Mill.) from orchards in the far north of the country (Ramsfield et al. 2009). Other isolates, which were from Northland and the Central Plateau of the North Island, were from soil and were not linked to plant disease. However, *P. kernoviae* has now been recovered from locations throughout the North Island, and from a few locations in the Westland and Nelson regions in the South Island of New Zealand (MA Dick, unpublished data). In addition to infecting *An. cherimola*, foliar necrosis of *Rhododendron* spp. and *Pinus radiata* D.Don (both exotics) has occasionally been recorded (Dick et al. 2014). To date, *Phytophthora kernoviae* has been isolated only twice from native New Zealand species, once from the dying lower foliage of a podocarp *Prumnopitys ferruginea* (D.Don) de Laub (Dick et al. 2014) and once from an asymptomatic fern *Blechnum novae-zelandiae* T.C. Chambers and P.A. Farrant (MA Dick, unpublished data).

Little is known of the life cycle and ecology of *Phytophthora kernoviae* in New Zealand. Apart from the disease outbreak on *An. cherimola,* there is no evidence of the destructive behaviour that has been reported in woodlands and heathlands in parts of Great Britain (Brasier et al. 2005; Beales et al. 2006, 2009; Denman et al. 2006a, 2009). Phylogenetically, *P. kernoviae* is located in *Phytophthora* clade 10 and hence has a likely Southern Hemisphere connection (Ramsfield et al. 2009). The recent identification of *P. kernoviae* from indigenous forest in Chile (Sanfuentes et al. 2014) may lend further support to this concept. The hypothesis that *P. kernoviae* is native to New Zealand is being tested using a population genetic study (D. Cooke, D. Studholme and R. McDougal, unpublished data). If this hypothesis is found to hold true, then *P. kernoviae* may be either acting as an occasional minor pathogen of indigenous plants or it could be endophytic and generally having no detrimental effect on its hosts.

This paper reports on several studies undertaken to develop an understanding of the life cycle of *P. kernoviae* in New Zealand and to determine the effect that it has on New Zealand indigenous plants. Firstly, the recovery of *P. kernoviae* from soil and litter throughout the year was examined at a forest location where this species had been recorded in the 1950s and 1960s (McAlonan 1970) and subsequently in 2006 (M. Bullians, unpublished data). Secondly, understorey vegetation (both exotic and indigenous plants) was evaluated and foliage with lesions of unknown cause collected for analysis each month for 8 months from September 2007 until April 2008. Plants of *R. catawbiense* Michx., which is known to be susceptible to *P. kernoviae* (Denman et al. 2005), were taken into one stand as indicator plants for *P. kernoviae* pathogenic activity.

In addition, glasshouse and laboratory studies included stem and foliar inoculations of a selection of arborescent plants representing major groups within the New Zealand flora, and on a number of ferns that are common in plantation forest environments. The ability of *P. kernoviae* to colonise these tissues asymptomatically after inoculation was tested through re-isolation. Lesion formation was also recorded. For comparison, three exotic host species of known susceptibility were also inoculated.

Methods

Field study

Site

Four plots (labelled 'a'–'d'), each 1.5 m^2, were established about 250 m apart within a plantation of 20-year-old *Pinus radiata* where soil had tested positive for *Phytophthora kernoviae* in 2006. The plantation was located at Pinedale (38° 03′ S, 175° 48′ E) in the central North Island.

Indicator plants

Rhododendron catawbiense plants were used both as indicator plants for *P. kernoviae* pathogenic activity and to provide bait leaves for soil bioassays.

A 1-m-tall *R. catawbiense* was planted in August 2007 at each of plots 'a' and 'b'. These plants were monitored monthly for any typical symptom development. Three leaves were collected each month from each plant, symptomatic leaves if present or healthy leaves if there were no signs of disease, for isolation in the laboratory. If dieback occurred, branch or twig samples were collected for isolation from affected tissue. Once dead, these plants were dug up and returned to the laboratory for analysis.

A potted 50-cm-tall *R. catawbiense* was placed at each of plots 'a' and 'b' in each month from October 2007 to April 2008. Each month, the previous month's plants were collected, placed in a shade house at the laboratory and monitored for development of leaf lesions and dieback. Plants were spatially separated and bottom-watered to ensure that no cross-infection could occur. Any diseased material was collected for isolation in the laboratory. Plants used for this trial were propagated by cuttings from a parent plant to ensure genetic uniformity.

Soil samples

For each of the four plots, soil and plant litter were collected with a trowel from three randomly selected places within the plot to a soil depth of 150 mm. The three samples from each plot were bulked and thoroughly mixed. The total amount of soil and litter was not less than 1.0 L per plot. Samples were collected at monthly intervals and analysed in the laboratory as described below. Soil sampling commenced in April 2007 and was carried out for 12 months.

Understorey plants

Foliage samples from the six understorey plants that dominated the plots and the immediate area surrounding the plots were collected each month from July 2007 until April 2008. These species were: *Bl. novae-zelandiae* (palm-leaf fern); *Cyathea medullaris* (G.Forst.) Sw. (tree fern), two ground ferns (*Pteris tremula* R.Br. and *Diplazium australe* (R.Br.) N.A. Wakef.), *Rubus fruticosus* agg. (blackberry) and *Metrosideros diffusa* (G.Forst.) Sm.) (climbing rata). Foliage with lesions of unknown cause was selected, but healthy leaves were also included. Leaves from a sole *Camellia* sp. bush (probably originating from garden rubbish that had been dumped near plot 'a') was also sampled as *Camellia* species are a recognised host of *Phytophthora kernoviae* in the United Kingdom. The samples were tested for *Phytophthora kernoviae* in the laboratory.

Isolations

Soil samples Each soil sample was flooded with distilled water to a depth of twice the depth of the soil and four *R. catawbiense* leaves floated on the surface of the water as bait. After 7 days, bait leaves were surface sterilised, cut into sections and plated onto amended synthetic Mucor agar (SMA) medium (prepared as described by Brasier et al. 2005) and incubated in the dark at 20 °C for up to 14 days. Any *Phytophthora*-like colonies were sub-cultured as they emerged onto carrot agar (CA) (Erwin and Robeiro 1996) for identification. Colony morphology, sporangial and gametangial characters were examined, and the temperature at which mycelial growth ceased was determined. DNA sequencing of selected isolates was carried out to confirm morphological identification of *P. kernoviae*.

Leaf samples

Small pieces (5 mm^2) of leaf were cut either from live/dead margins of lesions if present or from random pieces from healthy leaves. Leaf pieces were surface sterilised and placed on SMA or pimaricin, ampicillin, rifampicin and pentachloronitrobenzene (PARP; Jeffers and Martin 1986) media and incubated in the dark at 20 °C. *Phytophthora*-like colonies were examined as described for those obtained from soil samples.

Inoculations

Species selection

Eighteen native plant species for stem inoculations and 29 native species for foliar inoculations were selected from 17 genera. Species choice was influenced by availability of potted plants for the stem inoculations. Both canopy and understorey plants from the major forest types (broadleaf-podocarp and beech) were represented. Three exotic species (*Fagus sylvatica* L., *An. cherimola* and *R. catawbiense*) were also inoculated for comparison as these are known susceptible hosts of *P. kernoviae*. All host species are listed in Table 1. The plants were potted and were held in a shade house during inoculation and for the assessment period of 3 months.

Phytophthora kernoviae isolates

Four isolates of *P. kernoviae* (NZFS 2646, NZFS 3062, NZFS 3339 and NZFS 2707) obtained from the New Zealand Forest Research Institute Culture Collection were used for the inoculations. Details of the original host and collection date are provided in Table 2. All isolates were grown on CA medium at 18 °C in the dark for 14 days prior to use.

Stem inoculations

Two isolates (NZFS 2646 and NZFS 3062) were used for stem inoculations. Inoculum comprised a 5-mm-diameter mycelial plug taken from the edge of an actively growing colony. Carrot agar plugs without cultures were used as controls.

A flap of bark (10 mm long × 3 mm wide) was peeled back from a clear section of the stem, midway between the soil level and first branch of the host plant, and the inoculum placed with the mycelium-side against the

Table 1 Host plants used for stem and foliar inoculations with *Phytophthora kernoviae* isolates and number of replicates

Host	Family	Stem inoculations			Foliar inoculations				Foliar inoculations of wounded leaves		
		Isolate 2646	Isolate 3062	Control	Isolate 2646 (b)	Isolate 2707 (a)	Isolate 3339 (d)	Control	Isolate 3339 wounded	Isolate 3339 Unwounded inoculated control	Wounded Control
Ferns											
Blechnum novae-zelandiae T.C.Chambers and P.A.Farrant	Blechnaceae	–	–	–	–	10	–	5	5	5	5
Cyathea medullaris (G.Forst.) Sw.	Cyatheaceae	––	–	–	–	–	10	5	–	–	–
Paesia scaberula (A.Rich.) Kuhn	Dennstaedtiaceae	–	–	–	–	10	–	5	–	–	–
Asplenium bulbiferum G.Forst.	Pteridaceae	–	–	–	–		5	5	–	–	–
Pteris tremula R.Br.	Dryopteridaceae	–	–	–	–	–	15	5	–	–	–
Conifers											
Agathis australis (D.Don) Lindl.	Araucariaceae	5	5	5	–	–	21	10	–	–	–
Dacrycarpus dacrydioides (A.Rich.) de Laub.	Podocarpaceae	5	5	5	5	15	–	–	–	–	–
Podocarpus totara G.Benn. ex D.Don	Podocarpaceae	–	–	–	–	30	10	5	20	20	5
Angiosperms–monocotyledons											
Cordyline australis (G.Forst.) Endl.	Asparagaceae	5	–	2	–	–	10	5	–	–	–
Angiosperms–dicotyledons											
Annona cherimola Mill.[a]	Annonaceae	5	4	4	–	10	3	5	5	5	5
Beilschmiedia tawa (A.Cunn.) Benth. and Hook.f. ex Kirk	Lauraceae	5	5	5	6	18	31	5	5	5	5
Clianthus maximus Colenso	Fabaceae	–	–	–	–	–	15	5	–	–	–
Coprosma sp.	Rubiaceae	–	8	5	10	17	15	5	–	–	–
Corynocarpus laevigatus J.R.Forst. and G.Forst.	Corynocarpaceae	–	–	–	–	–	15	5	–	–	–
Dodonaea viscosa Jacq.	Sapindaceae	5	5	5	5	–	0	5	–	–	–
Fagus sylvatica L.[a]	Fagaceae	5	5	5	5	15	18	5	5	5	5
Fuchsia excorticata (J.R.Forst. and G.Forst.) L.f.	Onagraceae	5	5	5	5	25	23	5	5	5	5
Gaultheria depressa Hook.f.	Ericaceae	–	–	–	–	–	10	5	–	–	–
Griselinia littoralis Raoul	Griseliniaceae	5	5	5	5	14	5	5	–	–	–
Laurelia novae-zelandiae A.Cunn.	Atherospermataceae	5	5	5	–	20	20	5	–	–	–

Table 1 Host plants used for stem and foliar inoculations with *Phytophthora kernoviae* isolates and number of replicates *(Continued)*

Leptospermum scoparium J.R.Forst. and G.Forst.	Myrtaceae	5	5	5	–	–	15	5	–	–	–
Lophomyrtus bullata Burret	Myrtaceae	–	–	–	–	19	10	5	10	10	10
Melicytus ramiflorus J.R.Forst. and G.Forst.	Violaceae	5	5	5	5	15	15	5	–	–	–
Metrosideros excelsa Sol. ex Gaertn.	Myrtaceae	5	5	5	4	20	10	5	5	5	5
Myrsine australis (A.Rich.) Allan	Primulaceae	5	5	5	5	–	–	5	–	–	–
Nothofagus menziesii (Hook.f.) Oerst.	Nothofagaceae	5	5	5	5	15	14	5	–	–	–
Nothofagus truncata (Colenso) Cockayne	Nothofagaceae	5	5	5	5	5	12	5	–	–	–
Pittosporum eugenioides A.Cunn.	Pittosporaceae	5	5	5	5	11	5	5	–	–	–
Pittosporum ralphii Kirk	Pittosporaceae	5	5	5	5	–	–	5	–	–	–
Pseudopanax sp.	Araliaceae	–	–	–	–	–	10	5	–	–	–
Pseudowintera colorata (Raoul) Dandy	Winteraceae	5	5	5	5	20	5	5	5	5	5
Rhododendron catawbiense Michx.[a]	Ericaceae	–	4	3	–	18	–	5	5	5	5

[a]Species not native to New Zealand

sapwood. The flap was placed back in position over the inoculum, the wound wrapped in moist cotton wool, then with plastic sealing film (Parafilm M®) and then covered with aluminium foil. The number of replicates for each host species inoculated with each of the isolates and the controls is shown in Table 1.

After inoculation, plants were placed in a randomised block arrangement within the shade house. Where possible, each of the five blocks consisted of one plant of each species with a stem inoculation of each isolate and one control inoculation for each species. Plants were watered from below as required to maintain moist soil. Inoculations took place in the spring with night time temperatures ranging from 1 to 19 °C with means of 9, 10 and 12 °C for the months of September, October and November, respectively. Day time temperatures were marginally higher, ranging from 2 to 24 °C and means of 12, 13 and 16 °C for the 3 months as above.

Plants were examined weekly, and any wilting or other indications of decline were recorded. All of the plants were destructively sampled 3 months after inoculation. Bark was peeled back from the wound and lesions or discolouration in the wood measured. Isolations were made from the centre, from 10 mm above and from 10 mm below the centre of the wound. For those plants with lesions or discolouration of the sapwood beyond the inoculation wound, isolations were also made from all the margins of the discoloured tissue. Pieces of wood (approximately 4 mm^3) were excised from the host and were plated onto PARP. After 5 and 10 days, any *Phytophthora*-like colonies that had formed were transferred to

Table 2 Isolates of *Phytophthora kernoviae* used for inoculations

NZFS	ICMP	Host	Geographical region	Date collected
2646	16306	*Annona cherimola*	Northland	19 November 2005
2707		*Blechnum novae-zelandiae*	Waikato	27 July 2006
3062		*Rhododendron catawbiense*	Waikato	18 December 2007
3339		*Rhododendron* sp.	Wellington	21 December 2009

CA for identification using the method described in the soil sampling.

Foliar inoculations

Segments of agar from the actively growing margin of 7-day-old cultures of NZFS 2646, NZFS 2707 and NZFS 3339 (Table 2) were placed in sterile pond water and held at 20 °C with a 12/12-h light/dark cycle for 72 h to induce sporangial production. Zoospore release was promoted by placing the solutions at 4–6 °C for 20–60 min and then returning to 20 °C. Zoospore concentration was determined using a haemocytometer. Zoospore concentration ranged from 1.0×10^4 to 6.0×10^4 zoospores/mL. The number of replicates, for each host species, inoculated with each isolate and the controls is shown in Table 1.

Foliage inoculation 1 Leaves of 32 host species (Table 1) were inoculated on the day they were collected. Detached leaves were dipped (apical end) into a freshly prepared zoospore suspension and held there for 5 min. For host species with small leaves (e.g. *Leptospermum scoparium*), the apical end of a small branch with leaves attached was dipped in the inoculum. Replicates ranged from three to 30. Five or ten leaves of each species were dipped into sterile pond water as control treatments. The dipped material was then laid on moist paper towels in containers and covered in plastic wrap held in a culture cabinet at 20 °C with a 16/8-h light/dark schedule. Lighting was provided by cool white fluorescent tubes.

Foliage inoculation 2 Leaves of 10 host species, three exotic and seven indigenous (Table 1) were wounded by cutting off the apical end prior to dipping in a spore suspension of NZFS 3339 (2×10^4 zoospores/mL). Equal numbers of non-wounded leaves were dipped concurrently. The inoculation and monitoring procedure was as described above for inoculation 1. There were five replicates each of wounded and non-wounded leaves with the exception of the small-leaved *Podocarpus totara* (20 replicates) and *Lophomyrtus bullata* (10 replicates).

Leaves were monitored every 2 days for lesion development. After 7 days, all leaves (whether symptomatic or not) were surface sterilised in 70 % ethanol for 30 s, rinsed in sterile distilled water and plated onto PARP. Plates were incubated at 18 °C in the dark. *Phytophthora*-like colonies were transferred to CA for further examination and identification as for soil sampling.

Statistical analysis

All statistical computations were performed within the R software environment (R Core Team 2014). Generalised linear models using a binomial distribution and logit link were used to analyse the stem and foliar pathogen recovery data, which were recorded as binary responses. The occurrence of quasi-complete separation resulting from the null response seen in numerous species required the use of the Firth-bias reduction (Kosmidis 2013). The response variable was represented by a two-column matrix holding the number of 'successes' (positive recovery) and 'failures' (no recovery). Species identity was modelled as the sole explanatory variable. The large number of tested species (>20), made multiple comparisons ineffectual and, therefore, the responses of the native taxa were pooled and compared with the lumped response of the exotic species. The data aggregation involved with this approach resulted in proportions, which were modelled using generalised additive models for location scale and shape (Rigby and Stasinopoulos 2005).

Stem lesion length was modelled using generalised least-squares models fitted by restricted maximum likelihood (Pinheiro et al. 2014). As described above, species identity and later the pooled responses of native vs. exotic species were modelled as the sole explanatory variable. Graphical validation tools were used to check the model assumptions (residual plots for variance homogeneity and quantile–quantile plots for normality). Heteroscedasticity was detected and modelled with an exponential variance structure using the fitted values as variance covariate.

For all models, the significance of the main effect was tested using likelihood ratio tests.

Results

Field study

Indicator plants

Plot A The planted rhododendron remained healthy from August to February but subsequently wilted and died. *Phytophthora kernoviae* was not recovered from the three healthy leaves collected each month nor from any part of the dying plant, including the roots, at any time.

Potted rhododendrons did not exhibit leaf or shoot dieback, and *P. kernoviae* was not recovered from any asymptomatic foliage that was tested.

Plot B Some shoots on the planted rhododendron died back after 3 months (in December), and the plant continued to decline thereafter. *Phytophthora kernoviae* was isolated from the margins of dead shoot tissue in December and February. The three leaves collected each month tested positive for *P. kernoviae* in September and December although only the December leaves had lesions. The plant was dead by the end of March. No *P.*

kernoviae was isolated from roots (including fine roots) when the plant was uprooted and dissected.

Necrosis was present on the potted rhododendrons exposed during October, November, December, February and March, and *P. kernoviae* was recovered from the leaves in November, December, February and March. No further symptoms developed on the plants once removed from the site.

Soil samples

Phytophthora kernoviae was obtained from at least one of the four plots during the months of April to November inclusive, and from all four plots in July (mid-winter). *Phytophthora kernoviae* was not recovered during the months of December through March (summer). Plot results are given in Table 3.

Understorey plants

The most commonly occurring understorey plants throughout the forest are ferns, tree ferns and fern allies. Other understorey plants comprised both exotic and indigenous shrubs and ground cover plants. In total, 170 pieces of foliage from the seven most common understorey species were plated but no *P. kernoviae* was recovered.

Inoculations

Stem inoculations

Eight weeks after the plants were inoculated, two of the five *Nothofagus truncata* inoculated with NZFS 3062, two inoculated with NZFS 2646 and one of the five control plants had wilted and exhibited signs of root disease. These five plants were harvested at this time, 4 weeks before all other plants. Roots and soil of these *N. truncata* were tested for *Phytophthora* species, and *Phytophthora cinnamomi* Rands was recovered from both soil and roots of three plants. *Phytophthora cinnamomi* was considered likely to be the cause of the wilt of these three plants as no other plants had any indication of root disease. Although stem lesions had not formed at the inoculation point, *P. kernoviae* was recovered from the stem of two of the plants inoculated with NZFS 3062. Data from all plants was included in the analysis.

There had been little indication of ill-health in the foliage of any of the indigenous species (including the remaining *N. truncata* plants) at 3 months when all other plants were harvested for stem dissection. Of all the exotic plants inoculated, only one of the nine *An. cherimola* showed signs of wilt at this time.

Table 3 Recovery of *Phytophthora kernoviae* from soil samples collected monthly

Plot	Apr	May	Jun	July	Aug	Sep	Oct	Nov	Dec	Jan	Feb	Mar
a	+	+	+	+	+	+	–	+	–	–	–	–
b	+	+	+	+	+	+	+	–	–	–	–	–
c	–	–	+	+	–	–	–	–	–	–	–	–
d	–	–	–	+	–	–	+	–	–	–	–	–

Destructive sampling showed that there was no stem lesion development on any replicates of either *Fuchsia excorticata*, or *Cordyline australis* (Fig. 1). There was considerable variation in the number of replicates developing lesions for the other 14 native species, however. A visible lesion formed in only one of the 10 inoculated plants of each of the following species: *Coprosma* sp., *Dacrycarpus dacrydioides, Dodonaea viscosa, Pseudowintera colorata, Pittosporum eugenioides, Nothofagus menziesii*; and two of the *Beilschmiedia tawa, Lep. scoparium* and *Pittosporum ralphii.* Of the exotic species, the nine *An. cherimola*, eight of ten *F. sylvatica* plants and two of four *R. catawbiense* had a stem lesion. Indigenous hosts were more susceptible to stem lesion formation when inoculated with isolate NZFS 2646 (obtained from *An. cherimola*) than to NZFS 3062 (ex *R. catawbiense*) whereas the exotic *F. sylvatica* and *An. cherimola* were similarly susceptible to both isolates. *Rhododendron catawbiense* was only tested with NZFS 3062, and half of the plants developed lesions.

Mean lesion lengths (excluding the original wound length of 10 mm) are shown in Fig. 2. Results for the two *P. kernoviae* isolates have been combined for the figure. The longest lesion, 270 mm, was formed on one of the *An. cherimola* specimens, and the longest lesion on a native plant occurred on *Myrsine australis* with a lesion length of 80 mm. The wilting *An. cherimola* had been girdled at the inoculation point, and the lesion length was 40 mm (Fig. 2). All wounds on the control plants of every host species had healed, and there was no discolouration in the stem.

Recovery of *P. kernoviae* was more consistent from the exotic species than any of the indigenous hosts. Of the latter, *P. kernoviae* was not recovered from any plants of *Melicytus ramiflorus, N. menziesii, Pit. eugenioides, Agathis australis, Myr. australis, Cordyline australis* or *Coprosma* sp. but at varying rates from the other hosts (Fig. 3). *Phytophthora kernoviae* was occasionally recovered from *Fch. excorticata, Pit. ralphii* and *Ps. colorata* although lesions did not form. *Phytophthora kernoviae* was not recovered from any of the control plants.

Foliage inoculation 1

Symptom development Leaf-tip necrosis of leaves following dipping into a zoospore suspension was not common on the indigenous hosts. Where necrosis was recorded, it occurred only in low proportions, i.e. *B. tawa* (2 %), *Fch. excorticata* (4 %), *Corynocarpus laevigatus* (7 %), *D. dacrydioides* (8 %) and *Gaultheria depressa* (10 %). Symptoms were recorded on all three exotic

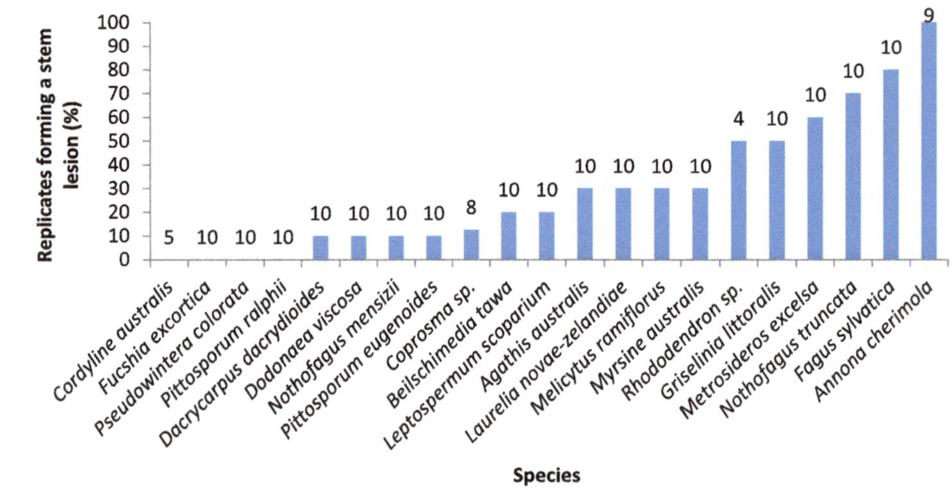

Fig. 1 Percentage of replicates within a species forming stem lesions. The number *above* the *bar* is the number of replicates of each host

species though percentages were also low with *F. sylvatica* (5 %), *An. cherimola* (15 %) and *R. catawbiense* (17 %). No lesions developed on any of the control treatments.

Pathogen recovery Although symptom development was uncommon, *P. kernoviae* was recovered from many of the asymptomatic leaves. *Phytophthora kernoviae* was recovered from leaves of 14 of the 27 native species tested though not from two of the species that exhibited symptoms (*Cor. laevigatus* and *G. depressa*). The proportion of pathogen recovery from leaves varied significantly among the tested species ($\chi^2 = 138.06$, $df = 31$, $P < 0.0001$). Recovery from foliage was highest for two of the exotic species, *R. catawbiense* (56 ± 26 %) and *F. sylvatica* (42 ±

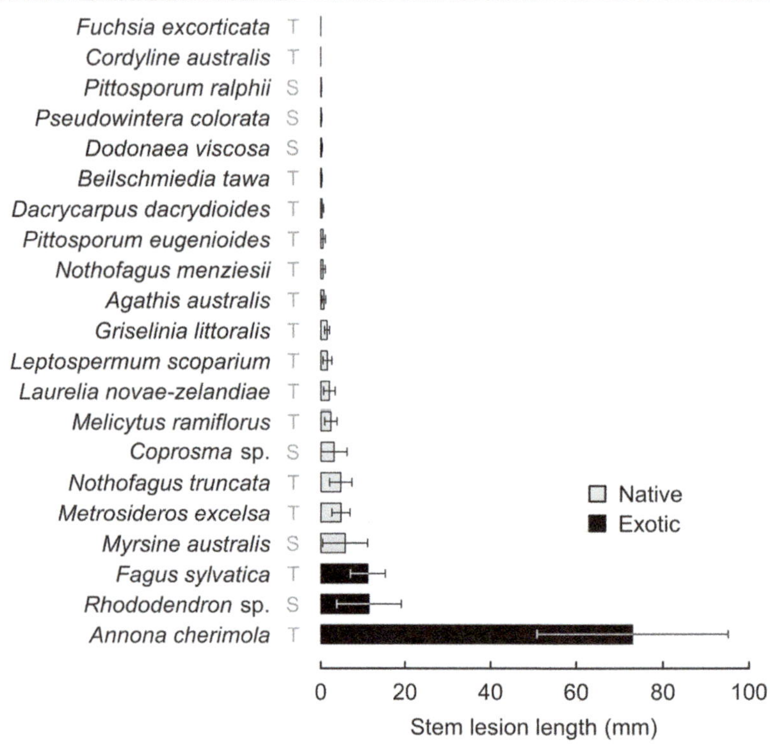

Fig. 2 Mean lengths of stem lesions induced by *Phytophthora kernoviae*. Means ± SE, $n = 2$–9. Key: T = tree; and S = shrub

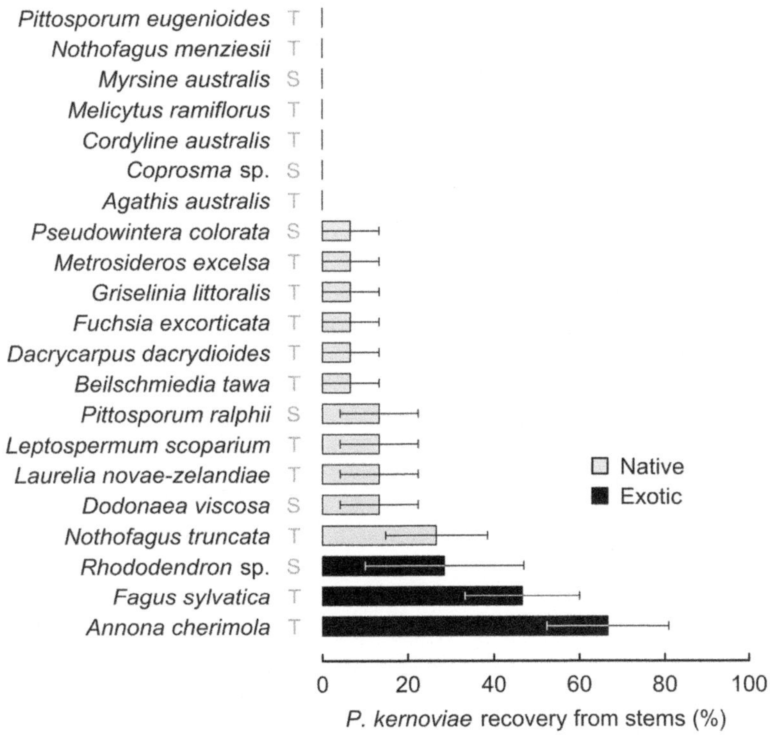

Fig. 3 Recovery of *Phytophthora kernoviae* from inoculated stems. *Bars* show the proportion of stems with pathogen recovery for each species (means ± SE, *n* = 2–9). Key: T = tree; and S = shrub

15 %), followed by the natives *Fch. excorticata* (25 ± 12 %) and *Lep. scoparium* (20 ± 20 %). In all other species, less than 20 % pathogen recovery was observed. Overall pathogen recovery was significantly higher in the exotic compared to the native taxa (37 ± 12.7 % vs. 5.6 ± 1.4 %; χ^2 = 8.32, *df* = 1, *P* = 0.004). Among the five native fern species, *P. kernoviae* was recovered only from *Bl. novae-zelandiae* fronds (12 ± 8 %). Results are illustrated in Fig. 4.

Foliage inoculation 2

Symptom development Complete results are given in Table 4. Lesions developed only on leaves (including both wounded and unwounded) of two of the three exotic species (*An. cherimola* and *R. catawbiense*). No lesions developed on leaves of any of the seven indigenous species tested.

Pathogen recovery Of the seven native plants tested, the pathogen was recovered from wounded but asymptomatic *Bl. novae-zelandiae* fronds (40 %) and *Loph. bullata* leaves (10 %). Recovery rates from the exotics were variable with high recovery (corresponding to symptom formation) from both wounded + inoculated and unwounded + inoculated for rhododendron with similar recoveries from most of the asymptomatic *F. sylvatica* leaves. However, the pathogen was recovered from just one of the five wounded + inoculated *An. cherimola*.

Discussion

The studies detailed here regarding the presence of *P. kernoviae* in the ecosystem, plus the tests on the susceptibility of New Zealand natives (through stem and foliar inoculations), have enhanced our knowledge of *P. kernoviae* occurrence and virulence in New Zealand.

Phytophthora kernoviae was recovered from the soil at the site from April to November but not during December to March (summer and early autumn) possibly because drier and warmer conditions during these seasons reduce activity of the oomycete. Webber (2008) reported that climatic analysis identified the western British Isles, with its mild temperatures and tendency to have high levels of precipitation, as suitable for establishment of *P. kernoviae* and very conducive for infection of aerial plant parts by the pathogen. Similar observations were made by Davidson et al. (2005) in their studies of *Phytophthora ramorum* Werres, De Cock and Man in't Veld (Brasier, Beales and S.A. Kirk), a species with comparable activity to *P. kernoviae*, in evergreen forests in California. Davidson et al. (2005) found that detectable levels of *P. ramorum* inoculum were present largely during the period of winter rains, detection ceased shortly after the rains stopped and did not occur during the hot, dry summer. The viability of

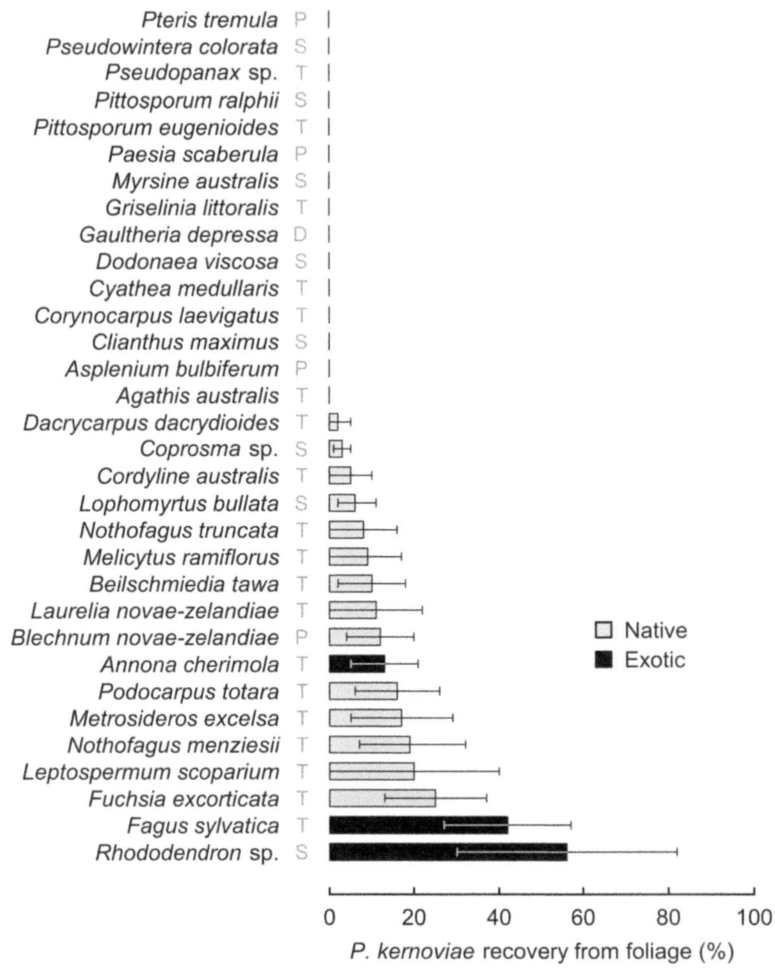

Fig. 4 Recovery of *Phytophthora kernoviae* from inoculated foliage for each woody species and four native fern species. *Bars* show the proportion of leaves with pathogen recovery per species (means ± SE, *n* = 2–9). Key: T = tree; S = shrub; D = dwarf shrub; and P = non-woody perennial

propagules of *P. kernoviae* in the soil, especially dormant oospores, remains in question. We do not know as yet whether the soil is a reservoir of inactive inoculum during the summer/early autumn. Colonisation of root systems of plants by *P. kernoviae* has not been studied in New Zealand.

No foliar infection of understorey plants was identified in this study; therefore, the role of understorey plants in ensuring survival of the fungus over time at this site remains unknown. Brasier et al. (2005) reported that, in England, *P. kernoviae* is primarily a foliar pathogen and that its presence in soil was likely due to decay of fallen leaves from infected shrubs in the litter layer. In studies in the same part of south-west England, Webber (2009) found that *P. kernoviae* persisted in soil and litter for at least 3 years following removal of infected rhododendrons from the site. Fichtner et al. (2009, 2011) studied colonisation of roots of plants in woodland and heathlands in this area and found that *P. kernoviae* could be isolated from roots of asymptomatic *Rhododendron ponticum* L. plants and seedlings and from asymptomatic roots and rhizomes of heathland *Vaccinium myrtillus* L. Root infectivity and in planta oospore production suggest mechanisms of *P. kernoviae* survival.

Variable results were obtained from the *Rhododendron* indicator plants. The two plots where the plants were located were both under a high canopy of 20-year-old *Pinus radiata* but with differences in the surrounding understorey vegetation. Plot 'a' was adjacent to a large healthy camellia shrub that provided a partial barrier to needle drop from the *Pinus radiata* canopy and from other understorey vegetation (mainly ferns and small tree ferns). In contrast, plot 'b' was relatively open to the canopy and surrounded by ferns and ground-covering shrubs. The *Pinus radiata* canopy is a potential inoculum source given the subsequent finding (Dick et al. 2014) that *Pinus radiata* needles can be infected by

Table 4 Foliage inoculation 2: Symptom expression and recovery of *P. kernoviae* (isolates NZFS 3062, 2646 and 3339)

	No. leaves treated			No. symptomatic leaves			Pathogen recovery (no. of leaves)		
	Wounded inoculated	Unwounded inoculated	Wounded control	Wounded inoculated	Unwounded inoculated	Wounded control	Wounded inoculated	Unwounded inoculated	Wounded control
Native hosts									
Blechnum novae-zelandiae	5	5	5	0	0	0	2	0	0
Beilschmiedia tawa	5	5	5	0	0	0	0	0	0
Fuchsia excorticata	5	5	5	0	0	0	0	0	0
Lophomyrtus bullata	10	10	5	0	0	0	1	0	0
Metrosideros excelsa	5	5	5	0	0	0	0	0	0
Podocarpus totara	20	20	5	0	0	0	0	0	0
Pseudowintera colorata	5	5	5	0	0	0	0	0	0
Exotic hosts									
Annona cherimola	5	5	5	3	1	0	1	0	0
Fagus sylvatica	5	5	5	0	0	0	4	3	0
Rhododendron catawbiense	5	5	5	5	3	0	5	3	0

Phytophthora kernoviae. Of the two planted rhododendrons, one (at plot 'a') remained asymptomatic whereas the other developed leaf lesions and branch tip dieback from which *Phytophthora kernoviae* was recovered. Likewise for the potted rhododendrons, there was neither symptom development nor recovery of *Phytophthora kernoviae* for any of the plants placed at plot 'a' but periodic results for plants from plot 'b'. As *Phytophthora kernoviae* recoveries from the soil were almost identical throughout the study at the two plots, and root infection was not recorded, the inoculum source would appear to be from other plants (albeit unidentified) either from the understorey or the canopy.

Susceptibility to *Phytophthora kernoviae* via stem inoculation was determined by lesion formation and also by recovery of the pathogen. The three exotic species inoculated via a wound on the lower stem all proved to be more susceptible than any of the indigenous species tested with the longest lesions recorded in *An. cherimola* stems. Widmer (2015) inoculated whole *An. cherimola* plants by spraying with a zoospore suspension and then examining the non-wounded plants for stem and foliar lesions. No stem lesions were recorded by Widmer (2015) using this method. One isolate employed by Widmer (2015) (NZFS 2646) was also used in this study and when applied to a stem wound resulted in lesion formation of all the *An. cherimola* and the death of one via girdling. An entry point to the stem would therefore appear to be a necessary precursor to stem invasion for this host. Foliage results for *An. cherimola* were similar to those we recorded in this study (<20 % symptom expression unless wounded) though Widmer (2015) also observed that he recorded significantly more leaf drop in inoculated as compared to non-inoculated plants. However in New Zealand, *An. cherimola* has been assumed to be highly susceptible to *P. kernoviae* in the wild. *Phytophthora kernoviae* was isolated from diseased plants obtained from two *An. cherimola* orchards in the far north of the country where there was extensive shoot and fruit blight, and branch cankering (Ramsfield et al. 2009). It is possible that additional adverse factors may also have contributed to the orchards becoming non-productive.

A similar pattern of infection was recorded for the less susceptible *F. sylvatica* paralleling information from the United Kingdom where foliage infection is negligible (Denman et al. 2006b, Anon 2012), and primarily stem cankers are recorded on this host (Brasier et al. 2005, Brown and Brasier 2007). We did record, however, a high recovery rate of *P. kernoviae* from asymptomatic *F. sylvatica* foliage in our inoculation tests. Denman et al. (2009) postulated that asymptomatic infections may be quiescent, awaiting a change in host tissue that will trigger tissue colonisation and symptom expression. The relationship may also be dependent on inoculum density, with disease expression occurring at higher inoculum levels as suggested by Hansen et al. (2005) and Tooley et al. (2004) from their work on the similarly behaved *P. ramorum*.

Mean lesion length on *R. catawbiense* stems equalled that on *F. sylvatica* in this study but foliage was comparatively more susceptible with the highest percentage symptom development and recovery of *P. kernoviae* recorded for any host. *Rhododendron* species have frequently been recorded as susceptible hosts from the Northern hemisphere (Brasier et al. 2005; Anon 2012; Denman et al. 2009; Beales et al. 2006; Fichtner et al. 2012).

Rhododendron ponticum was considerably more susceptible to *P. kernoviae* than other hosts tested by Widmer (2015). Although there was no difference in stem lesion rating between the United Kingdom isolates, the isolate from the republic of Ireland scored significantly lower. In contrast, the six isolates from New Zealand were very variable; two with comparable virulence to the United Kingdom isolates and the others significantly lower. At our study site, little disease developed on the *R. catawbiense* indicator plants suggesting that the resident *P. kernoviae* population was of low virulence to this host.

Phytophthora kernoviae was recovered from a high proportion of symptomatic tissue from both stem and foliar inoculations. Recovery rates were also high from asymptomatic tissue for some species, e.g. *Fch. excorticata* (34 % recovery and 4 % symptoms). In contrast, *Myr. australis* and *A. australis* formed lesions, but *P. kernoviae* was rarely recovered from the tissue. A few of the native plants tested (e.g. *Ps. colorata, Pit. eugenoides, Pit. ralphii, C. australis, D. dacrydioides*) showed little susceptibility to *P. kernoviae*, never forming lesions or persisting in asymptomatic tissues. In general, no marked susceptibility of the native plants was discovered. Some species could carry the oomycete asymptomatically in inoculated detached leaves, indicating an ability to survive attack from *P. kernoviae* without causing visible damage but whether these plants are a natural part of a *P. kernoviae* life cycle remains to be determined. Denman et al. (2009) demonstrated that this occurs for a number of understorey species in woodland in the Britain. In the current study, there was little indication that ferns could be a component in the life cycle of *P. kernoviae*. No symptoms were recorded on the fronds dipped into a zoospore suspension, and *P. kernoviae* was recovered only from *Bl. novae-zealandiae* (15 %). Although *P. kernoviae* was not recovered from *Bl. novae-zealandiae* during the understorey plant testing component of this study, it has been recorded as a host (MA Dick, unpublished data) and the isolate obtained (NZFS 2707) was one of those used in these inoculation experiments.

In 2014, *P. kernoviae* was found causing leaf necrosis and defoliation of *Drimys winteri* J.R.Forst. and G.Forst. in Chile (Sanfuentes et al. 2014). *Drimys winteri* is a native of Chile and Argentina where it is a dominant tree in the coastal temperate rain forests. This host is also planted in other parts of the world and is common in southern England where it has also been reported (Anon 2012; Webber 2008) to be susceptible to *P. kernoviae* (foliar necrosis and dieback). The New Zealand genus *Pseudowintera* is a member of the same family (Winteraceae). One of the three New Zealand species, *Ps. colorata*, is a common understorey plant in native forests and was tested in this study but showed very little response to inoculation. No symptoms were formed on either foliage or stems; *P. kernoviae* was not recovered from any of the leaves and from only one (10 %) of the inoculated stems.

Few gymnosperms have been found susceptible to *P. kernoviae* infection. The 2012 host list published by Department for Environment, Food and Rural Affairs (DEFRA) (Anon 2012) includes just two species: shoot tip wilt and foliar blight of *Podocarpus salignus* D.Don (Podocarpaceae) and leaf and stem necrosis of *Sequoiadendron giganteum* (Lindl.) J.Buchholz (Cupressaceae). Foliage of the New Zealand native *Podocarpus totara* (Podocarpaceae) was tested in this study without a positive result. However, subsequently in 2012, foliage death in the lower crowns of a group of the native *Prumnopitys ferruginea* (miro), also a member of the Podocarpaceae, in the lower North Island was reported (Dick et al. 2014) and *Phytophthora kernoviae* was consistently isolated from affected needles. Apart from this one collection, all disease records in New Zealand are of exotic plants. *Prumnopitys ferruginea* is found growing on both lowland terrain and on hill slopes throughout the country and is also common in urban areas. Considering the wide distribution of *Phytophthora kernoviae* in New Zealand, it is likely that this host and the pathogen cohabit at other locations. Reasons for the disease outbreak on this occasion are unknown.

Ramsfield et al. (2009) reviewed available records of *Phytophthora kernoviae* in New Zealand and overseas. They considered the possibility that *P. kernoviae* may be native to New Zealand based on its relatively widespread distribution, its presence in soil in native forests or recently converted native forests, and on genetic diversity in the New Zealand population based on a limited number of isolates. In Great Britain, *P. kernoviae* is recognised as a serious pathogen (e.g. Brasier et al. 2005; Brown and Brasier 2007; Webber 2008) with a wide, and increasing, host range of both trees and shrubs (e.g. Beales et al. 2009) whereas in New Zealand records of disease are few in spite of records dating back approximately 60 years (Ramsfield et al. 2009).

Phytophthora kernoviae is currently known only from the United Kingdom, the Republic of Ireland, Chile and New Zealand. Studies on *P. kernoviae* populations in Great Britain have shown considerable genetic uniformity. This, together with an initially limited geographic distribution, is consistent with a likely recent introduction (Brasier 2008). Concern about the potential spread to other countries has led to a number of studies examining potential host range (Fichtner et al. 2012; Widmer 2015). A range of isolates has been employed as information on differences in virulence between isolates helps to guide choices for regulatory agencies and researchers involved in monitoring programmes and for the conduct of host range tests. Hence Widmer (2015) compared the virulence of isolates of *P. kernoviae* from Great Britain and the Republic of Ireland with that of isolates from New Zealand. Leaf necrosis symptoms varied between isolates, but when results were combined within the geographical regions the isolates from the northern hemisphere resulted in a significantly higher percentage of necrosis than those from New Zealand.

The difference in disease expression between naturally infected vegetation in New Zealand and the northern hemisphere is striking. There is very little genetic variation in the population from Great Britain whereas population genetic studies have shown considerable variability exists within the New Zealand population (D. Cooke, D. Studholme and R. McDougal, unpublished data). Results from the inoculation study carried out by Widmer (2015) support these observations.

The goal of the studies reported in this paper was to gather further knowledge on interactions between *Phytophthora kernoviae* and native plants and environments. No evidence of disease of native plants was found in this study and there is only one record of a natural disease caused by *Phytophthora kernoviae* to a native host, that of *Prumnopitys ferruginea* in 2012.

Conclusions

Based on the results of the inoculations and the lack of historical records of disease of indigenous plants associated with *P. kernoviae*, this oomycete does not appear to be a damaging pathogen of New Zealand's indigenous flora. This may indicate that it has co-evolved with the flora. The recent finding of *P. kernoviae* in Chile associated with lesions on fallen leaves of *D. winteri* (Sanfuentes et al. 2014) may indicate a Gondwanan connection.

To gain a better understanding of the role of *P. kernoviae* in New Zealand ecosystems, further investigations need to be carried out. These include expanding host inoculation tests with a wider range of isolates, examining the ability of indigenous flora to support sporulation on asymptomatic foliage, and survival of *P. kernoviae* in roots and soil. The results from these preliminary studies suggest that exotic plant species are more likely to be potential susceptible hosts than indigenous species. New Zealand depends heavily on exotic plants to support its infrastructure and it is therefore encouraging that so few disease records have emerged in the past 60+ years.

Competing interests

The authors declare that they have no competing interests.

Authors' contributions

Field and laboratory work was carried out by JG and MD. MB carried out the statistical analysis. All authors read and approved the final manuscript.

Acknowledgements

Rita Tetenburg and Pam Smith are thanked for their technical assistance. We are also grateful to Peter Gadgil and Rebecca McDougal for discussion and advice on the manuscript. Funding was provided by the New Zealand Forest Owners' Association and the New Zealand Ministry for Primary Industries.

References

Anon (2012). Fera list of natural hosts of *Phytophthora kernoviae* with symptom and location. Resource document. The Food and Environment Research Agency. http://webarchive.nationalarchives.gov.uk/20140904082245/http://www.fera.defra.gov.uk/plants/plantHealth/pestsDiseases/phytophthora/documents/pK. Accessed 13 October 2015.

Beales, PA, Lane, CR, Barton, VC, & Giltrap, PG. (2006). *Phytophthora kernoviae* on ornamentals in the UK. *EPPO Bulletin, 36*(2), 377–379.

Beales, PA, Giltrap, PG, Payne, A, & Ingram, N. (2009). A new threat to UK heathland from *Phytophthora kernoviae* on *Vaccinium myrtillus* in the wild. *Plant Pathology, 58*(2), 393.

Brasier, C. (2008). *Phytophthora ramorum + P. kernoviae* = international biosecurity failure. In Frankel, SJ, Kliejunas, JT, & Palmieri, KM. (Tech. Coords.). *Proceedings of the Sudden Oak Death Third Science Symposium, March 5-9, 2007, Santa Rosa, California.* (USDA Forest Service General Technical Report PSW-GTR-214, pp. 133-139). Albany, CA, USA: US Department of Agriculture, Forest Service, Pacific Southwest Research Station.

Brasier, CM, Beales, PA, Kirk, SA, Denman, S, & Rose, J. (2005). *Phytophthora kernoviae* sp. nov., an invasive pathogen causing bleeding stem lesions on forest trees and foliar necrosis of ornamentals in the UK. *Mycological Research, 109*(8), 853–859.

Brown, AV, & Brasier, CM. (2007). Colonization of tree xylem by *Phytophthora ramorum*, P. kernoviae and other Phytophthora species. *Plant Pathology, 56*(2), 227–241.

Davidson, JM, Wickland, AC, Patterson, HA, Falk, KR, & Rizzo, DM. (2005). Transmission of *Phytophthora ramorum* in mixed evergreen forest in California. *Phytopathology, 95*(5), 587–596.

Denman, S, Kirk, SA, Brasier, CM, & Webber, JF. (2005). *In vitro* leaf inoculation studies as an indication of tree foliage susceptibility to *Phytophthora ramorum* in the UK. *Plant Pathology, 54*(4), 512–521.

Denman, S, Kirk, S, Whybrow, A, Orton, E, & Webber, JF. (2006a). *Phytophthora kernoviae* and *P. ramorum*: host susceptibility and sporulation potential on foliage of susceptible trees. *EPPO Bulletin, 36*(2), 373–376. doi:10.1111/j.1365-2338.2006.01014.

Denman, S, Brasier, CM, Brown, A, Kirk, SA, Orton, E, Webber JF. (2006b). Preliminary results of foliage susceptibility to *Phytophthora kernoviae* sp. nov.: a new pathogen of forest trees in the UK. [Poster]. Presented at the *Third International IUFRO working party S07.02.09. Freisling, Germany, 11-18 September 2004.*

Denman, S, Kirk, SA, Moralejo, E, & Webber, JF. (2009). *Phytophthora ramorum* and *Phytophthora kernoviae* on naturally infected asymptomatic foliage. *EPPO Bulletin, 39*(1), 105–111. doi:10.1111/j.1365-2338.2009.02243.x.

Dick, MA, Williams, NM, Bader, MK, Gardner, JF, & Bulman, LS. (2014). Pathogenicity of *Phytophthora pluvialis* to *Pinus radiata* and its relation with red needle cast disease in New Zealand. *New Zealand Journal of Forestry Science, 44*, 6. doi:10.1186/s40490-014-0006-7.

Erwin, DC, & Robeiro, O. (1996). *Phytophthora diseases worldwide*. St Paul, MN, USA: APS Press.

Fichtner, EJ, Rizzo, DM, Kirk, S, & Webber, J. (2009). Epidemiology of *Phytophthora kernoviae* in UK woodlands and heathland and risk to North American forests. *(Abstr.) Phytopathology, 99*, S35.

Fichtner, EJ, Rizzo, DM, Kirk, S, & Webber, J. (2011). Root infections may challenge management of invasive *Phytophthora* spp. in U.K. woodlands. *Plant Disease, 95*, 13–18.

Fichtner, EJ, Rizzo, DM, Kirk, SA, & Webber, JF. (2012). Infectivity and sporulation potential of *Phytophthora kernoviae* to select North American native plants. *Plant Pathology, 61*(2), 224–233.

Hansen, EM, Parke, JL, & Sutton, W. (2005). Susceptibility of Oregon forest trees and shrubs to *Phytophthora ramorum*: a comparison of artificial inoculation and natural infection. *Plant Disease, 89*(1), 63–70.

Jeffers, SN, & Martin, SB. (1986). Comparison of two media selective for *Phytophthora* and *Pythium* species. *Plant Disease, 70*(11), 1038–1043.

Kosmidis, I. (2013). brglm: Bias reduction in binomial-response Generalized Linear Models. *R package version*, 0, 5–9. http://www.ucl.ac.uk/~ucakiko/software.html

McAlonan, MJ. (1970). *An undescribed Phytophthora sp. recovered from beneath stands of Pinus radiata* (p. 64). New Zealand: Master of Science Thesis, Auckland University.

Pinheiro, J, Bates, D, DebRoy, S, Sarkar, D, & Core Team, R. (2014). nlme: Linear and Nonlinear Mixed Effects Models. *R package version, 3*, 1–117.

R Core Team. (2014). R: A language and environment for statistical computing. R Foundation for Statistical Computing, Vienna, Austria. URL http://www.R-project.org/

Ramsfield, T, Dick, MA, Beever, RE, Horner, IJ, McAlonan, MJ, & Hill, CF. (2009). *In Phytophthora kernoviae in New Zealand. Phytophthora in forests and natural ecosystems* (pp. 47–53). Monterey, California: Proceedings of the Fourth Meeting of IUFRO Working Party 507.02.09. August 26-31, 2007.

Rigby, RA, & Stasinopoulos, DM. (2005). Generalized additive models for location, scale and shape. *Applied Statistics, 54*(3), 507–554.

Sanfuentes, E, Fajardo, S, Sabag, M, Hansen, E, González, M. (2014) *Phytophthora kernoviae* detection in *Drimys winteri* (Winter's Bark) forest of southern Chile. Programme Abstracts of the Seventh meeting of IUFRO Working Party, 7.02.09, (p 45). November 10–14, 2014, Patagonia, Argentina. http://www.iufrophytophthora2014.org/downloads/proceedings.pdf

Tooley, PW, Kyde, KL, & Englander, L. (2004). Susceptibility of select ericaceous ornamental host species to *Phytophthora ramorum*. *Plant Disease, 88*(9), 993–999.

Webber, JF (2008). Status of *Phytophthora ramorum* and *P. kernoviae* in Europe. In Frankel, SJ, Kliejunas, JT & Palmieri, KM. (Tech. Coords.). *In Proceedings of the Sudden Oak Death Third Science Symposium, March 5-9, 2007, Santa Rosa, California*. (USDA Forest Service General Technical Report PSW-GTR-214, pp. 19-26). Albany, CA, USA: US Department of Agriculture, Forest Service, Pacific Southwest Research Station.

Webber, JF (2009). Management of *Phytophthora kernoviae* and *P. ramorum* in Southwest England. In Goheen, EM & Frankel, SJ (Tech. Coords.). *Phytophthoras in Forests and Natural Ecosystems. Proceedings of the Fourth Meeting of the International Union of Forest Research Organizations (IUFRO) Working Party 07.02.09. August 26-31, 2007, Monterey, California*. (USDA Forest Service General Technical Report PSW-GTR-221, pp. 177-183). Albany, CA, USA: US Department of Agriculture, Forest Service, Pacific Southwest Research Station.

Widmer, TL (2015) Differences in sporulation and virulence of *Phytophthora kernoviae* isolates originating from two distinct geographical regions. *Plant Disease, 99*(4), 460–466.

The effect of land use on slope failure and sediment generation in the Coromandel region of New Zealand following a major storm in 1995

Michael Marden[1*] and Donna Rowan[2]

Abstract

Background: Part of the Coromandel region (North Island, New Zealand) was subjected to a severe storm in March 1995. Analysis of relevant data provides a valuable opportunity to assess the type, extent, distribution and sediment generation rates by slope failures associated with steep-land forests and harvest practice.

Methods: Slope failures were mapped at 1:10 000 scale for stands of planted exotic forest, areas of exotic forest cutover, indigenous forest, indigenous secondary regrowth, and pasture. Slope failure dimensions and bulk density were used to calculate catchment-based sediment mass and generation rates by: (i) failure type, (ii) vegetation type, (iii) slope group, (iv) Land Use Capability unit (LUC), and by (v) catchment. The proportion of the total storm sediment load discharged as yield was estimated from records of flow and depth-integrated sediment samples.

Results: Storm-initiated slope failures generated ~0.5 Mt of sediment, predominantly by debris avalanche. Most were located within indigenous forest and secondary regrowth and generated ~78% of the total sediment mass. Few slope failures occurred within standing exotic forest and inclusive of areas disturbed by harvesting operations (cutover) generated ~21% of the mass and, 1% was derived from pastoral hill country. Sediment generation rates were greater from areas of exotic forest clearfelled three years before the storm and these were 2.5 times greater than from cutover clearfelled just before the storm. This result is explained by the progressive loss of strength from decaying tree roots that had not yet been countered by an effective root system under a new tree crop. Furthermore, rates were highest for slopes between 26 and 35° and, of the 11 LUC units, were highest for units VIe11 and VIIe2. For the combined Opitonui and Awaroa catchments, ~24% of the storm sediment load was discharged as yield with ~76% remaining as in-channel storage.

Conclusions: The erosion response was primarily controlled by rainfall variation and slope, which overrode the influence of vegetation cover. Time since clear-felling had a secondary influence. A re-evaluation of erosion susceptibility in steep-land terrain and a return to the identification, mapping and geomorphic interpretation of site-specific hazards, particularly from an operational perspective—in advance of harvesting—is recommended.

Keywords: Storm effects; Slope failures; Sediment mass and load; Geomorphic mapping

* Correspondence: mardenm@landcareresearch.co.nz
[1]Landcare Research, PO Box 445, Gisborne 4010, New Zealand
Full list of author information is available at the end of the article

Background

Internationally, much has been written on the relationship between forest-harvesting activities and landslide initiation. Studies have varied in scope from regional inventories (Rood 1984; Sauder et al. 1987; Millard 1999), watershed and other area-limited studies (Thomson 1987; Jakob 2000) to individual landslide reports. Similarly, studies have varied in objectives from those focussed primarily on causative factors (Schwab 1983; Millard 1999) to frequency of landslide occurrence (O'Loughlin 1972; Rood 1984; Guthrie 2002), recovery (Smith et al. 1986), and terrain types identified as particularly prone to landsliding (Howes 1987; Rollerson 1992; Rollerson et al. 1998).

In New Zealand, there has been a long history of concern about mass wasting (erosion)—predominantly by landsliding—that has largely focused on 'storm-effects' in areas of highly erodible hill country in the North Island (e.g. Committee of Inquiry 1939; Taylor 1938, Taylor, NH comp 1970; Blaschke et al. 1991; Dymond et al. 2006)[a]. Landslides triggered by major storm events are the most widespread and destructive erosion process (Glade 1998) and yet the documentation of incidences of storm damage has been poor. Glade and Crozier (1996) and Phillips et al. (2012) commented on the issues this raises not only for identifying relationships between land use/management practices and landsliding, but also for determining temporal trends in the incidence of landsliding, and developing improved hazard and risk analysis approaches to underpin land (including forest lands) management policy and practice.

The majority of documented storm-related studies have focussed on the impacts of landslide initiation following regional and localised storm events. They have also tended to compare relative differences in landslide densities across a range of forest types—both indigenous and exotic—and drawn comparisons with adjacent areas of pasture (Phillips et al. 1990; Hicks 1991; Marden et al. 1991; Marden and Rowan 1993; Bergin et al. 1995; Fransen and Brownlie 1995; Fahey and Marden 2000; Reid and Page 2002; Fahey et al. 2003, 2004; Dymond et al. 2006)[b]. These studies indicate a strong relationship between the incidence of landslides and land cover with fewer landslides occurring in areas of indigenous forest, and in exotic forest—once canopy closure has been attained between ~5–8 years after planting—than in areas of adjacent pastoral hill country (Selby 1967; Salter et al. 1983; Hicks 1989; Phillips et al. 1990; Marden et al. 1991; Marden and Rowan 1993; Dymond et al. 2006)[a]. Many assessments have been produced as client-based reports or as internal file notes. These have often not been officially published so cannot be cited as references. However, such reports may contain valuable information so details of relevant ones are provided as endnotes.

Much of New Zealand's steep hill country was originally covered in indigenous forest that was cleared for pasture in the nineteenth century. Pastoral farming failed (Poole 1960; Olsen 1970), in part, due to erosion. Attempts to reduce erosion have largely been through reforestation with the establishment of exotic plantations (Taylor, 1970; McKelvey 1992; Marden 2004, 2012; Phillips and Marden 2005)[c,d]. Of the current exotic forest estate, thirty-three per cent (0.6 million hectares) is located on steep hill country[e]. Most of this was originally planted as watershed protection/production forests with a soil conservation role and for most of the rotation such forests provide a high level of slope stability. Many are located in regions where climate is predicted to become dominated by long dry spells but with more frequent high intensity storms[f], including those in Northland, Coromandel, Bay of Plenty, Taranaki, Gisborne-East Cape, Nelson and Marlborough[g]. Within these regions, significant areas of forest have reached, or are nearing, maturity so harvesting is on the increase.

Landslides do occur in planted forests[h,i,j] but tend to be most extensive on slopes recently clearfelled following harvesting. The predominant planted tree species is *Pinus radiata* D.Don, for which a 30-year period between planting and harvest is common in New Zealand (McLaren 1993). Within planted forests, storm-initiated debris avalanches and soil slip failures are the main sources of sediment and woody debris entering stream channels—often resulting in major debris flows[k]. Their occurrence between 1994 and 1998 was more common in Northland, Auckland, Coromandel, Eastern Bay of Plenty, and Nelson-Marlborough, with fewer occurrences in the central North Island, East Coast, and Hawke's Bay[k]. In the last five years, however, there has been a high frequency of events, not only in the Bay of Plenty[l,m], Nelson[n,o], and Marlborough[p] but also in the East Coast in 2002 and 2011 and Hawke's Bay in 2011 (Phillips et al. 2012). In all these regions, storms producing more than 80 mm of rain in 24 hours usually lead to the saturation of shallow hill soils resulting in landslides (Crozier and Eyles 1980; Caine 1980).

Internationally and locally, it is widely accepted that forest harvesting accelerates sediment production, particularly during storm events. Often this increase is associated with road construction and subsequent road-related and landing failure (Mosley 1980; Pearce and Hodgkiss 1987; Fahey and Coker 1989, 1992; Beverley et al. 2001; Megahan et al. 1991, 2001). The susceptibility of forest cutover to storm-initiated mass wasting (predominantly landslides) has similarly been well documented (Bishop and Stevens 1964; Ziemer 1981a, b, c; Furbish and Rice 1983; Guthrie 2002; Marden et al. 2006, 2007)[h,i,n,q], with consequent on-and off-site impact following a storm often exacerbated by the presence of logging slash (woody debris)[e,k]. Large areas of exotic forest are approaching maturity. These are due to be harvested within the next decade, which will result in an increase in clearfelled land susceptible to erosion. There is also a high probability that the frequency of severe weather events will increase in the future due to climate change. The combination of these factors means that it is highly likely that erosion will occur in forested areas in the future.

There have been numerous reports of off-site impacts downstream of areas of exotic planted forest following high intensity storm cells in recent years (Horner 2012)[i,j,n,o] although not always coincident with harvesting. Several reports have specifically implicated forest cutover as a significant source of sediment delivered to stream channels in regions considered environmentally sensitive (and therefore susceptible to human-induced disturbances) (Basher et al. 2011; Horner 2012)[i,q]. Previous research has suggested that areas of cutover are at greatest risk and will remain susceptible to the impact of storms until the replacement trees attain canopy closure at ~8-years after establishment (Phillips et al. 1990; Marden et al. 1991; Marden and Rowan 1993)[r]. Surprisingly, however, only rarely has sediment production (either as a consequence of storm-initiated landsliding (Dymond et al. 2006)[a,h] or by other processes such as slope-wash erosion (Marden et al. 2006, 2007) been quantified for areas of exotic forest cutover.

Such reports place pressure on the forest industry to review their current harvesting and post-harvest practices to lessen the off-site impacts associated with future storm events[e,o]. The introduction of the Resource Management Act in 1991 has also made forest management more challenging due to its emphasis on avoiding, mitigating and remedying adverse 'effects' on the environment. Thus, harvest- and post-harvest-related impacts on steep and erodible hill country must be balanced against stewardship of the land (Marden and Saunders 1992). It is difficult to assess the potential for landslide initiation and to devise/evaluate appropriate counter measures due to the unpredictability of storm events and a paucity of quantitative data on storm-related impacts of exotic forest harvesting over its short history, particularly for areas of cutover.

This paper presents a retrospective assessment of sediment generated by slope failures following a single storm in March 1995. The distribution of slope failures, mass and rate of sediment generation are evaluated for indigenous and exotic forest, over a range of slopes and for specific areas of land. This study provides much needed process-based quantitative data on sediment generation from New Zealand's forested areas, particularly cutover, during such events.

Background on study area and environs

Location

The Whangapoua forest study area (Figure 1) is located on the eastern side of the Coromandel Peninsula within the Thames-Coromandel District. The steeper parts of this peninsula form the Coromandel Range, of which the uppermost slopes constitute part of the Coromandel Forest Park. Here, the vegetation is predominantly indigenous forest.

From its highest point near Castle Rock (521 m above sea level), this part of the Coromandel Range is deeply incised by short, steep streams—the Awaroa, Opitonui, Owera, Weiti, Waiatekatanga, and several minor streams—that have a high incidence of flood flows and the potential to deliver significant volumes of sediment to Whangapoua Harbour and Mercury Bay Estuary (Figure 1).

History of use

Whangapoua forest (10,500 hectares) spans the mid-slope reaches of the Awaroa, Opitonui, Owera and Weiti streams. This forest was planted between 1949 and 1985, predominantly with exotic *Pinus radiata*. Substantial areas of secondary indigenous vegetation were permitted to regenerate on steep banks (riparian) adjacent to stream channels following scrub clearance (by root raking and roller crushing). Pastoral farming is currently restricted to rolling hill country located in the lowermost parts of these catchments and to the coastal floodplain.

A storm in 1972 initiated shallow landslides that, in turn, caused significant sedimentation on farmland located downstream of this forest. This storm coincided with the early establishment phase of Whangapoua Forest and raised a number of issues of concern to residents, regional authorities, recreationists, and the forest owners well before harvesting began. These included the likelihood of impacts resulting from on-site forestry activities (road and landing construction, harvesting, logging traffic) and their potential off-site effects, particularly on streams and associated estuaries and harbours[s]. These harbours are noted habitats for estuarine birds and shellfish while the surrounding sea is internationally renowned for its game fishing. In addition, the Coromandel area is a popular tourist destination for both land and sea-based recreation.

Upgrading of the forest's road infrastructure began in 1991 and harvesting began in 1992. By 1995, ~6% of Whangapoua Forest had been harvested (~8% of the area assessed for post-storm damage). Cutover areas were aerially desiccated before replanting then oversown with a mix of introduced grasses (Yorkshire fog, *Holcus lanatus*; Punawai browntop, *Agrostis capillaris*) (Wardle 1991), and legumes (White clover, *Trifolium repens*; Birdsfoot trefoil, *Lotus angustissimus*) (Roy et al. 1998) to minimise surface erosion, predominantly by slope-wash, during the post-harvest period. Aerial spraying with herbicides has been

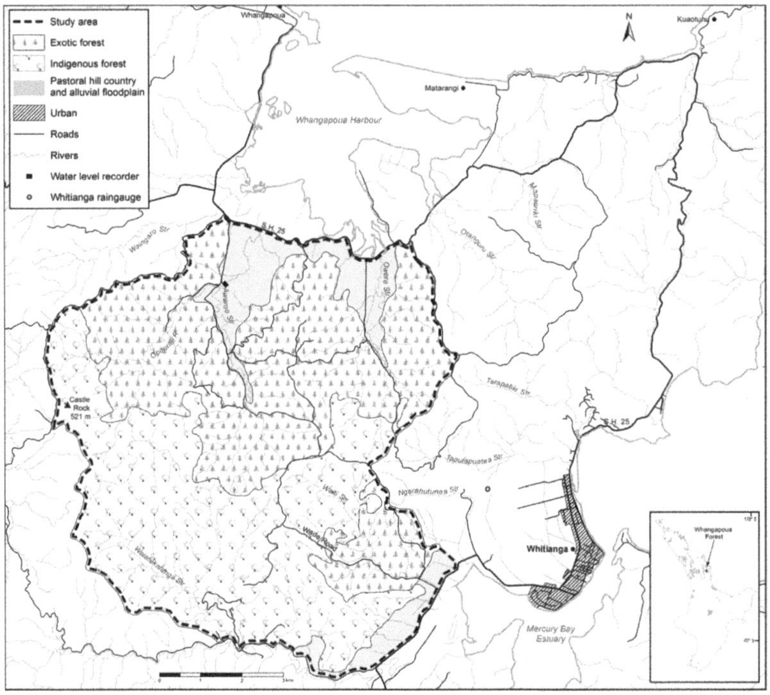

Figure 1 Location map of the Whangapoua forest study area showing the distribution of indigenous and exotic forest, pastoral hill country and floodplain, catchment boundaries, and location of rain gauges and water level recording site.

used to control competition from regenerating indigenous and wilding exotic pines.

Geology

The basement geology of the area consists of hydrothermally altered Whangapoua and Matarangi andesites of Miocene age (Skinner 1976). Soils are highly variable, with soil types being strongly related to parent materials, elevation and slope steepness. They have developed from deeply weathered andesite. Steep land soils, including Aroha (As) and Te Kie (Tks), occur in steep to very steep terrain in the uppermost parts of the major river catchments where exposed bluffs of weathered andesite are common. At lower elevations, on moderately steep to steep terrain, Waitakere and Rangiuru hill soils predominate (McCraw and Bell 1975); that is, Typic Orthic Brown Soils and Mottled Orthic Brown Soils (Hewitt 2010), respectively. Soils on easy to rolling terrain include Whitianga Silt loam and Waitekere Clay loam (McCraw and Bell 1975).

The weathered nature and high clay content of these soils predispose slopes to shallow, rapid slides (soil slip and debris avalanche) and flows involving soil and regolith. Soil slips typically have a small scar ≤ 1 m deep exposing a slip surface with debris being redeposited as a narrow debris tail downslope of the scar. A debris avalanche is a similar type of failure but tends to be larger, the scar is deeper (2–5 m) and, hence the depositional debris tail tends to occupy a significant length of the slope (Eyles 1985).

Climate

Though temperate, the climate at Whangapoua is known for its frequent, high-intensity localised storms, often of tropical origin that frequently result in severe flooding. The average annual rainfall is 1729 mm, with a distinct March to June 'wet season'[s]. The estimated 2-year return period rainfall is 127–133 mm in 24 hours (New Zealand Meteorological Service 1980). One of the largest storms in recent years occurred on 3rd March 1995. Visual inspection of landslide damage, rainfall, and stream flow data together suggest that the March 1995 storm encompassed an area of approximately 40 km^2 [h]. Rainfall records at Castle Rock, located in the headwaters of the Opitonui catchment, show that the duration of this storm was ~24-hours. During this period 94 mm of rainfall was recorded (1.3-year return period), half fell within a 2-hour period at an intensity of 25 mm/hour while the maximum hourly rainfall intensity reached 37 mm/hour (return period 1.8 years)[t] (from Table 1). However, eastward of this location, rainfall data collected at an automatic weather station located near Whitianga (Figure 1) showed that the 6-hour total of 112 mm (return period 8.4 years), and a 24-hour period total of 154 mm (return period of 3-years) was considerably greater than that recorded at Castle Rock. As the Whitianga rain gauge only records 6-hourly totals (Table 1) it is not possible to estimate return periods, based on rainfall intensities, for the early part of the storm and where local reports suggest that much of the rain fell within a less than 6-hour period[t], and likely centred over the most severely impacted Awaroa, Owera and Weiti catchments. Although storm-based, stream water-level records from a site located downstream of the Opitonui and Awaroa confluence (Figure 1), provide valuable insights into the impact of this storm they do not reflect the contrasting flow response between these two streams. For example, post-storm observations of their channel form and fauna indicate that flows in the Opitonui Stream had minimal impact on the in-stream habitat and macro invertebrate population whereas the flow response in the Awaroa Stream had considerable impact. Further east in the Owera and Weiti Streams the impact was extreme, and from the limited river flow data available from the Owera Stream, it is likely the channel was affected by debris flow(s)[u]. A 24-hr rainfall intensity of 12.8 mm hr^{-1} (which translates to an annual return period of 30–40 years (High Intensity Rainfall Design System HIRDS)) was recorded at the Whitianga rainfall gauge. This value equates well with the estimated return period of flow from the Owera Stream of between 20 and 50 years as reported by Quinn et al[v].

Methods

The study area (6,984 ha) includes Weiti, Opitonui, Awaroa, Owera, Waiatekatanga, and a number of smaller catchments (Figure 1). The first four catchments incorporate a range of land uses including indigenous and exotic forest, secondary indigenous regrowth, exotic forest cutover, and pastoral hill country. Predominantly indigenous forest, the Waiatekatanga and several smaller catchments are here collectively referred to as 'miscellaneous' catchments (Table 2).

Mapping

Landslides (debris avalanche and soil slip), locations of stream bank collapse (the result of bank undercutting during flood flows), and landing (a cleared area to which logs are hauled for sorting) failures were identified by stereoscopic analysis of 1:10,000 scale colour photographs taken within days of the March 1995 storm. Whole-catchment aerial photographic coverage was available for all but ~2 km^2 of the upper reaches of the Opitonui and Waiatekatanga catchments.

Each erosion feature was delineated on photographic prints using a fine mapping pen, assigned a unique number, and the failure type (debris avalanche, soil slip, stream bank collapse or landing failure) was noted. The first three types are considered as natural phenomena while the latter is associated with harvesting activities. Using an epidiascope to adjust for minor differences in scale, mapped areas were then transferred from the photographs onto a base map. A 1:10,000 scale base map was created from a 1:25,000 scale topo plot (New Zealand Map Series 270, part Sheet T 11) to match the scale of both the aerial photographs, and the forest compartment maps. The latter were geo-referenced to the New Zealand Map Grid 1949 projection.

A number of problems were encountered with the aerial photography, including:

(a) photographic prints were of variable contrast and exposure
(b) shadow obscured landslides located in the steeper, dissected, and more heavily vegetated areas, and within areas of secondary indigenous regrowth along steep-sided riparian stream banks
(c) low-altitude photography in very steep terrain produced considerable relief distortion.

These were countered by mapping landslides located within the central part of each photographic print. The

Table 1 Maximum rainfall intensities, totals and return periods recorded during the March 1995 storm

Maximum rainfall intensities and total rainfall	Castle Rock		Whitianga	
	Rainfall (mm)	Return period (years)	Rainfall (mm)	Return period (years)
Max 10 minute intensity	15.5	1.8		
Max 20 minute intensity	25	2.4		
Max 30 minute intensity	31	2.5		
Max 1 hour intensity	37	1.8		
Max 2 hour intensity	52	1.8		
Max 3 hour intensity	57	1.7		
Max 6 hour intensity	65.5	1.4	112	8.4
Max 12 hour intensity	83.5	1.4		
Max 24 hour intensity	94	1.3	154	3
Total rainfall over period	96		154	

Source: Williams[t].

Table 2 Details of slope failure number and per cent by area (ha) for each of the classification types used to define the study area

Classification	Area assessed		Slope failures		Failures per hectare
	ha	%	Number	%	
Land use					
Exotic Forest	2139	30.6	36	4	0.02
Indigenous Forest	2684	38.4	394	42	0.15
Secondary Indigenous Regrowth	859	12.3	210	23	0.24
Exotic Cutover 1992/93	115	1.6	86	9	0.75
Exotic Cutover 1993/94	213	3.1	109	12	0.51
Exotic Cutover 1994/95	263	3.8	83	9	0.32
Pasture (hill country)	375	5.4	13	1	0.03
Pasture (floodplain)	336	4.8	0	0	0
TOTAL	6984	100	931	100	
Slope Group					
A (0–3°)	306.5	4.4	0	0	0
D (16–20°)	361.6	5.2	9	1	0.02
E (21–25°)	1623.3	23.2	239	25.5	0.15
F (26–35°)	4493.9	64.3	679	73	0.15
G (>35°)	198.7	2.9	4	0.5	0.02
TOTAL	6984	100	931	100	
LUC unit					
IIs	306.5	4.4	0	0	0
IIIe	183.1	2.6	0	0	0
VIe2	148.8	2.1	2	0.2	0
VIe3	212.8	3.0	7	0.8	0.03
VIe8	465.9	6.7	36	3.9	0.08
VIe10	1403.7	20.1	123	13.2	0.09
VIe11	2071.1	29.7	478	51.3	0.23
VIIe2	1882.7	27.0	264	28.4	0.14
VIIe7	110.7	1.6	17	1.8	0.15
VIIe8	191.8	2.7	4	0.4	0.02
VIIe9	6.9	0.1	0	0	0
TOTAL	6984	100	931	100	
Catchment					
Opitonui	1538	22.0	83	9	0.05
Awaroa	1168	16.7	94	10	0.08
Owera	1365	19.6	402	43	0.29
Weiti	910	13.0	160	17	0.18
Miscellaneous	2003	28.7	192	21	0.09
TOTAL	6984	100	931	100	

~60% overlap provided by each print allowed all landslides to be captured with minimal distortion. Mapped landslides were then transferred to the base map.

In addition, numerous check marks (e.g., fences, stream edges, tracks, roads, quarries, trigonometric survey points) identified on the aerial photography were also present on both the base map and the forest compartment maps. These provided useful geo-referenced points with which to register the location of mapped landslides and thus were instrumental in minimising any spatial distortion (unquantified) that may have incurred during the transfer of erosion features from the photographs, and their registration to the base map.

For financial reasons, spatial data were not captured in geographic information system (GIS) at the time of mapping. Instead, a digital planimeter was used to measure the spatial extent and number of slope failures in each of eight vegetation types (exotic forest, indigenous forest, secondary indigenous regrowth, exotic cutover 1992/93, cutover 1993/94, cutover 1994/95, pasture (hill country) and pasture (floodplain) assessed from the forest compartment maps (Table 2). The study area and number of slope failures were also classified into five slope groups: A (0–3°); D (16–20°); E (21–25°); F (26–35°); and G (>35°), and into separate LUC units (IIs, IIIe, VIe2, VIe3, VIe8, VIe10, VIe11, VIIe2, VIIe7, VIIe8, and VIIe9) from the New Zealand Land Resource Inventory Worksheets (1975) (Table 2). The distribution of slope failures, by catchment, is presented in Table 2. Details of each LUC unit are provided in Additional file 1. The distribution of exotic and indigenous forest relative to slope group is presented in Table 3.

Sediment mass calculations

The eroded area (i.e. the source zone from which sediment was derived) was measured using a dot grid for each of the four failure types and this was termed 'scar'. For the larger slope failure types (debris avalanche and landing failure) the 'scar' and the depositional material extending downslope from the source area (termed 'debris tail') were measured separately. In May 1995, the dimensions (length, width and depth) of scars and debris tails for a variety of sizes of each of the failure types were measured in the field. This verified that for debris avalanche and landing failures their source area (scar) made up about one third of the total mapped eroded area (scar plus debris tail). Therefore a 'scar to debris tail' ratio of 1:3 was used along with a mean depth of 3 m for debris avalanches and 1.04 m for landing failures (scar areas only) to calculate sediment mass. For soil slips, the small size of the scar and minimal run-out distances on the gentler topography (i.e. limited development of a debris tail) precluded the mapping of areas of scar and debris tail separately. Scar depths approximate depths reported following an earlier storm (Salter et al. 1983). Consequently, sediment mass (Table 4) was calculated using the total measured area and a mean depth of 1.0 m. Similarly, sediment mass for stream bank failures is based on the measured eroded area (bank length times height) and a mean bank retreat distance of 0.86 m. Mass (t) (Additional files 2–5) was derived using a mean bulk density of 1.2 ± 0.11 t m^3 from 10 samples of weathered andesitic parent material dried for 24 hours at 105°C and weighed as described in Marden et al. (2006).

The density of the four slope failure types combined, relative to vegetation cover, slope group and LUC unit, is presented in Table 2.

Table 3 Distribution (percentage) of standing exotic and indigenous forest relative to slope group

Slope group	Exotic forest	Indigenous forest
D (16–20°)	8	0
E (21–25°)	40	20
F (26–35°)	49	60
G (>35°)	3	20

Contribution to stream sediment yield

An assessment of the relative contribution of sediment delivered to stream channels is based on the connectivity of each mapped feature to a watercourse. An erosion feature with a debris tail that reached a watercourse was recorded as 'connected' and likely contributed to stream sediment load. A subsequent Coromandel-based study determined that for landslides (debris avalanche and soil slips) initiated during storms in 1991 and 1992, and assessed as being 'connected' to a stream channel, approximately half the material generated at the time of failure was delivered to the stream and half was retained on slope (Marden et al. 2006). Based on this approach, a sediment delivery ratio (SDR) of 0.5 for storm initiated debris avalanche and soil slips was adopted and applied also to landing failures on similarly steep slopes. It was assumed that all the material derived from failures located on riparian stream banks entered the channel so this was assigned a SDR of 1.0. Using these criteria, the relative proportion of sediment mass delivered to streams as sediment load was assessed. Source area (ha), sediment mass (t), sediment generation rates (t ha^{-1} of scar area), and the mass (t) delivered to stream channels, are presented in Table 4.

Sediment generation rates and the annual sediment yield (primarily from the March 1995 storm) were compared with the longer term (1992–2004) record of flow and depth-integrated sediment samples collected from a permanent water level and sediment sampling site located on the Opitonui Stream downstream of the Opitonui/Awaroa confluence (Figure 1). For the March 1995 storm, the suspended sediment yield was calculated by summing the product of suspended sediment concentration and flow over the duration of the storm[w].

Results and Discussion

The March 1995 storm

The relative sediment generation rates (t ha^{-1}) were quantified by process (Figure 2; see also Additional files 2–5), by vegetation type (Figure 3; Additional file 2), by

Table 4 Scar area, sediment mass and generation rate, sediment load and assumed sediment delivery ratio for each of the four slope failure types

Characteristic	Failure type				Total
	Debris avalanche	Soil slip	Stream bank collapse	Landing failure	
Total scar area (ha)	9.9	11.6	0.3	0.2	22.0
Total scar area (%)	45.0	53.0	1.0	1.0	100.0
Sediment mass generated (Mt)	0.36	0.14	0.003	0.003	0.56
Total sediment mass generated (%)	71.1	27.8	0.6	0.5	100
Sediment generation rate (t ha^{-1})	51	20	0.5	0.3	71.8
Scar area (ha) with connection to stream	9	5.9	0.3	0.07	15.3
Assumed sediment delivery ratio (SDR)	0.5	0.5	1.0	0.5	
Mass delivered as sediment load (Mt)	0.18	0.07	0.003	0.001	0.254
Total mass delivered as load (%)	71	28	0.7	0.3	100

Totals have been rounded.

slope groupings (Figure 4; Additional file 3), by Land Use Capability unit (LUC) (Additional file 4), and for groupings of the four major catchments (Figure 5; Additional file 5).

In the Coromandel area, the return period for high intensity rainfall events is estimated to be only 2 years (New Zealand Meteorological Service 1980). However, comparison of the March 1995 storm that produced daily rainfalls in excess of 100 mm[t] with previous storms in this area (Coulter and Hessell 1980), and with calculated discharges (using slope area flow) suggests that its recurrence interval was in the range of 20–50 years[v]. Between 1995 and 1998, there were other storms but none seem to have been as intense and localised or produced as high peak discharges[x]. In contrast, there were at least six storms between 1998 and 2013 where the maximum daily discharge exceeded the 180 cumecs of the March 1995 storm. Of note is the initiation of numerous slips throughout Whangapoua Forest during the largest storm recorded over the past 21 years (31 May 2000) when discharge measured at the Opitonui site reached 218 cumecs[y].

The March 1995 storm is considered atypical of the more frequent storms that characterise the Coromandel climate but is characteristic of high intensity rainstorms recorded elsewhere throughout New Zealand within the last three decades. Some assessment of resultant landslide damage, has been made for some of these storms although the results are largely qualitative (Basher et al. 2011)[i,m,n,z].

Caine (1980) suggested a general threshold for shallow landslide failure on undisturbed slopes in terms of rainfall intensity and duration and, in terms of hourly rainfall intensities, equates to 25 mm hr^{-1}. Hourly rainfall intensities at Whangapoua during the March 1995 storm exceeded this threshold with half (52 mm) of the total storm rainfall (96 mm) occurring within a 2-hour period (i.e. ~26 mm/hour), and the recording of a 1-hour maximum rainfall intensity of 37 mm[t] (Table 1). This storm initiated 931 new erosion features which included 441 debris avalanches, 451 soil slips, 27 stream bank failures and 12 landing failures. The combined area of failed sites was ~42 ha (0.6% of the study area) of which ~22 ha was identified as scar (the location of individual failures), and ~20 ha as debris tail (depositional areas predominantly associated with debris avalanche and landing failures). Of the ~22 ha scar area, soil slips comprised ~11.6 ha, debris avalanches ~9.9 ha, stream bank failures ~0.3 ha, and landing failures ~0.2 ha (Table 4).

Collectively, slope failures generated ~0.5 Mt of sediment at a rate of ~72 t ha^{-1} (Table 4). Debris avalanches generated ~71% (~51 t ha^{-1}) of this sediment, more than twice the rate generated by soil slips ~28% (~20 t ha^{-1}), and which was an order of magnitude greater than was

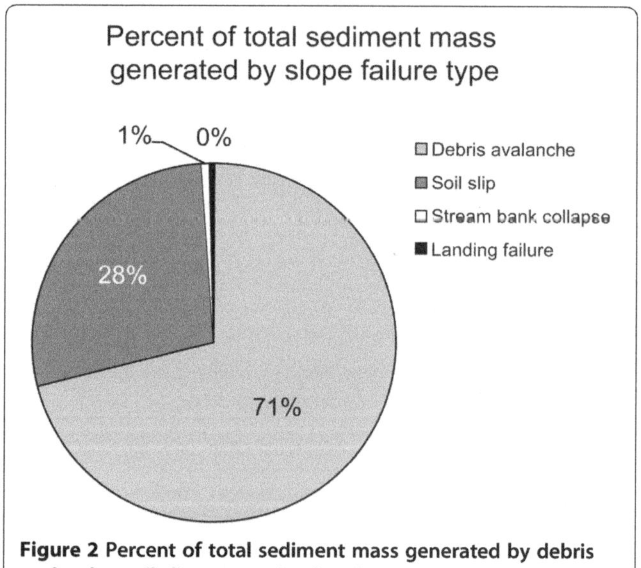

Figure 2 Percent of total sediment mass generated by debris avalanche, soil slip, stream bank collapse and landing failure.

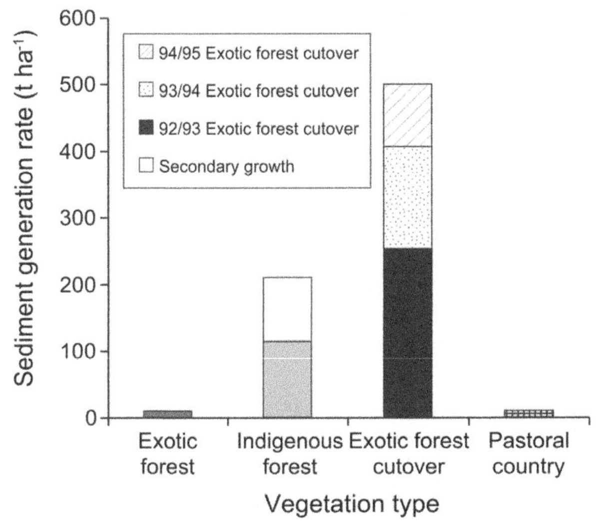

Figure 3 Comparative sediment generation rates (t ha^{-1}) by vegetation type for standing exotic forest and exotic forest cutover harvested 3, 2, and 1 year before the March 1995 storm. Indigenous vegetation was partitioned into areas of intact forest and secondary regrowth. Sediment generation rates from areas of pastoral land are from hill country areas only.

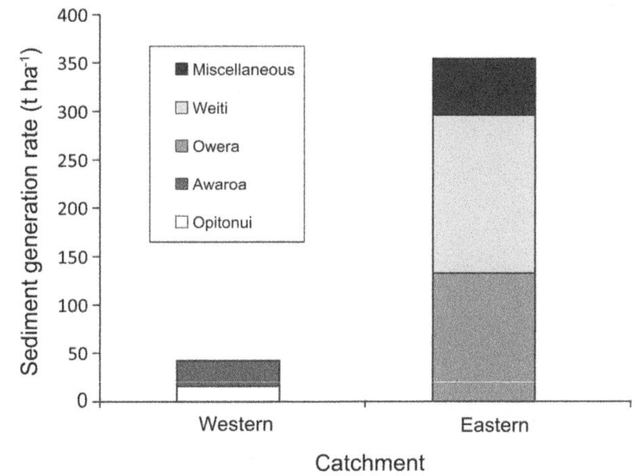

Figure 5 Contrasting sediment generation rates (t ha^{-1}) by all processes combined, for catchments located in the westernmost versus those located in the easternmost parts of the study area.

generated either by stream bank ~0.6% (~0.5 t ha^{-1}) or landing failures ~0.5% (~0.3 t ha^{-1}) (Table 4, Figure 2).

Approximately 38% of the study area was covered with indigenous forest (Table 2) primarily located in the uppermost and steepest parts of each of the study catchments. Here, 80% of the indigenous forest area coincides with slopes >26° (Table 3). The majority of slope failures (73.5%) were associated with this slope group (Table 2), thus most (42%) were coincidental with areas of indigenous forest predominantly in the Awaroa, Owera and Weiti catchments where rainfall amount and intensity is considered to have been significantly greater than for other parts (e.g. Opitonui and miscellaneous catchments) of the study area[t]. Here, slope failures were predominately the deeper and larger debris avalanches and collectively they generated the greatest proportion (~61%) of the total sediment mass at a rate of ~114 t ha^{-1} (Additional file 2). The density of slope failures (0.15 ha^{-1}) (Table 2), was similar (0.14 slips ha^{-1}) to that recorded for forested areas located in the Thames-Te Aroha area following a storm on 11-13th April 1981 with a recurrence interval of 15–20 years (Salter et al. 1983), suggesting that the March 1995 Whangapoua storm was indeed a severe event.

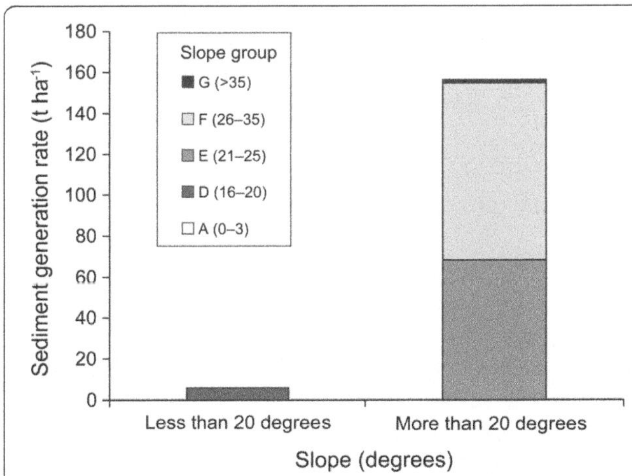

Figure 4 Contrasting sediment generation rates (t ha^{-1}) by all processes combined, for slopes less than and greater than 20 degrees.

While the cutting of trees does not increase sediment generation *per se*, the effects of the removal of the vegetation cover (i.e. a reduction in evapotranspiration, loss of tree root strength) generally increases the vulnerability of cutover to landslide initiation, and ultimately results in increased sediment yield (Phillips et al. 2005, Basher et al. 2011). At Whangapoua, the harvesting of exotic forest undoubtedly contributed to a significant increase in slope failures on cutover compared with mature forest. Thirty per cent (278) of all failures initiated during the March 1995 storm were on cutover, which covered only ~ 8% of the study area (Table 2) while just 4% (36) occurred on land covered in standing exotic forest, which accounted for nearly 32% of the study area. Of the slope failures on cutover, only 12 were associated with forest landings while the remainder occurred on natural slopes (data not shown). The incidence of slope failure was highest (0.75 ha^{-1}) on cutover harvested earliest 1992/93 (<2% of study area). Debris avalanches predominated in this vegetation type and resulted in a sediment generation rate ~2.5 times higher (254 t ha^{-1})

than that generated from cutover clearfelled in 1994/95 (<4% of study area), just before the storm (Figure 3). Here, the incidence of slope failures was less than half (0.32 ha^{-1}) that of the earlier cutover and generated just 93.8 t ha^{-1} with approximately equal contributions from both debris avalanches and soil slips (Additional file 2). This finding is consistent with that found after a storm in the Marlborough Sounds in July 1983 (O'Loughlin 1985). This result is explained by the progressive loss of strength from decaying roots that had not yet been counteracted by the development of an effective root system under the new tree crop. The root systems of *Pinus radiata* decay after harvesting, losing half their strength within 15-months. After 3 years, the large (>5 cm diameter) structural roots are in an advanced state of decay (Watson 1990). Thus, areas of exotic forest harvested in 1992/93 would have been at their most susceptible to the initiation of the deeper and larger debris avalanches at the time of the 1995 storm. Such areas have been shown to remain susceptible to storm damage for ≥6–8 years after replanting and until adequate root-soil reinforcement by the replacement trees has been re-established (Watson et al. 1995, 1999). The incidence of landslides decreases markedly after this time (Marden and Rowan 1993; Hicks 1989; Phillips et al. 1990; Marden et al. 1991)[r].

Slope failure densities were minimal within areas of pasture (<0.03 ha^{-1}) and mature exotic forest (0.02 ha^{-1}) (Table 2).

Fifty-two per cent of the mature exotic forest was on slopes >26° (Table 3) yet generated just ~4% of the total mass at a rate of ~9.5 t ha^{-1}. This was the lowest of all vegetation types except pasture (Figure 3). O'Loughlin and Ziemer (1982) found that total root biomass and tree root morphology, consisting of overlapping and intertwining root systems, appear to be the major tree root variables that influence slope stability and sediment production the most. Under a mature 25-year-old pine forest, total root biomass would likely approximate 100 tonnes per hectare (O'Loughlin 1985) with extensive lateral roots spreading to ~10 m from the stump and an average depth of 2.4 m (Watson 1990). Conversely, the root systems of indigenous tree species generally consist of a well-developed but surficial lateral root network, confined largely to the top 50–60 cm of soil and, a weakly developed vertical root system (Phillips and Watson 1994). Based on a 25% measured fraction of aboveground biomass across all New Zealand indigenous forests, the total root biomass of these forests was estimated at be ~76 t ha^{-1} (Holdaway et al. 2013), i.e. 60% of the value for mature pine forest. Therefore, indigenous forests are likely to impart a less substantial contribution to slope stability, particularly during severe storm events when pore-water pressure is greatest.

The majority (~99%) of slope failures were initiated on slopes >20° of which 73% occurred on slopes 26–35° (~64% of study area) at a density of 0.15 ha^{-1} (Table 2). Slope failures in this slope group collectively generated ~77% of the total sediment mass (Additional file 3) at a rate of ~86 t ha^{-1} (Figure 4). Previous slope-stability studies in other parts of New Zealand have shown that soils on slopes over 30° become unstable as they approach saturation if they do not have substantial reinforcement from competent root systems (O'Loughlin et al. 1982; Rogers and Selby 1980; Crozier and Eyles 1980). However, landslides were less frequent (0.02 ha^{-1}) on slopes >35° degrees (~3% of study area), and generated only ~0.05% of the total sediment mass at a rate of ~1.5 t ha^{-1} (Figure 4, Additional file 3). The initiation of slope failures on the steeper slopes was limited because much of the soil and colluvium had been stripped during previous storm events[h]. No slope failures were recorded on slopes < 16° (Additional file 3) (~4% of study area) (Table 2).

Of the 11 LUC units that occurred in the study area, slope failures were densest (0.23 ha^{-1}) for unit VIe11 (Table 2) with a correspondingly high sediment generation rate of ~100 t ha^{-1} (Additional file 4). The LUC units were defined using the New Zealand Land Resource Inventory Worksheets of 1975, i.e. 20 years before the March 1995 storm. Debris avalanche failure had not been identified in the mid-1970s as a significant erosion process associated with this unit (Additional file 1) and yet debris avalanches alone generated ~71% of the total sediment mass (from Additional file 4). The potential for the current 'moderate' erosion severity to increase to 'severe' was however acknowledged but only for sheet, wind and soil slip (National Water and Soil Conservation Organisation 1975). In fact, the density of slope failures on VIIe2 identified as prone to severe debris avalanches was half (0.14 ha^{-1}) that on VIe11 (0.23 ha^{-1}) (Table 2) yet both units actually generated about the same mass (at an equivalent rate of ~100 t ha^{-1}; Additional file 4). Slope failures from these two LUC units combined represented approximately 57% of the study area but generated ~80% of the total sediment mass (Additional file 4). Conversely, no debris avalanches were initiated on either VIIe8 or VIIe9 although both are recorded as being susceptible to severe debris avalanche failure with the potential to become very severe. In fact, only minor soil slip occurred on VIIe8 (Additional file 4).

These apparent inconsistencies in land classification highlight the difficulty of not only predicting the mode of failure on different units but also, as in the case of the March 1995 storm, of predicting the potential magnitude of the erosion response to a severe event. The variability in storm characteristics (particularly rainfall intensity and duration) associated with different storms also adds another level of complexity. Importantly, the results of this and previous assessments of storm damage in this region raise concerns for all areas that are currently or are likely to experience severe storm events in the future. In particular, the susceptibility of forested slopes between 26 and 31° to debris avalanche failures has likely been underestimated. Also, perhaps Class VIe11 has been misidentified and should be Class VII. Furthermore, there is potential for debris avalanche(s) initiated in indigenous and exotic forests to transform into debris flow, which will pose a significant risk to the environs downstream of forests. Therefore, the current erosion potential classification likely warrants re-evaluating.

As a consequence of the March 1995 storm event, debris avalanche and soil slip failure of natural slopes along with stream bank erosion collectively generated >99% of the total sediment mass and contributed >99% of the ~0.25 Mt delivered to stream channels as sediment load (Table 4). Conversely, landing failures generated <0.5% of the total sediment mass and delivered <0.5% to the stream sediment load. Only four of the 12 landing failures contributed sediment to the stream network. Half the resulting sediment (and any associated woody debris) mobilised by them—was redistributed downslope of the point of failure and the other half remained on-slope as a depositional debris tail.

Most of the hill country in the Coromandel region is susceptible to storm-initiated landsliding. In the event of a major storm coinciding with areas of indigenous forest little can be done to prevent landslide initiation particularly as slope failures are typically the deeper-seated debris avalanches that tend to generate the bulk of sediment. Within exotic forests, however, the risk of landslide damage could be alleviated by identifying the 'most-at-risk' areas and consider their exclusion from commercial forest use. As an example, Pearce and O'Loughlin[aa] suggested retiring slopes >35° in the Coromandel Region from productive uses and allowing reversion to indigenous forest. In support of this, the highest frequency of shallow landsliding initiated during a storm on 14–16 March 1985 in the northern Hawke's Bay was on slopes 38–40°, that is, predominantly Class VII land (Harmsworth et al. 1987). More recently, Page et al.[o] identified 30–45° slopes in the Nelson region as having the highest susceptibility to landsliding following forest harvesting. That said, most of New Zealand's exotic forest estate is likely located on Class VI land, and on slopes in the range 21–35°[e]. As identified in this paper, and previously by Salter et al. (1983), it is these slopes that are potentially of 'highest risk' to the initiation of landslides and potentially a significant source of landslide-generated sediment. Proposed management options for the 'most-at- risk' slopes have generally included: (i) scheduling the harvesting of the most at risk sites to periods of the year when storm events are least likely; (ii) limiting the size of harvest areas and spatially separating cutover so that areas clearfelled in successive years are not contiguous[s], and (iii) increasing the planting density to reduce the period that cutover is most susceptible to storm damage (Phillips et al. 2012).

Riparian areas have also been promoted as an effective means of reducing the delivery of sediment to streams (Quinn et al. 1993). Where ground-cover vegetation has remained intact after harvesting, slope-wash-transported sediment may be effectively filtered out with only a minimal amount reaching the stream network (Marden et al. 2006, 2007). However, a comparison of the rate of sediment generated by landslides and by slope-wash erosion on forest cutover, and of their relative contribution to the stream sediment load during a 2-year post-harvest period showed that: (i) the delivery of sediment to the stream network by either slope-wash erosion or landslides is highly dependent on the coupling of their respective source areas to the stream network; (ii) slope-wash erosion is the least significant of these erosion processes by several orders of magnitude (Marden et al. 2006)—a finding similar to that from a related study undertaken on cutover located in the central North Island (Marden et al. 2007); and (iii) episodic, storm-initiated landslides are the single most important sediment-generating process, providing the primary mechanism for mobilising and delivering most sediment to the stream network. The latter finding was most evident during the March 1995 storm when ~91% of debris avalanches and ~51% of soil slips (cf. 40% of landslides initiated during the 11-13th April Thames-Te Aroha storm in 1981 reported in Salter et al. 1983, and 50% at Otoi during a storm on 14–16 March of 1985 reported by Harmsworth et al. 1987) tracked sediment and debris through standing forest and into stream channels[h]. Furthermore, storms commonly initiate slope failures within riparian areas and these can potentially contribute significant sediment and woody debris to the stream network. For example, soil slips were prevalent (0.24 ha^{-1}) in areas of secondary indigenous regrowth (~12% of study area) at Whangapoua, which were primarily located along stream banks (riparian), and collectively generated ~17% of the total sediment mass at ~97 t ha^{-1} (Figure 3). Here, the absence of large trees with an extensive network of lateral roots likely increased the susceptibility of riparian slopes to the initiation of soil slips and bank undercutting during peak discharge with much of the sediment entering stream channels[h]. Nonetheless, it is evident that the existence of riparian areas does serve to reduce the incidence of slope failures that would otherwise have occurred in the absence of a woody vegetation cover and that the trapping efficiency of riparian vegetation is significant in the interval between storm events.

Sediment delivery and yield

Of the 931 documented slope failures initiated during this storm, 91% of debris avalanches, 51% of soil slips, 33% of landing failures, and 100% of stream bank failures contributed sediment directly to a permanent watercourse. Of the 0.25Mt delivered to stream channels as sediment load (Table 4), ~99.5% was derived from landslides initiated on natural slopes, while <0.5% originated from failures associated with harvesting landings. Seventy-eight percent of the total sediment load was derived from slope failures initiated within areas of indigenous vegetation (standing forest and secondary indigenous regrowth), ~21% from within standing and cutover exotic forest, and <1% from pastoral hill country (from Additional file 2).

For the two westernmost catchments, landslide number and sediment generation rates within the Opitonui (0.05 ha^{-1}; ~26 t ha^{-1}) and Awaroa (0.08 ha^{-1}; ~16 t ha^{-1}) catchments (Table 2 & Additional file 5) were an order of magnitude less than in the easternmost Owera (0.29 ha^{-1}; ~132 t ha^{-1}) and Weiti catchments (0.18 ha^{-1}; ~163 t ha^{-1}) as was that within the collective of miscellaneous catchments (0.09 ha^{-1}, ~58 t ha^{-1}) (Figure 5; Additional file 5). In both the westernmost catchments debris avalanche and soil slip failures generated similar amounts of sediment mass whereas in each of the easternmost catchments debris avalanche failures generated 2 to 4 times more mass than did soil slips (Additional file 5).Thus landslide distribution, the type of failure, and sediment generation rates and yield suggest that the March 1995 storm had its greatest impact on the easternmost catchments where rainfall intensity and amount was likely greater than in the westernmost catchment thereby suggesting a strong relationship with rainfall.

The combined Opitonui and Awaroa catchments (2900 ha) (Figure 1) experienced a period of relatively benign suspended sediment discharge flows in the three years before the March 1995 storm. Peak flow during this period was less than 50 m^3 s^{-1}, which was lower than in all years since the March 1995 storm[w]. The average annual yield for 1995 was estimated at 6758 t (~2.3 t ha^{-1}), and largely derived during the March storm[w]. Using a sediment delivery ratio of 0.5 only for the connected landslide and landing failures and 1.0 for stream bank failures, the sediment load delivered to these streams during the 1995 storm was estimated for the Opitonui at ~12,331 t (~8.0 t ha^{-1}) and for the Awaroa at ~15,516 t (~13 t ha^{-1}) (from Additional file 5). Approximately 24% (6758 t; ~2.3 t ha^{-1}) of the combined sediment load of 27,847 t (~9.6 t ha^{-1}) delivered to these streams during the March 1995 storm, was discharged as yield. This result implies that ~76% of the load remained as temporary in-channel storage. This high storage component likely reflects: (i) the capacity of stream channels upstream of the gauging site to store sediment behind debris dams, within in-filled pools, and widened reaches of both the Opitonui and Awaroa streams, and (ii) the relatively low peak storm discharge of ~0.04.0 m^3 s^{-1} ha^{-1}[u]. However, for every year following1995 through to 2004, the annual sediment loads at the Opitonui gauging site exceeded the pre–1995 load and reflect the probable reworking of stored bedload. Elevated yields following storms have also been observed in the Orewa catchment[u] and the Motueka River (Hicks and Basher 2008). Additionally, as less than ~0.5% of these catchments had been harvested before the storm, slope failures on cutover would have contributed little to either the sediment load or yield at the time of this storm.

By contrast, the annual sediment yield was likely significantly greater for the Owera and Weiti catchments because: (i) rainfall intensities and totals was highest, (ii) >30% of the total sediment load generated during the March 1995 storm was delivered to each of these streams (from Additional file 5), and (iii) severe peak specific storm discharges were estimated to be 10-times greater (0.34–0.69 m^3 s^{-1} ha^{-1}) for the Owera catchment than those recorded at the Opitonui site[v]. This resulted in substantial changes in channel morphology, sediment deposition and composition[u], and also to the macro invertebrate population suggesting that the slope-area gauging may have been affected by a debris flow. Additionally, harvesting had yet to begin in the Weiti catchment while a significant proportion of the total sediment load within Owera catchment was likely derived from cutover since ~30% of the catchment area had been harvested before the March 1995 storm.

While most storm damage assessments demonstrate a higher incidence of slope failures associated with pastoral hill country than with forested areas, the results of this study show the reverse. This suggests that rolling hill country in pastoral use (~5% of the study area) was peripheral to the storm centre with rainfall intensities below the 25 mm/hour threshold required to trigger shallow landslides (Caine 1980). Here, slope failures were minimal (0.03 ha^{-1}) (Table 2) and, generated ~1% of the total sediment mass at a rate of ~9.6 t ha^{-1} (Additional file 2).

Implications for the future

'Climate Change 2014: Impacts, Adaptation and Vulnerability' (IPCC 2014) reveals that the effects of climate change on New Zealand are likely to be more frequent floods, storms, landslides, and droughts. The country already appears to be experiencing more floods with consequent storm-initiated landsliding, and in many instances such events are coincident with areas of exotic forests. Each forest has its own unique set of physical characteristics (e.g. geology, slope, elevation, soils), a different mix of vegetation types at varying stages of maturity, and as each is located in a different geographical region the probabilities of storm frequency and associated rainfall intensity will be different, as will the geomorphic response and resultant aftermath. In the case of a planted forest, the extent of harvesting at the time of a storm will be critical. Furthermore, retrospective assessments of storm damage in many of the more highly erodible forested areas suggest that historical factors (physical or policy) may also have contributed, at least in part, to an increase in resultant damage—often a legacy of land clearance and stock retention policies for marginal land introduced between 1979–82 (Chudleigh et al., 1983), inherited forest infrastructure[m], declining health of steep land indigenous forests as a result of pre-European fires (Esler 1963), burning and grazing (Cunningham and Stribling 1978), increased storminess (Grant 1977), and the influence of introduced animals (James 1973) (remnants of which remain within catchments re-established as exotic forest), and inadequate due diligence on a landscape's susceptibility to forest disturbance (Marden et al., 2015).

To date, the standard response following a significant storm has been to undertake a post-storm damage assessment. Most assessments are produced as client-based reports or as internal file notes and few such reports are made public. Many such assessments are merely a cursory documentation, often non-quantitative, of the extent and type of slope failure. Some may include an assessment of possible causative factors as related to geology, vegetation maturity, planting density, slope angle and, storm-related factors such as rainfall intensity, amount, and duration of the event. However, similar studies need to be conducted across many more physiographic and climatic regions to provide more transparent information on the relationships between factors that contribute to landslide occurrence within planted forests. These also need to focus particularly on areas established in first-rotation exotic forest. To be of any value, the information collected must be standardised, quantitative, and in a format for analyses and interpretation of results to be comparable. If not, the information will have little value in supporting or modifying current practices aimed at reducing the likelihood of landslide occurrence[e].

Previously, a number of approaches (including terrain mapping, terrain stability mapping, and terrain assessments) have been credited with reducing the risk of landslide occurrence through improved forest practices on steep, potentially unstable terrain (Fannin et al. 2005). In New Zealand, such maps were produced both for existing forests and for areas designated for future planting as 'conservation forests' (essentially for erosion-control purposes) in geologically unstable terrain. The aim was to forearm harvest planners with knowledge on the relative stability of different parts of the landscape, and on the dominant slope failure process, and to signal potential difficulties likely to be encountered during roading and harvesting. Production of these maps was shelved at the time of the New Zealand Forest Service restructuring (Phillips et al. 1989) and the subsequent sale of the State's exotic forests. As a consequence, these maps were initially underutilised by the new owners of these former state-owned 'conservation' forests, and their real value was never fully appreciated until much later when difficulties with maintaining road infrastructure arose. Interestingly, Pearce (1977) commented that a landscape zoning scheme like this needed to be developed urgently for all New Zealand plantation forests if it was to provide landslide hazard information at the planning stage of forest activities and at a scale suitable for use at the operational level—35 years later recent erosion events suggest the need remains[e]. With the advent of new technology (LiDAR, GIS, slope-stability tools), used in combination with stereoscopic analysis of aerial photography that pre-dates planting, it is possible to anticipate with reasonable certainty the type and location of geomorphic responses during and following major storm events. Access to such information would forewarn harvest planners with the knowledge required to design a harvest strategy that avoids, or is at least cognisant of, the most vulnerable areas at times of greatest risk from the impact of storms.

The challenge ahead lies in managing New Zealand's exotic forests within environments where storms and landslide failures are a regular occurrence and where the failure of 'natural' slopes beyond the forest boundary is more often than not the greatest contributor to stream sediment load and yield. Thus the identification and avoidance of unstable terrain, especially in those forests located in areas recognised as being geologically fragile, are critical to reducing environmental, social, and economic costs incurred by slope failures that may occur at any stage throughout a forest's rotation. Although the knowledge gained from retrospective assessments of storm-related impacts is unlikely to help prevent slope failures during future storms, lessons learned will nonetheless assist in the identification of areas elsewhere where a similar potential for slope failure associated with forest activities—not just harvesting—exists. In addition, lessons learned have value in either underpinning or re-evaluating existing best management practices and environmental standards.

Conclusions

The extent and severity of landslide damage sustained across the study area during the March 1995 storm was consistent with a severe event with a likely recurrence interval of several decades.

The factors that most influenced differences in the extent and type of slope failure (and ultimately, in the severity of the resultant on- and off-site damage) were rainfall distribution and intensity, slope, and vegetation. Though based on limited rainfall, stream water level and sediment yield data, the general consensus is that this storm had its greatest impact where rainfall intensities and totals were highest. Additionally, high peak stream water levels and sediment yield data, together with a post-storm documentation of the distribution of slope failures (this paper), observations of channel characteristic, and impact assessments of in-stream fauna and the macro invertebrate population further substantiate that the March 1995 storm had its greatest impact within the Awaroa, Owera and Weiti catchments and, to a lesser degree, in the Opitonui and miscellaneous catchments.

The highest proportion (>99%) of the sediment load delivered to streams was generated by the failure of natural slopes (e.g. debris avalanche, soil slip and stream bank collapse) while failures associated with forest harvesting (e.g. landing failures) contributed <0.5%.

The clearfelling of exotic forest undoubtedly increased the susceptibility of cutover to storm-initiated slope failure, predominantly from the earliest and thus the most vulnerable of the cutover areas where the progressive loss of strength from decaying roots had not yet been counteracted by the development of an effective root system under the new tree crop.

While woody debris and sediment derived from forest cutover contributed to the resultant damage sustained on floodplain areas located downstream of Whangapoua Forest, it was exacerbated by the severity and extent of landslide-generated debris derived primarily from within areas of indigenous forest located upstream of the exotic forest estate. This finding is contrary to public opinion at the time of the storm.

Also, contrary to the findings of most storm damage-vegetation assessments, least sediment mass was generated from areas of pastoral hill country suggesting they lay at the outer edge of this storms influence and that rainfall intensity and totals were below the threshold required to initiate slope failure.

The initiation of landslides in steep-land forests during storms is inevitable, and valuable lessons can be learned from such events. In keeping with previous post-storm assessments of slope failure, the 'storm effect' at Whangapoua was highly variable because landslide occurrence and type were determined by specific combinations of topography, storm characteristics, vegetation cover, and in the case of areas of plantation forest, the extent of harvesting at the time of the March 1995 storm.

Recommendations

The findings of this study suggest that there is a need to reassess the erosion potential/susceptibility of steep land LUC units. Moreover, a return to terrain stability assessments based on the identification, mapping, and geomorphic interpretation of site-specific hazards, particularly from an operational perspective—in advance of harvesting—is recommended. Such maps would either support or identify the need to modify current practices aimed at reducing the likelihood of landslide (both shallow and deep-seated) occurrence and the consequent on- and off-site impacts of future storms, thereby permitting foresters to look forward toward prevention rather than backward toward correction.

Endnotes

[a]Hancox, GT, Wright, K. (2005). *Analysis of landsliding caused by the February 2004 rainstorms in the Wanganui-Manawatu hill country, southern North Island, New Zealand.* (GNS Report 2005/11). Wellington, New Zealand: Institute of Geological & Nuclear Sciences Limited.

[b]Fransen, PJB. (1998). *Slips and sedimentation in mature plantation forest and pastoral hill country, Hawke's Bay, New Zealand.* (LIRO Report PR 73). Rotorua, New Zealand: Liro Forestry Solutions.

[c]Bayfield, MA, Meister, AD. (1998). *East Coast Forestry Project Review.* (Report to Ministry of Agriculture and Forestry). Wellington, New Zealand.

[d]Bayfield, MA, Meister, AD. (2005). *East Coast Forestry Project Review.* (Report to Ministry of Agriculture and Forestry). Wellington, New Zealand.

[e]Amishev D, Basher L, Phillips C, Hill S, Marden M, Bloomberg M, Moore J. (2013). *New forest management approaches to steep hills* (MPI Technical Paper No: 2014/39 Prepared by Scion). Rotorua, New Zealand: Scion.

[f]Gray, W, Ibbitt, R, Turner, R, Duncan, M, Hollis, M. (2005). *A methodology to assess the impacts of climate change on flood risk in New Zealand.* (NIWA Client Report CHC2005-060). Christchurch, New Zealand: National Institute of Water and Atmospheric Research.

[g]Bloomberg M, Davies T, Visser R, Morgenroth J 2011. *Erosion Susceptibility Classification and Analysis of Erosion Risks for Plantation Forestry.* (Report for the Ministry for the Environment by University of Canterbury). Christchurch, New Zealand.

[h]Marden, M, Rowan, D. (1995). *Assessment of Storm Damage to Whangapoua Forest and its Immediate Environs Following the Storm of March 1995.* (Landcare Research Contract Report LC9495/172). Lincoln, New Zealand: Landcare Research New Zealand Limited.

[i]Jelinek, L. (2010). *The Tapawera-Baton 16th of May 2010 flood-a review of erosion mechanisms, resulting*

damage and learnings from forest operations planning. (Report for Nelson Forests). Nelson, New Zealand.

[j]Page, MJ. (2013). *Landslide and debris flows caused by the 15–17 June 2013 rain storm in the Marahau-Motueka area, and the fatal landslide at Otuwhero Inlet.* (Geological Sciences Report 2013/44). Wellington, New Zealand: Institute of Geological & Nuclear Sciences Limited.

[k]Baillie, BR. (1999). *Management of logging slash in streams – results of a survey.* (LIRO Project Report PR85). Rotorua, New Zealand: Liro Forestry Solutions.

[l]Douglas, J, Stokes, S, Waiora Soil Conservation. (2011). *Report on exotic forest debris management related to storm events in the Bay of Plenty.* (Operations Publication 2011/03 for the Bay of Plenty Regional Council). Whakatane, New Zealand: Eastern Bay of Plenty Regional Council.

[m]Phillips, CJ, Marden, M. (2011). *Houpotu Forest storm damage assessment and review.* (Landcare Research Contract Report LC777). Lincoln, New Zealand: Landcare Research New Zealand Limited.

[n]Basher, L. (2010). *Storm damage in the Tapawera area during the storm of 16th March 2010.* (Landcare Research Internal Report). Lincoln, New Zealand: Landcare Research New Zealand Limited.

[o]Page, MJ, Langridge, RM, Stevens, GJ, Jones, KE. (2012). *The December 2011 debris flows in the Pohara-Ligar Bay area, Golden Bay: causes, distribution, future risks and mitigation options.* (Geological Science Consultancy Report 2012/305). Wellington, New Zealand: Institute of Geological & Nuclear Sciences Limited.

[p]Gray, C, Spencer, M. (2011). *Some observations of erosion as a result of the 28 December 2010 storm event.* (Technical Report No. 11–024 for the Marlborough District Council). Blenheim, New Zealand: Marlborough District Council.

[q]Phillips, CJ, Marden, M. (1999). *Review of vegetation-slope stability in plantation forests and risk assessment of Ohui Forest to landsliding.* (Landcare Research Contract Report LC9899/66). Lincoln, New Zealand: Landcare Research New Zealand Limited.

[r]Marden, M, Rowan, D. (1988). *Protective value of different vegetation types in the East Coast Region after Cyclone Bola: a preliminary assessment.* (Internal Forest Research Institute Report). Lincoln, New Zealand: Landcare Research New Zealand Limited.

[s]Coker, R. (1988). *Assessment of the Environmental Impacts of Harvesting Whangapoua Forest.* (Contract report prepared for NZ Timberlands). Auckland, New Zealand: New Zealand Timberlands Limited.

[t]Williams, M. (1995). *Whitianga Storm Event - 4th March 1995.* (Environment Waikato Memorandum File 410003). Hamilton, New Zealand. Environment Waikato.

[u]Quinn, JM, Kemp, CLS. (1998). Effects of harvesting Whangapoua Forest on stream clarity and temperature: 1997 annual report and review of five years monitoring. NIWA Client Report ERN90201/1. Prepared for Ernslaw One Ltd.

[v]Quinn, JM, Stroud, MJ, Parkyn, SM, Burrell, GP. (1995). *Impact of an intense rain storm on 3–4 March 1995 on Whangapoua Forest landscape and streams.* NIWA Client Report ERN301/1. Prepared for Ernslaw One Ltd.

[w]Wild M, Hicks M. (2005). *Opitonui Stream suspended sediment analysis.* (NIWA Client Report CHC2005–115). Christchurch, New Zealand: National Institute of Water and Atmospheric Research Ltd.

[x]Morrisey DJ, Burt K, Funnell GA, Hume TM, Kemp CLS, Ovenden R, Smith RK, Swales A. (1999). *Ecological monitoring for potential effects of forest activity on the intertidal habitats of Whangapoua Harbour, Annual Report 1998–1999.* (NIWA Client Report ERN90202/1). Hamilton, New Zealand. New Zealand: National Institute of Water and Atmospheric Research

[y]Wright-Stow, AE, Quinn, J M. (2013). *Progressive catchment harvesting in Whangapoua Forest: impacts on stream clarity, dissolved oxygen.temperature, habitat and invertebrates:* Twenty-first annual report incorporating results from 1992–2013. NIWA Client Report HAM2013-039.

[z]Phillips, CJ, Marden, M. (1996). *Mohaka forest storm damage.* (Landcare Research Contract Report LC9596/121). Lincoln, New Zealand: Landcare Research New Zealand Limited.

[aa]Pearce, AJ, O'Loughlin, C. (1976). *Effects of forest management on water and soil conservation.* (Internal Report to Auckland Conservancy of the New Zealand Forest Service). Auckland, New Zealand: New Zealand Forest Service Limited.

Competing interests

The authors declare that they have no competing interests.

Authors' contributions

MM was the primary author, and undertook the stereoscopic analysis, mapping, and interpretation of the data. MM and DR undertook the field work, and compiled the data into spreadsheets. Both authors read and approved the manuscript.

Acknowledgements
This research was supported by funding from the Ministry of Business, Innovation, and Employment, Contract number CO4X1306 Growing Confidence in Forestry to Scion/Future Forest Research. Our thanks go to the staff of Ernslaw One Limited for commissioning the original 1995 contract, for supplying forest compartmental records and aerial photography of the storm-affected area, and latterly for granting permission to publish this work. Environment Waikato (Regional Council) is acknowledged for supplying rainfall, and flow data. Thanks to Drs Les Basher, Chris Phillips and two external reviewers for their helpful comments, and suggestion. Figures were drawn by Nicolette Faville, and Anne Sutherland. Anne Austin edited the script. Suggested improvements by the journal's Associate editor and Editor-in-chief are much appreciated.

Author details
[1]Landcare Research, PO Box 445, Gisborne 4010, New Zealand. [2]16 Elgin Place, Whangarei 0112, New Zealand.

References

Basher, LR, Hicks, DM, Clapp, B, & Hewitt, T. (2011). Sediment yield responses to forest harvesting and large storm events, Motueka River, New Zealand. *New Zealand Journal of Marine and Freshwater Research, 45*, 333–356.

Bergin, DR, Kimberley, MO, & Marden, M. (1995). Protective value of regenerating tea-tree stands on erosion-prone hill country, East Coast, North Island, New Zealand. *New Zealand Journal of Forest Science, 25*, 3–19.

Beverley, CW, Swanson, FJ, & Jones, JA. (2001). Forest roads and geomorphic process interactions, Cascade Range, Oregon. *Earth Surface Processes and Landforms, 26*, 191–204.

Bishop, DM, & Stevens, ME. (1964). *Landslide on logged areas in south-eastern Alaska.* (Paper NOR-1).Juneau, AL, USA: Northern Forest Experiment Station, Forest Service, U.S. Dept. of Agriculture.

Blaschke, PM, Trustrum, NA, & DeRose, RC. (1991). Ecosystem processes and sustainable land use in New Zealand steep lands. *Agriculture Ecosystems and Environment, 41*, 153–178.

Caine, N. (1980). Rainfall intensity-duration control of shallow landslide and debris flows. *Geografiska Annaler, 62A*(1–2), 23–29.

Chudleigh, PD, Greer, G. & Sheppard, RL. (1983). *Supporting the Agricultural Sector: Rationale and Policy*. Agricultural Economics Research Unit, Lincoln College, New Zealand. ISSN 0110-7720.

Committee of Inquiry. (1939). *Maintenance of vegetative cover in New Zealand, with special reference to land erosion.* (DSIR Bulletin 77). Wellington, New Zealand: DSIR.

Coulter, JD, & Hessell, JWD. (1980). *The frequency of high intensity rainfalls in New Zealand. Part III. Point estimates.* (New Zealand Meteorological Service Miscellaneous Publication 162). Wellington, New Zealand: New Zealand Meteorological Service.

Crozier, MJ, & Eyles, RJ. (1980). *Assessing the probability of rapid mass movement. Third Australia-New Zealand Conference on Geomechanics*, (Vol 2, pp. 2.47–2.53). Wellington, New Zealand: Institute of Engineers Proceedings of Technical Groups.

Cunningham, A, & Stribling, PW. (1978). The Ruahine Range – A Situation Review and Proposals for Integrated Management of the Ruahine Range and the Rivers Affected By It. *Water and Soil Technical Publication*, 13

Dymond, JR, Ausseil, AG, Shepherd, JD, & Buettner, L. (2006). Validation of a region-wide model of landslide susceptibility in the Manawatu–Wanganui region of New Zealand. *Geomorphology, 74*, 70–79.

Esler, AE. (1963). The Influence of the Pre-European Fires in the Tiritea Catchment, Northern Tararuas. *Proceedings of the New Zealand Ecological Society, 10*, 8–12.

Eyles, GO. (1985). *The New Zealand Land Resource Inventory Erosion Classification.* (Water and Soil Miscellaneous Publication 85). Wellington, New Zealand: National Water and Soil Conservation Authority.

Fahey, BD, & Coker, RJ. (1989). Forest road erosion in the granite terrain of southwest Nelson, New Zealand. *Journal of Hydrology (NZ), 28*(2), 123–141.

Fahey, BD, & Coker, RJ. (1992). Sediment production from forest roads in Queen Charlotte Forest and potential impact on marine water quality, Marlborough Sounds, New Zealand. *New Zealand Journal of Marine and Freshwater Research, 26*, 187–195.

Fahey, BD, & Marden, M. (2000). Sediment yields from a forested and a pasture catchment, coastal Hawke's Bay, North Island, New Zealand. *Journal of Hydrology (NZ), 39*(1), 49–63.

Fahey, BD, Marden, M, & Phillips, CJ. (2003). Sediment yields from plantation forestry and pastoral farming, coastal Hawke's Bay, North Island, New Zealand. *Journal of Hydrology (NZ), 42*(1), 27–38.

Fahey, B, Duncan, M, & Quinn, J. (2004). Impacts of Forestry. In J Harding, P Mosley, C Pearson, & B Sorrell (Eds.), *Freshwaters of New Zealand* (Chapter 33). Christchurch, New Zealand: Institute of Engineers Proceedings of Technical Groups.

Fannin, RJ, Moore, GD, Schwab, JW, & VanDine, DF. (2005). Landslide risk management in forest practices. In: Proceedings *International Conference on Landslide Risk Management* (pp. 299–320), Vancouver, B.C., Canada, 31 May—June 3.

Fransen, PJB, & Brownlie, RK. (1995). Historical slip erosion in catchments under pasture and radiata pine forest, Hawke's Bay hill country. *New Zealand Forestry, 40*(4), 29–33.

Furbish, DJ, & Rice, RM. (1983). Predicting landslide related to clearcut logging, North-western California. *U. S. A. Mountain Research and Development, 3*(3), 253–259.

Glade, T. (1998). Establishing the frequency and magnitude of landslide triggering rainstorm events in New Zealand. *Environmental Geology, 35*, 160–174.

Glade, T, & Crozier, MJ. (1996). Towards a national landslide information base for New Zealand. *New Zealand Geographer, 52*(1), 29–40.

Grant, PJ. (1977). *Recorded Channel Changes of the Upper Waipawa River, Ruahine Range*, New Zealand. Water & Soil Technical Publication, 6. Wellington, NZ: Ministry of Works and Development.

Guthrie, RH. (2002). The effects of logging on frequency and distribution of landslide in three watersheds on Vancouver Island, British Columbia. *Geomorphology, 43*, 273–292.

Harmsworth, GR, Hope, GD, Page, MJ, & Manson, PA. (1987). *An assessment of storm damage at Otoi in Northern Hawkes Bay.* (Publication 10).Palmerston North, New Zealand: Soil Conservation Centre.

Hewitt, AE. (2010). *New Zealand Soil Classification.* (Landcare Research Science Series 1 (3rd ed.). Lincoln, New Zealand: Manaaki Whenua Press.

Hicks, DL. (1989). *Storm damage to bush, pasture and forest during Cyclone Bola.* (Technical Report PN2). Palmerston North, New Zealand: DSIR Soil Conservation Centre.

Hicks, DL. (1991). Erosion under pasture, pine plantations, scrub and indigenous forest: a comparison from Cyclone Bola. *New Zealand Forestry, 36*(3), 21–22.

Hicks, DM, & Basher, LR. (2008). The signature of an extreme erosion event on suspended sediment loads: Motueka River Catchment, South Island, New Zealand. Sediment Dynamics in Changing Environments. In: *Proceedings of International Association of Hydrological Sciences Symposium* (325, 184–191), Christchurch, New Zealand.

High Intensity Rainfall Design System (HIRDS). http://www.niwa.co.nz/software/hirds. Accessed 05.08.2014

Holdaway, RJ, Easdale, TA, Mason, NWH, & Carswell, FE. (2013). *LUCAS National Forest Plot Analysis: Are New Zealand's Natural Forests a Source or Sink of Carbon?* Wellington, New Zealand: Ministry for the Environment.

Horner, M. (2012). After the storms-a case study in risk reduction. *New Zealand Journal of Forestry, 56*(4), 13–15.

Howes, DE. (1987).A method for predicting terrain susceptibility to landslide following forest harvesting: a case study from the southern Coast Mountains, British Columbia. Forest Hydrology and Watershed Management. In: *Proceedings of the International Association of Hydrological Sciences Symposium*, (pp. 143–154). XIX General Assembly of the International Union of Geodesy and Geophysics, August 12–22, University of British Columbia, Vancouver.

IPCC (2014). Summary for policymakers. In: *Climate Change 2014: Impacts, Adaptation, and Vulnerability. Part A: Global and Sectoral Aspects. Contribution of Working Group II to the Fifth Assessment Report of the Intergovernmental Panel on Climate Change* [Field, CB. VR. Barros, DJ. Dokken, KJ. Mach, MD. Mastrandrea, TE. Bilir, M. Chatterjee, KL. Ebi, YO. Estrada, RC. Genova, B. Girma, ES. Kissel, AN. Levy, S. MacCracken, PR. Mastrandrea, and LL. White (Eds.)]. Cambridge University Press, Cambridge, United Kingdom and New York, NY, USA, pp. 1–32.

Jakob, M. (2000). The impacts of logging on landslide activity at Clayoquot Sound, British Columbia. *Catena, 38*, 279–300.

James, IL. (1973). Mass Movement in the Upper Pohangina Catchment, Ruahine Range. *New Zealand Journal of Hydrology, 12*, 92–102.

Marden, M. (2004). Future-proofing erosion-prone hill country against soil degradation and loss during large storm events: have past lessons been heeded? *New Zealand Journal of Forestry, 49*, 11–16.

Marden, M. (2012). Effectiveness of reforestation in erosion mitigation and implications for future sediment yields, East Coast catchments, New Zealand: A review. *New Zealand Geographer, 68*, 24–35.

Marden, M, & Rowan, D. (1993). Protective value of vegetation on tertiary terrain before and during Cyclone Bola, East Coast, North Island, New Zealand. *New Zealand Journal of Forest Science, 23*(3), 255–263.

Marden, M, & Saunders, H. (1992). Balancing forestry economics and stewardship of the East Coast environment. *Conservation Quorum (Gisborne District Council), 8*, 8–9.

Marden, M, Phillips, CJ, & Rowan, D. (1991). Declining soil loss with increasing age of plantation forests in Uawa catchment, East Coast region, North Island, New Zealand. In *Proceedings, International Symposium on Sustainable Land Management, (pp. 358–361)* (pp. 18–23). New Zealand: Napier.

Marden, M, Rowan, D, & Phillips, C. (2006). Sediment sources and delivery following plantation harvesting in a weathered volcanic terrain, Coromandel Peninsula, North Island, New Zealand. *Australian Journal of Soil Research, 44*, 219–232.

Marden, M, Rowe, L, & Rowan, D. (2007). Slope wash erosion following plantation harvesting in pumice terrain and its contribution to stream sedimentation, Pokairoa catchment, North Island, New Zealand. *Journal of Hydrology, 46*, 73–90.

Marden, M, Basher, L, Phillips, C, & Black, R. (2015). Should detailed terrain stability or erosion susceptibility mapping be mandatory in erodible steep lands? *New Zealand Journal of Forestry, 59*(4), 42–52.

McCraw, JD, & Bell, JL. (1975). Soils of the Coromandel-Thames Counties. In *Land Inventory Survey of Coromandel-Thames Counties*. Department of Lands and Survey. 88 p, 6 maps.

McKelvey, P. (1992). The development of the concept of steep land protection forestry in New Zealand. *New Zealand Forestry*, 20–24.

McLaren, P. (1993). *Radiata Pine Growers Manual*. Forest Research Institute Bulletin 184, New Zealand Forest Research Institute Limited, Rotorua, New Zealand. 134 p.

Megahan, WF, Monsen, SB, & Wilson, MD. (1991). Probability of sediment yields from surface erosion on granite road fills in Idaho. *Journal of Environmental Quality, 20*(1), 53–60.

Megahan, WF, Wilson, M, & Monsen, SB. (2001). Sediment production from granitic cutslopes on forest roads in Idaho, USA. *Earth Surfaces Processes and Landforms, 26*, 153–163.

Millard, T. (1999). *Debris flow initiation in Coastal British Columbia gullies*. (Forest Research Technical Report). Nanaimo, British Columbia, Canada: British Columbia Ministry of Forests, Vancouver Forest Region.

Mosley, MP. (1980). The impact of forest road erosion in the Dart Valley, Nelson, New Zealand. *New Zealand Journal of Forestry, 23*, 184–198.

National Water and Soil Conservation Organisation. (1975). *New Zealand Land Resource Inventory Worksheets*. Wellington, New Zealand: New Zealand Government Printer.

O'Loughlin, CL. (1972). A preliminary study of landslide in the coast mountains of southwestern British Columbia. In S HO & MP HJ (Eds.), *Mountain Geomorphology, Geomorphological Processes in the Canadian Cordillera. (B. C. Geographical Series Vol. 14, pp. 101–111)*. Vancouver, British Columbia, Canada: Tantalus Research.

O'Loughlin, CL. (1985). Influences of exotic plantation forest on slope stability implications for exotic forest management in the Marlborough Sounds. In *Proceedings of the Soil Dynamics and Land-Use Seminar, Blenheim*.

O'Loughlin, CL, & Ziemer, RR. (1982). The importance of root strength and deterioration rates upon edaphic stability in steep land forests. In RH Waring (Ed.), *Carbon uptake and allocation in subalpine ecosystems as a key to management*. Corvallis: Proceedings of IUFRO Workshop.

O'Loughlin, CL, Rowe, LK, Pearce, AJ. (1982). Exceptional storm influences on slope erosion and sediment yields in small forest catchments, north Westland, New Zealand. In: EM, O'Loughlin, EJ, Bren (Eds.). *Institute of Engineers, Australian National Conference publication, 82/6*, 152 p.

Olsen, PF. (1970). Mangatu: A production forest with major protection value. *New Zealand Journal of Forestry, 15*(2), 169–183.

Pearce, AJ. (1977). *Landscape zoning, erosion control and forest management. What's New in Forest Research? 55*. Wellington, New Zealand: Forest Research Institute.

Pearce, AJ, & Hodgkiss, PD. (1987). Erosion and sediment yield from a landing failure after a moderate rainstorm, Tairua Forest. *New Zealand Journal of Forestry, 32*(2), 19–22.

Phillips, CJ, & Marden, M. (2005). *Reforestation schemes to manage regional landslide risk. In T, Glade, MG, Anderson, MJ, Crozier. (Eds.). Landslide Hazard and Risk. (pp. 731–752)*. Chichester, UK: John Wiley and Sons Ltd.

Phillips, CJ, Marden, M, & Rowan, D. (1989). Planning for forestry after Cyclone Bola: a comment. *New Zealand Forestry*, 16–17.

Phillips, CJ, Marden, M, & Pearce, AJ. (1990). Effectiveness of reforestation in the prevention and control of landsliding during large cyclonic storms. In *Proceedings of International Union of Forest Research Organisations XIX World Congress* (pp. 340–350), Montreal 1990.

Phillips, CJ, Marden, M, & Rowan, D. (2005). Sediment yield following plantation harvesting, Coromandel Peninsula, North Island, New Zealand. *Journal of Hydrology (NZ), 44*, 29–44.

Phillips, C, Marden, M, & Basher, L. (2012). Plantation forest harvesting and landscape response–what we know and what we need to know. *New Zealand Journal of Forestry, 56*, 4–12.

Poole, AL. (1960). Protection forests in New Zealand and a Poverty Bay example. *New Zealand Geographer, 16*(2), 115–130.

Quinn, JM, Cooper, AB, & Williamson, RB. (1993). Riparian zones as buffer strips: A New Zealand perspective. In *Proceedings of a National Workshop on research and management needs for riparian zones in Australia*. Held in association with the 32nd Annual Congress of the Australian Society for Limnology, Marcoola, Australia.

Reid, LM, & Page, MJ. (2002). Magnitude and frequency of landsliding in a large New Zealand catchment. *Geomorphology, 49*, 71–88.

Rogers, NW, & Selby, MJ. (1980). Mechanisms of shallow translational landsliding during summer rainstorms, North Island, New Zealand. *Geografiska Annaler, 62A*, 11–21.

Rollerson, TP. (1992). *Relationship between landscape attributes and landslide frequencies after logging; Skidegate Plateau, Queen Charlotte Islands* (Ministry of Forests Land Management Report 76). British Columbia, Canada: Victoria.

Rollerson, TP, Jones, C, Trainor, K, & Thomson, B. (1998). *Linking post-logging landslide to terrain variables: Coast Mountains, British Columbia-preliminary analyses* (In *Proceedings of the 8th International Congress of the International Association for Engineering Geology and the Environment*, (pp. 1973–1979)). Rotterdam: Balkema.

Rood, KM. (1984). *An aerial photograph inventory of the frequency and yield of mass wasting on the Queen Charlotte Islands, British Columbia. (Ministry of Forests Land Management Report 34)*. Victoria, British Columbia, Canada: Ministry of Forests.

Roy, B, Popay, I, Champion, P, James, T, & Rahman, A. (1998). *An Illustrated Guide to Common Weeds of New Zealand*. New Zealand Plant Protection Society (Inc) 282 p. ISBN 0-473-05296-2.

Salter, RT, Crippen, TF, & Noble, KE. (1983). *Storm damage assessment of the Thames-Te Aroha area following the storm of April 1981. (Publication I)*. Palmerston North, New Zealand: Soil Conservation Centre.

Sauder, EA, Krag, RK, & Wellburn, GV. (1987). *Logging and mass wasting in the Pacific Northwest with application to the Queen Charlotte Islands, B.C. (Land Management Report, 53)*. Victoria, British Columbia, Canada: British Columbia Ministry of Forests.

Schwab, JW. (1983). *Mass wasting: October-November 1978 storm, Rennell Sound, Queen Charlotte Islands, British Columbia. (Ministry of Forests Research Note 91)*. Victoria, British Columbia, Canada: Ministry of Forests.

Selby, MJ. (1967). Aspects of the geomorphology of the greywacke ranges bordering the lower and middle Waikato basins. *Earth Science Journal, 1*(1), 37–58.

Service, NZM. (1980). *Depth-Duration-Frequency Tables based on daily rainfalls*. (New Zealand Meteorological Service Miscellaneous Publication, 162, Supplement 1). Wellington, New Zealand: New Zealand Meteorological Service.

Skinner, DNB. (1976). *Geological Map of New Zealand 1:63360 Sheets N40 and part sheets N35, N36 and N39 Northern Coromandel*. Wellington, New Zealand: New Zealand Geological Survey.

Smith, RB, Commandeur, PR, & Ryan, MW. (1986). *Soils, vegetation, and forest growth on landslide and surrounding logged and old-growth areas on the Queen Charlotte Islands, British Columbia*. (Ministry of Forests Land Management Report 41). Victoria, British Columbia Canada: Ministry of Forests.

Taylor, NH. (1938). *Land deterioration in the heavier rainfall districts of New Zealand*. (DSIR Bulletin 62). Wellington, New Zealand: DSIR.

Taylor, NH (Ed.). (1970). *Wise land use & community development, report of Technical Committee of Inquiry into the problems of the Poverty Bay–East Cape District of New Zealand.* Wellington. Wellington, New Zealand: Water and Soil Division, Ministry of Works for the National Water and Soil Conservation Organisation.
Thomson, B. (1987). *Chapman Creek landslide inventory (1:20,000 scale map with legend).* Victoria, British Columbia, Canada: British Columbia Ministry of Environment, Lands and Parks.
Wardle, P. (1991). *Vegetation of New Zealand.* ISBN: 1-930665-58-X. The Blackwell Press, USA. 672 p.
Watson, AJ. (1990). *Contribution of tree roots to slope stability. What's New in Forest Research 196.* Wellington, New Zealand: Forest Research Institute.
Watson, A, Marden, M, & Rowan, D. (1995). Tree species performance and slope stability. In: DH, Barker. (Ed.) *Vegetation and Slopes, stabilization, protection and ecology.* (pp. 161–171). London, UK: Thomas Telford.
Watson, A, Phillips, C, & Marden, M. (1999). Root strength, growth, and rates of decay: root reinforcement changes of the two species and their contribution to slope stability. *Plant and Soil, 217,* 39–47.
Ziemer, RR. (1981a). Roots and the stability of forested slopes. *In Proceedings of the International Symposium on Erosion and Sediment Transport in Pacific Rim Steep lands. (pp. 343–361). (International Association of Hydrological Sciences Publication 132).* Christchurch, New Zealand: International Association of Hydrological Sciences.
Ziemer, RR. (1981b).The role of vegetation in the stability of forested slopes. In *Proceedings of the International Union of Forestry Research Organisations XVII World Forestry Congress,* (pp. 297–308). Kyoto, Japan.
Ziemer, RR. (1981c). Management of steep land erosion: An overview. Keynote address, Theme 4. *Journal of Hydrology* (NZ), *20,* 8–16.

21

Effects of stand density and seedlot on three wood properties of young radiata pine grown at a dry-land site in New Zealand

John R Moore[1*], Dave J Cown[1], Russell B McKinley[1] and Charles O Sabatia[1,2]

Abstract

Background: Manipulation of stand density and choice of genetic material are two key mechanisms through which forest managers can influence tree growth and wood properties. Past silvicultural practices in New Zealand have been characterised by early thinning to relatively wide spacing, while tree improvement programmes have primarily focussed on growth and form. The aim of this study was to quantify the impacts of stand density and past genetic selections on the wood properties of radiata pine (*Pinus radiata* D. Don).

Methods: Stress-wave velocity, wood density, microfibril angle (MFA) and modulus of elasticity (MOE) were measured on trees and wood samples taken from a 15-year-old silvicultural and tree breeds trial located in Canterbury, New Zealand. The focus of this study was comparison of seedlots with genetic ratings of GF6 and GF25. Data from pruned stands with final densities of 100, 200 and 400 stems ha^{-1} were compared as were data from unpruned stands with stand densities of 200, 400, 600 and 1000 stems ha^{-1}.

Results: Stress-wave velocity of trees was affected by final stand density, with the lowest values recorded from trees in the 100 and 200 stems ha^{-1} treatments. Values for wood MFA and MOE both showed the greatest amount of change when final stand density was less than 400 stems ha^{-1}. Trees from the GF25-rated seedlot had wood density that was approximately 34 kg m^{-3} lower than trees from the GF6-rated seedlot.

Conclusion: Very high stand densities are not required in order to influence wood properties, but early thinning to low stand densities should be avoided. Such results are of practical importance to forestry managers.

Keywords: Silviculture; Genetics; Wood properties; Wood density; Microfibril angle; Modulus of elasticity

Background

Control of the growing space available to trees and the selection of planting stock with improved traits are two of the main processes through which forest managers can influence the growth and wood properties of trees within a stand (Savill and Sandels 1983; Smith et al. 1997). Spacing affects the vigour and characteristics of individual trees within a stand and also the total volume of wood obtained from a stand (Assmann 1970; Carson et al. 1999; Kilpatrick et al. 1981; Whyte and Woollons 1990). Trees grown in widely spaced stands are typically more tapered and have larger branches that remain alive further down the stem than trees growing more closely together (Auty et al. 2012; Evert 1971; Hein et al. 2008; Maguire et al. 1991). Tree spacing can also affect a number of wood properties, such as density, microfibril angle (MFA) and stiffness (modulus of elasticity, MOE), that affect the performance of solid timber (Panshin and de Zeeuw 1980). Trees that are widely spaced (whether established at wide spacing or later thinned to a reduced stand density) generally have poorer wood properties than closely spaced trees (Macdonald and Hubert 2002).

Radiata pine (*Pinus radiata* D. Don) is the predominant commercial species grown in New Zealand (Ministry for Primary Industries 2012). Silvicultural regimes for this species are generally characterised by multiple thinning and pruning operations. Silviculture is driven by the primary goal of maximising profitability and this led to the development of regimes based on wide tree

* Correspondence: john.moore@scionresearch.com
[1]Scion, Private Bag 3020, Rotorua 3046, New Zealand
Full list of author information is available at the end of the article

spacing and heavy early thinning from the 1960s to the 1990s (James 1990). In the most extreme cases, the residual stand densities following final thinning were as low as 200 trees ha^{-1}. The rationale used during that period was that the best financial return came from growing large pruned logs (Fenton and Sutton 1968). Silvicultural regimes developed more recently do not generally involve such low stocking rates but are, nevertheless, still based on the desire to maximise profitability. Given that rotation lengths are generally around 30 years, trees will contain a relatively high proportion of corewood, which generally has poorer wood properties than outerwood (Burdon et al. 2004; Zobel and Sprague 1998).

Many wood properties are also under a moderate to high degree of genetic control, but these were often considered secondary traits in tree improvement programmes. Breeding programmes have instead focussed on improving growth, form and tolerance of a wide range of site conditions (Zobel and Jett 1995). However, competition from alternative materials has increased in recent years so the performance characteristics of wood products must continue to meet end-user expectations and requirements otherwise market share is likely to be lost (Roos et al. 2010; Vinden 2003). These economic drivers have led to an increased focus on: (a) breeding trees for improved wood quality; (b) non-destructive evaluation of wood properties (e.g. measuring tree stiffness using portable acoustic tools (Kumar 2004; Kumar and Burdon 2010; Kumar et al. 2006)); and (c) understanding the impact of past genetic selections on wood quality given that many desirable growth and wood quality traits are adversely correlated (Gapare et al. 2009; Kumar et al. 2008).

New Zealand also has a long history of tree breeding dating back to the 1950s when the first plus tree selections were made (Burdon 2008; Carson 1996; Jayawickrama and Carson 2000). The long-standing goal of various radiata-pine breeding programmes has been the rapid growth of large, well-formed trees, which has given rise to the Growth and Form (GF) breed (Jayawickrama and Carson 2000). Different selections of plus trees have been made within this breed; the main ones being the 850, 268 and 870 Series. The 850 Series was an early selection based on the results of an intensive search of radiata pine stands across New Zealand for trees with better-than-average growth, good stem form and light branching. The 268 Series was obtained from trees in the northern part of Kaingaroa Forest (latitude 38°41'S, longitude 176°34'E). The forest was divided into one-hectare blocks then the best tree in each block was identified and selected using the same criteria as for the 850 Series. The 870 Series was obtained by selecting trees from throughout New Zealand with long internodes, combined with good growth and stem form (Carson 1996; Jayawickrama et al. 1997). Numerous different seedlots have been produced within each series, initially through open pollination and more latterly through controlled pollination. Each seedlot is assigned a rating based on the breeding values for growth and form (GF), in which growth is given twice the weighting as form. This rating gives a strong indication of genetic worth (Vincent 1987).

A set of trials was established between 1987 and 1991 at 28 locations throughout New Zealand to compare the performance of these genetically improved breeds when grown at a range of stocking levels. This set of trials was designed to include all eight major forest growing regions. Up to four levels of site quality were tested within each region and each trial installation typically included five different stand densities. The trials are intended to run for an entire rotation, i.e. around 30 years. Early growth and yield results from those trials installed in 1987 were reported by Carson et al. (1999). One of the 1991 trials was located at Shellocks Forest, 40 km south of Christchurch. This trial contained five seedlots – two from the 268 Series, one from the 850 Series, one from the 870 Series plus one from an early "climbing select" series (i.e. mild mass selection from climbing trees in 16–20 year old stands). Unfortunately, the site of this trial was sold for conversion into farmland and the forest scheduled for harvesting in 2005 when the trees were only 15 years old. Although premature removal of this trial prevents the collection of 30-year data as intended and cross-comparison with other sites, this situation did provide an opportunity to examine the effects of stand density, pruning and genetic selection on various wood properties at a younger age. Thus, the aim of this study became to collect and analyse data on three wood properties (wood density, microfibril angle and modulus elasticity) from the 15-year-old trees felled at the Shellocks Forest site. The results reported here are a specific subset of the total data collected.

Methods

Site

The trial was established in 1991 at Shellocks Forest, Dunsandel on the Canterbury Plains, 40 km south of Christchurch, New Zealand (latitude 43°40'S, longitude 172°11'E, elevation 115 m a.s.l.). Soil at the site is free draining, of low fertility and is classified as Pallic Firm Brown soil belonging to the Lismore series (Hewitt 1998). Long-term average annual rainfall at the closest meteorological station to the site (Christchurch Airport) is approximately 630 mm (New Zealand Meteorological Service 1983) and seasonal water deficits are common (Watson et al. 2004; Watt et al. 2003).

Trial design

The experiment was designed to investigate the effects of seedlot, stand density and pruning on growth and wood properties. It contained five different radiata pine seedlots, four levels of post-thinning stand density (100, 200, 400, 600 stems ha^{-1}) plus one unthinned treatment (1000 stems ha^{-1}), and two levels of pruning (pruned or unpruned). Only 25 of the 50 possible factorial combinations of seedlot, stand density and pruning were actually included in the original experiment (Table 1). Furthermore, there was only a single replicate of each selected treatment combination, so that the experiment consisted of 25 plots, each assigned a different treatment combination. These plots were 45 m × 36 m in size except in the treatment combinations associated with the lowest level of stand density (100 stems ha^{-1}) where plot size was 45 m × 72 m.

Seedlots

Five radiata pine seedlots were used (88/201, 90/294, 88/105, 89/15 and 88/102). The origin and GF rating of each seedlot are given in Table 2. The seedlots will be referred to by their GF rating from here on for clarity. None of the seedlots were bred for improved wood properties.

Silvicultural treatments

Plots in the experiment were established at three levels of initial stand density (250, 500 and 1000 stems ha^{-1}). Thinning was carried out in 1999 when the mean stand height was 7.1 m and the trees were eight years old. The GF6-, GF13- and GF25-rated seedlots were planted at each combination of initial stand density and thinned to each final level of stand density as follows. All the plots established at 250 stems ha^{-1} or 500 stems ha^{-1} were thinned using a thinning ratio (ratio of initial stand density to post-thinning stand density) of 2.5:1. This ratio was also applied to a proportion of the 1000 stems ha^{-1} plots, which were thinned to 400 stems ha^{-1}. A thinning ratio of 1.7:1 was applied to another proportion of 1000 stems ha^{-1} plots to obtain a final stand density of 600 stems ha^{-1}. This resulted in four levels of post-thinning stand density (100, 200, 400, 600 stems ha^{-1}). The remainder of the 1000 stems ha^{-1} plots were left unthinned (thinning ratio of 1:1). The GF14- and GF16-rated seedlots were each planted at only two initial stand densities (500 and 1000 stems ha^{-1}). In each case, all the 500 stems ha^{-1} plots were thinned to 200 stems ha^{-1} while all the 1000 stems ha^{-1} plots were unthinned (Table 1).

Pruning was also carried out in 1999 when the mean stand height was 7.1 m and the trees were eight years old. Trees in 11 of the 25 plots were pruned (P) to leave an average of 4.2 m of green crown remaining; this corresponded to a mean pruned height of approximately 2.7 m. Trees in the remaining 14 plots were unpruned (UP) (Table 1).

Data collection

Field measurements

In the time between planting and data collection, it had been discovered that seedlot 89/15 from the long-internode 870 Series produced trees with poor stem form. Consequently, there was a lack of market acceptance for this seed, and it is not commonly planted commercially (Burdon, 2008). Therefore, no data were collected from trees in the seven plots planted with this seedlot. In each of the remaining 18 plots, 15 trees with good stem form were selected from across the diameter range (270 trees in total) and their diameter at a breast height of 1.4 m (DBH) and total height (H) were measured. Stress-wave (acoustic) velocity measurements were made on each tree using two different standing-tree tools: an IML Hammer (Instrumenta Mechanic Labor GmbH, Wiesloch, Germany), and a ST-300 (Fibre-gen, Christchurch, New Zealand). Two sets of

Table 1 Trial design for the silviculture and tree breeds trial located at Shellocks Forest

Silviculture					Seedlot rating				
Trt.	Pruning	Stand density (stems ha^{-1})		Thinning ratio	GF6	GF14	GF16	GF25	GF13
		Initial	Final						
1	P	250	100	2.5:1	x[1]			x	x
2	P	500	200	2.5:1	x	x	x	x	x
3	P	1000	400	2.5:1	x			x	x
4	UP	500	200	2.5:1	x			x	x
5	UP	1000	400	2.5:1	x			x	x
6	UP	1000	600	1.7:1	x			x	x
7	UP	1000	1000	1:1	x	x	x	x	x

Each of the 25 treatment combination present in the trial is indicated by an "x". None of the GF13 seedlot material was sampled in the current study.

Table 2 Description of the seedlots planted in the experiment

Breeding series	Seedlot number	Seedlot rating	Description
"268 series"	88/201	GF16	Open-pollinated seedlot from the Brightwater seed orchard
"268 series"	90/294	GF25	Control-pollinated mix from the top 16 clones in the "268 series"
"850 series"	88/105	GF14	Open-pollinated mix of plus trees from Gwavas, Kaingaroa and Waimihia forests
"870 series"	89/15	GF13 (LI25)	Open-pollinated mix from seven long-internode clones
"climbing select"	88/102	GF6	Climbing select from Kaingaroa and Rotoehu forests

measurements were made with each instrument; one on the north and one on the south side of each tree. These measurements were centred about breast height and the probes of the instruments were spaced approximately 1 m apart along the longitudinal axis of the stem. The acoustic velocity value for a treatment plot was computed as the arithmetic mean of the acoustic velocity measurements from the 15 sample trees in the plot.

The trees were felled once these measurements had been completed. The resulting logs were initially cross-cut at the point where the stem diameter was 10 cm. The length of this stem section was typically 9–12 m, but was shorter in some cases where the stem was forked. The stress-wave velocity was measured on this stem section using the HM-200 tool (Fibre-gen, New Zealand). A disc 50 mm in thickness was then cut out at the breast height position on the log to provide samples for measuring ring-level wood properties.

Wood property measurements

Ten discs were selected at random from the 15 discs obtained from each plot. Wood samples measuring 15 mm in both the tangential and longitudinal directions were taken from the pith to the cambium on the north side of the disc. Regions containing compression wood and/or knots were avoided. These samples were soaked in ethanol to prevent fungal decay and to minimise fibre collapse; four ethanol exchanges were carried out, each lasting 48 hours. From each sample a strip spanning the full radius of the tree was milled using a twin-blade sawing system. Each strip had dimensions of 7 mm in the longitudinal direction and 2 mm in the tangential direction, and was prepared in accordance with the requirements for the SilviScan system (Evans 1999). All strips were extracted with boiling acetone using Soxhlet apparatus to remove resins present in the wood that could affect subsequent measurements. Strips were then conditioned at 40% RH and 20°C in a climate-controlled room to reach an equilibrium moisture content of 7-8%.

Radial profiles of MFA were determined for each strip from the standard deviation of the 002 azimuthal diffraction pattern using the SilviScan-2 instrument (Evans 1999; Evans et al. 1999) located at CSIRO, Melbourne, Australia. X-ray diffraction measurements were made at 50 μm resolution and the average MFA was determined over successive 5 mm intervals. X-ray densitometry measurements were made on each strip at a resolution of 50 μm in order to identify annual growth rings. Ring-level average values of wood density and MFA were calculated from the radial profiles, and these were used to estimate MOE using the approach described in Evans and Ilic (2001).

Data analysis

The current investigation focused on two components of the trial (Table 1) that formed complete two-way factorial experiments. The two components were: 1) a 3 by 2 stand-density by seedlot experiment for pruned stands with final densities of 100, 200 and 400 stems ha^{-1} containing trees from seedlots with genetic ratings of GF6 and GF25, hereinafter referred to as the pruned experiment; and 2) a 4 by 2 stand-density by seedlot experiment in unpruned stands with final densities of 200, 400, 600 and 1000 stems ha^{-1} containing trees from seedlots with genetic ratings of GF6 and GF25, hereinafter referred to as the unpruned experiment. The two experiments were analysed separately.

Tree-level acoustic velocity

There was just a single mean value of acoustic velocity per treatment combination (Tables 3 and 4), so data from the pruned and unpruned experiments were subjected separately to simple two-way comparison analysis. Prior to analysis, the data were tested for stand-density by seedlot interaction using Tukey's single-degree-of-freedom test for non-additivity (Milliken and Johnson 1989; Tukey 1949) using the GLM procedure in SAS software (SAS Institute Inc. 2004–2009). There was no evidence to show there was a seedlot by stand density interaction in tree-level acoustic velocity in either the pruned or unpruned experiments ($p \geq 0.1359$ and $p \geq 0.0948$, respectively). For this reason, the analysis of tree-level acoustic velocity data was based on the main effects of stand density and seedlot. The seedlot by stand density interaction was used to provide an estimate of experimental error (analogous to the case of a randomised complete block design with fixed blocks). The following statistical model was used:

$$y_{ij} = \mu + S_i + G_j + \varepsilon_{ij} \quad (1)$$

Table 3 Average plot-level acoustic-velocity measurements for treatment combinations in the pruned experiment

Acoustic velocity tool	Seedlot genetic rating	Acoustic velocity (km s^{-1}) at final stand density		
		100	200 (stems ha^{-1})	400
ST-300	GF6	2.434	2.743	3.069
	GF25	2.461	2.755	3.125
IML hammer	GF6	1.903	2.094	2.316
	GF25	1.899	2.101	2.356
HM-200	GF6	1.848	2.098	2.345
	GF25	1.918	2.081	2.385

where y_{ij} is the average acoustic velocity measurement for the ith level of stand density and the jth seedlot, μ is the overall mean, S_i is the fixed effect of the ith level of stand density ($i = 100$, 200, 400 for the pruned experiment and $i = 200$, 400, 600, 1000 for the unpruned experiment), G_j is the fixed effect of seedlot with the jth genetic rating (j = GF6, GF25), and ε_{ij} is a random error associated with the jth genetic rating plot under the ith stand density ($\varepsilon_{ij} \sim N(0, \sigma_\varepsilon^2)$). The statistical model in Equation 1 was fitted using the GLM procedure in SAS software (SAS Institute Inc. 2004–2009). Multiple comparisons between the factor-level means were made using the Tukey honest significant difference (HSD) adjustment of comparison-wise p-values to maintain an experiment-wise cut-off p-value of 0.05.

Ring-level MFA, MOE and wood density

The radial trends in wood density, MFA and MOE were modelled using a power function of the form:

$$Y = \alpha X^\beta \tag{2}$$

where Y is the MFA, MOE or wood density, X is the ring number with the ring closest to the pith designated as ring number 1, and α and β are parameters to be estimated from the data. The parameter α is the value of the response variable for ring number 1 and β is a measure of the radial rate of change of the response variable. Each treatment combination in both the pruned and unpruned experiments was considered to be a separate group and was analysed separately using the following model:

$$Y_{ijkl} = \left(\alpha_{ij} + \mathrm{a}_{ijk}\right) R_{ijkl}^{\left(\beta_{ij} + \mathrm{b}_{ijk}\right)} + \varepsilon_{ijkl} \tag{3}$$

where Y_{ijkl} is the breast height MFA, MOE or wood density of the lth ring (R_{ijkl}) in the kth tree in the ith stand density by j^{th} seedlot treatment combination; α_{ij} is the pith ring MFA, MOE or wood density parameter, and β_{ij} is the radial rate of change parameter for these responses, in the ith stand density by j^{th} seedlot treatment combination; a_{ijk} and b_{ijk} are random effects of the kth tree in the ith stand density by j^{th} seedlot treatment combination ($\mathrm{a}_{ijk} \sim N(0, \sigma^2_{\mathrm{a}_{ij}})$, $\mathrm{b}_{ijk} \sim N(0, \sigma^2_{\mathrm{b}_{ij}})$); and ε_{ijkl} is a zero expectation random error due to the lth ring in the kth tree in the ith stand density by j^{th} seedlot treatment combination. The residuals from fitting the data to Equation 3 exhibited heteroscedasticity so this was modelled (separately for each treatment combination) as a power function of the predicted value of Y_{ijkl} at the given R_{ijkl}. Correlations among the errors ε_{ijkl} for rings within the same tree were not modelled as these were considered to have been accounted for by the random effect parameters a_{ijk} and b_{ijk}. The size of the model residuals was reduced and the estimates of the variance of the random effect parameters became negligible when the random effect and correlation parameters were simultaneously included in Equation 3. The the '*nlme*' package (Pinheiro et al. 2013) in the R software package (R Development Core Team 2014) was used for computation.

Table 4 Average plot-level acoustic-velocity measurements for treatment combinations in the unpruned experiment

Acoustic velocity tool	Seedlot genetic rating	Acoustic velocity (km s^{-1}) at final stand density			
		200	400 (stems ha^{-1})	600	1000
ST-300	GF6	2.577	2.963	2.907	2.832
	GF25	2.624	2.955	2.807	2.831
IML hammer	GF6	1.891	2.186	2.183	2.130
	GF25	1.966	2.124	2.075	2.105
HM-200	GF6	2.008	2.305	2.364	2.332
	GF25	1.994	2.254	2.223	2.349

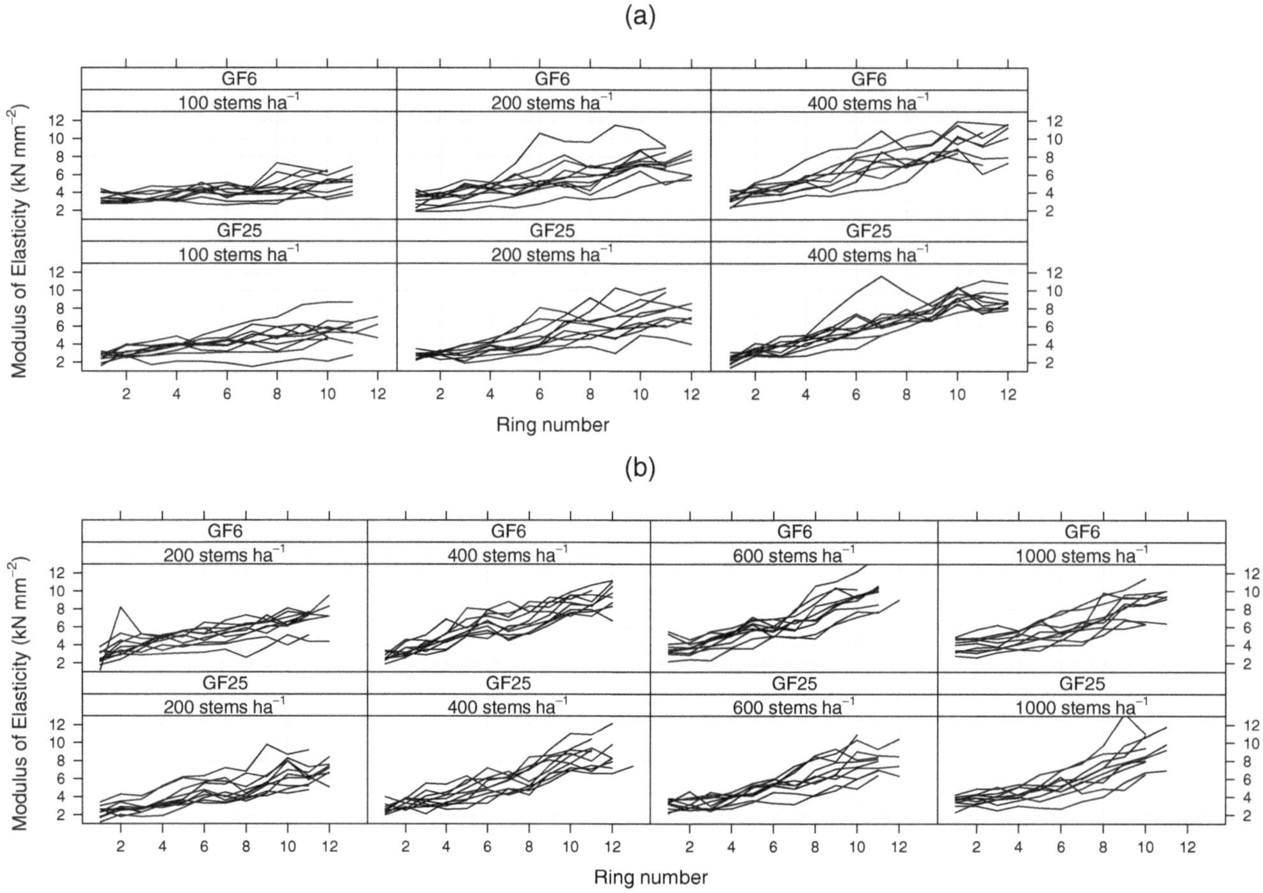

Figure 1 Trends in ring level values of modulus of elasticity at breast height for GF6- and GF25-rated trees at various final stand densities in (a) the pruned experiment; and (b) the unpruned experiment.

One GF25-rated tree in the pruned 400 stems ha^{-1} plot exhibited an unusual MOE trend (Figure 1a) and one GF6-rated tree from the unpruned 200 stems ha^{-1} plot exhibited an unusually high density for the first two rings (Figure 2b) so data from these two trees were excluded from the model. The effects of genetic rating and final stand density on breast-height ring-level MFA, MOE and wood density were inferred from comparison of the 95% confidence intervals of the parameter estimates of α_{ij} and β_{ij} across the stand-density and seedlot treatments. Equation 3 was also used to calculate the expected values of MFA, MOE and wood density for the ring 15 from the pith, which corresponded to the age of the trees when felled.

Results

Tree dimensions

Differences in tree height, DBH and height:DBH ratio were not formally compared between treatments using tests of significance, but mean values and general trends are reported to aid in interpreting the influence of treatment factors on wood properties. The higher the final stand density, the lower the DBH for unpruned trees of either seedlot examined. A similar trend was also observed in most cases for pruned trees (Figure 3). The height of unpruned GF6-rated trees grown at a final stand density of 400 stems ha^{-1} was almost approximately 2 m greater than those at a final stand density of 200 stems ha^{-1}, but higher final stand densities resulted in a reduction in tree height. Similar, but less pronounced, differences were observed for the unpruned GF25-rated seedlot and for pruned trees from both seedlots (Figure 3). The ratios of height to DBH were greater for trees grown at higher final stand densities, irrespective of seedlot or pruning treatment (Figure 3).

Tree-level acoustic velocity

Average tree-level acoustic-velocity measurements for the two-way factorial experiments for pruned and unpruned trees are given in Tables 3 and 4, respectively.

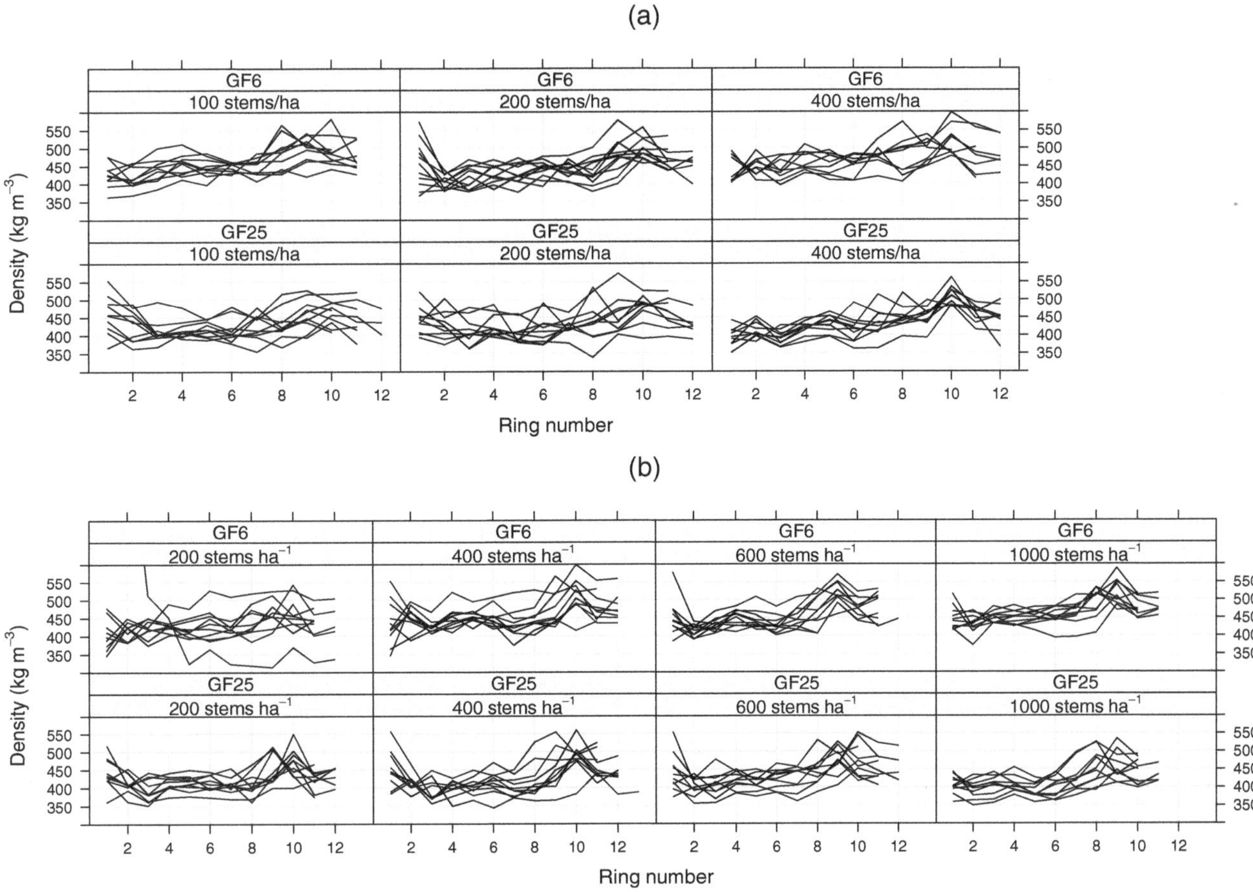

Figure 2 Trends in ring level values of wood density at breast height for GF6- and GF25-rated trees at various final stand densities in (a) the pruned experiment; and (b) the unpruned experiment.

Analysis of variance using the linear statistical model given in Equation 1 showed that final stand density had a significant effect on the acoustic velocity of trees in both the pruned and unpruned experiments. These results were irrespective of the type of tool used with the exception of the unpruned experiment measured using the IML hammer (Table 4). No significant differences in acoustic velocity were found between seedlots in either the pruned or the unpruned experiment, irrespective of the type of tool used. The test statistics and associated p-values for the overall analysis of variance in the two experiments are given in Table 5.

Marginal mean values of acoustic velocity across the range of final stand densities and between seedlots with different genetic ratings are compared in Table 6 for the two experiments. The effect of stand density was significant for mean acoustic velocity values (irrespective of the measurement tool used) from the pruned experiment (adjusted $p \leq 0.0401$). Results from the unpruned experiment were less clear cut. Mean acoustic velocity for trees in the 200 stems ha^{-1} treatment was significantly lower (adjusted $p \leq 0.0386$) than at the other final stand densities when measured using the ST-300 or HM-200 tools but not for the IML hammer. The mean acoustic velocities for trees grown at the other three final stand densities tested (400, 600 and 1000 stems ha^{-1}) were not significantly different from one another (adjusted $p \geq 0.1683$). For acoustic velocity values measured using the IML hammer, the mean acoustic velocity from trees across all four final stand densities in the unpruned experiment were not significantly different from each other (adjusted $p \geq 0.0757$).

Ring-level microfibril angle

Trends in the breast-height ring-level values of MFA are shown for each individual tree in Figures 4, 1 and 2, respectively. Neither the pith-ring MFA (parameter α_{ij}) nor the radial rate of change in MFA (parameter β_{ij}) differed between the seedlots tested for a given final stand density in either the pruned or the unpruned experiment (Figure 5). The only exception was in the 200 stems ha^{-1} treatment in the unpruned experiment where the pith ring MFA in GF25-rated trees was higher than that in GF6-rated trees, while the radial rate of change in MFA for the GF25-rated trees was slower than that of the

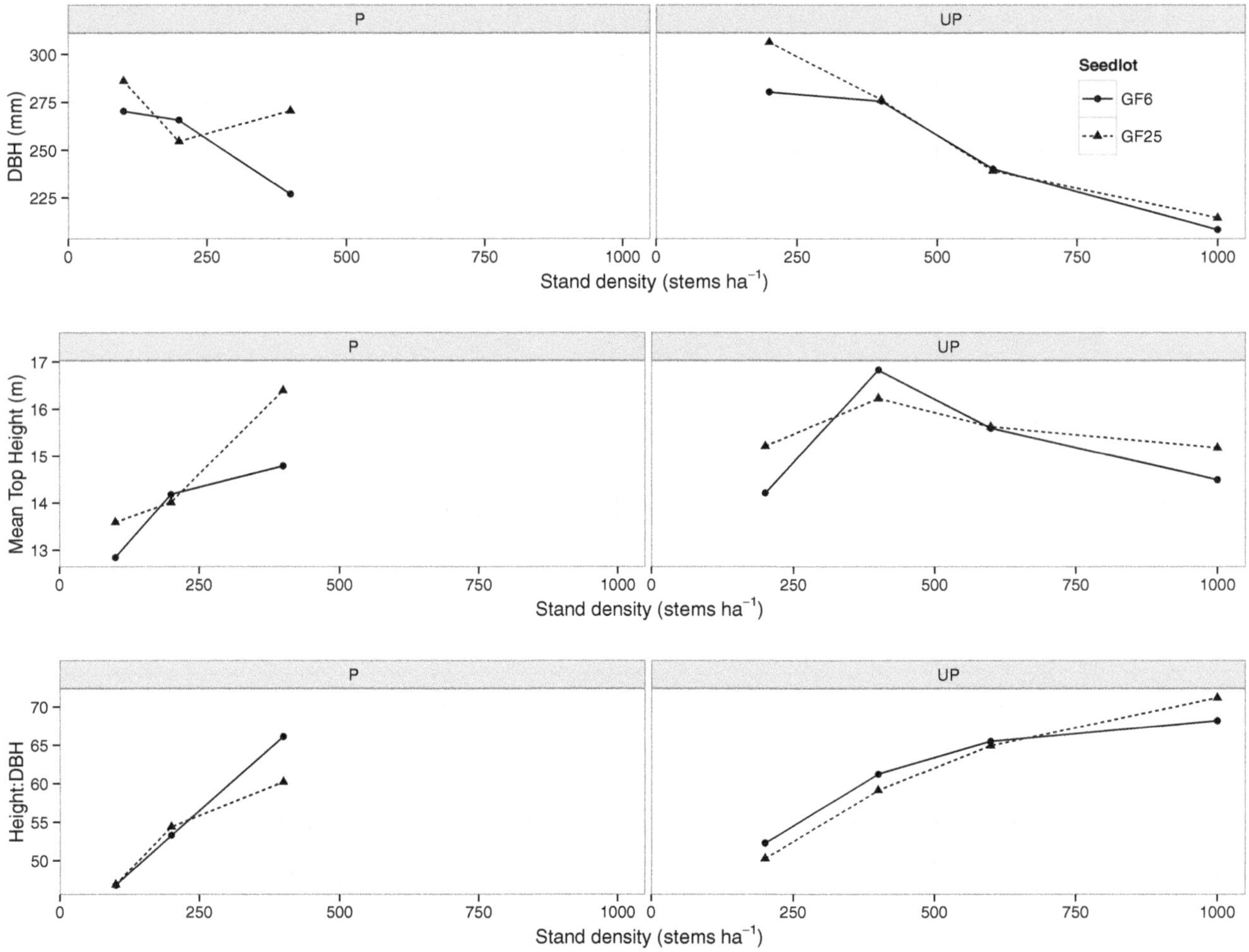

Figure 3 Trends in tree dimensions with stand density for each seedlot in the pruned experiment (P) and the unpruned experiment (UP).

GF6-rated trees. As expected, the radial rate of change in MFA was negative indicating a decrease in MFA the further away the ring is from the pith in all the treatment combinations across both the pruned and unpruned experiments.

In the pruned experiment, pith-ring MFA (parameter α_{ij}) was significantly higher ($p < 0.05$) in GF6-rated trees grown at a final stand density of 200 stems ha^{-1} compared with those grown at a final stand density of 100 stems ha^{-1} but did not change at higher final stand densities (Figure 5a). No statistically significant differences were found for trees from the GF25-rated seedlot. Radial rate of decrease in MFA was slowest in trees grown at a final stand density of 100 stems ha^{-1} and fastest in trees

Table 5 Analysis of variance values of the F-test statistics and associated probability > F (*p*) values for the effects of the stocking and genetic rating factors on acoustic velocity in the pruned and unpruned experiments

Factor	Test statistic	Experiment and type of tool used					
		Pruned			Unpruned		
		ST-300	IML hammer	HM-200	ST-300	IML hammer	HM-200
Final stand density (stems ha^{-1})	$F_{ndf,ddf}$	863.61	351.88	120.49	24.27	7.12	20.12
	p-value	0.001*	0.003*	0.008*	0.013*	0.071	0.017*
Genetic rating	$F_{ndf,ddf}$	6.19	1.10	1.46	0.26	0.59	1.90
	p-value	0.131	0.404	0.350	0.645	0.498	0.262

Note: For the stand density, $F_{ndf,ddf} = F_{2,2}$ for the pruned experiment and $F_{ndf,ddf} = F_{3,3}$ for the unpruned experiment. For the genetic-rating factor, the test statistics are $F_{1,2}$ and $F_{1,3}$ for the two experiments respectively.
*denotes significance at $p < 0.05$.

Table 6 Treatment level marginal means for the acoustic velocity measurements, obtained by the different tools for the different levels of final stand density and genetic rating treatments in the pruned and unpruned experiments

Factor	Factor level	Pruned experiment marginal mean (km s^{-1})			Unpruned experiment marginal mean (km s^{-1})		
		ST-300	IML hammer	HM-200	ST-300	IML hammer	HM-200
Final stand density (stems ha^{-1})	100	2.447^{a} (0.013)	1.901^{a} (0.002)	1.883^{a} (0.035)	-	-	-
	200	2.749^{b} (0.006)	2.097^{b} (0.003)	2.089^{b} (0.008)	2.600^{a} (0.024)	1.928^{a} (0.038)	2.001^{a} (0.007)
	400	3.097^{c} (0.028)	2.336^{c} (0.020)	2.365^{c} (0.020)	2.959^{b} (0.004)	2.155^{a} (0.031)	2.280^{b} (0.026)
	600	-	-	-	2.857^{b} (0.050)	2.129^{a} (0.054)	2.293^{b} (0.071)
	1000	-	-	-	2.831^{b} (0.001)	2.118^{a} (0.013)	2.341^{b} (0.009)
Genetic rating	GF6	2.748^{s} (0.183)	2.104^{s} (0.119)	2.097^{s} (0.144)	2.820^{s} (0.085)	2.097^{s} (0.070)	2.252^{s} (0.075)
	GF25	2.780^{s} (0.192)	2.118^{s} (0.132)	2.128^{s} (0.137)	2.804^{s} (0.068)	2.067^{s} (0.035)	2.205^{s} (0.075)

Note: The standard error of each mean is given in parentheses below each mean. Means in the same column, for a given factor, indicated by the same letter a, b, or c (for the stocking factor) and s and t (for the genetic rating factor) are not significantly different from each other at $p = 0.05$ using the Tukey honest significant difference p-value adjustment for multiple comparisons to maintain an experiment-wise error p-value of 0.05.

grown at a final stand density of 400 stems ha^{-1} with trees from both seedlots exhibiting a faster radial rate of decline in MFA at higher final stand densities (Figure 5a).

In the unpruned experiment, pith-ring MFA (parameter α_{ij}) was significantly higher ($p < 0.05$) in GF6-rated trees grown at a final stand density of 400 stems ha^{-1} compared with those grown at 200 stems ha^{-1} (Figure 5b) but there were no significant differences at higher final stand densities. For the GF25-rated seedlot, pith-ring MFA in trees grown at a final stand density of 1000 stems ha^{-1} was significantly lower than those grown at final stand densities of 200, 400 or 600 stems ha^{-1}. The trend

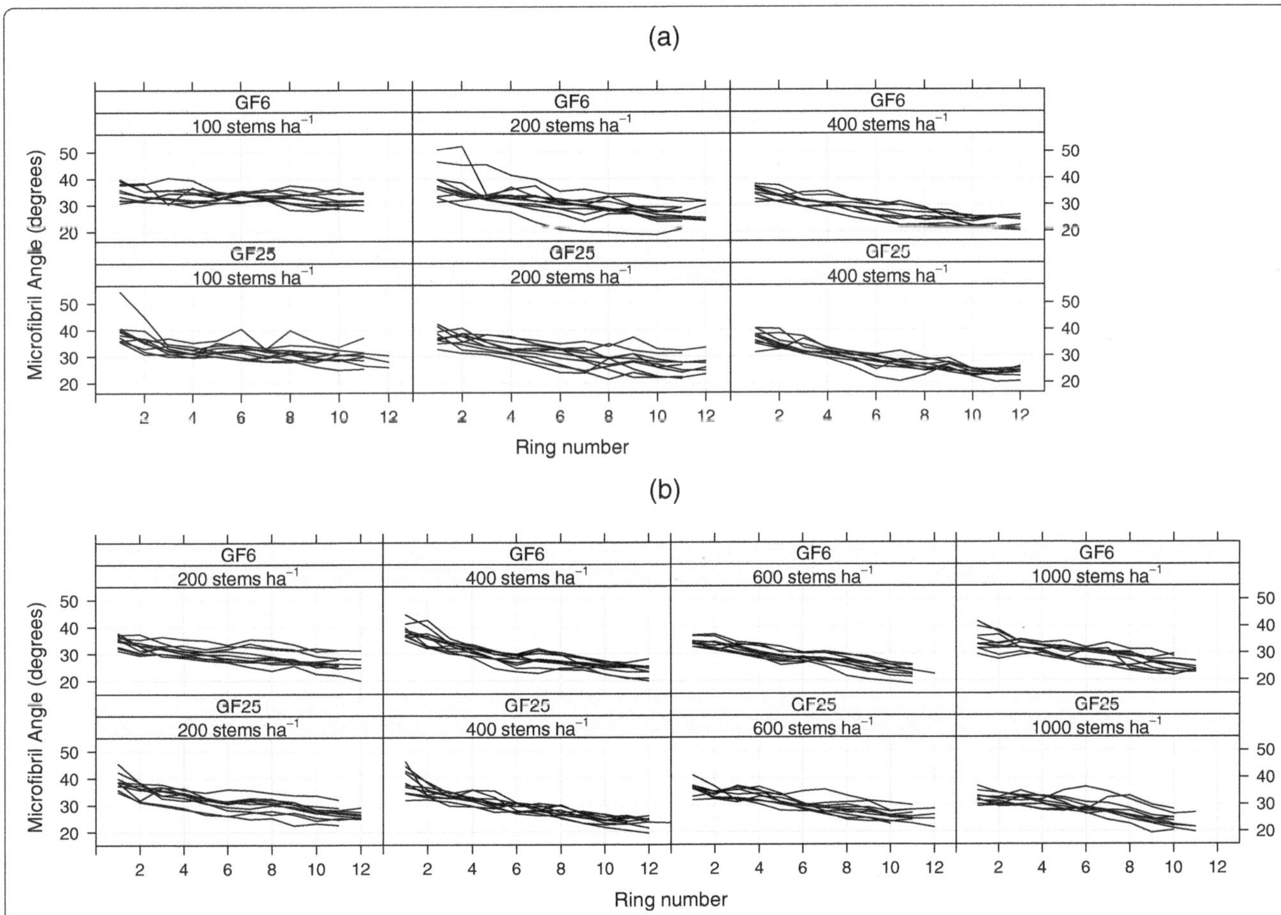

Figure 4 Trends in ring-level values of microfibril angle at breast height for GF6- and GF25-rated trees at various final stand densities in (a) the pruned experiment; and (b) the unpruned experiments.

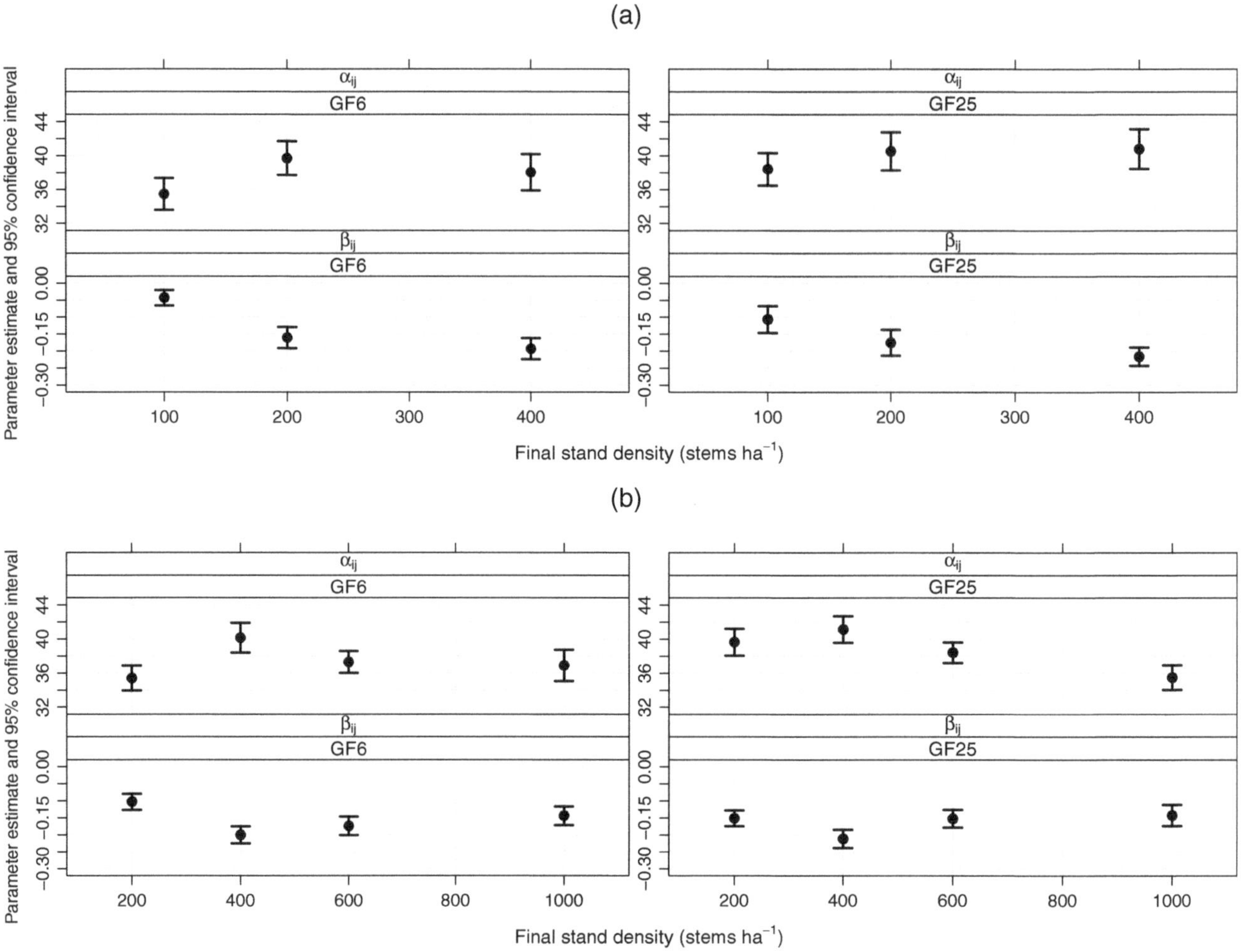

Figure 5 Plots of Equation [3] fixed parameter estimates and 95% confidence intervals for the breast height ring level microfibril angle trends for GF6- and GF25-rated trees at various final stand densities in (a) the pruned experiment; and (b) the unpruned experiments.

in radial rate of decrease in MFA with final stand density in the unpruned experiment differed from that observed in the pruned experiment. Except for the faster radial rate of decrease in MFA at a stand density 400 stems ha^{-1}, there was essentially no change in rate of decline in MFA with change in stand density for trees from the GF25-rated seedlot (Figure 5b). In trees from the GF6-rated seedlot, the radial rate of decrease in MFA at a stand density of 400 stems ha^{-1} was significantly higher than at a stand density of 200 stems ha^{-1}. However, the rate of decrease at stand densities of 600 and 1000 stems ha^{-1} did not differ from that observed at 400 stems ha^{-1}.

Overall, there was no difference in predicted MFA for ring 15 from the pith between the GF6- and GF25-rated seedlots (Table 7). However predicted MFA differed with stand density, and was 8° higher in pruned trees growing at 100 stems ha^{-1} than in pruned trees growing at 400 stems ha^{-1}. In the unpruned experiment, there was no change in MFA with further increases in stand density above a value of 400 stems ha^{-1} (Table 7).

Ring-level modulus of elasticity

In the pruned experiment, pith-ring MOE (parameter α_{ij}) in the GF25-rated seedlot trees was significantly lower than in GF6-rated seedlot trees (at $p = 0.05$) at final stand densities of 100 and 400 stems ha^{-1} (Figure 6a). There was no significant effect of final stand density on pith-ring MOE in the GF6-rated seedlot trees but a small reduction for GF25-rated seedlot trees. The estimated radial rate of change (parameter β_{ij}) was positive in all cases indicating that breast-height MOE for trees in the pruned experiment increased with increasing ring number from the pith. For both seedlots, there was a positive correlation between rate of radial increase in breast-height ring-level MOE and final stand density. Overall, MOE in ring 15 from the pith was twice as high in trees grown

Table 7 Treatment level marginal means for predicted values of wood density, MFA and MOE for ring 15 from the pith for the different levels of final stand density and genetic rating treatments in the pruned and unpruned experiments

Factor	Factor level	Pruned experiment marginal mean			Unpruned experiment marginal mean		
		Density (kg m^{-3})	MFA(°)	MOE (kN mm$_{-2}$)	Density (kg m^{-3})	MFA(°)	MOE (kN mm$_{-2}$)
Final stand density (stems ha^{-1})	100	465^{a} (9.2)	30^{a} (0.8)	5.39^{a} (0.23)	-	-	-
	200	455^{a} (9.2)	25^{b} (0.8)	8.11^{b} (0.23)	445^{a} (4.6)	27^{a} (0.7)	7.54^{a} (0.55)
	400	483^{a} (9.2)	22^{b} (0.8)	10.66^{c} (0.23)	459^{ab} (4.6)	23^{b} (0.7)	10.28^{b} (0.55)
	600	-	-	-	470^{b} (4.6)	24^{b} (0.7)	10.26^{b} (0.55)
	1000	-	-	-	470^{b} (4.6)	25^{ab} (0.7)	9.98^{b} (0.55)
Genetic rating	GF6	485^{s} (7.5)	26^{s} (0.6)	8.04^{s} (0.18)	478^{s} (3.3)	25^{s} (0.5)	9.72^{s} (0.39)
	GF25	451^{t} (7.5)	27^{s} (0.6)	8.07^{s} (0.18)	444^{t} (3.3)	25^{s} (0.5)	9.29^{s} (0.39)

See notes from Table 6 for assistance in interpreting this table.

at a final stand density of 400 stems ha^{-1} than those grown at 100 stems ha^{-1} irrespective of seedlot (Table 7).

In the unpruned experiment, generally only small differences were observed in either pith-ring MOE (parameter α_{ij}) or radial rate of change (parameter β_{ij}) between seedlots and among final stand densities (Figure 6b). The exception was at a final stand density of 200 stems ha $^{-1}$, where pith-ring MOE for trees from the GF25-rated seedlot was lower than those from the GF6-rated seedlot,

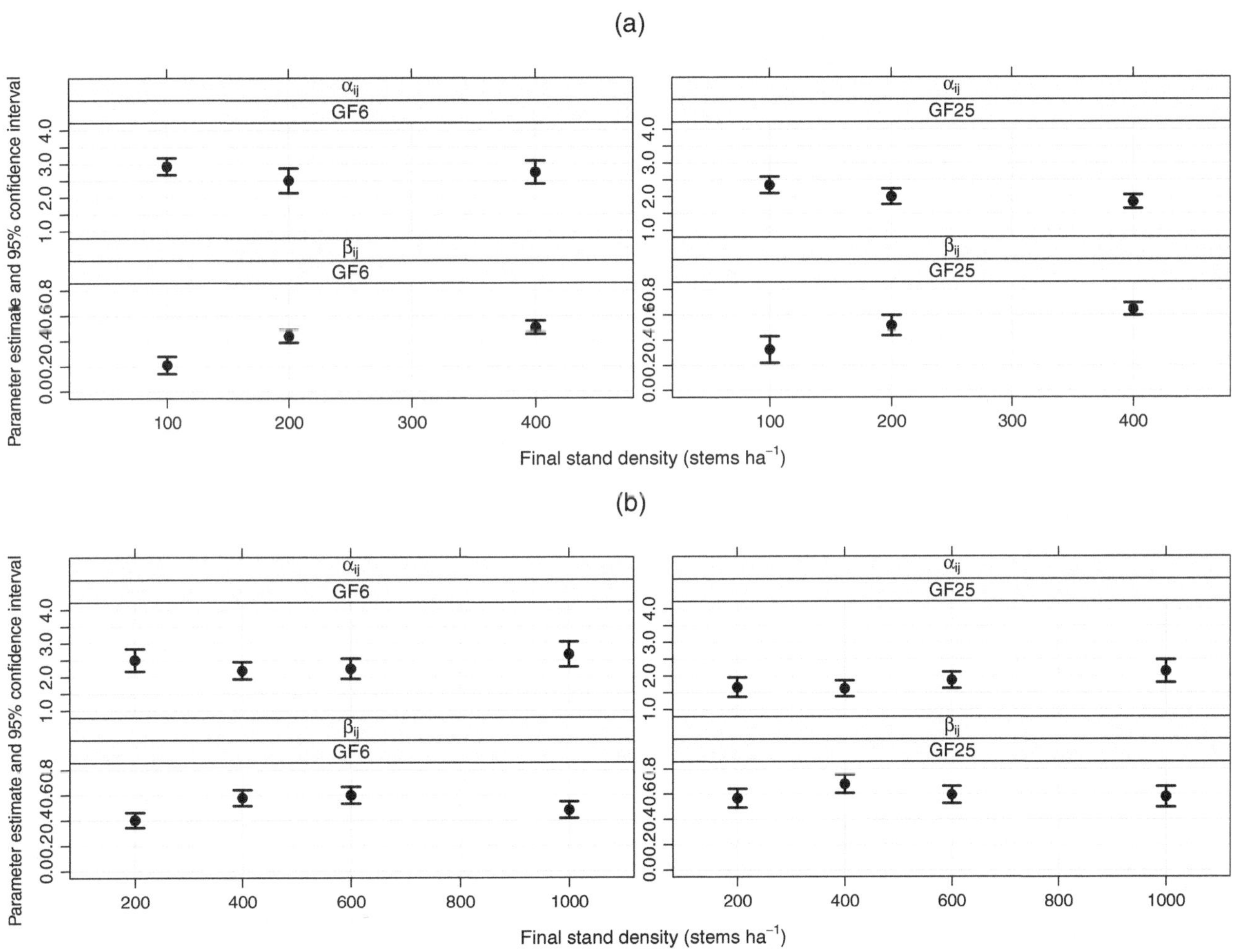

Figure 6 **Plots of Equation [3] fixed parameter estimates and 95% confidence intervals for the breast height ring level modulus of elasticity trends for GF6- and GF25-rated trees at various final stand densities in (a) the pruned experiment; and (b) the unpruned experiments.**

and the rate of radial increase in MOE for GF25-rated seedlot trees was faster than for GF6 seedlot trees. The optimal final stand density for MOE in ring 15 from the pith was 400 stems ha^{-1} irrespective of seedlot.

Ring-level wood density

Some differences in ring-level wood density between seedlots were observed in both the pruned and unpruned experiments. There were some differences in pith-ring wood density (parameter α_{ij}) between trees from the GF6- and GF25-rated seedlots, but generally no differences across final stand densities for trees from a given seedlot. In the pruned experiment, pith-ring wood density in trees from the GF25-rated seedlot trees growing at a final stand density of 400 stems ha^{-1} was significantly lower than equivalent GF6-rated seedlot trees (Figure 7a). A similar trend was observed for trees grown at a final stand density of 1000 stems ha^{-1} in the unpruned experiment (Figure 7b).

The rate of radial increase in wood density (parameter β_{ij}) did not change with final stand density for the GF6-rated seedlot in either experiment, but was higher at greater stand densities for the GF25-rated seedlot in the pruned experiment (Figure 7a). However, no significant differences (at $p = 0.05$) were observed in the rate of radial increase in ring-level wood density with changes in stand density in the unpruned experiment (Figure 7b). Overall, the predicted value of wood density in ring 15 was 34 kg m^{-3} lower for GF25-rated seedlot trees than those from the GF6-rated seedlot (Table 7). In the unpruned experiment, wood density for ring 15 was 24 kg m^{-3} higher for trees grown at a final stand density of 1000 stems ha^{-1} compared with 200 stems ha^{-1}.

Discussion

The direct positive correlation between MOE and final stand density was consistent with observations from other studies on radiata pine (Lasserre et al. 2005; Lasserre

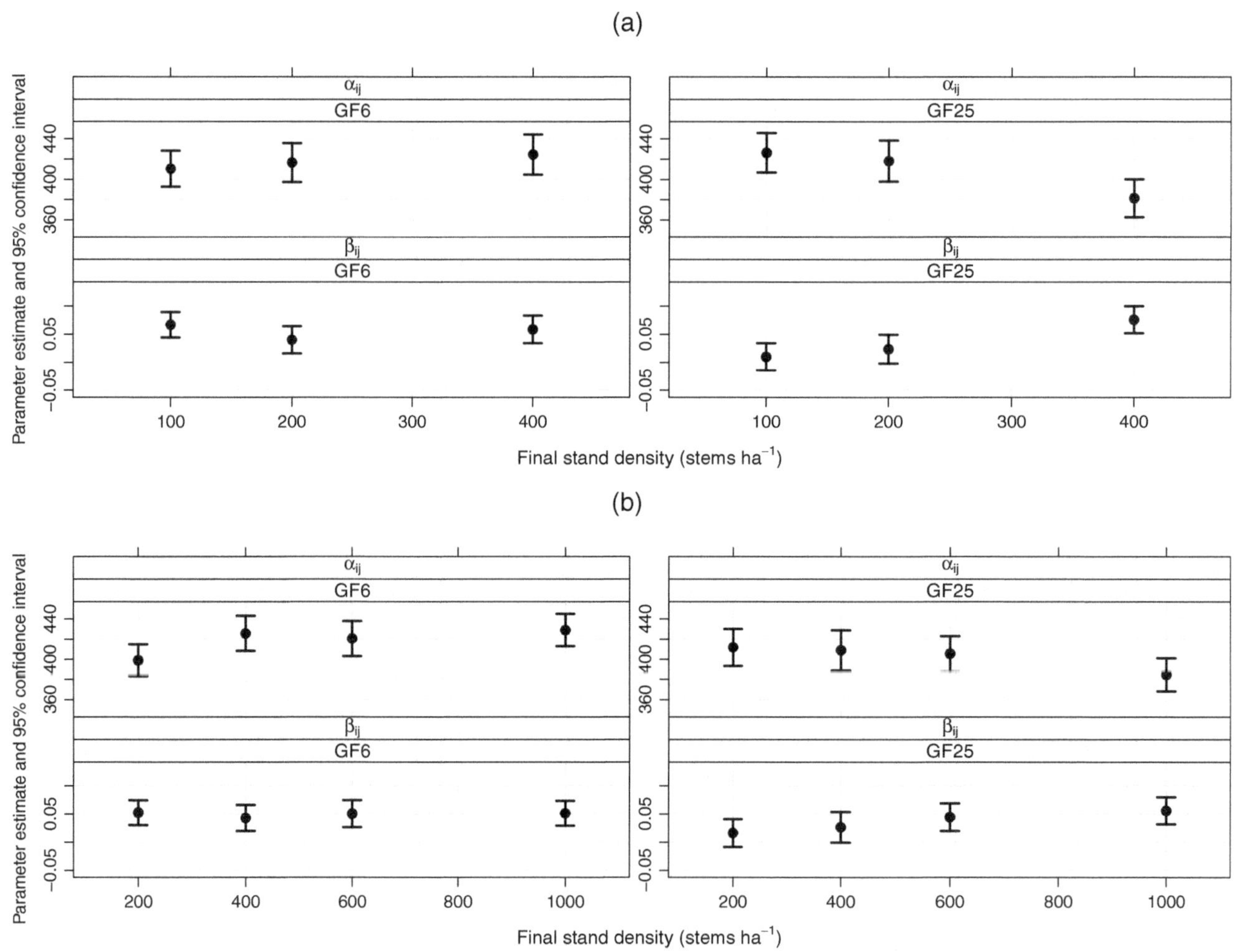

Figure 7 Plots of Equation [3] fixed parameter estimates and their 95% confidence intervals for the breast height ring level wood density trends for GF6- and GF25-rated trees at various final stand densities in (a) the pruned experiment; and (b) the unpruned experiments.

et al. 2008; Lasserre et al. 2009; Raymond et al. 2008; Waghorn et al. 2007; Watt et al. 2011) and a range of different species (Brazier and Mobbs 1993; Chuang and Wang 2001; Clark III et al. 2008; Moore et al. 2009; Roth et al. 2007; Wang and Ko 1998; Wang et al. 2005; Zhang et al. 2002; Zhang et al. 2006). The smallest MOE values occurred when final stand density was below 400 stems ha^{-1}, with higher MOE values occurring at higher final stand densities. A similar effect was observed in a nearby Nelder-spacing experiment with radiata pine. Outerwood MOE was 5.4 kN mm^{-2} in trees grown at a stand density of 209 stems ha^{-1} compared with 6.6 kN mm^{-2} for trees at a stand density of 489 stems ha^{-1}. This latter value of MOE was not significantly different from that for trees grown at a stand density of 2551 stems ha^{-1}(Waghorn et al. 2007).

In this study wood density differences were mostly related to differences between the two seedlots examined rather than to differences in final stand density. This result is in contrast to several previous studies in radiata pine that have found a significant effect of stand density on wood density (Cown 1973, 1974; Grace and Evans 2012; Watt et al. 2011), but consistent with others that have shown little or no effect (Lasserre et al. 2009; Sutton and Harris 1974). In 24-year-old radiata pine stands, Watt et al. (2011) observed a 92 kg m^{-3} increase in wood density over a similar range of stand densities to those examined in this study (i.e. 200 stems ha^{-1} up to 1100 stems ha^{-1}). There was a large amount of tree-to-tree variation observed in the radial profiles of wood density in the current study as well as a number of distinct annual shifts (Figure 2), which meant that only around 18% of the variation in wood density could be explained using the fixed effects included in the model tested. It is possible that these shifts in density between years, particularly the one observed in 2002 (i.e. at ring 10), were due to climatic factors as similar responses were observed across trees from both seedlots and at all final stand densities. In a study on loblolly pine (*Pinus taeda* L.) Cregg et al. (1988) observed that climatic signals had more influence on wood density than thinning treatments. Final stand density (irrespective of thinning) can still have an effect on average stem or log density even if there is no effect on wood density on an equivalent annual-ring basis. This effect occurs because growing space affects radial growth. Thus, trees in thinned stands or those planted at a wide initial spacing will reach a target diameter in a shorter period of time than more densely planted stands. Consequently, they have a higher proportion of low density corewood and lower average density (Cown 1974).

Fewer studies examining the effect of stand density on MFA have been published than those investigating the effect on wood density. However, two such studies in radiata pine found that MFA was lower at higher initial stand densities (Lasserre et al. 2009; Watt et al. 2011). The study by Watt et al. (2011) was over a similar range of stand densities to those reported in this study (i.e. 200 stems ha^{-1} to 1100 stems ha^{-1}), but the study by Lasserre et al. (2009) was in stands grown at 833 stems ha^{-1} up to 2500 stems ha^{-1}. In a separate study, Downes et al. (2002) found that thinning (combined with fertilisation) resulted in larger MFA values for radiata pine. Several studies in other species have also found a positive relationship between growth rate and MFA, i.e. faster growth rates are associated with higher MFA (Auty et al. 2013; Herman et al. 1999; Lindström et al. 1998). Therefore, it was somewhat unexpected that the largest difference in MFA observed in the current study occurred between trees grown in stands with final densities of 100 stems ha^{-1} and 400 stems ha^{-1}, neither of which are considered high. It is also pertinent to note that the largest difference in DBH occurred between stand densities of 400 stems ha^{-1} and 1000 stems ha^{-1}.

The almost 5° improvement in MFA and >2 kN mm^{-2} reduction in MOE between final stand densities of 400 stems ha^{-1} and 200 stems ha^{-1} respectively are highly significant from a wood utilisation perspective (Moore 2012; Walker and Butterfield 1995). The two MOE values are different enough to represent a change of one grade in the New Zealand machine-strength grading system, which could have a substantial impact for a sawmill producing structural lumber. The results presented here indicate that it is not necessary to maintain extremely high stand densities in order to have good wood quality in radiata pine. Also, early thinning has a number of other benefits as it will result in lower mortality later in the rotation and also produces larger diameter logs that generally have higher value (Smith et al. 1997). Another important finding was that the GF25-rated seedlot used for this trial produced trees with a lower wood density and MOE compared with those from the GF6-rated seedlot. These results indicate that, in some cases, past gains in growth and form have come at the expense of wood quality. This phenomenon is well recognised by tree breeders who have focussed on improving the wood density of radiata pine for some time (Jayawickrama and Carson 2000). More recently, improvements in the wood stiffness of radiata pine have also been sought (Dungey et al. 2006; Kumar et al. 2006).

The Shellocks trial is part of a much larger trial series, which has installations across different site qualities within New Zealand. These trials will reach maturity in the next 5–8 years, which will enable researchers to determine whether the trends observed here also hold at other sites and in mature trees where the contrast in levels of competition between spacing treatments is expected to be more marked. Results from an earlier

24-year-old thinning trial where the treatments had been applied at age 5 years indicate that differences in properties could be expected to increase with increasing stand age (Watt et al. 2011). The experimental design for the national trial series also enables the site by silviculture and site by seedlot interactions to be investigated. The earlier installations in this trial series have two replicates of each combination of seedlot and final stand density, which will allow the interaction between these factors to also be tested. However, two previous studies that have examined this interaction have found that it was not significant (Lasserre et al. 2009; Waghorn et al. 2007).

Conclusion

The results obtained in this study indicate, for this site at least, that early thinning down to a final stand density of 400 stems ha^{-1} had no detrimental effect on wood stiffness. However, there are substantial adverse impacts on wood MFA and MOE at final stand densities below this level. Thus, forest managers are able to influence the wood properties of radiata pine trees through manipulating stand density. The greatest change in wood properties occurred over a range of final stand densities where there was a significant change in tree height, but only a modest change in tree diameter. With the exception of wood density, tree spacing had a much stronger effect on wood properties than differences in seedlot. This result indicates that extremely high stand densities are not necessary in order to improve wood properties, but thinning to low residual stand densities should be avoided. Assessing the remaining trials in this series will enable the conclusions obtained from this trial to be validated across a wider range of sites and in mature trees.

Competing interests

The authors declare that they have no competing interests.

Authors' contributions

JRM was assisted with planning the study, field data collection, data analysis and writing of the manuscript. DJC and RBMcK planned the study, undertook field and laboratory analyses and assisted with the interpretation of the results. COS analysed the data and assisted with writing the manuscript. All authors read and approved the final version of the manuscript.

Acknowledgements

The authors wish to thank WQI Ltd for funding the study and giving permission to publish the results. Don McConchie, Marco Lausberg and Craig Treloar assisted with the field work. Funding to support the additional analysis and publication of the results was provided by Future Forests Research Ltd. Drs Ruth Falshaw and Ruth Gadgil provided numerous helpful comments on an earlier version of the manuscript.

Author details

[1]Scion, Private Bag 3020, Rotorua 3046, New Zealand. [2]Present address: Department of Forestry, Mississippi State University, P.O. Box 9681, Mississippi State, MS 39762, USA.

References

Assmann, E. (1970). *The Principles of Forest Yield Study*. New York: Pergamon Press.

Auty, D, Gardiner, BA, Achim, A, Moore, JR, & Cameron, AD. (2013). Models for predicting microfibril angle variation in Scots pine. *Annals of Forest Science, 70*, 209–218. doi:10.1007/s13595-012-0248-6.

Auty, D, Weiskittel, AR, Achim, A, Moore, JR, & Gardiner, BA. (2012). Influence of early re-spacing on Sitka spruce branch structure. *Annals of Forest Science, 69*, 93–104. doi:10.1007/s13595-011-0141-8.

Brazier, JD, & Mobbs, ID. (1993). The influence of planting distance on structural wood yields of unthinned Sitka spruce. *Forestry, 66*, 333–352.

Burdon, RD. (2008). Branching habit in radiata pine - breeding goals revisited. *New Zealand Journal of Forestry, 52*(4), 20–23.

Burdon, RD, Kibblewhite, RP, Walker, JCF, Megraw, RA, Evans, R, & Cown, DJ. (2004). Juvenile versus mature wood: a new concept, orthogonal to corewood versus outerwood, with special reference to *Pinus radiata* and *P. taeda*. *Forest Science, 50*(4), 399–415.

Carson, SD. (1996). Greater specialisation of improved seedlots in New Zealand: New developments for efficient selection of parents and evaluation of performance. *New Zealand Journal of Forestry, 41*(1), 12–17.

Carson, SD, Kimberley, MO, Hayes, JD, & Carson, MJ. (1999). The effect of silviculture on genetic gain in growth of *Pinus radiata* at one-third rotation. *Canadian Journal of Forest Research, 29*(12), 1979–1984. doi:10.1139/x99-152.

Chuang, ST, & Wang, SY. (2001). Evaluation of standing tree quality of Japanese cedar grown with different spacing using stress wave and ultrasonic-wave methods. *Journal of Wood Science, 47*, 245–253.

Clark III, A, Jordan, L, Schimleck, L, & Daniels, RF. (2008). Effect of initial planting spacing on wood properties of unthinned loblolly pine at age 21. *Forest Products Journal, 58*(10), 78–83.

Cown, DJ. (1973). Effects of severe thinning and pruning treatments on the intrinsic wood properties of young radiata pine. *New Zealand Journal of Forestry Science, 3*, 379–389.

Cown, DJ. (1974). Comparison of the effects of two thinning regimes on some wood properties of radiata pine. *New Zealand Journal of Forestry Science, 4*, 540–551.

Cregg, BM, Dougherty, PM, & Hennessey, TC. (1988). Growth and wood quality factors in relation to stand density and climatic factors. *Canadian Journal of Forest Research, 18*, 851–858.

Downes, GM, Nyakuengama, JG, Evans, R, Northway, R, Blakemore, P, Dickson, RL, & Lausberg, M. (2002). Relationship between wood density, microfibril angle and stiffness in thinned and fertlized *Pinus radiata*. *IAWA Journal, 23*(3), 253–265.

Dungey, HS, Matheson, AC, Kain, D, & Evans, R. (2006). Genetics of wood stiffness and its component traits in *Pinus radiata*. *Canadian Journal of Forest Research, 36*(5), 1165–1178.

Evans, R. (1999). A variance approach to the X-ray diffractometric estimation of microfibril angle in wood. *Appita Journal, 52*(4), 283–289. 294.

Evans, R, Hughes, M, & Menz, D. (1999). Microfibril angle variation by scanning X-ray diffractometry. *Appita Journal, 52*(5), 363–367.

Evans, R, & Ilic, J. (2001). Rapid prediction of wood stiffness from microfibril angle and density. *Forest Products Journal, 51*(3), 53–57.

Evert, F. (1971). *Spacing Studies: A Review. Information Report FMR-X-37*. Ottawa: Canadian Forest Service, Forest Management Institute.

Fenton, RT, & Sutton, WRJ. (1968). Silvicultural proposals for radiata pine on high quality sites. *New Zealand Journal of Forestry, 13*, 220–228.

Gapare, WJ, Baltunis, BS, Ivković, M, & Wu, HX. (2009). Genetic correlations among juvenile wood quality and growth traits and implications for selection strategy in *Pinus radiata* D. Don. *Annals of Forest Science, 66*(6), 606–606. doi:10.1051/forest/2009044.

Grace, JC, & Evans, R. (2012). Impact of thinning and pruning on selected wood properties in individual radiata pine trees in New Zealand. *New Zealand Journal of Forestry Science, 42*, 117–129.

Hein, S, Weiskittel, AR, & Kohnle, U. (2008). Effect of wide spacing on tree growth, branch and sapwood properties of young Douglas-fir [*Pseudotsuga menziesii* (Mirb.) Franco] in south-western Germany. *European Journal of Forest Research, 127*, 481–493.

Herman, M, Dutilleul, P, & Avella-Shaw, T. (1999). Growth rate effects on intra-ring and inter-ring trajectories of microfibril angle in Norway spruce (*Picea abies*). *IAWA Journal, 20*(1), 3–21.

Hewitt, AE. (1998). *New Zealand Soil Classification* (2nd ed.). Lincoln: Landcare Research Science Series.

James, RN. (1990). Evolution of Silvicultural Practice Towards Wide Spacing and Heavy Thinnings In New Zealand. In RN James & GL Tarlton (Eds.), *New Approaches to Spacing and Thinning in Plantation Forestry. FRI Bulletin 151* (pp. 13–20). Rotorua, New Zealand Ministry of Forestry, Forest Research Institute.

Jayawickrama, KJS, & Carson, MJ. (2000). A breeding strategy for the New Zealand Radiata Pine Breeding Cooperative. *Silvae Genetica, 49*(2), 82–90.

Jayawickrama, KJS, Shelbourne, CJA, & Carson, MJ. (1997). New Zealand's long internode breed of *Pinus radiata. New Zealand Journal of Forestry Science, 27*, 126–141.

Kilpatrick, DJ, Sanderson, JM, & Savill, PS. (1981). The influence of five early respacing treatments on the growth of Sitka spruce. *Forestry, 54*(1), 17–29.

Kumar, S. (2004). Genetic parameter estimates for wood stiffness, strength, internal checking, and resin bleeding for radiata pine. *Canadian Journal of Forest Research, 34*(12), 2601–2610.

Kumar, S, & Burdon, RD. (2010). Genetic improvement of stiffness of radiata pine: synthesis of results from acoustic assessments. *New Zealand Journal of Forestry Science, 40*, 185–197.

Kumar, S, Burdon, RD, Stovold, GT, & Gea, LD. (2008). Implications of selection history on genetic architecture of growth, form, and wood-quality traits in *Pinus radiata. Canadian Journal of Forest Research, 38*(9), 2372–2381.

Kumar, S, Dungey, HS, & Matheson, AC. (2006). Genetic parameters and strategies for genetic improvement of stiffness in radiata pine. *Silvae Genetica, 55*(2), 77–84.

Lasserre, J-P, Mason, EG, & Watt, MS. (2005). The effects of genotype and spacing on *Pinus radiata* [D. Don] corewood stiffness in an 11-year old experiment. *Forest Ecology and Management, 205*(1–3), 375–383. doi:10.1016/j.foreco.2004.10.037.

Lasserre, JP, Mason, EG, & Watt, MS. (2008). Influence of the main and interactive effects of site, stand stocking and clone on *Pinus radiata* D. Don corewood modulus of elasticity. *Forest Ecology and Management, 255*(8–9), 3455–3459.

Lasserre, JP, Mason, EG, Watt, MS, & Moore, JR. (2009). Influence of initial planting spacing and genotype on microfibril angle, wood density, fibre properties and modulus of elasticity in *Pinus radiata* D. Don corewood. *Forest Ecology and Management, 258*(9), 1924–1931. doi:10.1016/j.foreco.2009.07.028.

Lindström, H, Evans, JW, & Verrill, SP. (1998). Influence of cambial age and growth conditions on microfibril angle in young Norway spruce (*Picea abies* [L.] Karst.). *Holzforschung, 52*(6), 573–581.

Macdonald, E, & Hubert, J. (2002). A review of the effects of silviculture on timber quality of Sitka spruce. *Forestry, 75*(2), 107–137.

Maguire, DA, Kershaw, JA, Jr, & Hann, DW. (1991). Predicting the effects of silvicultural regime on branch size and crown wood core in Douglas-fir. *Forest Science, 37*, 1409–1428.

Milliken, GA, & Johnson, DE (1989). *Analysis of messy data: nonreplicated experiments* (Vol. 2) Boca Raton, FL: CRC Press.

Ministry for Primary Industries. (2012). *National Exotic Forest Description as at 1 April 2012*. Wellington, New Zealand: Ministry for Primary Industries.

Moore, J. (2012). Growing fit-for-purpose structural timber: what is the target and how do we get there? *New Zealand Journal of Forestry, 57*(3), 17–24.

Moore, J, Achim, A, Lyon, A, Mochan, S, & Gardiner, B. (2009). Effects of early re-spacing on the physical and mechanical properties of Sitka spruce structural timber. *Forest Ecology and Management, 258*(7), 1174–1180.

New Zealand Meteorological Service. (1983). *Summaries of Climate Observations To 1980*. Wellington: New Zealand Meteorological Service Miscellaneous Publication 177.

Panshin, AJ, & de Zeeuw, C. (1980). *Textbook of Wood Technology* (4th ed.). New York: McGraw-Hill.

Pinheiro, JC, Bates, DM, DebRoy, S, Sarkar, D, & R Development Core Team. (2013). *nlme: Linear and Nonlinear Mixed Effects Models. R package version 3* (pp. 1–113).

R Development Core Team (2014). R: a language and environment for statistical computing. Vienna: R Foundation for Statistical Computing.

Raymond, CA, Joe, B, Anderson, DW, & Watt, DJ. (2008). Effect of thinning on relationships between three measures of wood stiffness in *Pinus radiata*: Standing trees vs. logs vs. short clear specimens. *Canadian Journal of Forest Research, 38*(11), 2870–2879.

Roos, A, Woxblom, L, & McCluskey, D. (2010). The influence of architects and structural engineers on timber in construction – perceptions and roles. *Silva Fennica, 44*, 871–884.

Roth, BE, Li, X, Huber, DA, & Peter, GF. (2007). Effects of management intensity, genetics and planting density on wood stiffness in a plantation of juvenile loblolly pine in the southeastern USA. *Forest Ecology and Management, 246*(2–3), 155–162.

SAS Institute Inc. (2004–2009). *SAS 9.2 Help and Documentation*. Cary, NC: SAS Institute Inc.

Savill, PS, & Sandels, AJ. (1983). The influence of early respacing on the wood density of Sitka spruce. *Forestry, 65*(2), 109–120.

Smith, DM, Larson, BC, Kelty, MJ, & Ashton, PMS (1997). *The practice of silviculture: applied forest ecology* (Ninth ed.) New York: John Wiley & Sons.

Sutton, WRJ, & Harris, JM. (1974). Effect of heavy thinning on wood density in radiata pine. *New Zealand Journal of Forestry Science, 4*, 112–115.

Tukey, JW. (1949). One degree of freedom for non-additivity. *Biometrics, 5*(3), 232–242. doi:10.2307/3001938.

Vincent, TG. (1987). *Certification System for Forest Tree Seed and Planting Stock. (FRI Bulletin 134)*. Rotorua: New Zealand Forest Research Institute.

Vinden, P. (2003). Changes in markets, technology and resources: prospects for wood-based products. *Australian Forestry, 66*, 51–54.

Waghorn, MJ, Watt, MS, & Mason, EG. (2007). Influence of tree morphology, genetics, and initial stand density on outerwood modulus of elasticity of 17-year-old *Pinus radiata. Forest Ecology and Management, 244*(1–3), 86–92. doi:10.1016/j.foreco.2007.03.057.

Walker, JCF, & Butterfield, B. (1995). The importance of microfibril angle for the processing industries. *New Zealand Journal of Forestry, 40*(4), 34–40.

Wang, S-Y, & Ko, C-Y. (1998). Dynamic modulus of elasticity and bending properties of large beams of Taiwan-grown Japanese cedar from different plantation spacing sites. *Journal of Wood Science, 44*, 62–68.

Wang, S-Y, Lin, C-J, & Chiu, C-M. (2005). Evaluation of wood quality of Taiwania trees grown with different thinning prunning treatments using ultrasonic-wave testing. *Wood and Fiber Science, 37*(2), 192–200.

Watson, AJ, Davie, TJA, Bowden, W, & Payne, JJ. (2004). Drainage to groundwater under a closed-canopy radiata pine plantation on the Canterbury Plains, South Island, New Zealand. *Journal of Hydrology (NZ), 43*, 111–123.

Watt, MS, Whitehead, D, Richardson, B, Mason, EG, & Leckie, AC. (2003). Modelling the influence of weed competition on the growth of young *Pinus radiata* at a dryland site. *Forest Ecology and Management, 178*, 271–286.

Watt, MS, Zoric, B, Kimberley, MO, & Harrington, J. (2011). Influence of stocking on radial and longitudinal variation in modulus of elasticity, microfibril angle, and density in a 24-year-old *Pinus radiata* thinning trial. *Canadian Journal of Forest Research, 41*(7), 1422–1431. doi: 10.1139/x11-070.

Whyte, AGD, & Woollons, RC. (1990). Modelling stand growth of radiata pine thinned to varying densities. *Canadian Journal of Forest Research, 20*, 1069–1076.

Zhang, SY, Chauret, G, Ren, QH, & Desjardins, R. (2002). Impact of initial spacing on plantation black spruce lumber grade yield, bending properties, and MSR yield. *Wood and Fiber Science, 34*(4), 460–475.

Zhang, SY, Chauret, G, Swift, DE, & Duchesne, I. (2006). Effects of precommercial thinning on tree growth and lumber quality in a jack pine stand in New Brunswick, Canada. *Canadian Journal of Forest Research, 36*(4), 945–952. doi:10.1139/x05-307.

Zobel, BJ, & Jett, JB. (1995). *Genetics of Wood Production*. Berlin: Springer.

Zobel, BJ, & Sprague, JR. (1998). *Juvenile wood in forest trees*. Berlin: Springer.

22

Growth, biomass, leaf area and water-use efficiency of juvenile *Pinus radiata* in response to water deficits

Matthew J Waghorn[1*], David Whitehead[2], Michael S Watt[3], Euan G Mason[1] and Jonathan J Harrington[3]

Abstract

Background: As the frequency and severity of drought events are expected to increase globally, drought-induced reductions in plantation productivity are likely to become more important. Such reductions will be of concern to forest managers looking to improve forest productivity during the establishment and initial growth phases of plantation-grown *Pinus radiata* D. Don. The objective of this research was to assess how growth, biomass, leaf area and water-use efficiency in juvenile *Pinus radiata* responded to the timing and duration of water deficits.

Methods: Two-year-old *Pinus radiata* seedlings (cultivated in a polyhouse) were subjected to various water deficit treatments. Needle water potential, tree growth, biomass partitioning, leaf characteristics and water-use efficiency were measured to assess the impact that the timing and duration of water deficits had on productivity.

Results: Cyclical re-watering of the early- and late-season drought treatments led to large fluctuations in needle water potential. The summer drought treatment resulted in a sustained low needle water potential over the summer months. Total water stress integral (S_{ψ}) was 41.4, 66.8, 55.2 and 97.6 MPa-days for the well-watered, early-season cyclical drought, late-season cyclical drought and summer drought treatments, respectively. In general, water deficits decreased tree growth, reduced crown size, reduced biomass accumulation and leaf area, and resulted in more enriched values of $\delta^{13}C$, all of which were significantly ($P < 0.05$) affected by treatment. Summer drought reduced height, diameter and basal area by 24.7%, 33.1% and 52.3%, respectively. Total biomass was reduced by 64.7% and total leaf area by 40.0%, compared with the well-watered treatment. The reduction in stem diameter growth was larger for the late-season treatment than the early-season treatment when compared with the well-watered trees suggesting that late-season drought is more detrimental than early-season drought to tree growth.

Conclusion: The results provide insight for forest managers of *Pinus radiata* into the importance of managing water deficits in order to maximise forest production of juvenile trees. Results suggest that late-season drought is more detrimental to absolute growth and biomass accumulation than early-season drought.

Keywords: Biomass; Clone; Growth; Leaf area; *Pinus radiata*; Productivity; Water deficits; Water stress; Water-use efficiency

Background

Pinus radiata D. Don productivity is an important issue for forest managers in New Zealand. Increased or indeed diminished growth of *P. radiata* is directly related to a site's ability to provide resources to a forest crop. One of the major factors potentially limiting forest production is water availability (Linder 1987; Johnson 1990; Gower et al. 1992) and this is of particular importance for juvenile conifer growth (Sands and Nambiar 1984). Water deficits or drought can greatly influence the amount of foliage produced by a stand and consequently have a direct effect upon the fractional interceptance of radiation (Benson et al. 1992) and thus, net carbon gain and the potential growth rate of the stand.

During drier periods, reductions in available soil water result in closure of stomata, which in turn constrains growth of *P. radiata* and other conifers. While this response to water deficits helps conserve water by

* Correspondence: mjwaghorn@hotmail.com
[1]School of Forestry, University of Canterbury, Private Bag 4800, Christchurch, New Zealand
Full list of author information is available at the end of the article

limiting evaporative losses, it also reduces carbon dioxide uptake, thereby reducing photosynthetic rate, tree growth and wood production (Waring and Franklin 1979; Benecke 1980; Whitehead 1985; Thompson and Wheeler 1992; Yunusa et al. 1995a). Many plant traits are related to water uptake including hydraulic conductance, resistance to embolism, leaf to root ratios and root distribution (Hacke et al. 2000; Pinol and Sala 2000; Martinez-Vilalta et al. 2004). Earlier work by Rook et al. (1977) showed that greater levels of water stress caused transpiration and photosynthesis to decline to near zero, which induced shedding of older foliage.

With such pronounced physiological responses to water deficits in *P. radiata*, variation in morphology is expected. In response to water deficits, Rook et al. (1977) found that diameter and root growth were among the first processes to be affected. As a generalisation, morphological features such as diameter, height, basal area and leaf area are detrimentally impacted by water deficits (Jackson et al. 1976; Bongarten and Teskey 1987; Linder et al. 1987; Johnson 1990; Albaugh et al. 1998), which is of concern to forest managers looking to optimise forest growth.

Water deficits may also influence juvenile conifer productivity by altering partitioning of biomass between above- and below-ground tree components. Previous studies of *P. radiata* subjected to varying degrees of water availability have clearly shown that decreased availability of water reduces above-ground biomass accumulation (Snowdon and Benson, 1992; Woods et al. 1992; Yunusa et al. 1995a; Watt et al. 2003; Espinoza et al. 2013). However, the partitioning of biomass to above-ground components is often contradictory between studies (Snowdon and Benson, 1992; Watt et al. 2003).

Tree water stress can be interpreted from the ratio of the two stable carbon isotopes ^{12}C and ^{13}C found in plant material, known as $\delta^{13}C$. The $\delta^{13}C$ ratio is the result of discrimination against the heavier $^{13}CO_2$ during diffusion through the stomata and consequent carboxylation in favour of the lighter $^{12}CO_2$ (Farquhar et al. 1989; Brienen et al. 2011). Discrimination is linearly related to the ratio of leaf intercellular CO_2 concentration (C_i) to atmospheric CO_2 concentration (C_a) (i.e. C_i/C_a), which reflects the balance between the rate of inward CO_2 diffusion, mediated by stomatal conductance, and the rate of CO_2 assimilation in photosynthesis (Warren et al. 2001). Hence, drought stress related decreases in stomatal conductance and relative CO_2 concentration in the leaf are often correlated with increases (less negative values) in $\delta^{13}C$ of newly formed photosynthates (Farquhar et al. 1989). Water-use efficiency (the ratio of assimilation to transpiration) is also related to C_i/C_a, thus the $\delta^{13}C$ ratio and water-use efficiency are positively related. The $\delta^{13}C$ signature of tree material is widely regarded as a reliable proxy for water-use efficiency (Kruse et al. 2012).

Past research has shown that forest managers can influence soil water availability through choices in site, stand density and level of weed control (Nambiar 1990; Stogsdili et al. 1992; Yunusa et al. 1995b; Watt et al. 2003; Watt et al. 2006). Furthermore, it has been shown that growth of *P. radiata* is very sensitive to water stress during the juvenile phase (Sands and Nambiar 1984). In addition to the impact of forest management practices on water availability, global climate change is increasingly affecting seasonal rainfall distribution (Basher 2000; Kenny 2001). Therefore possible changes in water availability due to changes in weather with increasingly erratic seasonal rainfall and more frequent and severe drought events, coupled with increased evaporative demand brought about by higher temperatures have the potential to alter forest growth and reduce primary production (Kirschbaum and Fischlin 1996). At present, soil water deficits are common across many eastern regions of New Zealand during summer (Palmer et al. 2009) and many of these areas are predicted to receive less rainfall in the future (Mullen et al. 2005; Kirschbaum et al. 2012).

It is important for forest managers to be able to understand the response of growth and canopy dynamics of juvenile *P. radiata* to water deficits. In general, forest managers and researchers have a suitable understanding of the impacts of extended summer drought on accumulative growth and general productivity of juvenile trees (e.g. Snowdon and Waring 1991). However, this knowledge does not extend to the consequences of early-season (spring) or late-season (autumn) drought. Given that morphological and physiological processes can be different between spring and autumn and that different types of wood are being formed (earlywood vs. latewood), it is necessary to examine how either early- or late-season droughts compare with extended summer drought and well-watered conditions. As such, the early- and late-season drought results were of primary interest in this study.

The primary objective of this study was to assess how key growth characteristics in juvenile *P. radiata* respond to the timing and duration of water deficits over the course of one growing season. Specifically, this study sought to: (i) examine the effect that the timing and duration of water deficits has on productivity; (ii) quantify how early- and late-season water deficits influences above-ground biomass accumulation; and (iii) investigate water-use efficiency in response to the timing and duration of water deficits.

Methods

Experimental design and treatments

The experiment was undertaken in a polyhouse using *P. radiata* clonal stock. The nursery-raised trees were planted in the polyhouse for a two-year period with experimental treatments applied during the second year. The clone used was a high wood density, high acoustic velocity clone (Clone 35, Forest Genetics CellFor Ltd) produced using somatic embryogenesis. The trees were planted in a free-draining loamy sand soil within 200 litre pots to reduce any potential for water logging and root binding.

Soil particle size was determined using a hydrometer for smaller particles and sieve for larger particles prior to the experiment. Particles were separated into <0.002, 0.0021-0.06, 0.061-0.2, 0.21-0.6 and 0.61-2.0 mm size fractions. A comprehensive set of soil chemical measurements were made (Table 1). Six bulked soil samples were analysed for pH in water, total carbon (C), total nitrogen (N), Olsen phosphorus (P), exchangeable bases and cation exchange capacity (CEC), following the methods described by Blakemore et al. (1987).

The clonal tree material grown in the polyhouse was subjected to four water deficit treatments laid out in six blocks of five rows for one growing season (start of August 2009 to end of August 2010). Within each block of five rows, all four water deficit treatments were present plus one buffer row. Each treatment, which occupied one row, consisted of four trees. The treatments within each block were randomly assigned. The buffer row within each block was placed systematically on the next row to the north of rows subject to the summer drought treatment so that shading of the drought trees and thus any confounding influences were minimised. Trees were spaced 1.4 m apart so that there was no physical interaction between their crowns. Twenty four trees per treatment were used giving a total of 96 trees for the experiment.

Table 1 Mean soil physical and chemical properties

Variable	Value
Soil physical properties (%)	
Coarse sand	59
Medium sand	17
Fine sand	8
Silt	8
Clay	8
Soil chemical properties	
Total C (%)	6.2
Total N (%)	0.26
C:N ratio	23.9
pH	6.4
CEC (cmol/kg)	22
Exchangeable Na (cmol/kg)	0.53
Exchangeable K (cmol/kg)	1.68
Exchangeable Mg (cmol/kg)	2.77
Exchangeable Ca (cmol/kg)	12.6
Base saturation (%)	82
Olsen P (mg/kg)	69

Four water deficit treatments were applied to the experimental trees. Irrigation was applied using drip sprinklers until soil moisture capacity was reached. The first treatment was 'well-watered' with applications of water applied weekly between October and May (spring to autumn), and fortnightly between June and September (winter). The second treatment simulated 'extended summer drought' whereby water was withheld between November and April to allow drying of the soil profile. The drought was interrupted by the small, single application of water on March 1 to prevent tree mortality. Although predawn needle water potentials were not particularly low at this time, the trees had started showing large physical changes to continued drought. During the remainder of the year, water was applied at the same times as the well-watered treatment. The third and fourth treatments simulated 'early-season cyclical drought' and 'late-season cyclical drought', respectively. The early-season cyclical drought treatment was applied between October and January (spring/early summer), while the late-season cyclical drought treatment was applied between February and May (late summer/autumn). These treatments were timed to coincide with periods of expected early- and latewood development. Trees in these treatments were subjected to a series of drying cycles interrupted by water replenishment when needle water potential had fallen to approximately –0.5 MPa or lower. The early-season drought trees were re-watered on November 7, December 1, December 26 and January 16, while the late-season drought trees were re-watered on February 20, March 13 and April 19. This water potential was sufficient to generate a depressed growth response as previously observed in non-experimental trees. When not subject to the cyclical watering treatments between the stated months, water for both treatments was applied at the same times as that for the well-watered treatment. The length of time between watering for the cyclical drought treatments depended upon the time of year and air temperature. The time between watering varied from three to seven weeks during the October to May period. These cycles were expected to create large fluctuations in needle water potential.

Climatic variables

Climatic variables within the polyhouse were measured throughout the duration of the experiment. Variables measured daily were air temperature, relative humidity and daytime photosynethically active irradiance (PAR). These variables were measured using sensors connected to a data logger (HOBO weather station, ONSET Computer Corporation, Bourne, MA, USA). Vapour pressure deficit (D) was calculated from measurements of air temperature and relative humidity (Allen et al. 1998).

Measurements of tree growth

Measurements relating to tree growth were taken during year two. Monthly measurements of total tree height and diameter (at a tree height of 0.5 m) were made. Ground-line diameter was also measured every second month. Both measures of diameter were made on a single axis for all trees using a pair of electronic callipers. Whole tree litterfall from both current and second-year needles was collected monthly, dried at 70°C until constant weight was achieved and aggregated over the year for each treatment.

Predawn needle water potential

Measurements of predawn needle water potential (ψ_e) were made throughout the experiment using a pressure chamber. Three trees per treatment were sampled on each occasion. The timing of sampling depended on the season with weekly sampling occurring during spring, summer and autumn and fortnightly sampling occurring during winter. From each tree, three fascicles were collected from the youngest, well developed foliage. The fascicles were collected before dawn and stored in test tubes on ice until ψ_e was measured, which was always within one hour of collection. Following Myers (1988), water stress integral (S_ψ) was determined as the cumulative integral of predawn needle water potential over the period which water deficits were imposed (1 October to 31 May). In addition to total values of S_ψ, monthly S_ψ was also calculated. Values of S_ψ are expressed as an absolute sum, so that larger values of S_ψ represent greater accumulated water deficits.

Measurements of final tree growth, biomass and leaf area

At the conclusion of the experiment at the end of August 2010, prior to destructive sampling, final measurements of tree height, diameter at 0.5 m up the stem and ground-line diameter were made for all trees. Stem slenderness was calculated as the ratio of tree height to ground-line diameter. Basal area was calculated using Husch et al. (2003). Crown dimensions were measured for all trees. The number of whorls, internode length, number of branches per whorl, branch diameter and branch length were measured. The largest branch diameter in each quadrant of the whorl was measured and branch diameter was defined as the mean of the largest branch in each of the four quadrants.

Total above-ground and component biomasses were determined by destructively harvesting 48 trees at the end of the experiment (12 trees per treatment). Trees were separated into stem, branch and foliage components. All biomass components were dried to 70°C until constant mass was achieved, then weighed.

Leaf area was measured for 24 randomly selected trees (six per treatment) at the conclusion of the experiment. Leaf area by year was obtained by removing all foliage and separating into year-one and year-two growth based on measurements of height. Although treatments were only imposed during year two, year-one leaf area was assessed in order to examine subsequent retention of drought affected leaves. Following Beets (1977), specific leaf area was determined from measurements of fascicle diameter, length, and dry mass based on a subsample of 20 fascicles from both year-one and year-two age classes. Values for leaf area are presented on the basis of half the total surface area. The number of fascicles per tree was estimated as total tree foliage dry weight divided by leaf area fascicle subsample dry weight.

Foliar $\delta^{13}C$ isotope composition

Foliage stable carbon isotope ratios ($\delta^{13}C$) were measured for six trees per treatment (24 trees total) at the conclusion of the experiment. Oven-dried foliage samples from the year-two foliage of each sampled tree were ground in a Wiley mill to < 200 μm. Using 3 mg of ground sample material, carbon isotope composition was determined using a Dumas Elemental Analyser (Europa Scientific ANCA-SL) interfaced to a Stable Isotope Mass Spectrometer (Europa Scientific Tracermass, Europa Scientific Ltd, Crewe, UK). The $^{13}C/^{12}C$ ratios are expressed as delta (δ) values in thousandths (‰). All analyses were run against the CSIRO sucrose standard ($\delta^{13}C = -10.8‰$) and sample $\delta^{13}C$ was calculated relative to the Pee Dee Belemnite standard. The standard deviation of repeated $\delta^{13}C$ measurements of the laboratory samples was 0.3‰. The carbon isotope composition was calculated from:

$$\delta^{13}C\,(‰) = \left(R_{sample}/R_{standard} - 1\right) \times 1000 \quad (1)$$

where R represents the ratio of $^{13}C/^{12}C$ of samples and standards, respectively (Brienen et al. 2011).

Statistical analysis

All analyses were carried out using R software (R Core Team 2013). Prior to the commencement of the experiment at the start of year two, measures of growth were made and analysed to ensure no location bias existed within the polyhouse. There were no statistically significant treatment differences in any of the measured variables at the start of the experiment. The effect of water deficits on tested variables were examined using a mixed-effects model that included treatment as a fixed effect, with block as a random effect. A repeated-measures

ANOVA was used for measures of diameter, height and needle water potential. Tukey's HSD test was used, where applicable, to distinguish between treatment mean values. Differences were considered significant at $P < 0.05$.

Results

Climate

Air temperature fluctuated between −1 and 35°C with a mean and standard deviation of 14.3 ± 5.9°C. Vapour pressure deficit fluctuated between 0.01 and 3.89 kPa with a mean and standard deviation of 0.41 ± 0.53 kPa. The maximum value of daytime photosynthetically active irradiance was 1023 μmol m^{-2} s^{-1} while the minimum was 1.3 μmol m^{-2} s^{-1}. Annual variation in (a) average daily air temperature, (b) average daily vapour pressure deficit, and (c) average daytime photosynthetically active irradiance (PAR) are shown in Figure 1.

Predawn needle water potential

The implementation of the early-season drought, late-season drought and summer drought treatments had a pronounced effect on ψ_e and induced consistently more negative values of ψ_e than for the well-watered trees (Figure 2). The early-season and late-season drought treatments resulted in large fluctuations of ψ_e. The lowest value of ψ_e achieved for the early- and late-season drought treatments occurred during the first drying-rewetting cycle. Compared with these treatments, the summer drought treatment induced consistently more negative values of ψ_e during the period between mid-December and April. Values of total S_ψ were 41.4, 66.8, 55.2 and 97.6 MPa-days for the well-watered, early-season drought, late-season drought and summer drought treatments, respectively.

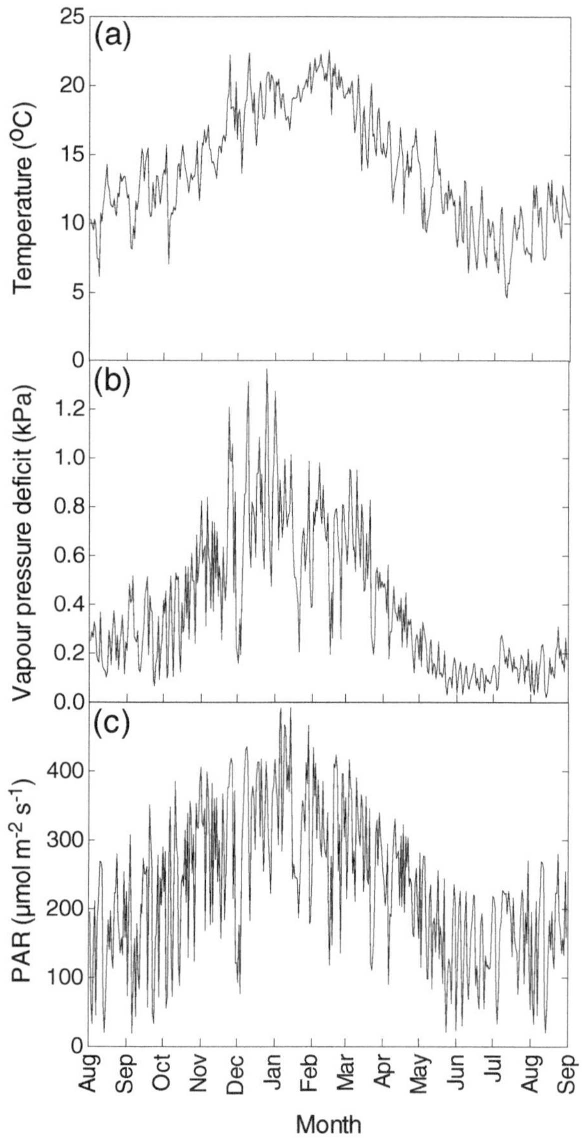

Figure 1 Climatic variation over the duration of the experiment showing annual variation in (a) average daily air temperature, (b) average daily vapour pressure deficit, and (c) average daytime photosynthetically active irradiance (PAR).

Tree characteristics

Tree height plus both measures of diameter and basal area were all significantly influenced by treatment ($P < 0.001$). Stem slenderness was also significantly influenced by treatment ($P = 0.012$), but only due to the variation in slenderness between the well-watered and early-season drought treatments. Basal area was the most sensitive indicator of water deficits as well-watered trees had double the basal area of the summer-drought trees (Table 2). With the imposition of water deficits, marked divergence in monthly diameter and height growth become immediately apparent (Figure 3). Decreased growth for all water deficit treatments coincided with more negative predawn needle water potential values. Diameter growth for the summer-drought treatment virtually ceased in March (0.04 mm growth for the month). However, the well-watered and early-season drought treatments experienced their period of greatest incremental diameter growth (3.16 and 3.36 mm, respectively) during the same month. After exceedingly different growth trajectories during the preceding months, diameter growth had converged for all treatments by mid-winter (July). Height growth started increasing towards the end of the experiment in July even though this month had the lowest daily temperatures. Rapid increases in height growth occurred the following month. Diameter growth started increasing in August. Compared with the well-watered treatment, the summer drought treatment reduced height, diameter and basal area by 24.7%, 33.1% and 52.3%, respectively.

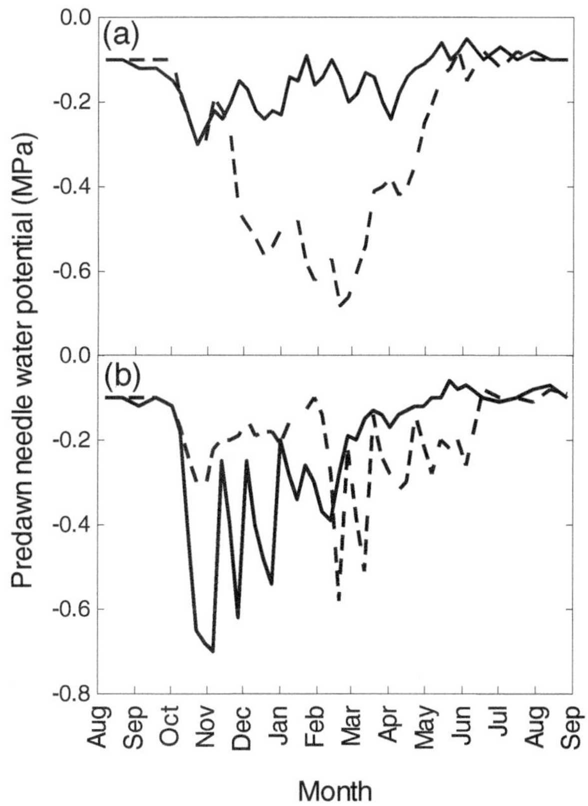

Figure 2 Seasonal changes in predawn needle water potential for (a), the well-watered (line) and summer drought (dash) treatments, and (b), early-season drought (line) and late-season drought (dash) treatments.

There were no significant differences ($P > 0.05$) in height, diameter and basal area between early- and late-season drought trees at the end of the experiment (Table 2), although total S_ψ was larger in the early season drought trees than in the late season drought trees. Monthly diameter growth was greater in early-season drought treatment trees even though they were subjected to higher levels of water stress than those in the late-season drought treatment. The relationship between monthly S_ψ and diameter growth is illustrated in Figure 4. Diameter growth was similar for both treatments irrespective of season during those periods when not subjected to water deficits and when monthly S_ψ values were comparable.

Trees exposed to early-season drought between October and January increased in diameter by 5.36 mm during that period compared with 10.60 mm for the late-season drought trees and 10.17 mm for the well-watered trees. When water deficits were imposed between February and May for the late-season drought trees, their diameter increased by 2.37 mm compared with 8.90 mm for the early-season drought and 9.41 mm for the well-watered treatment trees. The reduction in stem diameter growth was higher for the late-season treatment than the early-season treatment when compared with the well-watered trees (Figure 3). Furthermore, the early-season drought trees recovered rapidly after water stress, and even surpassed the well-watered trees for monthly stem height growth. This suggests that late-season drought is more detrimental than early-season drought to absolute tree growth.

The primary effect of water deficits on crown characteristics were reductions in branch diameter and branch length ($P < 0.001$). The magnitude of these effects increased with the severity of the water deficit treatment with branch length varying by two-fold between the well-watered and summer drought treatments (Table 3). The principal difference between the early- and late-season drought treatments was the number of fascicles per tree, otherwise crown variables were similar. Although summer drought produced a very similar number of whorls and branches per whorl as the other treatments, branch diameter and length were considerably different. This suggests that summer drought does not significantly inhibit formation of lateral branching but the size they grow to.

Needle litterfall was significantly ($P < 0.01$) affected by treatment. The impact of cyclical drought on the early-season treatment ceased early enough to minimise needle loss, however, the late-season treatment showed significant needle loss, especially during the last three

Table 2 Treatment variation in tree dimensions at the end of the experiment

Treatment	Tree height (mm)	Diameter (mm)	Ground-line diameter (mm)	Basal area (mm^2)	Slenderness
Well-watered	3544 (55) a	38.3 (0.8) a	44.8 (1.0) a	1590 (76) a	79.6 (1.2) a
Early-season drought	3220 (53) b	32.4 (0.6) b	37.1 (0.6) b	1086 (35) b	87.0 (1.0) b
Late-season drought	2997 (59) b	30.6 (0.6) b	35.8 (0.6) b	1015 (38) b	83.8 (1.3) ab
Summer drought	2606 (42) c	25.6 (0.4) c	31.0 (0.4) c	759 (21) c	84.1 (1.2) ab
Analysis of variance					
Treatment	37.3***	44.3***	45.3***	34.3***	5.12*

Each value is the treatment mean with standard error in parentheses for 24 trees. The Tukey HSD test was performed between treatments. Means followed by the same letter are not significantly different at $P < 0.05$. Significance of treatment effects is shown as F value followed by P category: *significant at $P < 0.05$; ***significant at $P < 0.001$.

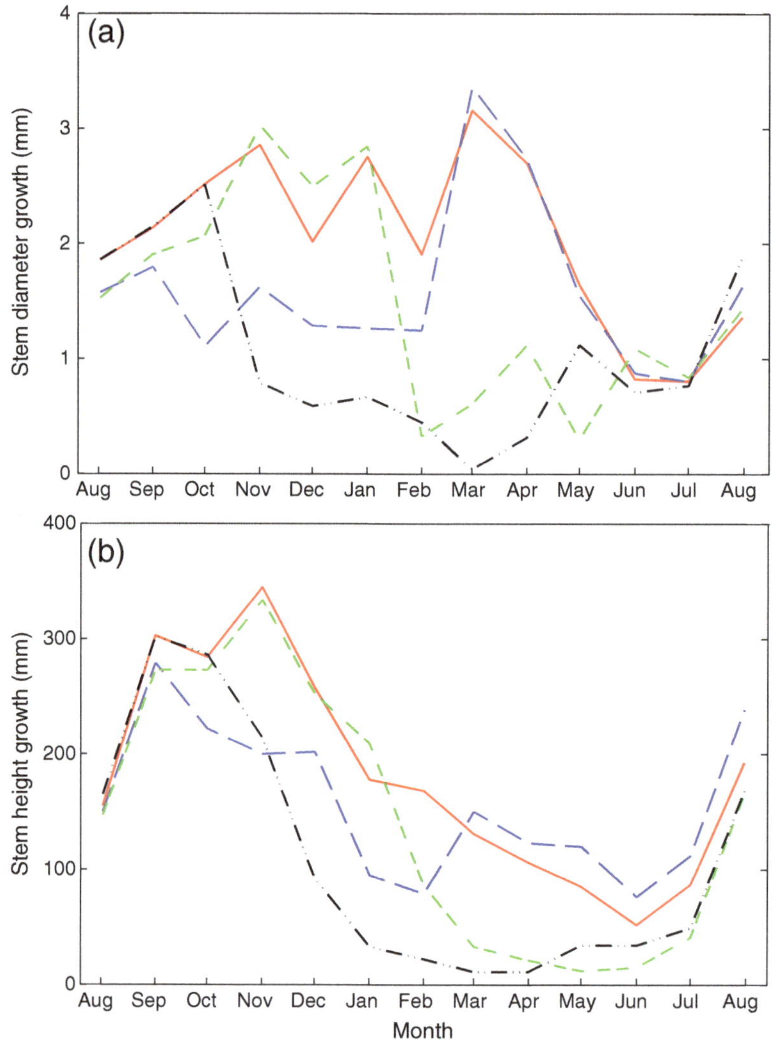

Figure 3 Average monthly incremental stem diameter growth (a) and height growth (b) for the well-watered (solid red line), early-season drought (long blue dash), late-season drought (short green dash) and summer drought (black dash-dot-dot) treatments.

months of the growing season when 5.18 grams of the total 6.88 grams fell. The large amount of litterfall from well-watered trees was most likely the result of tree size rather than a water deficit response.

Biomass

As expected, above-ground biomass component weights were influenced by treatment ($P < 0.001$). Stem and foliage biomasses of the early-season and late-season drought treatments were not significantly different ($P > 0.05$), (Table 4). The difference in biomass between the well-watered and summer drought treatments was significant for all components but was most apparent for branch biomass which differed by a factor of just over 4, compared with 3.5 for stem and 2.6 for foliage. Total biomass for the early-season, late-season and summer drought treatments was reduced by 36.3%, 44.8% and 68.3% compared with the well-watered treatment. The relationship between woody biomass and foliage biomass was highly correlated ($r^2 = 0.91$), indicating a strong functional relationship between conducting tissue of the tree and foliage.

Leaf characteristics

Leaf area varied considerably among treatments. A significant treatment effect was observed for year-one leaf ($P < 0.001$), year-two leaf ($P < 0.01$) and total leaf area ($P < 0.0001$). Although water deficits were only imposed during the second year, the effect on both the growth and development of the year-two foliage and the continued growth and retention of the year-one foliage was pronounced (Figure 5). The leaf area of year-two foliage for the well-watered treatment was double that of the summer drought treatment. The year-one, year-two

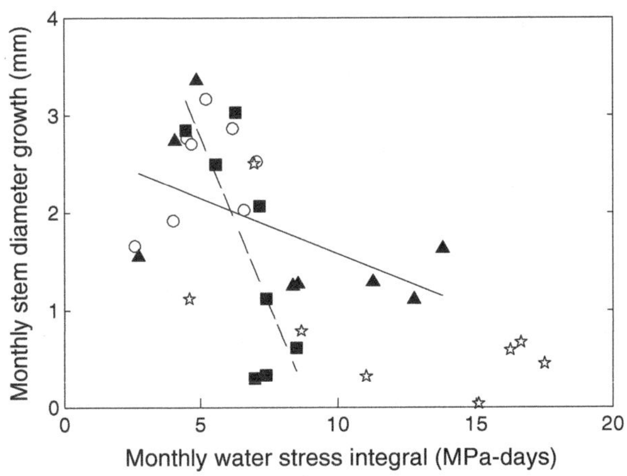

Figure 4 The relationship between monthly water stress integral (S_ψ) and monthly stem diameter growth for the period between October 2009 and May 2010 for the well-watered (open circles), early-season drought (filled triangle), late-season drought (filled square) and summer drought (open star) treatments. Best fit lines have been included for the early-season drought (solid line, $r^2 = 0.33$) and late-season drought (dashed line, $r^2 = 0.57$) treatments. The slopes were significantly different ($P = 0.0356$).

and total leaf areas were not significantly different between the early-season and late-season drought treatments ($P > 0.05$). Values of total leaf area of the early-season, late-season and summer drought treatments were 71%, 73% and 60%, respectively, of that for the well-watered treatment.

Estimates of water-use efficiency

Significant treatment differences in $\delta^{13}C$ ($P = 0.0375$) were observed. The well-watered trees showed significantly lower $\delta^{13}C$ values over the entire growth period compared with those for the summer drought trees because of higher soil water availability. The summer drought treatment had the most enriched values of $\delta^{13}C$, with an average of −25.93‰ compared to −26.57‰, −26.65‰ and −27.20‰ for the late-season, early-season and well-watered treatments, respectively. The variability among treatments (1.27‰) was smaller than the variation within the early-season drought treatment (3.06‰). Across all samples, foliar $\delta^{13}C$ ranged from −25.29‰ to −28.63‰ (range of 3.14‰).

Discussion

Previous experiments on dry sites that included irrigation have shown that water deficits limit tree growth by regulating the period during which tree growth is possible (Benson et al. 1992). Growth of trees in the well-watered treatment was largely unaffected by water stress as confirmed from measurements of predawn needle water potential, which did not fall below −0.30 MPa. Although differences among treatments in predawn needle water potentials were often quite small, Myers (1988) showed that small differences maintained over long periods can lead to large changes in growth. Improved water availability led to increased growth, and distinct differences were observed between the well-watered, cyclical drought and summer drought treatments in the current study.

The higher slope for the late-season drought treatment in Figure 4 means that similar monthly water stress integral had a more pronounced effect on monthly stem diameter growth compared with the early-season drought treatment, whilst demonstrating that the early-season drought trees experienced greater levels of water stress (total S_ψ = 66.8 vs. 55.2 MPa-days). In addition, the early-season drought trees rapidly recovered from water stress and exceeded the well-watered treatment for monthly stem height growth, whilst the late-season drought trees remained suppressed through to the end of the experiment and into the new growing season. This result demonstrates the complex interactions between the phenology of foliage production, partitioning of photosynthates and the availability of resources during different seasons. Although the early- and late-season

Table 3 Crown characteristics at the end of the experiment

	Well-watered	Early-season drought	Late-season drought	Summer drought	*Analysis of variance* treatment
No. of whorls	2.3 (0.1) a	2.3 (0.1) a	2.3 (0.1) a	2.2 (0.1) a	0.49^{ns}
Internode length (cm)	61.8 (3.1) a	58.3 (3.1) ab	60.0 (2.2) ab	50.6 (3.0) b	3.29^{ns}
Branches per whorl	7.6 (0.2) ab	6.8 (0.2) a	7.6 (0.2) ab	7.7 (0.2) b	3.89*
Branch diameter (cm)	7.8 (0.2) a	6.3 (0.2) b	6.4 (0.2) b	4.9 (0.1) c	45.8***
Branch length (cm)	40.3 (1.9) a	29.2 (1.3) b	31.3 (1.7) b	20.8 (1.3) c	44.2***
Needle length (cm)	12.9 (0.4) a	12.1 (0.2) a	12.1 (0.5) a	10.3 (0.3) b	10.6***
Fascicles per tree	2643 (404) a	2297 (167) a	1704 (123) a	1547 (290) b	3.61*
Litterfall (grams)	5.0 (0.7) ab	1.9 (0.3) b	6.9 (1.1) a	7.1 (1.3) a	9.31**

Each value is the treatment mean with standard error in parentheses for 24 trees, except for needle length which is for 6 trees. The Tukey HSD test was performed between treatments. Means followed by the same letter are not significantly different at $P < 0.05$. Significance of treatment effects is shown as F value followed by P category: ^{ns}non-significant; *significant at $P < 0.05$; **significant at $P < 0.01$; ***significant at $P < 0.001$.

Table 4 Above-ground biomass by treatment

Treatment	Tree component biomass (g tree^{-1}) Stem	Branches	Foliage	Total
Well-watered	350.1 (31.8) a	104.9 (12.9) a	291.4 (27.0) a	746.4 (67.8) a
Early-season drought	212.8 (14.5) b	56.7 (5.0) b	206.5 (10.0) b	475.9 (27.1) b
Late-season drought	187.1 (9.0) b	55.3 (4.8) b	169.9 (10.0) b	412.3 (20.5) b
Summer drought	100.2 (8.5) c	25.2 (3.3) c	111.6 (9.8) c	236.9 (19.9) c
Analysis of variance				
Treatment	36.4***	49.5***	52.9***	51.4***

Each value is the treatment mean with standard error in parentheses for 12 trees. The Tukey HSD test was performed between treatments. Means followed by the same letter are not significantly different at $P < 0.05$. Significance of treatment effects is shown as F value followed by P category: ***significant at $P < 0.001$.

drought treatments were implemented at different times of the year when different factors may have been influencing tree growth, such as air temperature and vapour pressure deficit, comparisons with the well-watered trees as the control treatment allow for direct comparisons between the relative effects of early- and late-season drought on physical measures of tree growth.

Diameter growth rates of the well-watered trees remained high until late autumn. However, there was a decline in incremental diameter growth for the summer months of December and February, which was possibly caused by atmospheric conditions. The highest values of vapour pressure deficit occurred in early summer (December) and the highest daily average temperatures in late summer (February). When the vapour pressure deficit increases, stomatal conductance decreases (Beneke 1980), thus reducing photosynthesis. It has also been shown that cell diameters decrease in the radial direction when the temperature is above 15°C (Jenkins 1975). These reductions in rate of diameter growth, more so for December than February, coincide with the main period of needle elongation. The foliage would thus have been a strong competing sink for assimilates, reducing the amount of resources available for diameter growth (Linder et al. 1987). The high vapour pressure deficit also affected the late-season drought treatment in December resulting in reduced diameter growth.

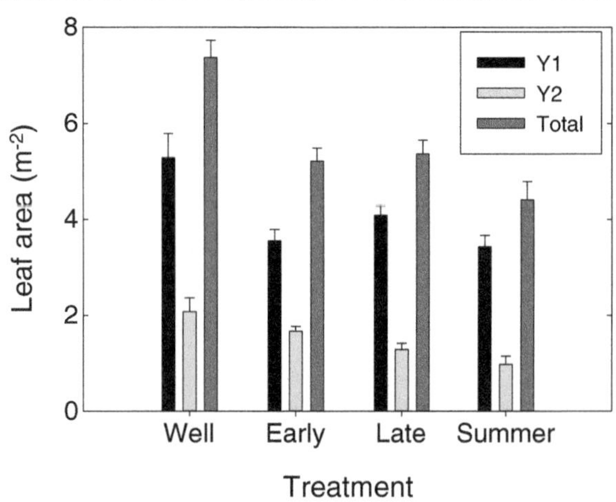

Figure 5 Leaf area for year one, year two and total tree foliage for the well-watered, early-season drought, late-season drought and summer drought treatments.

Distinct differences in biomass were observed among the well-watered, cyclical drought and the summer drought treatments. The well-watered trees in this study had over three-fold the total biomass of the summer drought trees and a four-fold difference in branch biomass which exceeded the difference for any other tree components between these treatments. This was driven by large differences in branch diameter and length. As the water stress integral increased, above-ground biomass was preferentially allocated to stem and branches over foliage. The strong positive relationship between woody biomass and foliage biomass across all treatments and tree sizes in this experiment conforms to the pipe model theory advocated by Shinozaki et al. (1964a, b). These results suggest that, in juvenile *P. radiata*, estimates of canopy leaf area can be made from the sapwood area at the base of the tree.

Water deficits resulted in smaller needles, reduced leaf area and considerably smaller crowns. Smaller crowns were the result of reduced branch length and diameter, and reduced number of fascicles. Pronounced differences existed for year-one, year-two and total leaf area across all treatments. Water deficits would have impacted on the growth and retention of year one foliage as well as the formation and growth of the year-two foliage. Early-season drought was expected to retard continued formation and growth of needles during the latter part of the growing season, however, this did not occur. It has been shown that needles on later season flushes can continue to develop until all above-ground growth sinks are halted and at the same rate as those earlier in the growing season (Rook et al. 1987; Dougherty et al. 1994).

The magnitude of $\delta^{13}C$ in plant foliage is determined by both stomatal conductance and photosynthetic capacity (Farquhar et al. 1989) so photosynthetic activity throughout the period the leaf tissue was synthesised impacts on $\delta^{13}C$ (McNulty and Swank 1995; Barbour et al. 2002; Dawson et al. 2002). In response to water deficits, stomatal conductance decreases to prevent excessive water losses, leading to reduced values for C_i and lower ^{13}C discrimination (Farquhar and Sharkey 1982). Since stomatal conductance is proportional to water loss (given similar ambient conditions), $\delta^{13}C$ is positively correlated with water-use efficiency (Farquhar and Richards 1984). Thus, given that the summer drought treatment had the most enriched values of $\delta^{13}C$, the results imply that the summer drought treatment had the greatest water-use efficiency and, as such, fixed the most carbon per unit amount of water transpired. Under water stress, diameter and height growth, above-ground biomass, and foliage area were significantly decreased, while water-use efficiency was significantly increased.

Conclusion

Strong positive correlations between water availability and various aspects of growth of juvenile *P. radiata* were observed in this experiment. Implementation of water deficit treatments resulted in pronounced reductions in tree dimensions, above-ground biomass accumulation, leaf area and increases in water-use efficiency compared with the well-watered treatment. Treatment effects were significant for all variables tested ($P < 0.05$). At a similar monthly water stress integral, late-season drought trees experienced a greater reduction in monthly stem diameter growth compared with the early-season drought trees, suggesting that late-season drought is more detrimental to tree growth and biomass accumulation in juvenile *P. radiata* than early-season drought. The results herein clearly demonstrate the benefit to forest managers of improved water availability to juvenile *P. radiata* to ensure sound establishment and vigorous growth and can be useful for improving both silvicultural and environmental approaches to forest management.

Competing interests

The authors declare that they have no competing interests.

Authors' contributions

MJW was the primary author and undertook the majority of analyses. DW assisted with some analyses, interpretation of data and experimental conception and design. MSW and EGM helped draft the manuscript and along with JJH assisted with experimental conception and design. All authors read and approved the final manuscript.

Acknowledgements

The lead author is grateful to Timberlands Ltd for in-kind support and providing the clonal tree stock. The lead author was supported by the University of Canterbury Doctoral Scholarship and funding from Scion.

Author details

[1]School of Forestry, University of Canterbury, Private Bag 4800, Christchurch, New Zealand. [2]Landcare Research, PO Box 40, Lincoln, New Zealand. [3]Scion, PO Box 29237, Fendalton, Christchurch, New Zealand.

References

Albaugh, TJ, Allen, HL, Dougherty, PM, Kress, LW, & King, JS. (1998). Leaf area and above- and belowground growth responses of loblolly pine to nutrient and water additions. *Forest Science, 44*, 317–328.

Allen, RG, Pereira, LS, Raes, D, & Smith, M. (1998). *Crop Evaporation - Guidelines for Computing Crop Water Requirements. FAO Irrigation and Drainage Paper 56. Chapter 3 - Meteorological Data.* Rome: Food and Agriculture Organization.

Barbour, MM, Walcroft, AS, & Farquhar, GD. (2002). Seasonal variation in $\delta^{18}C$ and $\delta^{13}C$ of cellulose from growth rings of *Pinus radiata. Plant, Cell and Environment, 25*, 1483–1499.

Basher, RE. (2000). The Impacts of Climate Change on New Zealand. In A Gillespie & WCG Burns (Eds.), *Climate Change in the South Pacific: Impacts and Responses in Australia, New Zealand and Small Island States* (pp. 121–142). Dordrecht, Netherlands: Kluwer Academic Publishers.

Beets, P. (1977). Determination of the fascicle surface area for *Pinus radiata. New Zealand Journal of Forestry Science, 7*, 397–407.

Benecke, U. (1980). Photosynthesis and transpiration of *Pinus radiata* D. Don under natural conditions in a forest stand. *Oecologia, 44*, 192–198.

Benson, ML, Myers, BJ, & Raison, RJ. (1992). Dynamics of stem growth of *Pinus radiata* as affected by water and nitrogen supply. *Forest Ecology and Management, 52*, 117–137.

Blackmore, LC, Searle, PL, & Daly, BK. (1987). *Methods for Chemical Analysis of Soils. New Zealand Soil Bureau Scientific Report No. 80.* Wellington, New Zealand: Department of Scientific and Industrial Research.

Bongarten, BC, & Teskey, RO. (1987). Dry weight partitioning and its relationship to productivity in loblolly pine seedlings from seven sources. *Forest Science, 33*, 255–267.

Brienen, RJW, Wanek, W, & Hietz, P. (2011). Stable carbon isotopes in tree rings indicate improved water use efficiency and drought responses of a tropical dry forest tree species. *Trees, 25*, 103–113.

Dawson, TE, Mambelli, S, Plamboeck, AH, Templer, PH, & Tu, KP. (2002). Stable isotopes in plant ecology. *Annual Review of Ecology and Systematics, 33*, 507–559.

Dougherty, PM, Whitehead, D, & Vose, JM. (1994). Environmental influences on the phenology of pine. *Ecological Bulletins, 43*, 64–75.

Espinoza, SE, Magni, CR, Martinez, VA, & Ivkovic, M. (2013). The effect of water availability on plastic responses and biomass allocation in early growth traits of *Pinus radiata* D. Don. *Forest Systems, 22*, 3–14.

Farquhar, GD, & Sharkey, TD. (1982). Stomatal conductance and photosynthesis. *Annual Review of Plant Physiology and Plant Molecular Biology, 33*, 317–345.

Farquhar, GD, & Richards, RA. (1984). Isotopic composition of plant carbon correlates with water use efficiency of wheat genotypes. *Australian Journal of Plant Physiology, 11*, 539–552.

Farquhar, GD, Ehleringer, JR, & Hubick, KT. (1989). Carbon isotope discrimination and photosynthesis. *Annual Review of Plant Physiology and Plant Molecular Biology, 40*, 503–537.

Gower, ST, Vogt, KA, & Grier, CC. (1992). Carbon dynamics of Rocky Mountain Douglas-fir: influence of water and nutrient availability. *Ecological Monographs, 62*, 43–65.

Hacke, UG, Sperry, JS, Ewers, BE, Ellsworth, DS, Schafer, KVR, & Oren, R. (2000). Influence of soil porosity on water use in *Pinus taeda. Oecologia, 124*, 495–505.

Husch, B, Beers, TW, & Kershaw, JA, Jr. (2003). *Forest Mensuration* (4th ed.). New Jersey, USA: John Wiley & Sons, Inc.

Jackson, DS, Gifford, HH, & Chittenden, J. (1976). Environmental variables influencing the increment of *Pinus radiata*: (2) Effects of seasonal drought on height and diameter increment. *New Zealand Journal of Forestry Science, 5*, 265–286.

Jenkins, PA. (1975). Influence of temperature change on wood formation in *Pinus radiata* grown in controlled environments. *New Zealand Journal of Botany, 13*, 579–592.

Johnson, JD. (1990). Dry-matter partitioning in loblolly and slash pine: effects of fertilisation and irrigation. *Forest Ecology and Management, 30*, 147–157.

Kenny, G. (2001). *Climate Change: Likely Impacts on New Zealand Agriculture.* Wellington, New Zealand: Ministry for the Environment.

Kirschbaum, MUF, & Fischlin, A. (1996). Climate Change Impact on Forests. In RT Watson, MC Zinyowera, & RH Moss (Eds.), *Climate Change 1995. Impacts, Adaptations and Mitigation of Climate Change: Scientific-Technical Analyses* (pp. 95–130). Cambridge, UK: Cambridge University Press.

Kirschbaum, MUF, Watt, MS, Tait, A, & Ausseil, AE. (2012). Future wood productivity of *Pinus radiata* in New Zealand under predicted climatic changes. *Global Change Biology, 18*, 1342–1356.

Kruse, J, Hopmans, P, Rennenberg, P, & Adams, M. (2012). Modern tools to tackle traditional concerns: Evaluation of site productivity and *Pinus radiata* management via $\delta^{13}C$- and $\delta^{18}O$- analysis of tree-rings. *Forest Ecology and Management, 285*, 227–238.

Linder, S. (1987). Responses to Water and Nutrients in Coniferous Ecosystems. In ED Schulze & H Zwolfer (Eds.), *Ecological Studies* (Potentials and limitations of ecosystem analysis, Vol. 61, pp. 180–202). Berlin, Germany: Springer Berlin Heidelberg.

Linder, S, Benson, ML, Myers, BJ, & Raison, RJ. (1987). Canopy dynamics and growth of *Pinus radiata*. I. Effects of irrigation and fertilisation during a drought. *Canadian Journal of Forest Research, 17*, 1157–1165.

Martinez-Vilalta, J, Sala, A, & Pinol, J. (2004). The hydraulic architecture of Pinaceae - a review. *Plant Ecology, 171*, 3–13.

McNulty, SG, & Swank, WT. (1995). Wood $\delta^{13}C$ as a measure of annual basal area growth and soil water stress in a *Pinus strobus* forest. *Ecology, 76*, 1581–1586.

Mullan, B, Porteous, A, Wratt, D, & Hollis, M. (2005). *Changes in Drought Risk With Climate Change. A Report Prepared for the Ministry for the Environment and Ministry of Agriculture and Forestry*. Wellington, New Zealand: NIWA.

Myers, BJ. (1988). Water stress integral – a link between short term stress and long term growth. *Tree Physiology, 4*, 315–323.

Nambiar, EKS. (1990). Interplay between nutrients, water, root growth and productivity in young plantations. *Forest Ecology and Management, 30*, 213–232.

Palmer, DJ, Watt, MS, Hock, BK, Lowe, DJ, & Payn, TW. (2009). *A Dynamic Framework for Spatial Modelling Pinus Radiata Soil Water Balance (SWatBal) Across New Zealand. Forest Research Bulletin No. 234*. Rotorua, New Zealand: Scion.

Pinol, J, & Sala, A. (2000). Ecological implications of xylem cavitation for several Pinaceae in the Pacific Northern USA. *Functional Ecology, 14*, 538–545.

R Core Team. (2013). R: A language and environment for statistical computing. R Foundation for Statistical Computing. http://www.R-project.org/.

Rook, DA, Swanson, RH, & Cranswick, AM. (1977). Reaction of Radiata Pine to Drought. In *Proceedings of Soil and Plant Water Symposium* (pp. 55–68). Wellington, New Zealand: New Zealand Department of Scientific and Industrial Research Information Series No. 126. Department of Scientific and Industrial Research.

Rook, DA, Bollmann, MP, & Hong, SO. (1987). Foliage development within the crowns of *Pinus radiata* trees at two spacings. *New Zealand Journal of Forestry Science, 17*, 297–314.

Sands, R, & Nambiar, EKS. (1984). Water relations of *Pinus radiata* in competition with weeds. *Canadian Journal of Forest Research, 14*, 233–237.

Shinozaki, K, Yoda, K, Hozumi, K, & Kira, T. (1964a). A quantitative analysis of plant form - the pipe model theory.I. Basic analyses. *Japanese Journal of Ecology, 14*, 97–105.

Shinozaki, K, Yoda, K, Hozumi, K, & Kira, T. (1964b). A quantitative analysis of plant form - the pipe model theory. II. Further evidence of the theory and its application in forest ecology. *Japanese Journal of Ecology, 14*, 133–139.

Snowdon, P, & Benson, ML. (1992). Effects of combinations of irrigation and fertilisation on the growth and above-ground biomass production of *Pinus radiata*. *Forest Ecology and Management, 52*, 87–116.

Snowdon, P, & Waring, AM. (1991). Effect of irrigation and artificial drought on the growth and health of *Pinus radiata* near Canberra, A.C.T. *Australian Forestry, 54*, 174–186.

Stogsdili, WR, Wittwer, RF, Hennessey, TC, & Dougherty, PM. (1992). Water use in thinned loblolly pine plantations. *Forest Ecology and Management, 50*, 233–245.

Thompson, WA, & Wheeler, AM. (1992). Photosynthesis by mature needles of field grown *Pinus radiata*. *Forest Ecology and Management, 52*, 225–242.

Waring, RH, & Franklin, JF. (1979). Evergreen coniferous forests of the Pacific Northwest. *Science, 204*, 1380–1386.

Warren, CR, McGrath, JF, & Adams, MA. (2001). Water availability and carbon isotope discrimination in conifers. *Oecologia, 127*, 476–486.

Watt, MS, Whitehead, D, Mason, EG, Richardson, B, & Kimberley, MO. (2003). The influence of weed competition for light and water on growth and dry matter partitioning of young *Pinus radiata*, at a dryland site. *Forest Ecology and Management, 183*, 363–376.

Watt, MS, Moore, JR, Facon, J, Downes, GM, Clinton, PW, Coker, G, Davis, MR, Simcock, R, Parfitt, RL, Dando, J, Mason, EG, & Bown, HE. (2006). Modelling the influence of stand structural, edaphic and climatic influences on juvenile *Pinus radiata* dynamic modulus of elasticity. *Forest Ecology and Management, 229*, 136–144.

Whitehead, D. (1985). A Review of Processes in the Water Relations of Forests. In JJ Landsberg & W Parsons (Eds.), *Research for Forest Management* (pp. 94–124). Canberra, Australia: CSIRO.

Woods, PV, Nambiar, EKS, & Smethurst, PJ. (1992). Effect of annual weeds on water and nitrogen availability to *Pinus radiata* trees in a young plantation. *Forest Ecology and Management, 48*, 145–163.

Yunusa, IAM, Mead, DJ, Lucas, RJ, & Pollock, KM. (1995a). Process studies in a *Pinus radiata* – pasture agroforestry system in a subhumid temperate environment. II. Analysis of dry matter yields in the third year. *Agroforestry Systems, 32*, 185–204.

Yunusa, IAM, Mead, DJ, Pollock, KM, & Lucas, RJ. (1995b). Process studies in a *Pinus radiata* – pasture agroforestry system in a subhumid temperate environment. I. Water use and light interception in the third year. *Agroforestry Systems, 32*, 163–183.

23

Comparing and contrasting flooded and unflooded forests in the Peruvian Amazon: seed rain

Randall W Myster

Abstract

Background: In order to understand how the seed rain differs among Amazonian forests, I sampled the seed rain in six different forest types across the Amazonian landscape.

Methods: I set up seed traps and took seed samples in three unflooded forests (*terra firme*, white sand-varillal, palm) and three white-water flooded forests (high restinga, low restinga, tahaumpa) in Peru over a period of one year.

Results: I found (1) all forest types had dispersed seed of many unique species, with a few species in common, (2) total seed load peaked in the early part of the year – near the end of the rainy season – and then decreased monotonically over the remainder of the year for all forest types, (3) species richness was greater in unflooded forests compared to flooded forests and the largest number of species were found in *terra firme*, (4) seeds were more evenly distributed among species in the unflooded forests compared to the flooded forests, and (5) Fisher's α diversity of seeds was greater in high and low restinga compared to tahaumpa.

Conclusions: I conclude for the unflooded forests that seed species number and richness increased with soil fertility but for the flooded forests seed species number and richness decreased with months under water. Furthermore, when taken together, results suggest that for forests across the Amazonian landscape differences in flooding regime may have a greater effect on both seed rain load and seed species richness than differences in availability of soil nutrients.

Keywords: High restinga; Low restinga; Palm; Tahuampa; *Terra firme*; White sand

Background

The Amazonian rainforest covers over 6 million square kilometres (Walter 1973; Myster 2009) and is the most productive (Daly and Prance 1989) and diverse terrestrial ecosystem on earth, containing more than 10% of its species (Pires and Prance 1985). This rainforest influences the entire world's weather patterns and climate (Keller et al. 2004), interacting intimately with its carbon cycle both as a "sink" by taking in large amounts of CO_2 through photosynthesis, but also as a "source" when, for example, its plants decay or burn. Within the Amazonian rainforest landscape are unflooded forests with structure similar to unflooded rainforests throughout the rest of the Neotropics (Kalliola et al. 1991; Everham et. al. 1966; Pitman et al. 2001; Myster and Santacruz 2005; Myster 2009) but with some differences which may be largely due to soil characteristics (e.g., *terra firme* on clay or loam soils, white sand on soils with large amounts of quartz, palm on permanently waterlogged soils: Tuomisto et. al. 2003; Honorio 2006).

Also defining the Amazonian rainforest are extensive flooded forests (Junk 1997; Parolin et. al. 2004) mainly derived from the nutrient rich "white" water from the Andes (generally called várzea) with the rest flooded by nutrient-poor forest runoff (generally called igapó: Junk 1984). Along with the variation in soils, flooding differences in factors such as water quality, frequency, duration, depth, and spatial variation (the flooding regime: Myster 2001) help create complex and diverse forest associations throughout the Amazon basin (Myster 2009).

Among the many factors controlling the regeneration of tropical forests, the seed rain has long been seen as key to understanding recruitment (Mesquita et. al. 2001; Muller-Landau et. al. 2008) because recruitment from the seed bank is low (Loiselle et. al. 1996) and root and stump sprouting is rare, even in forest clearings (Gorchov et. al. 1993). In addition, the seed rain plays a

Correspondence: myster@okstate.edu
Biology Department, Oklahoma State University, Oklahoma City, OK 73107, USA

Table 1 Number and species of seeds taken from seed traps located in unflooded forests (*terra firme*, white sand varillal, palm) and flooded forests (high restinga, low restinga, tahuampa) of Peru

Species	*Terra firme*	White sand varillal	Palm	High restinga	Low restinga	Tahuampa
Attalea butyracea	–	–	–	20	–	–
Campsiandra augustifolia	–	–	–	–	–	10
Caraipa punctulata	–	20	–	–	–	–
Casha moeria	33	–	–	–	–	–
Dicymbe puncticulosa	–	14	–	–	–	–
Drypetes amazonia	–	–	–	21	–	–
Eschweilera coriacea	30	–	–	–	–	–
Eschweilera bracteosa	36	–	–	–	–	–
Euterpe precatoria	–	–	20	–	–	–
Guarea macrophylla	–	–	–	–	6	–
Haploclathra paniculata	–	12	–	–	–	–
Hevea nitida	–	34	19	–	–	–
Iryanthera juruensis W.	–	–	–	5	–	8
Leonia glycycarpa	–	–	–	19	–	–
Macrolobium acifolium	–	–	–	51	67	–
Maquira coriacea	22	–	–	13	6	6
Mauritia flexuosa	–	–	25	–	–	–
Miconia tripliuaria	15	–	–	–	2	–
Minguartia guianensis	10	–	–	–	–	–
Miristhicaceae (family)	–	–	–	–	–	4
Ocotea sp.	20	–	–	–	–	–
Oenocarpus batana	4	–	–	–	–	–
Oxandra espintana	2	–	–	–	–	–
Pachira brevipes	–	15	–	–	–	–
Pourouma acurinata	–	–	–	–	12	–
Pouteria caimito	5	–	–	–	–	–
Qualea paraensis	–	–	–	25	–	–
Schefflera sp.	19	–	–	–	–	–
Virola paronis	22	19	12	–	10	–
Unknown	39	15	19	–	–	–

There were 10 traps sampled six times for each of the six forest type.

critical role in determining plant-plant replacements and, therefore community patterns such as diversity (Myster 2012b). Consequently, the seed rain is critical to explaining the structure, function, and dynamics of tropical forests.

This study was motivated by the importance of Amazon forests and the lack of seed rain studies known to be key to its dynamics. In addition I wanted to better understand the spatial heterogeneity (six different forest types common in the Amazonian landscape) and temporal heterogeneity (over a one year period) of the seed rain throughout the Amazon. Therefore I sampled seed rain in three unflooded forests and in three flooded forests of the Western Amazon in Peru. I address these questions: (1) What species of seed are dispersed in these common Amazon forest types, and are there species in common?, (2) How does the number of seeds vary over a one year time period within the three unflooded forests, within the three flooded forests, and between the unflooded forests and the flooded forests?, (3) Does the seed rain of common unflooded forests in the Amazon differ significantly in total number of seeds, species richness, species evenness, and Fisher's α diversity over a one year sampling?, (4) Does the seed rain of common flooded forests in the Amazon differ significantly in total number of seeds, species richness, species

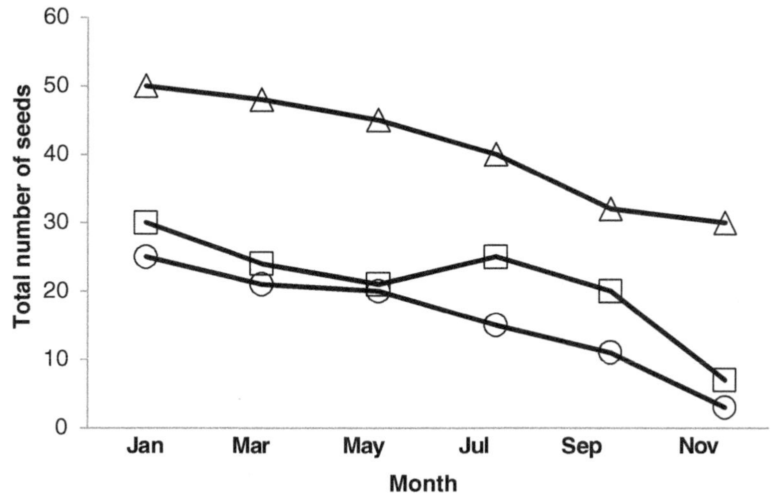

Figure 1 Mean number of seeds (over the 10 traps in each forest type) plotted over the one year collection period by month for the three unflooded forests: *terra firme* (triangle), white sand – varillal (rectangle), and palm (circle).

evenness, and Fisher's α diversity over a one year sampling?, and (5) Does the seed rain of the unflooded forests taken together, and the seed rain of the flooded forests taken together, differ significantly in total number of seeds, species richness, species evenness, and Fisher's α diversity over a one year sampling?

Methods

The study site was the Centro de Investigacion de Jenaro Herrera (CIJH) biological station operated by the Instituto de Investigaciónes de la Amazónia Peruana (www.iiap.org.pe/jenaro.htm) located 2.5 km from the town of Jenaro Herrera and 200 km south of Iquitos on the east margin of the Ucayali River in the Province of Requena, Department of Loreto, Peru (4° 54' S, 73° 40' W: Spichiger et. al. 1996; Honorio 2006). CIJH comprises an area of 4,336 km^2 with forest types relatively close to each other (Honorio 2006). The mean annual temperature is 26.0°C with a range between 25.1°C and 26.5°C and the mean annual rainfall is 2724 mm with a rainy season between the months of November and April (Kalliola et al. 1991). Dominating the unflooded portions of the CIJH are broad leaf forest (*terra firme*), white sand-varillal forest and low terrace palm forest. Common genera in the *terra firme* forest include *Eshweilera, Pouteria, Oenocarpus, Miconia*, and *Protium*, common genera in white sand-varillal forest include *Pachira, Haplolathra*, and *Macrolobium* and common species in palm forest include *Mauritia flexuosa, Oenocarpus bataua, Euterpe precatoria* and *Socratea exorrhiza* (Honorio 2006).

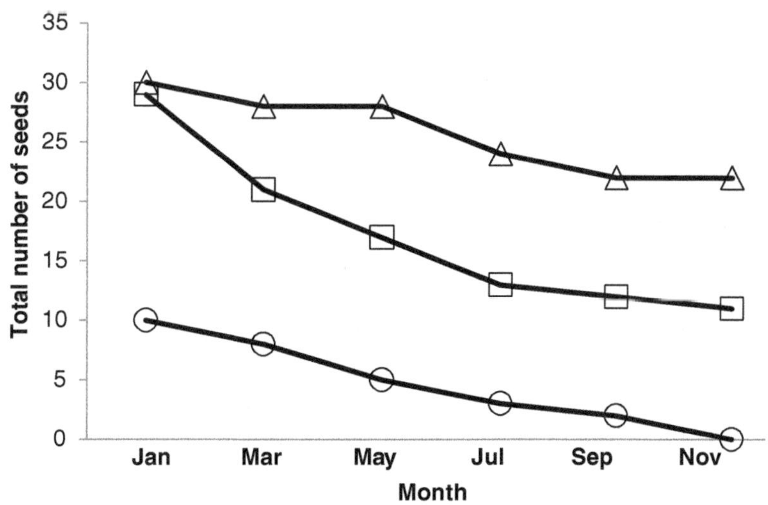

Figure 2 Mean number of seeds (over the 10 traps in each forest type) plotted over the one year collection period by month for the three flooded forests: high restinga (triangle), low restinga (rectangle), and tahnampa (circle).

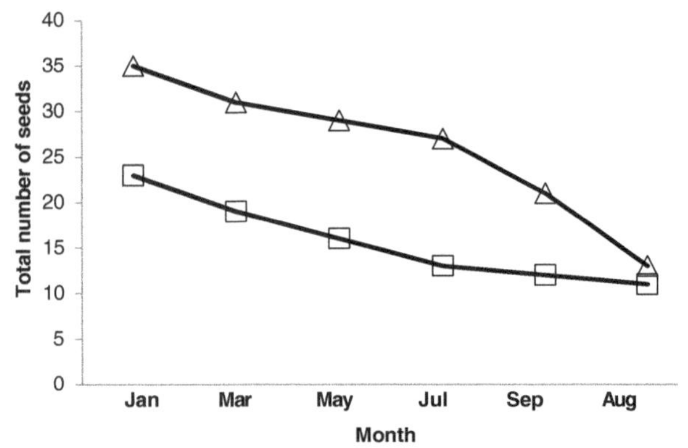

Figure 3 Mean number of seeds (over the 30 traps in the three unflooded forests and the 30 traps in the flooded forests) plotted over the one year collection period by month for the unflooded forests (triangle) and the flooded forests (rectangle).

Clay soils dominate *terra firme* forest and white sand forest soils consist mainly of quartz with individual trees having slender boles and roots concentrated at the soil surface (Klinge et al. 1990). Palm swamps occur in depressions or low-lying patches with poor drainage (Montufar and Pintaud 2006).

There are also forests that border CIJH white-water rivers and flood during the rainy season. These include forests which are underwater on average 1–2 months per year (called here high restinga), forests which are underwater on average 3–5 months per year (called here low restinga), and forests which are underwater at least 6 months per year (called here tahuampa). Common tree genera in Peru high restinga forests are *Chorisia*, *Eschweilera*, *Hura*, *Spondias*, and *Virola*, and common tree species in Peru low restinga forests and tahuampa forests include *Calycophyllum spruceanum*, *Ceiba samauma*, *Inga* spp., *Cedrela odorata*, *Copaifera reticulata*, *Phytelephas macrocarpa* with understorey palms such as *Scheelea* spp., *Guazuma rosea*, and *Piptadenia pteroclada* (Myster 2007a; Myster 2010).

In May 2010, I established a 50 m transect in each of six unlogged unflooded forests (two *terra firme* forests, two white sand-varillal forests, two palm forests) and in each of six unlogged flooded forests (two high restinga forests, two low restinga forests, two tahaumpa forests). Every 10 m along each transect I placed a cloth seed trap of 1 m^2 collecting area and 0.5 m depth, for a total of 5 traps per forest. Setting traps on transects at regular intervals, rather than in subjectively chosen microsites, better capture the variation in forest floor conditions that determine recruitment after dispersal, such as light, litter, and water availability (Dalling et. al. 1998; Myster 2012a). The traps were constructed with 5 cm^2 wire mesh on top, cloth underneath for collecting the seeds, and suspended on 1 m plastic poles for keeping the traps off the ground to reduce predation. This design has proven effective in several past Neotropic seed rain studies (e.g., Myster 2004; Myster 2007b) and predation is very rare for seed traps of this size (Muller-Landau et. al. 2008). No seed traps were flooded during the sample period.

Seeds were collected in July 2010, September 2010, November 2010, January 2011, March 2011 and May 2011. All seeds of trees and shrubs were removed from each trap during a collection period and no evidence of seed decay was observed. The design allowed an investigation into both the natural temporal heterogeneity of the seed rain over time and the natural spatial heterogeneity of the seed rain along the forest floor. Seeds were identified to species, or to genus in rare cases, using Gentry (1993), Romoleroux et al. (1997) and the on-site CIJH herbarium.

For each forest type (10 traps x 6 sampling periods per forest type), total number of seeds, species richness (total number of seed species), and seed species evenness (using Pielou's J index: Ludwig and Reynolds 1988) were calculated. Because richness trends could have been undermined by dissimilar densities (i.e., some plots could have had more species solely due to the larger number of seeds sampled there), Fisher's α diversity (Fisher et al. 1943) - an index independent of sample size and suggested for species-rich forests including those in the Amazon (Valencia et al. 1994; Valencia et al. 2004) - was also computed, using the iterative procedure described in Rosenzweig (1995).

These data were then used for one-way analysis of variance (ANOVA: SAS 1985) tests for each of the four seed community parameters (1) among the three unflooded forest types using all seed traps (10 microsites x 6 sampling periods = 60 replicates for each of the three forest types), (2) among the three flooded forest types using all seed traps (10 microsites x 6 sampling periods = 60 replicates for each of the three forest types) and (3) between all unflooded forests vs. all flooded forests (180

Table 2 F statistic summary table for the one-way ANOVAs using the response variables of total number of seeds, species richness (total number of seed species), seed species evenness (using Pielou's J index: Ludwig & Reynolds 1988) and Fishers α diversity

	Total number of seeds	Species richness	Species evenness	Fishers α diversity
Among unflooded forests	40.29***	5.63*	1.34	1.76
Among flooded forests	26.73***	0.11	1.98	7.58**
Unflooded forests vs. flooded forests	8.55**	6.01*	4.43*	2.19

The p-value is indicated as *if 0.05 < p < 0.01, as **if 0.01 < p < 0.001, and as ***if p < 0.001.

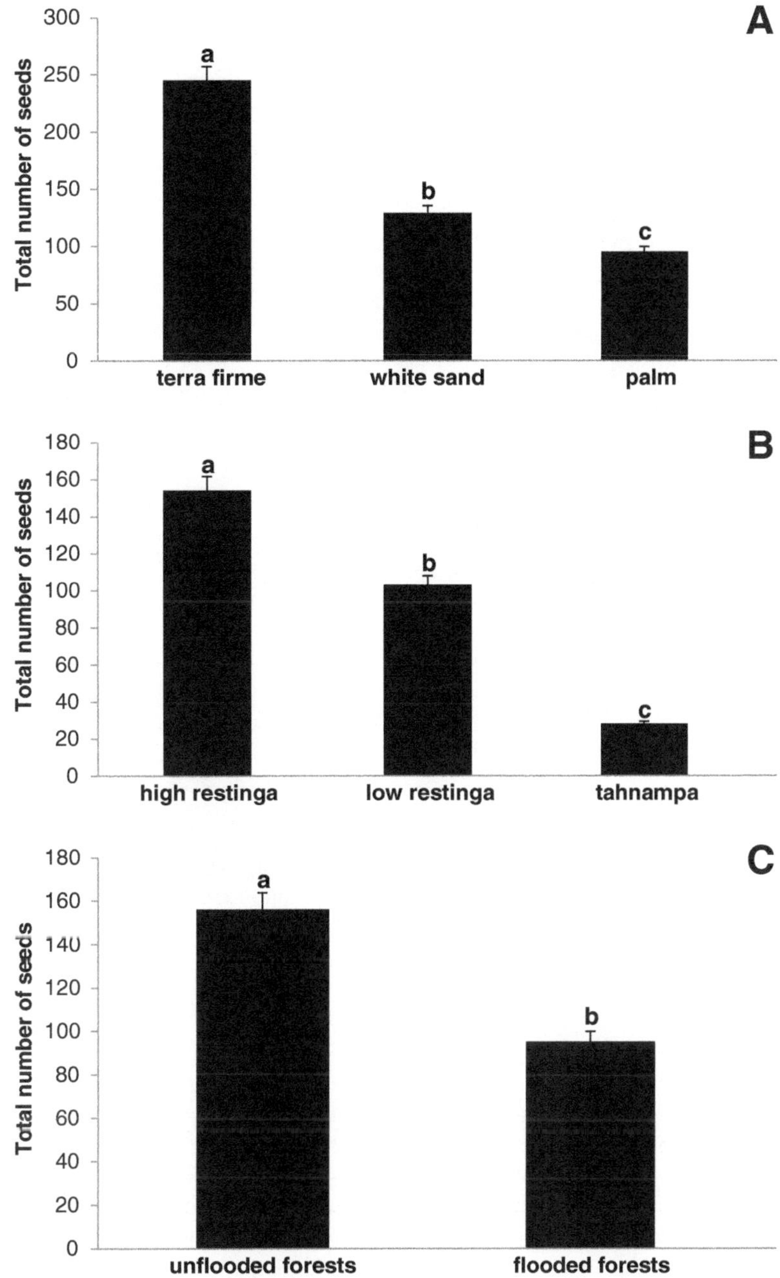

Figure 4 Mean and standard error bar graphs for the significant results from Table 2 involving total number of seeds (A) among unflooded forests, (B) among flooded forests, and (C) unflooded vs. flooded forests. Means testing results are indicated by lowercase letters which are different, if groups were significantly different, as given in Table 2.

replicates for flooded forests and 180 replicates for unflooded forests). Because seed traps were emptied completely after each collection period, and therefore could not influence the next sample collected in the same trap, repeated measure ANOVA was not necessary. If significance was found, means tests were conducted with the Tukey procedure (SAS 1985). All data were examined and found to be normally distributed.

Results

All forest types had dispersed seed of tree species that were not found in any of the other forest types: *terra firme* (*Casha moeria, Eschweilera coriacea, Eschweilera bracteosa, Minguartia guianensis, Ocotea* sp., *Oenocarpus batana, Oxandra espintana, Pouteria caimito*), white sand (*Caraipa punctulata, Dicymbe puncticulosa, Haploclathra paniculata, Schefflora* sp., *Pachira brevipes*),

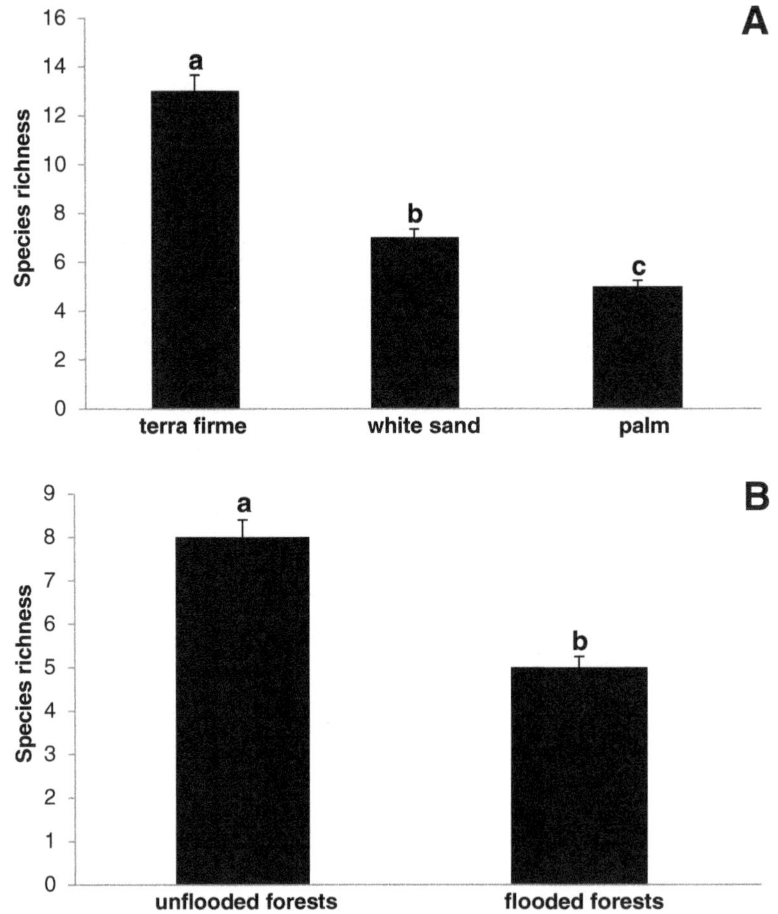

Figure 5 Mean and standard error bar graphs for the significant results from Table 2 involving species richness (A) among unflooded forests and (B) unflooded vs. flooded forests. Means testing results are indicated by lowercase letters which are different, if groups were significantly different, as given in Table 2.

palm (*Euterpe precatoria, Mauritia flexuosa*), high restinga (*Attalea butyracea, Drypetes amazonia, Leonia glycycarpa, Qualea paraensis*), low restinga (*Guarea macrophylla, Pourouma acurinata*) and tahuampa (*Campsiandra augustifolia* and seeds of the family Miristhicaceae*:* Table 1). There were few seed species shared among forest types. No species was found in all six forest types. The most common species were *Maquira coriacea* and *Virola paronis*, which were found in four of the six.

Total seed load peaks in the early part of the year and decreases monotonically over time for all forest types (Figures 1, 2 and 3). Within that general pattern, *terra firme* forest had significantly more seeds than the other unflooded forests (Table 2, Figure 4A), high restinga and low restinga had significantly more seeds than tahuampa (Table 2, Figure 4B) and unflooded forests had significantly more seeds than flooded forests (Table 2, Figure 4C).

Species richness was also significantly different among unflooded forests, and also between unflooded vs. flooded forests (Table 2). *Terra firme* forest had significantly more seed species as the other unflooded forests (Figure 5A) and unflooded forests had significantly more seed species than flooded forests (Figure 5B). The distribution of seed abundances among the species (species evenness) was significantly more in the unflooded vs. the flooded forest (Figure 6A). Finally, Fishers α diversity was significantly greater in high and low restinga compared to tahuampa (Figure 6B).

Discussion

The seed rain sampling was similar to other Neotropical seed rain studies (e.g. Dalling et. al. 2002; Muller-Landau et. al. 2008) in showing only a modest correspondence between the seed rain sampled in a forest and the trees sampled close by in 1 ha permanent plots (Myster 2013, Myster in press). This suggests that rare species, seed rain in other years, and/or other regeneration mechanisms and tolerances (Myster 2012b, Myster 2014) may play major roles in forest regeneration (Muller-Landau

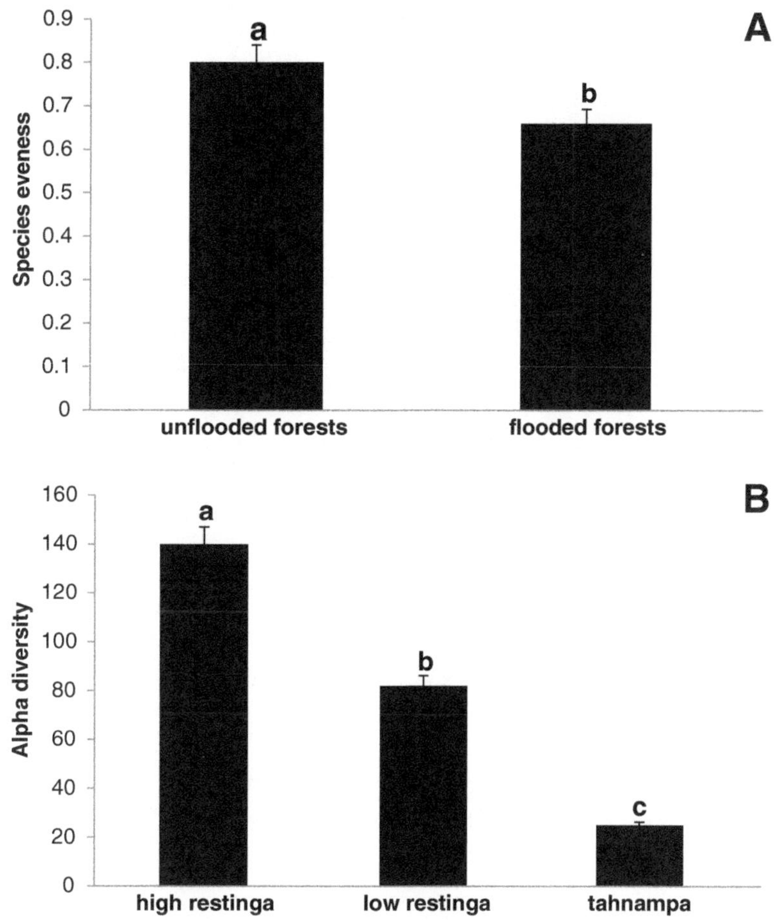

Figure 6 Mean and standard error bar graphs for the significant results from Table 2 involving species evenness (A) unflooded vs. flooded forests, and involving Fishers α diversity (B) among flooded forests. Means testing results are indicated by lowercase letters which are different, if groups were significantly different, as given in Table 2.

et. al. 2008). This result, along with studies of forest seed predation and seedling competition (e.g., conducted on site at CIJH: Notman et. al. 1996, author unpub. data), shows the complexities of forest recruitment and the major challenge for modelers who wish to predict the plant-plant replacements (Myster 2012b) which result in forest community patterns. As an example of this complexity, this is the first major sampling of the seed rain in flooded forests where secondary dispersal by fish is common (Parolin et. al. 2004).

Results agree with these other studies: (1) *terra firme* has much larger numbers of seed compared to the more open and less diverse white sand and palm forests (Honorio 2006), (2) less flooded forests have larger numbers of seeds compared to the more flooded – and open but less diverse – forests (Myster 2007a; Myster 2010) and (3) seed rain peaked after the start of the rainy season (Parolin et. al. 2004). Results also show, however, how different these forests are in seed species composition when compared to other Neotropical forests. For example, there were no species or genera in common with one sampling of a large plot in Panama (Dalling et. al. 2002) or Costa Rica (Loiselle et. al. 1996) and only a few similar genera with another Panama sampling (Muller-Landau et. al. 2008). Seed rain density was only slightly less in this *terra firme* forest compared to the same forest type in Panama.

There were much less seeds in areas recovering from agriculture both in Ecuador and Puerto Rico (Myster 2004; Myster 2007b), especially wind-dispersed tree species. After logging on site (CIJH: Gorchov et. al. 1993) there were also large declines in seed load from forest to center of logging strips with many tree succession genera also seen after agriculture (e.g., *Cecropia*, *Miconia*, *Piper* : Myster 2004). Lower montane *terra firme* forest in Puerto Rico and Ecuador has similar number of seeds, but less species with *Miconia* sp. and *Guarea* sp. found in common. Fisher's α diversity and species richness trends were very similar to past sampling in these kinds of forests (Myster 2010).

Whereas species diversity is lower in flooded forests compared to unflooded, adaptations in flooded trees may be remnants of preadaptions from the unflooded *terra firme* species where floodplain trees originated (Parolin et. al. 2004). Results show that increased flooding and decreased soil fertility both lead to reductions in seed rain which reflect loss of forest diversity and complexity (Myster 2007a; Myster 2010), and results further suggest that seed rain may be more reduced by loss of soil fertility than by increased flooding. Because differences among unflooded forests are due mainly to soil fertility (Tuomisto et. al. 2003; author unpub. data), results may help illustrate these two major structuring forces in Amazonian forests: flooding and soil fertility.

Conclusions

I conclude for the unflooded forests that seed species number and richness increased with soil fertility but for the flooded forests seed species number and richness decreased with months under water. I also conclude that seed rain peaks right after the start of the rainy season. Finally, when taken together, results suggest that for forests across the Amazonian landscape differences in flooding regime may have a greater effect on both seed rain load and seed species richness than differences in availability of soil nutrients.

Competing interests

The author declares that he has no competing interests.

Acknowledgments

I thank Eurdice Honorio and Julio Irarica of CIJH for their help in the logistics of this sampling.

References

Dalling, JW, Hubbell, SP, & Silvera, K. (1998). Seed dispersal, seedling establishment and gap partitioning among tropical pioneer trees. *Journal of Ecology, 86*, 674–689.

Dalling, JW, Mueller-Landau, HC, Wright, SJ, & Hubbell, SP. (2002). Role of dispersal in the recruitment limitation of Neotropical Pioneer species. *Journal of Ecology, 90*, 714–727.

Daly, DC, & Prance, GT. (1989). *Brazilian Amazon* (pp. 401–426). Bronx, NY: In Floristic inventory of tropical countries. NY Botanical Garden.

Everham, EM, III, Myster, RW, & VanDeqnachte, G. (1996). Effects of light, moisture, temperature, and litter on the regeneration of five tree species in the tropical montane wet Forest of Puerto Rico. *American Journal of Botany, 83*, 1063–1068.

Fisher, RA, Corbet, AS, & Williams, CB. (1943). The relation between the number of species and the number of individuals in a random sample of an animal population. *Journal of Animal Ecology, 12*, 42–58.

Gentry, A. (1993). *A field guide to woody plants of northwest South America (Colombia, Ecuador, Peru)*. Washington, DC: Conservation International.

Gorchov, DL, Cornejo, F, Ascorra, C, & Jaramillo, M. (1993). The role of seed dispersal in the natural regeneration of rain forest after strip-cutting in the Peruvian Amazon. *Vegetatio, 107*(108), 339–349.

Honorio, EN. (2006). *Floristic relationships of the tree flora of Jenaro Herrera, an unusual Area of the Peruvian Amazon*. M.S. thesis: University of Edinburgh, Edinburgh, UK.

Junk WJ. (1997) The Central Amazon Floodplain – Ecology of a Pulsing System. Springer -Verlag, Berlin. Ecological studies 126.

Kalliola, RS, Jukka, M, Puhakkaa, M, & Rajasilta, M. (1991). New site formation and colonizing vegetation in primary succession on the western Amazon floodplains. *Journal of Ecology, 79*, 877–901.

Keller, M, Alencar, A, Asner, AGP, Braswell, B, Bustamante, M, Davidson, E, Feldpausch, T, Fernandes, E, Goulden, M, Kabat, P, Kruijt, B, Luizao, F, Miller, S, Markewitz, D, Nobre, AD, Nobre, CA, Filho, NP, Da Rocha, H, Dias, PS, VonRandoq, C, & Vourlitis, GL. (2004). Ecological research in the large-scale biosphere atmosphere experiment in Amazonia: Early results. *Ecological Applications, 14*, S3–S16.

Klinge, H, Junk, WJ, & Revilla, CJ. (1990). Status and distribution of forested wetlands in tropical South America. *Forest Ecology and Management, 33/34*, 81–101.

Loiselle, BA, Ribbens, E, & Vargas, O. (1996). Spatial and temporal variation of seed rain in a tropical lowland wet forest. *Biotropica, 28*, 82–95.

Ludwig, JA, & Reynolds, JK. (1988). *Statistical ecology: a primer on methods and computing*. NY: Wiley & sons.

Mesquita, RC, Ickes, GK, Ganade, G, & Williamson, GB. (2001). Alternative successional pathways in the Amazon Basin. *Journal of Ecology, 89*, 528–537.

Montufar, R, & Pintaud, J. (2006). Variation in species composition, abundance and microhabitat preferences among western Amazonian *terra firme* palm communities. *Botanical Journal of the Linnean Society, 151*, 127–140.

Muller-Landau, HC, Wright, SJ, Calderon, O, Condit, R, & Hubbell, SP. (2008). Interspecific variation in primary seed dispersal in a tropical forest. *Journal of Ecology, 96*, 653–667.

Myster, RW. (2001). Mechanisms of plant response to gradients and after disturbances. *Botanical Review, 67*, 441–452.

Myster, RW. (2004). Regeneration filters in post-agricultural fields of Puerto Rico and Ecuador. *Plant Ecology, 172*, 199–209.

Myster, RW. (2007a). Interactive effects of flooding and forest gap formation on composition and abundance in the Peruvian Amazon. *Folia Geographica, 42*, 1–9.

Myster, RW. (2007b). Early successional pattern and process after sugarcane, banana and pasture cultivation in Ecuador. *NZ Jornal Botanical, 45*, 101–110.

Myster, RW. (2009). Plant communities of western Amazonia. *Botanical Review, 75*, 271–291.

Myster, RW. (2010). Flooding duration and treefall interactive effects on plant community richness, structure and alpha diversity in the Peruvian Amazon. *Ecotropica, 16*, 43–49.

Myster, RW. (2012a). Spatial and temporal heterogeneity of light and soil water along a *terra firma* transect in the Ecuadorian Amazon. *Canadian Journal of Forest Research, 42*, 1–4.

Myster, RW. (2012b). Plants replacing plants: the future of community modeling and research. *Botanical Review, 78*, 2–9.

Myster, RW. (2013). The effects of flooding on forest floristics and physical structure in the Amazon: results from two permanent plots. *Forest Research, 2*, 112. doi:10.4172/2168-9776.1000112.

Myster, RW. (2014). Interactive effects of flooding and treefall gap formation on *terre firme* forest seed and seedling mechanisms and tolerances in the Amazon. *Comm Ecol, 15*, 212–221.

Myster, R. W. in press. Várzea forest vs. *terra firme* forest floristics and physical structure in the Ecuadorean Amazon. Ecotropica

Myster, RW, & SantaCruz, PG. (2005). Una comparación de campo de insectos de suelo -morar de Amazonas: Tierra firme y bosques de tierras inundadas vs. espacios abiertos en el Parque Nacional Yasuní, Ecuador. *Revista Pontanti Universida Católica, 76*, 111–124.

Notman, E, Gorchov, DL, & Cornejo, F. (1996). Effect of distance, aggregation, and habitat on levels of seed predation for two mammal-dispersed Neotropical rain forest tree species. *Oecologia, 106*, 221–227.

Parolin, P, Desimobne, O, Haase, K, Waldhioff, D, Rottenberger, S, Kuhn, U, Kesselmeier, J, Kleiss, B, Schmidt, W, Piredade, MTF, & Junk, WJ. (2004). Central Amazonian floodplain forests: tree adaptations in a pulsing system. *Botanical Review, 70*, 357–380.

Pires, JM, & Prance, GT. (1985). *The vegetation types of the Brazilian Amazon* (pp. 109–145). In: Amazonia Pergamon Press, Oxford.

Pitman, NC, Terborgh, JW, Silman, MR, Nunez, PV, Neill, DA, Ceron, CE, Palacios, WA, & Aulestia, M. (2001). Dominance and distribution of tree species in upper Amazonian *terra firme* forests. *Ecology, 82*, 2101–2117.

Romoleroux K., Foster R., Valencia R., Condit R., Balslev H. & Losos E. (1997): Especies lenosas (dap = > 1 cm) encontradas en dos hectareas de un bosque de la Amazonia ecuatoriana. Pp 189–215 in Valencia, R. and H. Balslev (eds) Estudios Sobre Diversidad y Ecologia de Plantas. Pontificia Universidad Catolica del Ecuador, Quito, Ecuador.

Rosenzweig, ML. (1995). *Species diversity in space and time*. Cambridge UK: Cambridge University Press.
SAS. (1985). *User's Guide: Statistics, Ver. 5* (p. 956). Cary, NC: SAS Institute Inc.
Spichiger, R, Loizeau, P, Latour, C, & Barriera, G. (1996). Tree species richness of a South -Western Amazonian forest (Jenaro Herrera, Peru, 73°40'W/4°54'S). *Candollea, 51*, 559–577.
Tuomisto, H, Ruokolainen, K, & Yli-Halla, M. (2003). Dispersal, environment and floristic variation of Western Amazonian forests. *Science, 299*, 241–244.
Valencia, R, Balslev, H, Guillermo, P, & Mino, C. (1994). High tree alpha diversity in Amazonian Ecuador. *Biological Conservation, 3*, 21–28.
Valencia, R, Foster, RB, Villa, G, Condit, TR, Svenning, J, Hernandez, C, Romoleroux, K, Losos, E, Magard, E, & Balslev, SI. (2004). Tree species distributions and local habitat variation in the Amazon: large forest plot in eastern Ecuador. *Journal of Ecology, 92*, 214–229.
Walter, H. (1973). *Vegetation of the earth and the ecological systems of the geo-biosphere*. Berlin: Springer-Verlag.

The evaluation of aerially applied triclopyr mixtures for the control of dense infestations of wilding *Pinus contorta* in New Zealand

Stefan Gous[1*], Peter Raal[2] and Michael S Watt[3*]

Abstract

Background: Wilding conifers are a major threat to large areas of land throughout New Zealand where they compete with native vegetation, modify ecosystems, reduce available grazing land, limit future land-use options and visually change landscapes. Wilding conifers can also enhance damaging fires and in water limited catchments can affect surface flows and aquifer recharge.

A range of herbicide treatments was aerially applied by boom to a field trial established within a wilding *Pinus contorta* (Dougl.) (height range 1.5 – 12 m) infestation. Measurements of mortality taken two years post herbicide application were used to examine variation in efficacy of these herbicides using a mixed-effects model that tested the main and interactive effects of tree height and herbicide treatment on mortality. For a treatment to be considered effective, a mortality rate of over 85% should be achieved.

Findings: Herbicide treatment was found to significantly affect mortality. Application of herbicides that included triclopyr (18 kg a.i. ha^{-1}) and dicamba (5 kg a.i. ha^{-1}) applied in combination or these active ingredients applied with picloram (2 kg a.i. ha^{-1}) were found to effectively control *P. contorta* across the treated height range inducing respective mortality rates of 88 and 89%. Tree mortality rates for these two herbicides were significantly greater than for two treatments with only triclopyr (37%) or triclopyr and picloram (66%) used at the previously stated rates. Neither height nor the interaction of height and treatment affected mortality over the tested range of heights.

Conclusions: This research has shown that dense infestations of *P. contorta* can be successfully controlled using aerial broadcast application of triclopyr-based herbicides applied in a high volume (400 L ha^{-1}) mixture, with extremely large droplets (400 – 500 μm). The Department of Conservation has adopted the most effective formulation for operational use.

Keywords: Aerial boom spraying; Herbicides; Wilding conifers

Background

Exotic conifer species were planted for erosion control, research, shelter, landscaping and production forests from the late 1880s onwards (Ledgard 2001) throughout New Zealand. Most of these conifers are primary colonisers and have naturally regenerated from these plantings. These conifers, known as wildings, have spread extensively since then, and the total area in which wilding conifers occur in the South Island of New Zealand is estimated to be in excess of 500,000 ha (Raal and Gous 2010). Wilding conifers occur within more than 200,000 hectares of land administered by the Department of Conservation (DOC), of which approximately two thirds is invaded by *Pinus contorta* (Dougl.) (Ledgard 2001).

Wilding conifers are controlled either by mechanical (brush cutters and chainsaws), physical (hand pulling and slashing) or chemical (herbicides) methods. For dense infestations (greater than 80% canopy cover), aerial boom spray applied chemical control is a cost effective and

* Correspondence: stefan.gous@scionresearch.com; michael.watt@scionresearch.com
[1]Scion, Forest Biosecurity and Protection, Private Bag 3020, Rotorua, New Zealand
[3]Scion, PO Box 29237, Fendalton, Christchurch, New Zealand
Full list of author information is available at the end of the article

practical solution (Raal and Gous 2010). However, the feasibility of boom sprayed chemical control for invasive wilding conifers in New Zealand has had variable success to date (Raal and Gous 2010).

Potential herbicides for controlling wildings have been identified through pot trials (Gous 2010a, 2010b). The selective and systemic herbicides triclopyr ester (3,5,6-trichloro-2-pyridyloxyacetic acid) and picloram (4-amino-3,5,6-trichloropicolinic acid), applied alone (triclopyr ester) or in combination (triclopyr ester and picloram) and the non-selective systemic combination of glyphosate (*N*-(phosphonomethyl)glycine) and metsulfuron (2-[[[(4-methoxy-6-methyl-1,3,5-triazin-2-yl)amino]-oxomethyl]sulfamoyl]benzoic acid methyl ester) provided best control of small (height of ~0.3 m) *P. contorta*, *P. mugo* (Turra) and *Pseudotsuga mensiesii* (Mirb.).

Although these herbicides have been shown to be effective on young juvenile conifers (Gous 2010a, 2010b), little research has been conducted using these herbicides on mature conifers, growing in dense infestations where trees can reach heights of up to 15 m. Recent research shows that these herbicides have low efficacy (≤64% mortality) on large wilding *P. contorta* (height range 1 – 15 m) and *P. mugo* (height range 0.5 – 5 m) when they are mixed with water and aerially applied at low rates (150 L ha^{-1}) (Gous et al. 2014). Gous et al. (2014) concluded that application rates of water had to be high enough to ensure complete coverage of the foliage. As a consequence of these results, the objective of this research was to examine the efficacy of a range of systemic herbicides, applied in high water rates (400 L ha^{-1}), with wetting agents, to control dense field infestations of *P. contorta*.

Methods

Sites, treatments and application

The site selected for the *P. contorta* trial was located at Ferintosh Station (latitude 44° 6′ 00″ S; longitude 170° 07′ 29″), near Twizel, New Zealand. The herbicide treatments were applied as water-based mixtures at 400 L ha^{-1} using a calibrated helicopter boom spray and an extremely coarse (ASABE 2009) droplet spectrum of approximately 400 - 500 μm, that was determined by the manufacturer using tests with water. Application was undertaken with a MD 520N helicopter, flying at a ground speed of 33 knots with a release height of 10 m above the tree canopy. Sufficient spray was carried to treat a single plot (ca. 200 litres) which was within the maximum payload of 400 – 500 litres of the MD 520N helicopter.

The helicopter was fitted with 45 × AIC 11010 nozzles (Spraying Systems Co 2006), orientated straight back, at an angle of 0°, evenly spaced along the boom to within 80% of the rotor diameter. At the ground speed at which the helicopter was flown, the required flow rate was 163 L min^{-1}. This is equivalent to a flow rate of 3.6 L min^{-1} nozzle^{-1} and is delivered at a pressure of ca. 2.5 bar. Flying height and the flight lane separation of 4 m was measured and maintained by global positioning system (GPS) and spray-system navigation equipment.

Four treatments (Table 1) were selected based on results from previous herbicide screening and field trials (Gous 2010a, 2010b; Gous et al. 2014). These treatments were based on triclopyr ester and included combinations of this active ingredient with picloram and dicamba (Table 1) (AgriMedia Ltd 2010). Ammonium sulfate, Kwickin spray oil (Surface Science Technology, Auckland, New Zealand) and Companion surfactant (Surface Science Technology, Auckland, New Zealand) were added to the spray solution to enhance herbicide uptake (Nalewaja and Matysiak 1993) and increase canopy coverage. Each treatment was applied to three replicate 0.5 ha treatment plots using a randomised complete block design. A control of three replicate 0.5 ha plots, that received no treatment was also included in the design.

Treatments were applied on 14 December 2010, during the period of active growth to promote the uptake and translocation of herbicide throughout the trees (Radosevich

Table 1 Description of the herbicide treatments (and corresponding active ingredients) applied to *Pinus contorta* by broadcast aerial application within a total spray volume of 400 L ha^{-1}

Treatment	Herbicide product	Active ingredient
Code[1]	(quantity ha^{-1})	(kg a.i. ha^{-1})
T	30 L Grazon,	18 kg triclopyr
	20 L Kwickin oil	
	0.5 L Companion surfactant	
	4 kg Ammonium sulphate	
TP	20 L Grazon	18 kg triclopyr
	20 L Conquest	2 kg picloram
	20 L Kwickin oil	
	0.5 L Companion surfactant	
	4 kg Ammonium sulphate	
TD	20 L Grazon	18 kg triclopyr
	25 L Banvel	5 kg dicamba
	20 L Kwickin oil	
	0.5 L Companion surfactant	
	4 kg Ammonium sulphate	
TPD	20 L Grazon	18 kg triclopyr
	20 L Conquest	2 kg picloram
	25 L Banvel	5 kg dicamba
	20 L Kwickin oil	
	0.5 L Companion surfactant	
	4 kg Ammonium sulphate	

[1]Treatment codes: T is triclopyr, P is picloram and D is dicamba.

and Bayer 1979). Mean tree heights at the time of treatment ranged from 1.5 – 12.0 m (Table 2). The range in tree height was similar for all treatments (Table 2).

Damage assessment

Thirty-three trees were randomly selected from each treatment plot and marked for subsequent assessment of damage. Tree mortality was recorded two years following the treatment as the percentage dead foliage in increments of 10%. The crown of each tree was visually assessed from the ground by dividing it into three equal sections from top to bottom, and each section scored individually, then averaged to obtain a whole-tree score. A tree with 100% dead foliage was scored as dead. A treatment was considered effective when a mortality rate of over 85% was reached. Tree height was measured using a Suunto clinometer (Suunto, Vantaa, Finland).

Analysis

All analyses were undertaken using SAS software (SAS Institute Inc. 2008). Percentage mortality was the dependent variable used in analyses and this variable was transformed for analysis using an arcsine square root transformation to meet the underlying assumptions of the models used. Within each plot, mean mortality in 2 m height classes was used within analyses so that the effects of treatment and tree height on mortality could be examined.

Using this data, a mixed-effects model was used to test the effects of treatment, height class and the interaction between treatment and height class on transformed mortality. Replicate was included in the model as a random effect while treatment and height class were treated as fixed effects. All variables were included within the model as class-level variables. Where the treatment effect was significant, multiple comparisons were undertaken by examining the significance of least square differences using a t-test.

Findings

No mortality (0%) was recorded for the control treatment. Analyses showed herbicide treatment to have a strong significant effect on mortality (Table 3). Neither height class nor the interaction between height class and treatment were significant (Table 3) indicating that changes in mortality across height classes were relatively similar between treatments.

Table 3 Analysis of variance showing the main and interactive effects of height class and treatment on percentage mortality

Effect	Num DF	Den. DF	*F*-value	*p* value
Treatment (*T*)	3	30	14.4	<0.0001
Ht class (*H*)	4	30	1.29	0.297
H x *T*	10	30	0.32	0.967

Also shown is the numerator (Num. DF) and denominator (Den. DF) degrees of freedom.

The treatment containing triclopyr, dicamaba and picloram induced the highest mortality of the four tested herbicide combinations (Table 4). However, it is worth noting that there was no significant difference in mortality between this treatment and the treatment with only triclopyr and dicamba (89 vs 88%) suggesting that addition of picloram had little impact on tree mortality. Triclopyr, dicamba and picloram have little effect on monocotyledon plants, therefore the use of these herbicides may promote re-establishment of treated areas into native tussocks in areas where grassland is the dominant vegetation. The inclusion of picloram within the mixture may be beneficial as this active ingredient has long persistence and activity in the soil compared with triclopyr and dicamba. Previous research has shown that picloram has residual soil activity of up to 450 days after treatment (MacDiarmid 1975) which may provide an extended period over which germinating conifer seed and the wilding conifer can be killed.

Either an application of triclopyr alone or a mixture of triclopyr and picloram without dicamaba was significantly less effective at controlling *P. contorta* (inducing respective mortality of 66 and 37%) than the treatment containing triclopyr, picloram and dicamaba, where mortality was 89% (Table 4). These results suggest that dicamba is a far more effective mixing partner than picloram in herbicide mixtures with triclopyr.

In conclusion, this research showed that dense infestations of *P. contorta* can be successfully controlled using aerial broadcast application of triclopyr-based herbicides

Table 2 Mean tree height and range (in brackets) at the time of spray application

Treatment	Tree height (m)
T	6.3 (1.5 - 9.0)
TP	6.5 (1.5 - 10.0)
TD	6.8 (2.0 - 12.0)
TPD	6.7 (2.0 - 12.0)

Treatment codes: T is triclopyr, P is picloram and D is dicamba.

Table 4 Mean mortality by treatment

Treatment code[1]	Mortality (%)
T	36.5 (8.62) A
TP	66.0 (5.48) B
TD	88.1 (3.97) C
TPD	89.1 (4.48) C

[1]Treatment codes: T is triclopyr, P is picloram and D is dicamba.
Shown are treatment means followed by standard errors. Treatment means followed by the same letter are not significantly different at $p < 0.05$.

applied in a high volume (400 L ha^{-1}) mixture, with extremely large droplets (400 – 500 μm). On the basis of this research, the Department of Conservation is operationally using the triclopyr, picloram and dicamaba formulation.

Competing interests
The authors declare that they have no competing interests.

Authors' contributions
SG was the primary author and co-designed the study. PR undertook all field measurements and co-designed the trial. MSW was the secondary author and undertook all analyses. All authors have read and approved the final manuscript.

Acknowledgements
This project was funded by the FRST contract Beating Weeds 2 (Contract No. C09X0905). Thanks to Peter Willemse, Department of Conservation, Twizel, for taking care of all logistics, i.e. site selection, herbicide delivery and helicopter availability.

Author details
[1]Scion, Forest Biosecurity and Protection, Private Bag 3020, Rotorua, New Zealand. [2]Department of Conservation, Otago Conservancy, PO Box 5244, Dunedin, New Zealand. [3]Scion, PO Box 29237, Fendalton, Christchurch, New Zealand.

References
AgriMedia Ltd. (2010). *New Zealand Novachem Agrichemical Manual*. New Zealand: Christchurch.

ASABE (2009). ASABE - S572.1, Spray nozzle classification by droplet spectra. American Society of Agricultural and Biological Engineers.

Gous, SF, Watt, MS, Richardson, B, & Kimberley, MO. (2010a). Herbicide screening pot trial for wildling conifer control (*Pinus contorta, P. mugo* and *Pseudotsuga menziesii). New Zealand Journal of Forestry, 55*, 11–14.

Gous, SF, Watt, MS, Richardson, B, & Kimberley, MO. (2010b). Herbicide screening trial to control dormant *Pinus contorta, P. mugo* and *Pseudotsuga menziesii* during winter. *New Zealand Journal of Forestry Science, 40*, 153–159.

Gous, S, Raal, P, & Watt, MS. (2014). Dense wilding conifer control with aerially applied herbicides in New Zealand. *New Zealand Journal of Forestry Science, 44*, 4.

Ledgard, N. (2001). The spread of lodgepole pine (*Pinus contorta*, Dougl.) in New Zealand. *Forest Ecology and Management, 141*, 43–57.

MacDiarmid, BN. (1975). Soil residues of picloram applied aerially to New Zealand brushweeds. *Proceedings of the New Zealand Weed and Pest Control Conference, 28*, 109–114.

Nalewaja, JD, & Matysiak, R. (1993). Influence of diammonium sulfate and other salts on glyphosate phytotoxicity. *Pesticide Science, 38*(2–3), 77–84. doi:10.1002/ps.2780380203.

Raal, P, & Gous, SF. (2010). *Literature Review of Herbicides to Control Wilding Conifers. (Scion report no. 17750)*. Rotorua, New Zealand: Scion.

Radosevich, SR, & Bayer, DE. (1979). Effect of temperature and photoperiod on triclopyr, picloram and 2,4,5-T translocation. *Weed Science, 27*, 22–27.

SAS Institute Inc. (2008). *SAS/STAT 9.2 User's Guide*. Cary NC: SAS Institute Inc.

Spraying Systems Co. (2006). *Teejet Catalog 50-M*. IL, USA: Wheaton.

25

Use of LiDAR to estimate stand characteristics for thinning operations in young Douglas-fir plantations

Michael S Watt[1*], Andrew Meredith[2], Pete Watt[2] and Aaron Gunn[3]

Abstract

Background: Light Detection and Ranging (LiDAR) has been successfully used to describe a wide range of forest metrics at local, regional and national scales. However, little research has used this technology in young Douglas-fir stands to describe key stand characteristics used as criterion for operational thinning. The objective of this research was to develop models of Douglas-fir mean top height, basal area, volume, mean diameter (at breast height), green crown height and stand density from LiDAR and stand information.

Methods: Data for this study were obtained from four widely separated young (age range of 9 to 17 years) Douglas-fir plantations in the South Island, New Zealand. LiDAR was acquired for the entire area and stand metrics were measured within 122 plots established across the study area. Spatially synchronous stand and LiDAR metrics were extracted from the plots. Using this dataset, multiple regression models were developed for each of the six stand metrics.

Results: The final models constructed for mean top height, green crown height, total stem volume, mean diameter, basal area, and stand density had R^2 values of 0.85, 0.79, 0.86, 0.86, 0.84 and 0.55, respectively, with root mean square errors of 1.02 m, 0.427 m, 20.2 m^3 ha^{-1}, 13.9 mm, 3.81 m^2 ha^{-1} and 355 stems ha^{-1}, respectively. With the exception of stand density, all relationships were relatively unbiased. Variables with the greatest contribution (with the partial R^2 in brackets) to models of mean top height, green crown height, volume, mean diameter and basal area included the 75th (0.85), 1st (0.76), 10th (0.83), 95th (0.74), and 10th (0.72) LiDAR height percentiles. The LiDAR height interquartile distance was the most important contributor (partial $R^2 = 0.33$) to the model of stand density.

Conclusion: With the exception of stand density, the final models for stand metrics were sufficiently precise to be used for scheduling thinning operations. This study demonstrates the utility of LiDAR to accurately estimate key structural attributes of young Douglas-fir and to assist with forest management over a widely dispersed resource.

Keywords: Airborne laser scanning; ALS; LiDAR; Forest inventory; Laser; *Pseudotsuga menziesii*

Background

High productivity and superior wood properties have made Douglas-fir (*Pseudotsuga menziesii* [Mirb.] Franco) one of the premier and most widely planted species throughout temperate forest areas. Globally, there are *ca.* 15 million hectares of Douglas-fir plantations with substantial areas occurring in Europe, South America, New Zealand, Australia and western North America (Hermann and Lavender 1999). Douglas-fir forest plantations are some of the most productive in the world (McMurtrie 1993).

Thinning is an important management operation and has a major impact on Douglas-fir stand development. Douglas-fir responds well to thinning and this operation is necessary to avoid stand stagnation and achieve merchantable sized logs over economically viable time frames. Typically, one to two thinnings are undertaken during a rotation and depending on the size of felled trees these are classed as either pre-commercial or commercial thinnings. A number of relatively simple (Reukema 1975; Emmingham and Green 2003) and

* Correspondence: michael.watt@scionresearch.com
[1]Scion, Fendalton, PO Box 29237, Christchurch, New Zealand
Full list of author information is available at the end of the article

more complex growth modelling methods (MacLaren and Knowles 2005) have been developed that simulate the impact of different thinning regimes on future growth and rotation end stand dimension. Important metrics describing the state of the current stand that affect thinning decisions include height, stand density (stocking), tree diameter at breast height of 1.4 m, volume, basal area and green crown height (Reukema 1975; Emmingham and Green 2003).

Stand-level inventory is the most widely used method for determining current characteristics and the need for thinning in Douglas-fir plantations. However, over the last two decades, traditional field based assessment has been progressively replaced or supplemented with Light Detection and Ranging (LiDAR) in a range of plantation species. LiDAR is a laser-based remote sensing technology that has been widely used to accurately predict stand height (Watt and Watt 2013; Coops et al. 2007; Means et al. 2000; Næsset 2002; Means et al. 1999) and volume (Watt and Watt 2013; Coops et al. 2007; Means et al. 2000; Næsset 2002; Means et al. 1999). Correlations of moderate to high strength have been found between LiDAR metrics and basal area (Næsset 2004b, 2005, 2002; Nord-Larsen and Schumacher 2012; Means et al. 1999; Means et al. 2000), stem diameter (Næsset 2002) and green crown height (Næsset and Okland 2002). However, stand density is typically only predicted with a moderate degree of precision from LiDAR cloud metrics (Næsset and Bjerknes 2001; Hall et al. 2005; Næsset 2002).

In contrast to traditional stand level inventory, LiDAR is often less expensive and more accurate (Næsset et al. 2004). Predictions from LiDAR metrics can be used to characterise spatial variation in tree characteristics allowing stratification of within-stand thinning operations. Despite the relatively wide adoption of LiDAR within the forestry sector, little research has used the technology to characterise a wide variety of stand characteristics in young Douglas-fir plantations.

The objective of this research was to investigate the utility of discrete LiDAR for predicting variation in important stand characteristics in young Douglas-fir plantations.

Methods

LiDAR data and stand measurements

The data used in this study were obtained from four Douglas-fir plantation forests in the South Island, New Zealand (Figure 1). These four forests named Doon, Pentland, Saddle Peak and Shag River had areas of 1393, 1526, 561 and 2685 ha, respectively. Each forest was divided into a number of stands, based on age, with stand ages respectively ranging from 9–14, 14–17, 9–12 and 14–16 years old at Doon, Pentland, Saddle Peak and Shag River.

The LiDAR survey was conducted using a fixed wing aircraft between May and August 2012 using a small footprint (~0.20 m) Optech ALTM 3100EA system and a swath overlap of 50%. The LiDAR and flight parameters used to achieve first-return densities that averaged 15.6 points m^{-2} (range 5.7 to 28.2 points m^{-2}) are summarised in Table 1. The system also utilised an Applanix 510 Position and Orientation System (POS) that uses the GPS and inertial measurement unit (IMU) sensors, and a GPS-based computer controlled navigation system.

Measurements of stand characteristics were made from July to October 2012 across the resource in plots ranging in area from 0.015 to 0.1 ha. High grade differential GPS units were used to locate the plots. After exclusion of plots with species other than Douglas-fir and four plots in which the digital terrain model was poorly defined, there were 122 plots available for the modelling. Of these 122 plots, 28, 29, 11 and 54 plots were respectively allocated to the Doon, Pentland, Saddle Peak and Shag River forests. Within each plot diameter (at breast height; standard height of 1.4 m used in New Zealand) and green crown height were recorded for all trees and heights were recorded for at least 10 trees. Stand density and basal area were determined for each plot and mean top height, H_m (height of the 100 largest diameter trees per hectare) was derived from plot mean height, H_{mean}, and stand density, S, using the following function,

$$H_m = (1/H_{mean}(1-0.106(1-\exp(-0.228(S-100)/100))))^{-1} \quad (1)$$

Total volume, V, was determined from H_m and basal area, B, using the following function,

$$V = B(0.928 + 0.3208\,H_m) \quad (2)$$

Analyses

Models were developed using SAS software (SAS Institute Inc. 2008), to estimate mean top height, green crown height, total stem volume, mean diameter, basal area and total stand density. Explanatory variables used in the modelling included LiDAR metrics, stand age and stand density (determined from plot measurements). The LiDAR metrics included a range of LiDAR height percentiles from the first (H_{01}) to the ninety-ninth (H_{99}), several metrics describing the LiDAR height distribution through the canopy (skewness, coefficient of variation (CV), standard error, kurtosis, elevation interquartile distance (IQ)) and the percentage of returns reaching within 0.5 m of the ground (PC_{zero}).

Initial exploratory analyses were undertaken to assess the strength of linear relationships between all dependant stand characteristics and the explanatory variables. These models were then extended through the development of

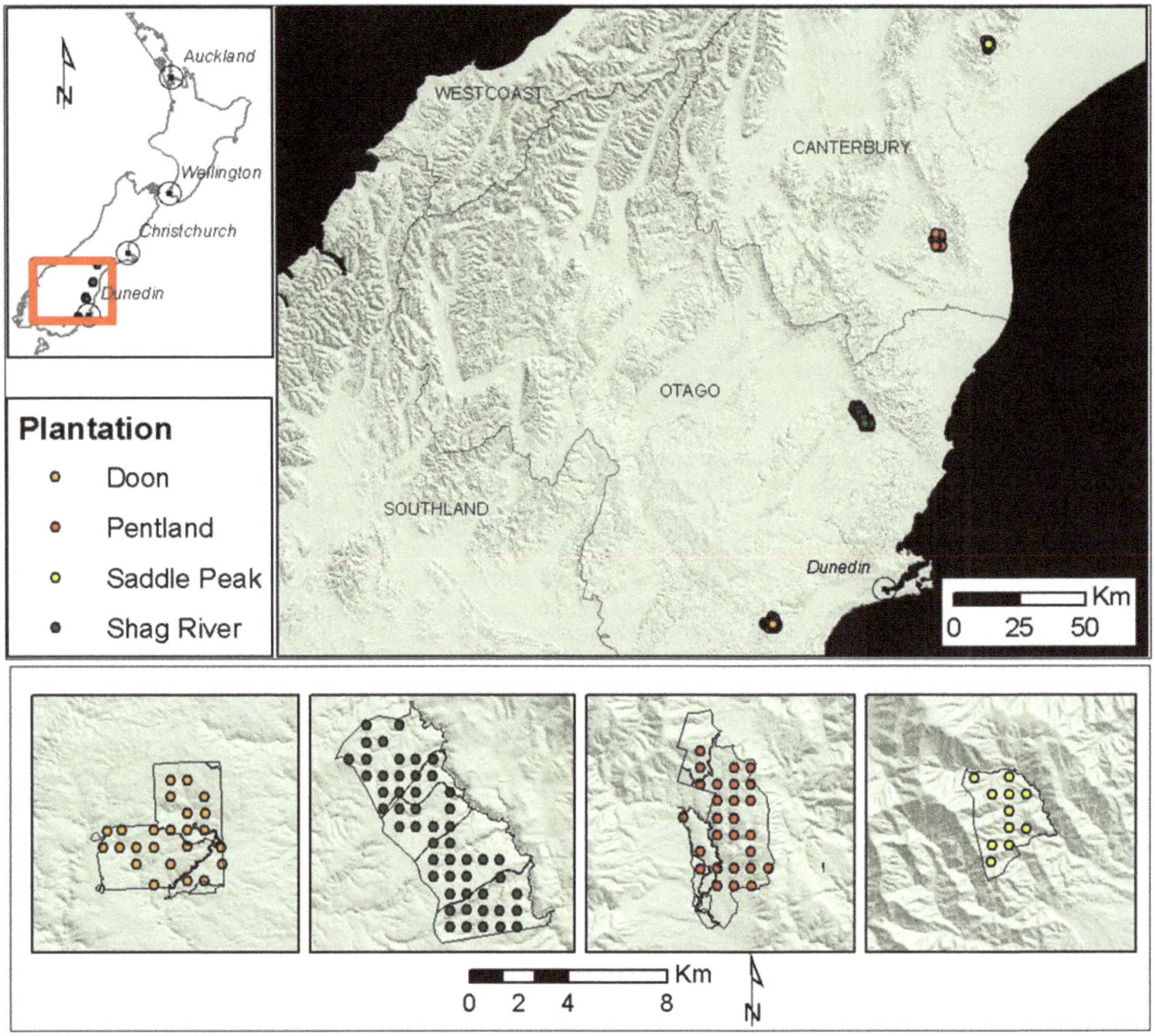

Figure 1 Map of the study area showing the distribution of the 122 plots used in the study.

Table 1 Summary of LiDAR attributes

Variable	Value
Wavelength (nm)	1064
Scan angle (deg.)	+/− 6.1
Pulse frequency (kHz)	70
Scan frequency (Hz)	70
Footprint diameter (m)	0.19
Ground speed (knots)	116.6
Flying height (m)	750

multiple regression models to ensure that robust models sensitive to a broad range of LiDAR metrics were created. For each multiple regression model, explanatory variables were sequentially introduced into the model starting with the variable that exhibited the strongest correlation with the dependent variable and subsequently the residuals from the model, until further additions were not significant, or did not improve the model coefficient of determination by more than 0.01. Variable selection was undertaken manually, one variable at a time, and plots of residuals were examined prior to variable addition to ensure that the variable was included in the model using the least biased functional form.

Model precision was determined using the coefficient of determination (R^2) and the root mean square error

(RMSE). Model bias was examined by plotting actual values against predictions and predicted values against residual values and dependant variables included in each model. The variance inflation factor was used to ascertain the degree of multicollinearity within the multiple regression models with values over 10 signifying problematic multicollinearity. Predictions from these models were not used to spatially estimate forest characteristics in this study.

Results

Ranges in LiDAR, stand metrics and slope

There was marked variation in LiDAR and stand metrics among plots. Mean top height, green crown height, volume, mean diameter, basal area, and stand density, respectively, ranged, six-fold, 17-fold, 226-fold, six-fold, 73-fold and 12-fold across the plot series (Table 2). Ranges in LiDAR metrics were similarly marked, with the 5th, 50th and 95th LiDAR height percentiles ranging 41-fold, 10-fold, and seven-fold, respectively (Table 2). Variation was also wide for PC_{zero}, which averaged 5.99%, ranging from 0.06 – 66.5% (Table 2). Plot slope averaged 16° and ranged from 2 to 41°.

Table 2 Variation in LiDAR and stand metrics

Variable	Mean	Range
LiDAR metrics		
H_{01} (m)	1.69 (0.124)	0 – 6.08
H_{05} (m)	2.58 (0.145)	0.17 – 6.98
H_{25} (m)	4.10 (0.166)	0.77 – 8.70
H_{50} (m)	5.17 (0.180)	1.00 – 9.90
H_{75} (m)	6.23 (0.194)	1.28 – 11.2
H_{95} (m)	7.67 (0.216)	1.77 – 13.0
H_{99} (m)	8.55 (0.230)	2.07 – 14.5
Elev kurtosis	2.83 (0.033)	1.95 – 4.31
Elev. skewness	−0.01 (0.025)	−0.74 – 0.75
Elev. CV (m)	0.323 (0.009)	0.17 – 0.70
Elev. std error (m)	1.54 (0.043)	0.36 – 3.15
Elev. IQ (m)	2.14 (0.066)	0.49 – 5.12
PC_{zero} (%)	5.99 (0.974)	0.06 – 66.5
Stand metrics		
Mean top height (m)	9.55 (0.242)	2.56 – 16.0
Green crown height (m)	1.32 (0.084)	0.255 – 4.31
Volume (m^3 ha^{-1})	92.0 (4.82)	1.01 – 228
Mean diameter at breast height (mm)	148 (3.33)	37.7 – 229
Basal area (m^2 ha^{-1})	21.6 (0.836)	0.580 – 42.1
Stand density (stems ha^{-1})	1289 (47.4)	250 – 3118
Stand age (years)	13.8 (0.192)	9 – 17

Values shown include the mean, with standard error in brackets, and the range.

Bivariate correlations

Green crown height was most strongly related to H_{01} with correlation coefficients diminishing as LiDAR height percentile increased (Table 3). The strongest relationships between green crown height and other LiDAR metrics were with the LiDAR height coefficient of variation and skewness (Table 3). Green crown height also exhibited strong positive relationships with both mean top height and volume (Table 3).

Mean top height was strongly correlated with the upper LiDAR height percentiles peaking with the seventy-fifth percentile, H_{75} (Table 3). In contrast, both basal area and total volume were more strongly correlated with the lower LiDAR height percentiles with the highest correlation coefficients occurring at the tenth percentile H_{10} for both characteristics (Table 3). For mean diameter, the strength of correlation coefficients increased with LiDAR height percentile reaching a peak value of 0.86 at the ninety-ninth percentile H_{95}. Stand density exhibited insignificant to weakly significant correlations with LiDAR height percentiles. The strongest bivariate relationship for stand density was observed with LiDAR height interquartile distance (Table 3).

Simple models describing the strongest linear relationships between LiDAR metrics and each of the stand characteristics were constructed (Figure 2). The positive linear relationships between LiDAR metrics and mean top height, green crown height, total stem volume, mean diameter and basal area had respective coefficients of determination (R^2) of 0.85, 0.71, 0.83, 0.74 and 0.72. The negative linear relationship between total stand density and LiDAR height interquartile distance had an R^2 of 0.33 (Figure 2).

Multiple regression models

Model formulation

Mean top height was best described by a simple linear relationship with H_{75} (Table 4). This relationship accounted for 85% of the variance in the data (Table 4).

Green crown height was best described by H_{01} and skewness in the LiDAR height distribution (Table 4). Inclusion of H_{01} in a quadratic form accounted for 76% of the variation in the data while addition of skewness as a negative linear term increased the R^2 by 0.03 to 0.79 (Table 4).

The best model for volume included H_{10}, PC_{zero} and stand age (Table 4). The strongest contributor to the model was H_{10} which when included using a quadratic form had a partial R^2 of 0.84. Inclusion of PC_{zero} as a negative linear term added a further 0.01 to the model while the addition of age as a positive linear term contributed a further 0.01 to the overall model. The final model had a R^2 of 0.86 (Table 4).

Table 3 Correlation coefficients describing the strength of linear relationships between stand metrics (mean top height, green crown height, volume, mean diameter, basal area, and stand density) with LiDAR and stand metrics

Variable	Mean top Height	Crown Height	Volume	Mean Diameter	Basal Area	Stand Density
LiDAR height percentiles						
H_{01}	0.71***	**0.85*** **	0.85***	0.49***	0.78***	0.14^{ns}
H_{05}	0.77***	0.82***	0.89***	0.58***	0.83***	0.10^{ns}
H_{10}	0.82***	0.82***	**0.91*** **	0.65***	**0.85*** **	0.06^{ns}
H_{20}	0.86***	0.81***	0.90***	0.72***	0.83***	-0.02^{ns}
H_{25}	0.87***	0.80***	0.89***	0.74***	0.82***	-0.05^{ns}
H_{30}	0.88***	0.80***	0.89***	0.76***	0.81***	-0.08^{ns}
H_{40}	0.91***	0.78***	0.87***	0.79***	0.79***	-0.13^{ns}
H_{50}	0.92***	0.77***	0.86***	0.81***	0.76***	-0.17^{ns}
H_{60}	0.92***	0.75***	0.84***	0.82***	0.74***	−0.20*
H_{70}	0.92***	0.73***	0.82***	0.84***	0.72***	−0.23*
H_{75}	**0.92*** **	0.72***	0.81***	0.84***	0.71***	−0.24**
H_{80}	0.92***	0.71***	0.80***	0.85***	0.70***	−0.25**
H_{90}	0.92***	0.70***	0.79***	0.86***	0.68***	−0.28**
H_{95}	0.92***	0.68***	0.77***	**0.86*** **	0.66***	−0.30***
H_{99}	0.91***	0.66***	0.74***	0.86***	0.63***	−0.33***
Other LiDAR metrics						
PC_{zero}	−0.39***	−0.26**	−0.47***	−0.35***	−0.59***	−0.42***
Elev. kurtosis	0.08^{ns}	0.23*	0.30***	0.01^{ns}	0.32***	0.20*
Elev. skewness	−0.53***	−0.29**	−0.43***	−0.56***	−0.45***	0.10^{ns}
Elev. CV	−0.51***	−0.61***	−0.72***	−0.35***	−0.75***	−0.32***
Elev. SD	0.60***	0.18*	0.24**	0.69***	0.14^{ns}	−0.56***
Elev. IQ	0.53***	0.12^{ns}	0.15^{ns}	0.63***	0.04^{ns}	**−0.58*** **
Stand metrics						
Mean top height	-	0.78***	0.85***	0.88***	0.74***	−0.23**
Crown height	0.78***	-	0.89***	0.60***	0.80***	0.03^{ns}
Volume	0.85***	0.89***	-	0.70***	0.96***	0.12^{ns}
Mean diameter at breast height	0.88***	0.60***	0.70***	-	0.63***	−0.46***
Basal area	0.74***	0.80***	0.96***	0.63***	-	0.29**
Stand density	−0.23**	0.03^{ns}	0.12^{ns}	−0.46***	0.29**	-
Stand age	0.62***	0.37***	0.49***	0.74***	0.45***	−0.39***

Values shown include the correlation coefficient, followed by the *P* category. ***, **, * denote significance at $P = 0.001$, 0.01 and 0.05, respectively; ^{ns} = not significant at $P = 0.05$. Values shown in bold represent the LiDAR variable that had the highest correlation coefficient with the associated stand metric.

The best model for mean diameter included H_{95}, stand density, stand age and PC_{zero} (Table 4). The largest contributor to the model was H_{95} and when included in the model as a positive linear term accounted for 74% of the variance in the dataset. Stand density and PC_{zero} were included in the model as negative linear terms and accounted for a further 4 and 5%, respectively, of the variance in the data. Inclusion of age as a positive linear term added a further 0.03 to the model R^2 and the final model R^2 was 0.86.

Basal area was best described by a model that included H_{10}, PC_{zero}, stand density and stand age (Table 4). The positive linear relationship with H_{10} accounted for 72% of the variance, while a negative linear relationship with PC_{zero} added a further 6%. Stand density and age were both included in the model as positive linear terms and contributed a further 2 and 3%, respectively, to the model. The final R^2 was 0.84.

The best model of stand density included LiDAR height interquartile distance, H_{IQ}, (height difference between the 25th and 75th height percentiles), PC_{zero} and stand age (Table 4). All variables were included in the model as negative linear terms and H_{IQ}, PC_{zero}

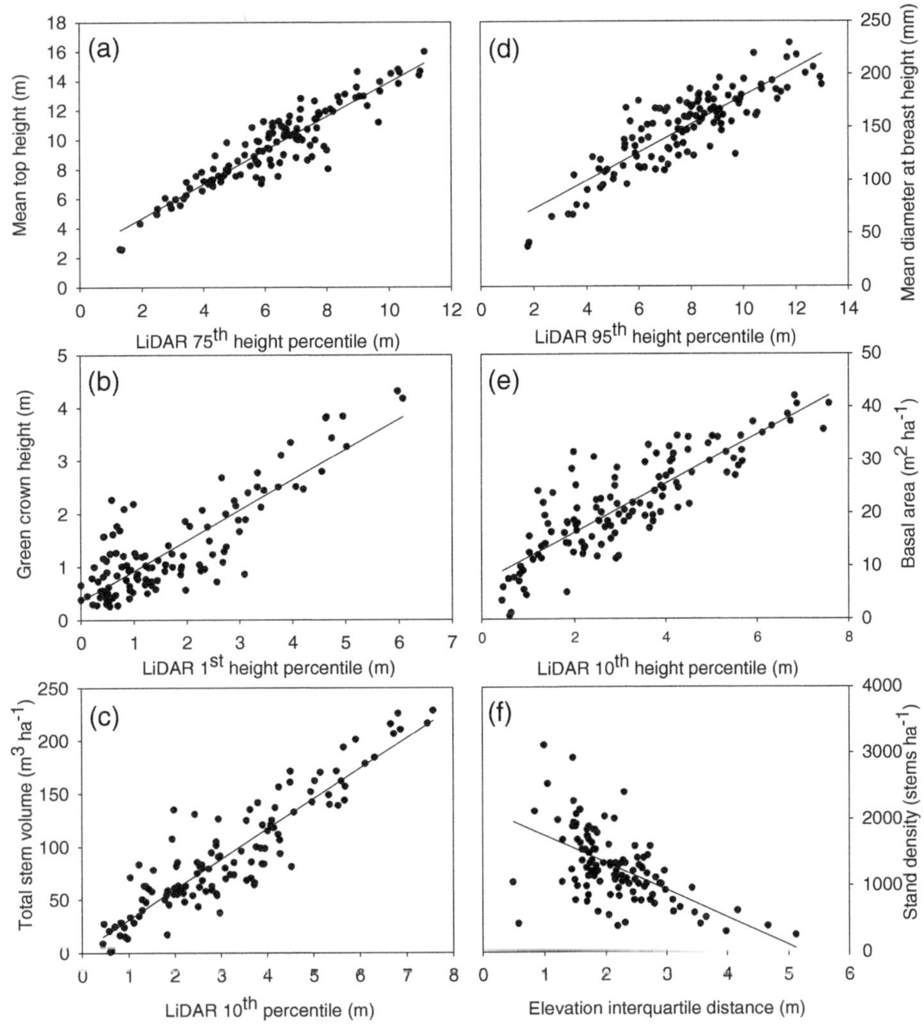

Figure 2 Bivariate relationships between LiDAR metrics and (a) mean top height, (b) green crown height, (c) total stem volume, (d) mean diameter at breast height, (e) basal area and (f) stand density. In all figures linear lines of the form $y = a + bx$ have been fitted. For these equations values of a, b, followed by R^2 and RMSE are **(a)** 2.381, 1.151; $R^2 = 0.85$, RMSE = 1.02 m, **(b)** 0.3602, 0.5699; $R^2 = 0.71$, RMSE = 0.497 m, **(c)** 2.501, 28.56; $R^2 = 0.83$, RMSE = 22.1 m^3 ha^{-1}, **(d)** 46.74, 13.24; $R^2 = 0.74$, RMSE = 18.9 mm, **(e)** 7.061, 4.632; $R^2 = 0.72$, RMSE = 4.87 m^2 ha^{-1}, **(f)** 2172, -413.8; $R^2 = 0.33$, RMSE = 430 stems ha^{-1}.

and stand age had respective partial R^2 values of 0.33, 0.17 and 0.05. The final model had an R^2 of 0.55.

Model precision and bias

With the exception of stand density, the coefficients of determination for the final developed models were high ranging from 0.79 to 0.86. The total amount of variance accounted for by the multiple regression model of stand density was markedly lower at 0.55. The RMSE of the final models was relatively low for all variables apart from stand density. Values of RMSE for the final models of mean top height, green crown height, stem volume, mean diameter, basal area and stand density were, respectively, 1.02 m, 0.427 m, 20.2 m^3 ha^{-1}, 13.9 mm, 3.81 m^2 and 355 stems ha^{-1}. All variables included within the six models were highly significant (Table 4). The highest variance inflation factor in for any of the terms in the multiple regression model was 1.97 indicating that multicollinearity was well within acceptable limits.

Plots of predicted against actual values for all models, apart from stand density, showed little apparent bias (Figure 3). For these five models there was also little pattern in the residuals when these values were plotted against predicted values or variables included in the final models (data not shown). For stand density, values of stand density were underpredicted at very low and very high stand densities (Figure 3f).

Discussion

With the exception of stand density, the models developed in this study had a high degree of precision.

Table 4 Multiple regression models for the stand characteristics described in this study

Dependant	Model coefficients		Model statistics		
Variable	Model term	Estimate	RMSE	Partial R^2	Overall R^2
Mean top height	Intercept	2.381			
	H_{75}	1.151	1.02 m	0.85***	0.85
Green crown height	Intercept	0.789			
	H_{01}	0.0125	0.497 m	0.71***	
	H_{01}^2	0.108	0.462 m	0.76***	
	Elev skewness	−0.680	0.427 m	0.03***	0.79
Volume	Intercept	−1.442			
	H_{10}	9.017	22.1 m^3 ha^{-1}	0.83***	
	H_{10}^2	2.215	21.8 m^3 ha^{-1}	0.01*	
	PC_{zero}	−0.7102	20.9 m^3 ha^{-1}	0.01**	
	Age	2.994	20.2 m^3 ha^{-1}	0.01**	0.86
Mean	Intercept	76.09			
Diameter at breast height	H_{95}	8.465	18.9 mm	0.74***	
	stand density	−0.02407	17.2 mm	0.04***	
	Age	3.189	15.8 mm	0.03***	
	PC_{zero}	−0.9579	13.9 mm	0.05***	0.86
Basal area	Intercept	−9.353			
	H_{10}	3.683	4.87 m^2 ha^{-1}	0.72***	
	PC_{zero}	−0.1117	4.32 m^2 ha^{-1}	0.06***	
	Stand density	4.975 x 10^{-3}	4.12 m^2 ha^{-1}	0.02***	
	Age	0.9880	3.81 m^2 ha^{-1}	0.03***	0.84
Stand density	Intercept	2986			
	Elev IQ	−314.9	430 stems ha^{-1}	0.33***	
	PC_{zero}	−22.94	371 stems ha^{-1}	0.17***	
	Age	−64.26	355 stems ha^{-1}	0.05***	0.55

Shown are the estimated coefficients for each term included in the six models. Model statistics shown include the root mean square error (RMSE), partial R^2 and significance category for each variable included in the model, with asterisks ***, **, * denoting significance at $P = 0.001$, $P = 0.01$ and 0.05 respectively. Also shown is the overall model R^2.

Despite the scattered nature of the resource from which data were acquired the models appeared to be quite robust. The use of LiDAR provides a viable approach for characterising the resource in young stands where there is little prior information (such as those sampled in this study).

The strength of relationships between LiDAR metrics and key stand characteristics were sufficiently high to delineate and prioritise areas for thinning. Thinning within the forests in the study area is typically undertaken when stands reach a mean top height of 14 m and the green crown reaches 3 m in height. Both mean top height and green crown height were predicted with a high degree of precision suggesting that LiDAR information could be useful for scheduling operations.

Of the key measurements used in forest inventory, mean top height is always predicted with the most precision using LiDAR. Although the coefficient of determination for the final mean top height model developed in this study was at the lower end of previously reported values, which range from 0.82 to 0.98 (Watt and Watt 2013; Coops et al. 2007; Means et al. 2000; Næsset 2002; Means et al. 1999; González-Ferreiro et al. 2012; Stone et al. 2011), the RMSE was similar to or lower than reported values for Douglas-fir and other coniferous species (Means et al. 2000; Watt and Watt 2013; Means et al. 1999). Previous research found LiDAR-derived estimates of height for Douglas-fir to be considerably less precise than those estimated for *Pinus ponderosa* Douglas ex. C. Lawson (Andersen et al. 2006). LiDAR may predict height less precisely in Douglas-fir as the pronounced conical shape of the species is more likely to be missed by LiDAR than the wider crowns of other coniferous species.

Total stem volume was predicted with a high degree of precision. The coefficient of determination for the final model was within the range cited by previous

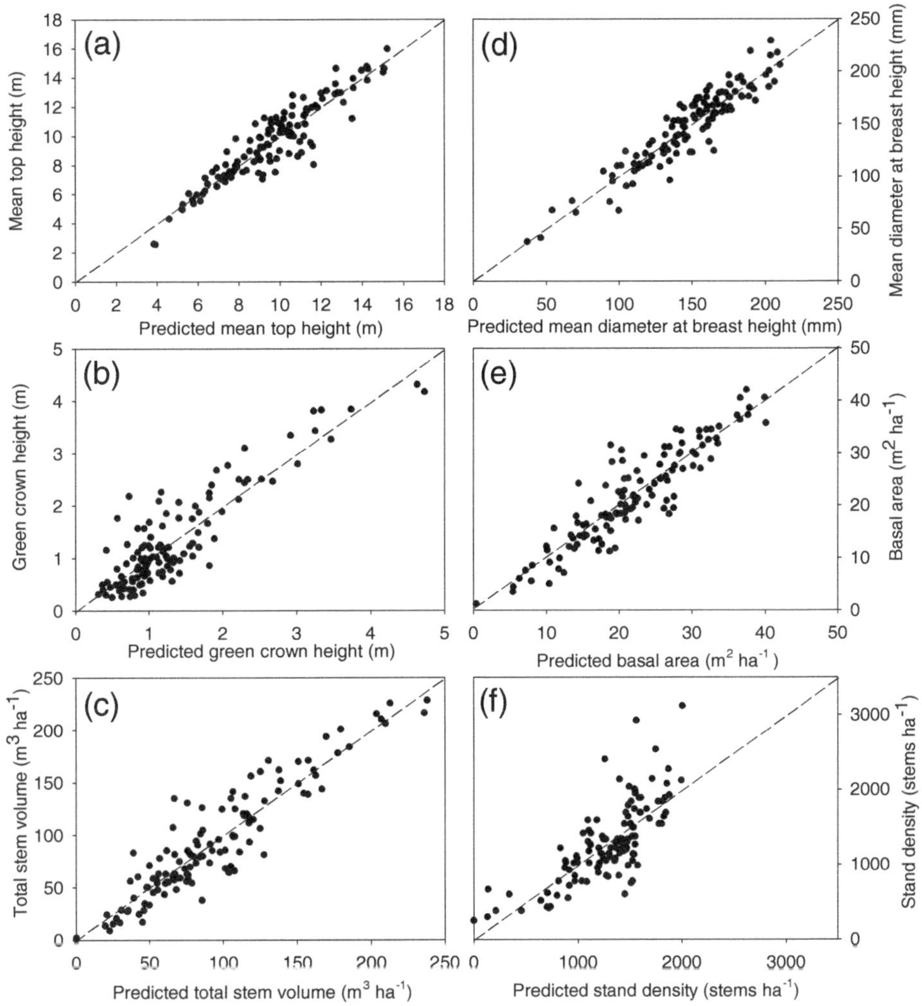

Figure 3 Relationship between predicted and actual (a) mean top height, (b) green crown height, (c) total stem volume, (d) mean diameter at breast height, (e) basal area and (f) stand density.

studies for coniferous species that vary from 0.46 to 0.97 (Means et al. 2000; Næsset 1997, 2002; Watt and Watt 2013). Compared to previous research, the RMSE of 20.2 m^3 ha^{-1} for the final model compares favourably to previous values that include 18.3 – 31.9 m^3 ha^{-1} (Holmgren and Jonsson 2004), 26.1 – 82.8 m^3 ha^{-1} (Næsset 1997), 71.6 m^3 ha^{-1} (Watt and Watt 2013) and 73 m^3 ha^{-1} (Means et al. 2000).

Model predictions of basal area had a comparable precision to similar studies within coniferous forests in boreal and temperate regions. Coefficients of determination ranged from 0.62 to 0.94 for models predicting basal area in coniferous forests in the United States of America (Means et al. 1999), Norway (Næsset 2004b, 2005, 2002) and Denmark (Nord-Larsen and Schumacher 2012). LiDAR-derived predictions of basal area in young and old growth Douglas-fir stands had a coefficient of determination of 0.95 (Means et al. 2000) that exceeds the value reported in this study. This high R^2 may reflect the relatively large plot size (0.25 ha) and wide range in basal area covered by the dataset as increases in both factors often result in increases in the coefficient of determination (see Zhao et al., 2009 for details on effects of plot size on precision).

Although little previous research has developed plot-level models of either stem diameter or green crown height from discrete-return LiDAR, both stand metrics were found to be strongly correlated to LiDAR metrics in the current study. Stem diameter was most strongly related to H_{95}, which most likely acts as a surrogate for mean top height, reflecting the strong relationship between mean diameter and mean top height in the dataset (Table 3). The strength of these relationships is broadly consistent with research on coniferous species within Norway where coefficients of determination ranged from 0.39 to 0.78 (Næsset 2002). Green crown height was most strongly related to H_{01} which is likely to represent the transition between green needles that intercept light to dead or absent needles at the crown base that allow greater light transmission. The degree of precision of the final model of green crown height exceeds that of models previously developed for stands dominated by *Picea abies* (L.) Karst. (Norway spruce) and *Pinus sylvestris* L. (Scots pine) where the coefficient of determination ranged from 0.6 – 0.71 (Næsset and Okland 2002).

Stand density was only predicted with a moderate degree of precision from LiDAR and stand level metrics. Previous research has also shown the correlation between LiDAR cloud metrics and stand density to be weaker than that for other stand characteristics reporting correlations that are only of moderate strength (Næsset and Bjerknes 2001; Hall et al. 2005; Næsset 2002).

Inclusion of stand age and stand density as predictive variables significantly improved a number of the developed models. Age featured in the final model for four of the six stand characteristics while stand density was a significant contributor to the models of mean diameter and basal area. These findings are broadly consistent with previous research that showed significant improvements in LiDAR models of stem volume through addition of stand density as an explanatory variable (Watt and Watt 2013; Watt et al. 2013).

The results of this study suggest that an accurate stand density map would provide a useful source of ancillary information to support use of LiDAR in predictions of stand metrics. As direct estimation of stand density from LiDAR cloud metrics is typically only moderately precise, the use of individual tree delineation may provide the most accurate method for deriving a stand density map. Although there have been difficulties in precisely counting trees (Vauhkonen et al. 2012), recent research has shown that plantation stand density can be accurately estimated from LiDAR imagery through use of a calibrated tree detection methodology (D. Pont, Scion pers. comm.).

The very low percentage of ground returns recorded in this dataset provided a robust test of the utility of LiDAR data in predicting stand characteristics. Compared to previous research, the mean percentage of ground returns recorded in this study (3.82%) was markedly lower that the typical range found for other coniferous species (Næsset 2002; Næsset and Bjerknes 2001; Næsset 2004a) but was similar to that recorded for high density unthinned hinoki cypress (*Chamaecyparis obtusa* Sieb. et Zucc.) and sugi (*Cryptomeria japonica* D. Don) plantations (Takahashi et al. 2006). The low penetration rates in this study are likely to be attributable to the dense closed canopy of the unthinned plantations and the topographically complex terrain over which the study sites were located. In older thinned stands of Douglas-fir, predictive precision of stand characteristics is likely to be higher due to greater canopy penetration by LiDAR and consequently more accurate definition of the digital terrain model and canopy height model. Further research should be undertaken to investigate how well LiDAR data can be used to predict stand characteristics in older thinned stands of Douglas-fir.

Conclusion

This research demonstrates that strong relationships can be developed for mean top height, green crown height, volume, mean diameter and basal area from LiDAR and stand metrics. Although the final model developed for stand density was only of moderate strength, these results are consistent with previous research linking LiDAR cloud metrics to stand density. Further research should investigate how well stand characteristics are predicted at older ages from LiDAR. It would also be useful to compare the accuracy and cost efficacy of predictions from LiDAR data with standard inventory methods for the scheduling of production thinning operations and later more detailed inventories in stands of Douglas-fir.

Competing interests

The authors declare that they have no competing interests.

Authors' contributions

MSW was the primary author and undertook all analyses. AM processed the LiDAR data, extracted metrics and produced the maps. PW and AG assisted with the writing, provision of technical detail. AG developed the specification for the LiDAR and provided the plot data. All authors read and approved the final manuscript.

Acknowledgements

We are grateful to Blakely Pacific Ltd for providing the LiDAR dataset on which this study is based. We also acknowledge useful comments made by two anonymous reviewers that helped improve the manuscript.

Author details

[1]Scion, Fendalton, PO Box 29237, Christchurch, New Zealand. [2]Indufor Asia-Pacific Ltd, PO Box 105039, Auckland, New Zealand. [3]Blakely Pacific Ltd, PO Box 13-980, Christchurch 8141, New Zealand.

References

Andersen, HE, Reutebuch, SE, & McGaughey, RJ. (2006). A rigorous assessment of tree height measurements obtained using airborne lidar and conventional field methods. *Canadian Journal of Remote Sensing, 32*(5), 355–366.

Coops, NC, Hilker, T, Wulder, MA, St-Onge, B, Newnham, G, Siggins, A, & Trofymow, JA. (2007). Estimating canopy structure of Douglas-fir forest stands from discrete-return LiDAR. *Trees-Structure and Function, 21*, 295–310.

Emmingham, WH, & Green, D. (2003). Thinning systems for western Oregon Douglas-fir stands EC1132. *The Woodland Workbook*. Corvallis, OR, USA: Oregon State University.

González-Ferreiro, E, Diéguez-Aranda, U, & Miranda, D. (2012). Estimation of stand variables in *Pinus radiata* D. Don plantations using different LiDAR pulse densities. *Forestry, 85*, 281–292.

Hall, SA, Burke, IC, Box, DO, Kaufmann, MR, & Stoker, JM. (2005). Estimating stand structure using discrete-return lidar: an example from low density, fire prone ponderosa pine forests. *Forest Ecology and Management, 208*(1–3), 189–209.

Hermann, RK, & Lavender, DP. (1999). Douglas-fir planted forests. *New Forests, 17*(1–3), 53–70, doi:10.1023/A:1006581028080.

Holmgren, G, & Jonsson, T. (2004). Large scale airborne laser scanning of forest resources in Sweden. *International Archives of Photogrammetry, Remote Sensing and Spatial Information Sciences, XXXV1 - 8/W2*, 157–160.

MacLaren, JP, & Knowles, RL. (2005). Version 2 calculators: upgrading the business of farm forestry. *New Zealand Tree Grower, 26*(2), 24.

McMurtrie, RE. (1993). Modelling of canopy carbon and water balance. In DO Hall, JMO Scurlock, HR Bolhar-Nordenkampf, RC Leegood, & SP Long (Eds.), *Photosynthesis and production in a changing environment: a field and laboratory manual* (pp. 220–231). London, UK: Chapman and Hall.

Means, JE, Acker, SA, Harding, DJ, Blair, JB, Lefsky, MA, Cohen, WB, Harmon, ME, & McKee, WA. (1999). Use of large-footprint scanning airborne LiDAR to estimate forest stand characteristics in the Western Cascades of Oregon. *Remote Sensing of Environment, 67*(3), 298–308.

Means, JE, Acker, SA, Fitt, BJ, Renslow, M, Emerson, L, & Hendrix, CJ. (2000). Predicting forest stand characteristics with airborne scanning LiDAR. *Photogrammetric Engineering and Remote Sensing, 66*(11), 1367–1371.

Næsset, E. (1997). Estimating timber volume of forest stands using airborne laser scanner data. *Remote Sensing of Environment, 61*(2), 246–253.

Næsset, E. (2002). Predicting forest stand characteristics with airborne scanning laser using a practical two-stage procedure and field data. *Remote Sensing of Environment, 80*(1), 88–99.

Næsset, E. (2004a). Effects of different flying altitudes on biophysical stand properties estimated from canopy height and density measured with a small-footprint airborne scanning laser. *Remote Sensing of Environment, 92*, 243–255.

Næsset, E. (2004b). Practical large-scale forest stand inventory using a small-footprint airborne scanning laser. *Scandinavian Journal of Forest Research, 19*(2), 164–179. doi:10.1080/02827580310019257.

Næsset, E. (2005). Assessing sensor effects and effects of leaf-off and leaf-on canopy conditions on biophysical stand properties derived from small-footprint airborne laser data. *Remote Sensing of Environment, 98*(2–3), 356–370.

Næsset, E, & Bjerknes, KO. (2001). Estimating tree heights and number of stems in young forest stands using airborne laser scanner data. *Remote Sensing of Environment, 78*(3), 328–340.

Næsset, E, & Okland, T. (2002). Estimating tree height and tree crown properties using airborne scanning laser in a boreal nature reserve. *Remote Sensing of Environment, 79*(1), 105–115.

Næsset, E, Gobakken, T, Holmgren, J, Hyyppa, H, Hyyppa, J, Maltamo, M, Nilsson, M, Olsson, H, Persson, A, & Soderman, U. (2004). Laser scanning of forest resources: The Nordic experience. *Scandinavian Journal of Forest Research, 19* (6), 482–499. doi:10.1080/02827580410019553.

Nord-Larsen, T, & Schumacher, J. (2012). Estimation of forest resources from a country wide laser scanning survey and national forest inventory data. *Remote Sensing of Environment, 119*, 148–157. doi:10.1016/j.rse.2011.12.022.

Reukema, DL. (1975). Guidelines for precommercial thinning of Douglas-fir. *USDA Forest Service General Technical Report PNW, 30*. Portland, OR, USA: United States Department of Agriculture.

SAS Institute Inc. (2008). *SAS/STAT 9.2 user's guide*. Cary NC, USA: SAS Institute Inc.

Stone, C, Penman, T, & Turner, R. (2011). Determining an optimal model for processing lidar data at the plot level: results for a *Pinus radiata* plantation in New South Wales, Australia. *New Zealand Journal of Forestry Science, 41*, 191–205.

Takahashi, T, Yamamoto, K, Miyachi, Y, Senda, Y, & Tsuzuku, M. (2006). The penetration rate of laser pulses transmitted from a small-footprint airborne LiDAR: a case study in closed canopy, middle-aged pure sugi (*Cryptomeria japonica* D. Don) and hinoki cypress (*Chamaecyparis obtusa* Sieb. et Zucc.) stands in Japan. *Journal of Forest Research, 11*(2), 117–123.

Vauhkonen, J, Ene, L, Gupta, S, Heinzel, J, Holmgren, J, Pitkanen, J, Solberg, S, Wang, YS, Weinacker, H, Hauglin, KM, Lien, V, Packalen, P, Gobakken, T, Koch, B, Næsset, E, Tokola, T, & Maltamo, M. (2012). Comparative testing of single-tree detection algorithms under different types of forest. *Forestry, 85*(1), 27–40. doi:10.1093/forestry/cpr051.

Watt, P, & Watt, MS. (2013). Development of a national model of *Pinus radiata* stand volume from LiDAR metrics for New Zealand. *International Journal of Remote Sensing, 34*, 5892–5904.

Watt, MS, Adams, T, Marshall, H, Pont, D, Lee, J, Crawley, D, & Watt, P. (2013). Modelling variation in *Pinus radiata* stem volume and outerwood stress-wave velocity from LiDAR metrics. *New Zealand Journal of Forestry Science, 43*.

Zhao, KG, Popescu, S, & Nelson, R. (2009). Lidar remote sensing of forest biomass: a scale-invariant estimation approach using airborne lasers. *Remote Sensing of Environment, 113*(1), 182–196. doi:10.1016/j.rse.2008.09.009.

26

The influence of LiDAR pulse density and plot size on the accuracy of New Zealand plantation stand volume equations

Michael S Watt[1*], Thomas Adams[2], Susana Gonzalez Aracil[3], Hamish Marshall[3] and Pete Watt[4]

Abstract

Background: LiDAR is an established technology that is increasingly being used to characterise spatial variation in important forest metrics such as total stem volume. The cost of forest inventory and LiDAR acquisition are strongly related to the inventory plot size and the LiDAR pulse density, respectively. It would therefore be beneficial to understand how reductions in these variables influence the strength of relationships between LiDAR and stand metrics. Although relatively high pulse densities are required for creating Digital Terrain Models (DTMs), once a DTM has been developed there is scope for reducing pulse density on subsequent flights to estimate stand metrics from LiDAR. This study used an extensive national dataset (for which the DTM had been characterised) obtained within New Zealand's planted forests. Using this dataset, the objective of this research was to investigate how variation in both pulse density and plot size influence the precision of relationships between LiDAR metrics and total stem volume.

Methods: LiDAR metrics were thinned to pulse densities ranging from 0.01 to 4 pulses m^{-2} across plot sizes ranging from 0.01 to 0.06 ha. For each pulse density/plot size combination regressions between LiDAR mean height and total stem volume were fitted using parameters fixed at values for the unthinned dataset or separately fitted for each pulse density/plot size combination.

Results: Using the unthinned dataset (plot size = 0.06 ha; pulse density = 4 pulses m^{-2}) the relationship between the mean LiDAR height and total stem volume had a coefficient of determination (R^2) of 0.77. Thinning of the data had little effect on R^2 above plot sizes of 0.03 ha and pulse densities of 0.1 pulses m^{-2}. As pulse densities decreased below 0.1 pulses m^{-2} within plots of less than 0.025 ha, there was a sharp decline in R^2 reaching values as low as 0.48 in plots of 0.01 ha with pulse densities of 0.01 pulses m^{-2}. Simulations where parameters were fixed yielded almost identical R^2 values to those where they were refitted for each plot size/pulse density combination. The number of pulses per plot integrates the effect of these two factors, with little change in the precision of the volume function until a threshold of 100 pulses per plot was reached.

Conclusions: This study showed that the precision of LiDAR-volume equations was relatively insensitive to reductions in pulse density and plot size when an accurate DTM was available. Acquisition of LiDAR information at lower pulse densities is likely to markedly improve the cost efficacy of this information for inventory purposes.

Keywords: Airborne laser scanning; ALS; LiDAR; Forest inventory; Laser; *Pinus radiata*

* Correspondence: michael.watt@scionresearch.com
[1]Scion, PO Box 29237, Fendalton, Christchurch, New Zealand
Full list of author information is available at the end of the article

Background

LiDAR is an established technology that is increasingly being used to derive spatial stand metrics, which is increasing the accuracy and efficiency of operational forest inventory. Since the first application of LiDAR in forestry almost three decades ago (Nelson et al. 1984), the technology has been widely used to quantify spatial variation in stand height, basal area (Watt 2005; Means et al. 1999), diameter, volume (Næsset 1997), canopy properties (Næsset and Okland 2002) and species composition (Donoghue et al. 2007). The strength of LiDAR lies in its ability to provide accurate distance information and to penetrate the canopy, thereby providing information on both the horizontal and vertical distribution of vegetation structure.

As stem volume is a key determinant of forest value, the use of LiDAR to provide an unbiased and accurate estimate of this stand dimension is of considerable interest to forest managers. Volume is typically determined from inventory measurements that are often time-consuming and expensive. Metrics derived from LiDAR have been shown to exhibit moderate to strong relationships with stem volume across a range of different coniferous species (Næsset 1997; Means et al. 2000), demonstrating the potential of LiDAR for inventory purposes. Compared to traditional inventory, with only plot-based measurements, LiDAR has a number of potential advantages that include increased fine scale spatial resolution, lower costs and increased precision (Næsset 2002; Holmgren and Jonsson 2004).

Field inventory costs increase markedly with plot size, and LiDAR costs increase with pulse density. Consequently a key issue requiring investigation is how far these elements can be reduced without compromising precision of the regression between LiDAR metrics and volume. Development of accurate LiDAR height metrics are reliant on an accurate digital terrain model (DTM) as the canopy returns must be converted from elevation above sea-level to height above the DTM. However, once an accurate DTM has been obtained it should be possible to reduce the pulse density for the estimation of volume.

The influence of pulse density on the precision of relationships between stand and LiDAR metrics has been widely investigated. In stands dominated by Norway spruce (*Picea abies* (L.) Karst.) and Scots pine (*Pinus sylvestris* L.), Gobakken and Næsset (2008) found that point densities could be reduced from 1.13 point to 0.25 points m^{-2} with little effect on the quality of inventory results. In a mixed conifer forest, where pulse densities were reduced from 9 to 0.01 pulses m^{-2}, it was found that correlations between LiDAR metrics and key stand metrics (tree height, diameter at breast height, and total basal area) were relatively unaffected until pulse densities were reduced below 1 pulse m^{-2} (Jakubowski et al. 2013). In mixed conifer-hardwood stands, it was found that pulse density could be reduced from 3.2 to 0.5 pulses m^{-2} with little effect on the precision of relationships between LiDAR and stand metrics (Treitz et al. 2012).

Fewer studies have investigated how plot size and the interaction between pulse density and plot size influence the precision of regressions between field measurements and canopy metrics. In general the homogeneity of canopies in even-aged forests allows for reductions of plot sizes to 0.02 ha, without substantial deterioration of the LiDAR-volume relationships (Gobakken and Næsset 2008). There are, however, substantial differences in this homogeneity between different forest types that may necessitate variation in plot sizes to accommodate this. This homogeneity ranges from high (in very dense unthinned planted stands with few gaps) through moderate (in thinned planted stands with greater ground area) to low (in natural forests).

Research is required that investigates how plot size and pulse density influence the quality of LiDAR-volume relationships across planted forest stands that include a wide variation in stand densities. Ideally, other important factors that influence estimation of canopy metrics, such as the laser system, flight specifications, data characteristics and data processing steps should be maintained constant to mitigate their influence. The characteristics described were part of a recently acquired LiDAR dataset that sampled the entire extent of New Zealand's predominantly *Pinus radiata* D. Don (radiata pine) resource. Using this dataset, the objective of this research was to investigate how pulse density and plot size influence the precision of LiDAR metric-volume relationships.

Methods

Dataset used

The dataset used was from a national inventory of planted forests (LUCAS) undertaken to measure and monitor temporal change in national carbon stocks. This inventory was undertaken by New Zealand as part of its obligations under the Kyoto Protocol and the United Nations Framework Convention on Climate Change (Beets et al. 2010). This dataset has been fully described previously (Beets et al. 2012; Stephens et al. 2012; Stephens et al. 2007).

Table 1 Summary of key LiDAR settings for the dataset collected

Variable	Value
Wavelength (nm)	1064
Scan angle (deg.)	± 6
Pulse frequency (kHz)	70
Scan frequency (Hz)	53
Footprint diameter (m)	0.27
Ground speed (knots)	105
Flying height (m)	1200

In stands established after 1 January 1990, circular plots with an area of 0.06 ha were installed between June and September 2008 using a regular systematic 4 km grid. In stands established prior to 1990, plots of 0.06 ha were installed during 2010 on an 8 km grid. Field plot centres were located using a 12-channel differentially corrected global positioning system (GPS) to within ± 3 m.

The LiDAR survey was undertaken aerially using a Cessna 207 aircraft in February 2008 (for pre-1990 forests) and 2010 (for post-1989 forests) using a small footprint (~0.20 m) Optech ALTM 3100EA system integrated with a Rollei AIC digital camera. The LiDAR settings used to achieve first-return pulse densities of at least 3 returns m^{-2} are summarised in Table 1. The system also utilised an Applanix 510 Position and Orientation System (POS) that uses GPS and inertial measurement unit (IMU) sensors, and a GPS-based computer controlled navigation system.

From this dataset, LiDAR metrics and stand information for 374 plots were extracted for the analysis. These data were taken across a range of plantation species that included plot representation of the following species: 350 plots (94%) in *Pinus radiata*, 13 plots (3%) in *Pseudotsuga menziesii* var *menziesii* (Mirb.) Franco (Douglas-fir), 7 plots (2%) in *Eucalyptus nitens* H.Deane & Maiden (shining gum), and the remaining four plots within stands with minor species. The distribution of these plots is shown in Figure 1.

Figure 1 Distribution of plot locations used for the modelling.

Table 2 Summary statistics describing the dataset

Variable	Mean	Range
Stand metrics		
Stem volume (m^3 ha^{-1})	240 (9.9)	0.05 – 1020
Crop stocking (stems ha^{-1})	464 (18)	17 – 2283
LiDAR metrics		
Canopy cover (%)	69.7 (1.1)	1.4 – 99.1
Height of 30th percentile (m)	9.2 (0.3)	0.6 – 30.7
Mean height (m)	11.3 (0.3)	0.7 – 31.2
Height of 95th percentile (m)	17.8 (0.5)	1.0 – 43.1

Shown are the mean and standard error (in brackets) along with the data range. LiDAR metrics shown include values from all returns.

Diameter at breast height (dbh) was recorded for each plot tree. A sample of at least 16 trees per crop type (i.e. silvicultural regime, age, species etc.) within each plot also had height measured. A plot-specific height-diameter relationship was used to determine height of any remaining trees in the plot. The bearing and distance from each tree to the plot centre was recorded. Stand slope was measured using an inclinometer and defined as the average of the maximum gradient at the plot centre and the gradient at ninety degrees to that.

Within the dataset there was a wide range in stand and LiDAR metrics (Table 2). Stand density ranged from 17 – 2283 stems ha^{-1} which exceeds the normal operational range found within planted forests in New Zealand. Total stem volume ranged from 0.048 to 1020 m^3 ha^{-1}. LiDAR metrics also ranged widely with most variation noted in the 30th height percentile which ranged 51-fold from 0.6 to 30.7 m (Table 2).

Processing of LiDAR and field data

Following the method described in Gonzalez Aracil (2011), which uses the equations described in Goulding (1995), individual tree volumes were determined from diameter and measurements (or estimations) of height. Summation of these volumes yielded a plot level estimate of total stem volume. Measurements of distance and bearing allowed total stem volume to be recalculated under a range of plot sizes. Where the plots were on sloping ground, a slope-corrected radius was used (Hayes 2005), both for the original plots and the simulated reduced size plots.

Terra-Scan software was used to classify ground and non-ground returns using the algorithm described by Axelsson (2000). Manual identification of breaklines by skilled operators was undertaken to improve DTM quality. These classified ground returns were used to construct a DTM by connecting them into a Triangulated Irregular Network (TIN), and then linearly interpolating them onto a regular grid. Pulses were extracted individually from the data and assimilated using the raw coordinates. The DTM grid size was 1 m, but linearly interpolated to the exact horizontal return location for more accurate ground characterisation.

LiDAR data were extracted for each plot assuming a circular plot shape. Although field plots were often elliptical, previous research demonstrates the assumption of circularity to have little effect on values for key LiDAR height metrics (<0.22%) within this dataset (Adams and Pont 2012). LiDAR metrics extracted for the analysis included the 30th, 70th and 95th height percentiles, mean height and canopy cover (percentage of returns from above a cut-off of 0.5 m to remove the understorey) as these are widely used for development of volume equations. The described LiDAR metrics were generated from all returns.

Thinning of LiDAR dataset

For the unthinned dataset (plot size = 0.06 ha; pulse density = 4 returns m^{-2}) the best model describing total stem volume was,

$$V = \alpha H_m + \beta \tag{1}$$

where V is the total stem volume (m^3 ha^{-1}), H_m is the mean LiDAR height (m), and α, β are the regression coefficients, which had respective values of 28.02 and −76.45. This relationship had a R^2 of 0.77. Addition of further variables to this model did not substantially improve model accuracy.

Models of V were developed under a wide range of plot size and pulse densities using the independent variable (H_m) described in Equation 1. LiDAR metrics were thinned to 84 different pulse densities ranging from 4 pulses m^{-2} to 0.01 pulses m^{-2} for plot sizes ranging from 0.01 to 0.06 ha (in 0.005 ha intervals). The raw data was thinned to remove a specified proportion of pulses (and all corresponding returns). Each iteration removed a random set of pulses from the original dataset, so successive thinnings would not contain the same pulses.

Statistics were extracted for two sets of simulations. In the first set, coefficients in the volume equation were refitted for each pulse density/plot size combination. In the second set, coefficients in the volume equation were fixed at those used for the highest plot size (0.06 ha) and pulse density (4 pulses m^{-2}). For both sets of simulations, the coefficient of determination (R^2) was obtained for each pulse density/plot size combination. For plot size/pulse density combinations in the first simulation the percentage variation in coefficients of the regression equation was determined as the mean percentage change in each of the two regression coefficients, compared to their original values (for a plot size of 0.06 ha and pulse density of 4 returns m^{-2}).

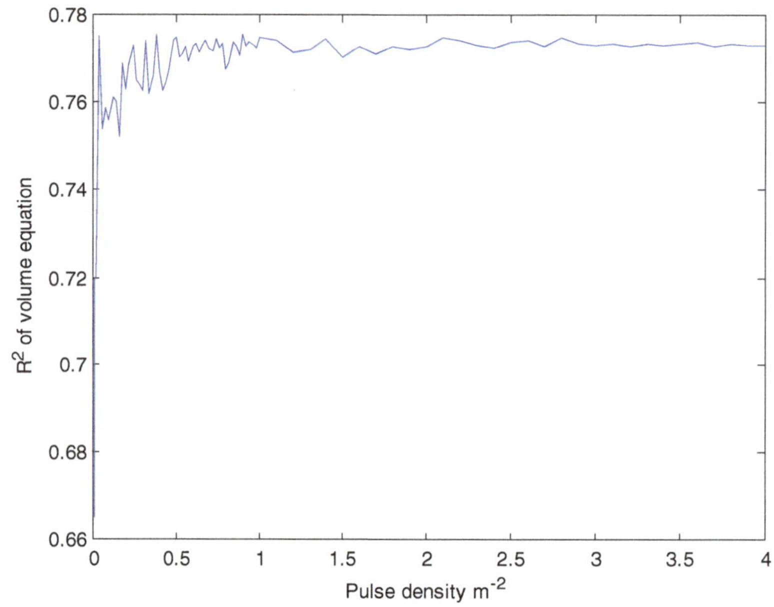

Figure 2 Relationship between pulse density and the coefficient of determination (R^2) for simulations where a fixed plot size of 0.06 ha was used.

Results

Influence of pulse density on model precision

For the first set of simulations, where coefficients were allowed to vary, the model performance (R^2) remained relatively constant in response to reductions in pulse density until values were reduced below 0.5 pulses m^{-2} (Figure 2). Following this there was a marked decline in R^2. The coefficients in the regression were almost constant above 1 pulse m^{-2} (varying by less than 0.4%), and only varied by more than 5% for the β coefficient (see Equation 1) with pulse densities less than 0.25 pulses m^{-2} (Figure 3). This stability was also apparent in the very low variation in model precision between simulations with fixed and varying coefficients. Differences in the coefficient of determination (R^2)

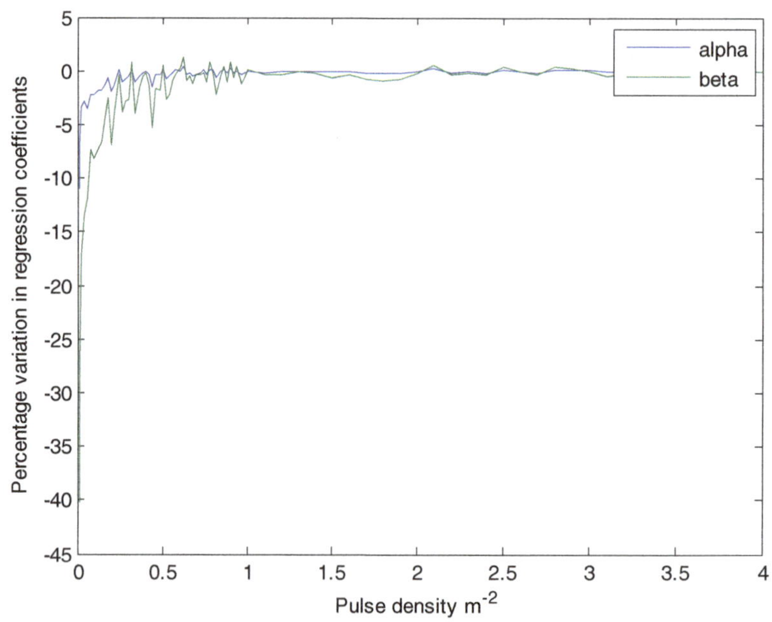

Figure 3 Relationship between pulse density and percentage variation in coefficients of the regression equation (described in Eqn. 1) for simulations where a fixed plot size of 0.06 ha was used.

between the two simulations, for a given pulse density, did not exceed 10^{-15}.

Influence of plot size and pulse density on model precision

For the first set of analyses where the coefficients were allowed to vary, the coefficient of determination (R^2) exhibited little change from the base value of 0.77 for plot sizes greater than 0.03 ha where the pulse density exceeded *ca.* 0.1 pulses m^{-2} (Figure 4). As long as pulse densities were maintained above 0.1 pulses m^{-2}, reductions in plot size to 0.01 ha resulted in only moderate losses in R^2 of up to *ca.* 0.03 (from ca. 0.77 to 0.74). Similarly, at plot sizes exceeding 0.025 ha, pulse densities could be reduced to 0.01 pulses m^{-2}, with only moderate losses in R^2 to 0.60 (Figure 4). As pulse densities declined below 0.1 pulses m^{-2} within plots of less than 0.025 ha, there was a sharp decline in R^2 reaching values as low as 0.48 in plots of 0.01 ha with pulse densities of 0.01 pulses m^{-2} (Figure 4).

At pulse densities exceeding 0.5 pulses m^{-2}, coefficients in the volume function were very stable across plot sizes greater than 0.04 ha, exhibiting less than 1% variation from the base coefficients (Figure 5). The average variation

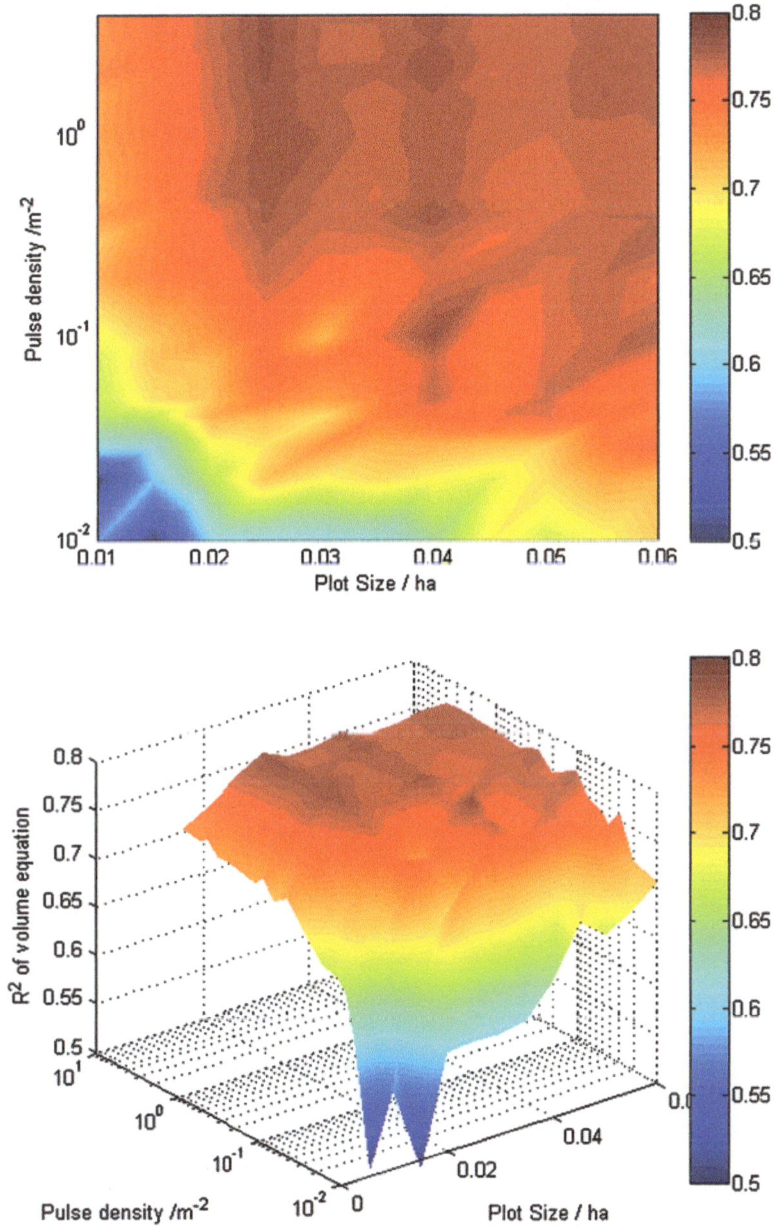

Figure 4 Relationship between coefficient of determination (R^2) for the volume equation and plot size and pulse density. Values are shown as a checkerboard plot (top) and as a 3D surface (bottom). The colour scale shows variation in R^2 for the volume equation.

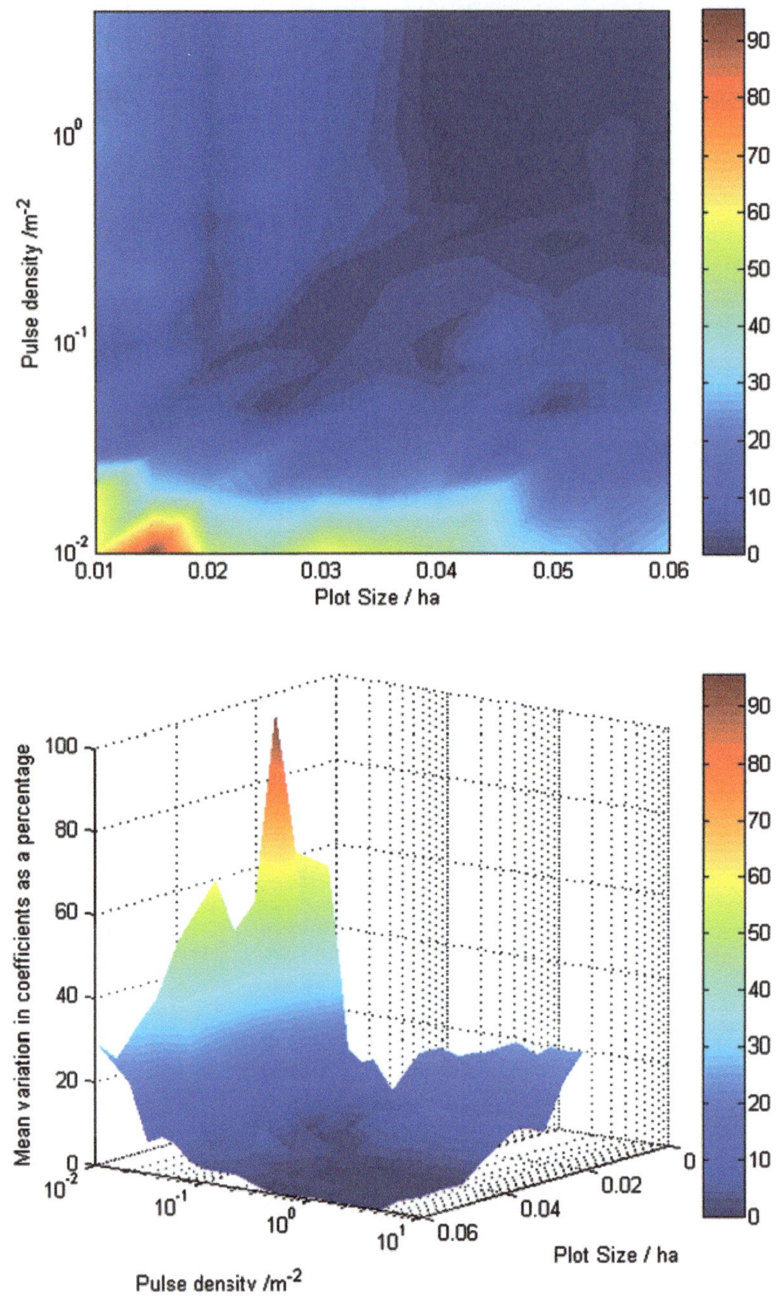

Figure 5 The relationship between mean percentage variation in regression coefficients for the LiDAR volume equation and plot size and pulse density. Values are shown as a checkerboard plot (top) and 3D surface (bottom). The colour scale shows percentage variation in regression coefficients.

of coefficients was never more than 13% for all plots greater than 0.03 ha under pulse densities greater than 0.01 pulses per m^2. There was a marked increase in average coefficient variation at very low pulse densities that was particularly pronounced as plot sizes declined below 0.025 ha (Figure 5).

Findings were very similar for simulations run with regression coefficients fixed at values found for the highest plot size (0.06 ha) and pulse density (4 pulses m^{-2}). These simulations show a very similar decline in R^2 to those where parameters were refitted (i.e. Figure 4), with absolute differences in R^2 not exceeding 10^{-15}. As R^2 values for these simulations were indistinguishable from those in Figure 4 these data are not shown.

Regression of the number of pulses per plot against model precision integrated the effect of the plot size and pulse density, described previously. The coefficient of determination did not markedly decline until a threshold value of 100 pulses per plot was reached (Figure 6).

Discussion

The results from this study are broadly consistent with previous research investigating the effects of plot size and pulse density on precision of relationships between LiDAR and canopy metrics. Most studies have found that LiDAR pulse densities can be reduced to between 0.06 – 0.5 pulses m^{-2} with little consequent effect on precision of canopy metrics (Goodwin et al. 2006; Næsset 2004a; Gobakken and Næsset 2008; Treitz et al. 2012; Jakubowski et al. 2013). Similarly, previous research has shown that reductions in plot size are possible to 0.02 ha without negative impacts on the precision of regressions between LiDAR metrics and volume (Gobakken and Næsset 2008).

This study extends previous research by examining the combined effect of plot size and pulse density on model precision. When these two variables were examined in combination, little deterioration in the precision of the volume function was noted for plot sizes greater than 0.03 ha where the pulse density exceeded *ca.* 0.1 pulses m^{-2}. Analyses indicate that the number of pulses per plot integrates

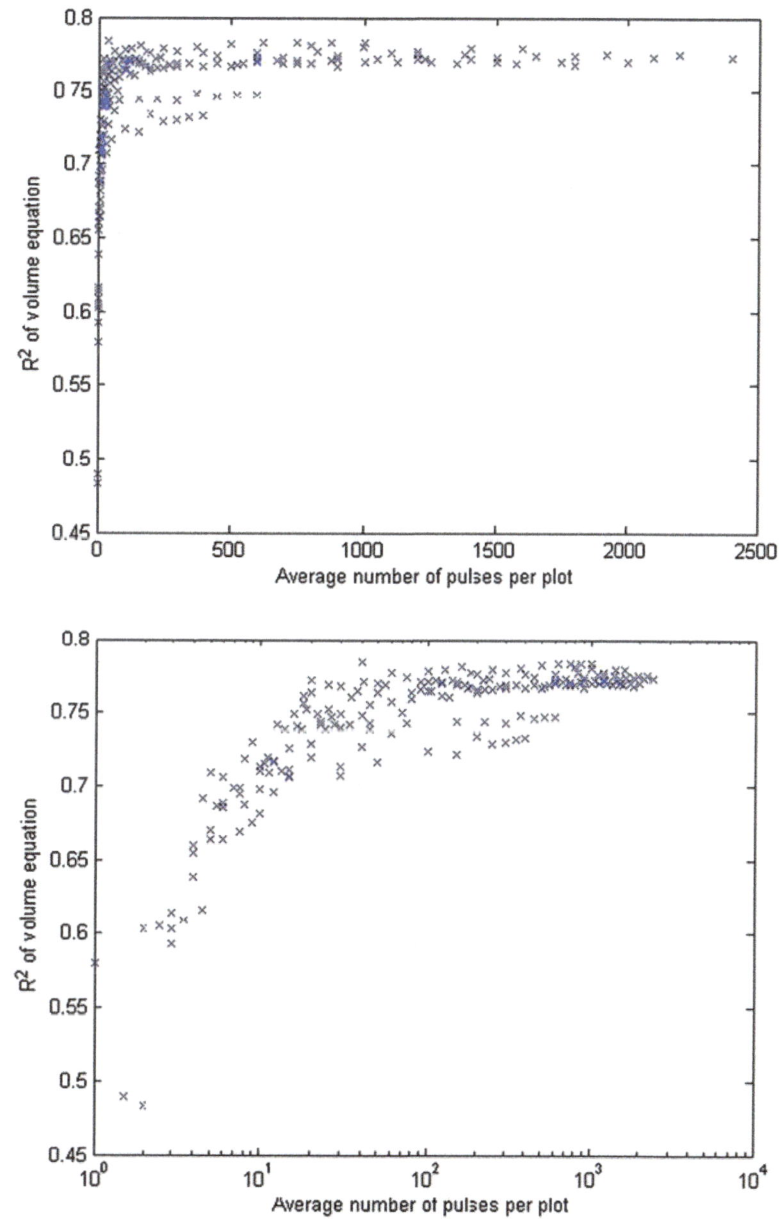

Figure 6 The relationship between the coefficient of determination (R^2) of the volume equation and number of pulses per plot, with values on the *x*-axis plotted using a linear scale (top) and log scale (bottom).

the effect of these two factors, with little change in the precision of the volume function until a threshold of 100 pulses per plot is reached. Once this point is reached there are too few pulses to enable the calculation of robust LiDAR metrics. The volume equation was found to be stable across most pulse density and plot size combinations. The effect of data thinning on R^2 hardly varied between simulations where the parameters were held constant or allowed to change.

The relatively low sensitivity of model precision to pulse density demonstrated in this study has implications for the cost effectiveness of LiDAR as an inventory tool. Currently most LiDAR is acquired within forests at a pulse density of between 1–2 pulses m^{-2} in New Zealand (New Zealand Aerial Mapping pers. comm.). Reduction of the pulse density to well below these levels, under situations where an accurate DTM is available, would markedly increase the cost efficacy of LiDAR as a tool for describing spatial variation in canopy metrics.

In an operational setting, pulse density and LiDAR acquisition cost can be reduced through flying faster, reducing the swath overlap or flying at higher altitude. Increases in aircraft speed or a reduction in the swath overlap will reduce the pulse density without influencing other specifications such as footprint diameter and ability of the laser to penetrate the canopy (Magnusson et al. 2010).

In contrast, flying at a higher altitude affects several features of the data. The footprint diameter and swath width increases while the reflected energy, ability of the laser to penetrate the canopy, and maximum pulse repetition frequency decline (Magnusson et al. 2010). As a result of these changes flying at higher altitude results in a change in the distribution of echo categories, an upward shift in the canopy-height distribution and a lower proportion of multiple returns (Goodwin et al. 2006; Næsset 2004b, 2009). Although changes to these features do not significantly affect the precision of regression equations for common stand metrics, such as height and stand volume, coefficients in the regression equations are likely to change between acquisitions at different altitudes (Næsset 2004b, 2009).

Conclusions

This study demonstrates the relative insensitivity of precision in LiDAR-derived models of volume to changes in plot size and pulse density. Little change in R^2 was noted until data were reduced to plot sizes of 0.03 ha and pulse densities of 0.1 pulses m^{-2}. Simulations where model parameters were fixed yielded almost identical results to those where parameters were refitted to each combination of plot size/pulse density. Acquisition of LiDAR data at lower pulse densities will markedly improve the cost-effectiveness of this technology for inventory purposes.

Competing interests

The authors declare that they have no competing interests

Authors' contributions

MSW was the primary author. TA processed the LiDAR data, extracted metrics and undertook analyses. SGA, HM and PW assisted with the writing and provision of technical detail. All authors read and approved the final manuscript.

Acknowledgements

We are very grateful to the Ministry for the Environment (MfE) for providing permission to use the data described in this paper. This project was funded within the Intensive Forest Systems project of Future Forests Research Ltd. The authors are grateful to Interpine for provision of the LUCAS dataset. We thank Marie Heaphy for producing Figure 1.

Author details

[1]Scion, PO Box 29237, Fendalton, Christchurch, New Zealand. [2]MetService, 30 Salamanca Road, Kelburn, Wellington 6021, New Zealand. [3]Interpine, 99 Sala Street, PO Box 1209, Rotorua 3010, New Zealand. [4]Indufor Asia-Pacific Ltd, PO Box 105039, Auckland, New Zealand.

References

Adams, T, & Pont, D. (2012). Projecting plots in LiDAR - Correcting for slope in remotely sensed data. *New Zealand Journal of Forestry, 57*, 25–31.

Axelsson, P. (2000). DEM generation from laser scanner data using adaptive tin models. *International Archives of Photogrammetry and Remote Sensing, 33*, 111–118.

Beets, PN, Brandon, A, Fraser, BV, Goulding, CJ, Lane, PM, & Stephens, PR. (2010). National Forest Inventories Report: New Zealand. In E Tomppo, T Gschwantner, M Lawrence, & RE McRoberts (Eds.), *National Forest Inventories: Pathways for Common Reporting* (pp. 391–410). Heidelberg, Germany: Springer.

Beets, PN, Brandon, AM, Goulding, CJ, Kimberley, MO, Paul, TSH, & Searles, N. (2012). The inventory of carbon stock in New Zealand's post-1989 planted forest for reporting under the Kyoto protocol. *Forest Ecology and Management, 262*, 1119–1130.

Donoghue, DNM, Watt, PJ, Cox, NJ, & Wilson, J. (2007). Remote sensing of species mixtures in conifer plantations using LiDAR height and intensity data. *Remote Sensing Environment, 110*, 509–522.

Gobakken, T, & Næsset, E. (2008). Assessing effects of laser point density, ground sampling intensity, and field sample plot size on biophysical stand properties derived from airborne laser scanner data. *Canadian Journal of Forest Research-Revue Canadienne De Recherche Forestiere, 38*(5), 1095–1109. doi:Doi 10.1139/X07-219.

Gonzalez Aracil, S. (2011). *The impact of plot size on LiDAR volume relationships in forest inventories - Masters Thesis*. Madrid, Spain: Universidad se Alcala.

Goodwin, NR, Coops, NC, & Culvenor, DS. (2006). Assessment of forest structure with airborne LiDAR and the effects of platform altitude. *Remote Sensing of Environment, 103*(2), 140–152.

Goulding, CJ. (1995). Individual tree volume, taper, bark, and breakage equations. In D Hammond (Ed.), *NZIF 1995 Forestry Handbook* (3rd ed.). Christchurch, New Zealand: New Zealand Institute of Forestry (Inc.).

Hayes, JD. (2005). Sample plots techniques and tables. In M Colley (Ed.), *Forestry Handbook*. Rotorua, New Zealand: New Zealand Institute of Forestry.

Holmgren, G, & Jonsson, T. (2004). Large scale airborne laser scanning of forest resources in Sweden. *International Archives of photogrammetry, remote sensing and spatial information sciences, XXXV1*, 8/W2.

Jakubowski, MK, Guo, QH, & Kelly, M. (2013). Tradeoffs between lidar pulse density and forest measurement accuracy. *Remote Sensing of Environment, 130*, 245–253. doi:DOI 10.1016/j.rse.2012.11.024.

Magnusson, M, Fransson, JES, & Holmgren, J. (2010). Effects of estimation accuracy of forest variables using different pulse density of laser data. *Forest Science, 53*, 619–626.

Means, JE, Acker, SA, Fitt, BJ, Renslow, M, Emerson, L, & Hendrix, CJ. (2000). Predicting forest stand characteristics with airborne scanning LiDAR. *Photogrammetric Engineering and Remote Sensing, 66*(11), 1367–1371.

Means, JE, Acker, SA, Harding, DJ, Blair, JB, Lefsky, MA, Cohen, WB, et al. (1999). Use of large-footprint scanning airborne LiDAR to estimate forest stand

characteristics in the Western Cascades of Oregon. *Remote Sensing of Environment, 67*(3), 298–308.
Næsset, E. (1997). Estimating timber volume of forest stands using airborne laser scanner data. *Remote Sensing Of Environment, 61*(2), 246–253.
Næsset, E. (2002). Predicting forest stand characteristics with airborne scanning laser using a practical two-stage procedure and field data. *Remote Sensing of Environment, 80*(1), 88–99.
Næsset, E. (2004a). Effects of different flying altitudes on biophysical stand properties estimated from canopy height and density measured with a small-footprint airborne scanning laser. *Remote Sensing of Environment, 91*(2), 243–255.
Næsset, E. (2004b). Effects of different flying altitudes on biophysical stand properties estimated from canopy height and density measured with a small-footprint airborne scanning laser. *Remote Sensing of Environment, 92*, 243–255.
Næsset, E. (2009). Effects of different sensors, flying altitudes, and pulse repetition frequencies on forest canopy metrics and biophysical stand properties derived from small-footprint airborne laser data. *Remote Sensing of Environment, 113*(1), 148–159. doi:DOI 10.1016/j.rse.2008.09.001.
Næsset, E, & Okland, T. (2002). Estimating tree height and tree crown properties using airborne scanning laser in a boreal nature reserve. *Remote Sensing of Environment, 79*(1), 105–115.
Nelson, R, Krabill, W, & MacLean, G. (1984). Determining forest canopy characteristics using airborne laser data. *Remote Sensing of Environment, 15*, 201–212.
Stephens, PR, Watt, PJ, Loubser, D, Haywood, A, & Kimberley, MO. (2007). Estimation of carbon stocks in New Zealand planted forests using airborne scanning LiDAR. *Workshop on Laser Scanning, September 12–14* (pp. 389–394). Finland: IAPRS XXXVI, Part 3/W52.
Stephens, PS, Kimberley, MO, Beets, PN, Paul, TSH, Searles, NB, Bell, A, Brack, C, & Broadley, J. (2012). Airborne scanning LiDAR in a double sampling forest carbon inventory. *Remote Sensing of Environment, 117*, 348–357.
Treitz, P, Lim, K, Woods, M, Pitt, D, Nesbitt, D, & Etheridge, D. (2012). LiDAR sampling density for forest resource inventories in Ontario, Canada. *Remote Sensing, 4*(4), 830–848. doi:Doi 10.3390/Rs4040830.
Watt, PJ. (2005). *An evaluation of LiDAR and optical satellite data for the measurement of structural attributes in British upland conifer plantation forestry. Doctoral thesis*. England: Department of Geography, University of Durham.

Permissions

All chapters in this book were first published in NZJFS, by Springer; hereby published with permission under the Creative Commons Attribution License or equivalent. Every chapter published in this book has been scrutinized by our experts. Their significance has been extensively debated. The topics covered herein carry significant findings which will fuel the growth of the discipline. They may even be implemented as practical applications or may be referred to as a beginning point for another development.

The contributors of this book come from diverse backgrounds, making this book a truly international effort. This book will bring forth new frontiers with its revolutionizing research information and detailed analysis of the nascent developments around the world.

We would like to thank all the contributing authors for lending their expertise to make the book truly unique. They have played a crucial role in the development of this book. Without their invaluable contributions this book wouldn't have been possible. They have made vital efforts to compile up to date information on the varied aspects of this subject to make this book a valuable addition to the collection of many professionals and students.

This book was conceptualized with the vision of imparting up-to-date information and advanced data in this field. To ensure the same, a matchless editorial board was set up. Every individual on the board went through rigorous rounds of assessment to prove their worth. After which they invested a large part of their time researching and compiling the most relevant data for our readers.

The editorial board has been involved in producing this book since its inception. They have spent rigorous hours researching and exploring the diverse topics which have resulted in the successful publishing of this book. They have passed on their knowledge of decades through this book. To expedite this challenging task, the publisher supported the team at every step. A small team of assistant editors was also appointed to further simplify the editing procedure and attain best results for the readers.

Apart from the editorial board, the designing team has also invested a significant amount of their time in understanding the subject and creating the most relevant covers. They scrutinized every image to scout for the most suitable representation of the subject and create an appropriate cover for the book.

The publishing team has been an ardent support to the editorial, designing and production team. Their endless efforts to recruit the best for this project, has resulted in the accomplishment of this book. They are a veteran in the field of academics and their pool of knowledge is as vast as their experience in printing. Their expertise and guidance has proved useful at every step. Their uncompromising quality standards have made this book an exceptional effort. Their encouragement from time to time has been an inspiration for everyone.

The publisher and the editorial board hope that this book will prove to be a valuable piece of knowledge for researchers, students, practitioners and scholars across the globe.

List of Contributors

Horacio E. Bown
Faculty of Forestry & Conservation, University of Chile, P.O. Box 9206, Santiago, Chile

Jean-Pierre Lasserre
Forestal Mininco S.A., Los Canelos 71, San Pedro, Concepción, Chile

John R. Moore, Dave J. Cown and Russell B. McKinley
Scion, Private Bag 3020, Rotorua, New Zealand

J. C. Grace and R. K. Brownlie
Scion, Private Bag, 3020 Rotorua, New Zealand

S. G. Kennedy
Coillte, Kilmacurra Park, Kilbride, County Wicklow, Ireland

B. R. Christensen
(Threats)-Conservation-Related Fire Research, Department of Conservation, 99 Sala Street, Rotorua, New Zealand

Chris J. Phillips
Landcare Research, PO Box 69040, Lincoln 7640, New Zealand

Michael Marden
Landcare Research, PO Box 445, Gisborne 4010, New Zealand

Suzanne M. Lambie
Landcare Research, Private Bag 3127, Hamilton 3240, New Zealand

Rebecca J Ganley, Cathy L Hargreaves and Lloyd A Donaldson
Scion, New Zealand Forest Research Institute Ltd., 49 Sala Street, Rotorua 3046, New Zealand

Jinquan Feng, Allan Woolf, Xingyun Yang and Shane Olsson
The New Zealand Institute for Plant & Food Research Limited, 120 Mt Albert Rd, Auckland 1142, New Zealand

Don Brash, Alistair Hall and Bruce Bycroft
The New Zealand Institute for Plant & Food Research Limited, Batchelar Rd, Palmerston North 4474, New Zealand

Hop Tran, Kerry C Harrington and Alastair W Robertson
Institute of Agriculture and Environment, Massey University, Palmerston
North 4442, New Zealand

Michael S Watt
Scion, Fendalton, PO Box 29237, Christchurch 8540, New Zealand

Gavin Melville
Trangie Agricultural Research Centre, Mitchell Hwy, Trangie 2823, Australia

Christine Stone
NSW Department of Primary Industries, Forest Research, Level 12, 10 Valentine Ave, Parramatta 2124, Australia

Russell Turner
Remote Census Pty Ltd, PO Box 527, Morisset 2264, Australia

Glen Murphy
Waiariki Institute of Technology, Rotorua, New Zealand

Dave Cown
New Zealand Forest Research Institute Ltd, Rotorua, New Zealand

Farzam Tavankar
Department of Forestry, Khalkhal Branch, Islamic Azad University, P.O. Box:
56817-31367, Khalkhal, Iran

Amireslam Bonyad
Department of Forestry, Faculty of Natural Resources, University of Guilan, P.O. Box: 1144, Somehsara, Iran

Enrico Marchi
Department of Agriculture, Food and Forest Systems, Florence University, Via San Bonaventura, 13 50145 Florence, Italy

Rachele Venanzi and Rodolfo Picchio
Department of Science and Technology for Agriculture, Forest, Nature and Energy (DAFNE), Tuscia University, Via S. Camillo De Lellis, 01100 Viterbo, Italy

Mark O. Kimberley, Dave J. Cown, Russell B. McKinley, John R. Moore and Leslie J. Dowling
Scion, Private Bag 3020, Rotorua, New Zealand

Karen M. Bayne, Barbara K. Höck, Tim W. Payn and Tim D. Barnard
New Zealand Forest Research Institute Ltd (Scion), University of Canterbury, Ilam, Christchurch, PO Box 29237, Forestry Road, Fendalton, Christchurch 8540, New Zealand

Harley R. Spence and Kirsten A. Crawford
Coastline Consultants, Rotorua 3010, New Zealand

Judy Frances Gardner, Margaret Anne Dick and Martin Karl-Friedrich Bader
New Zealand Forest Research Institute (Scion), Private Bag 3020, Rotorua 3046, New Zealand

Michael Marden and Donna Rowan
Landcare Research, PO Box 445, Gisborne 4010, New Zealand

John R Moore, Dave J Cown, Russell B McKinley and Charles O Sabatia
Scion, Private Bag 3020, Rotorua 3046, New Zealand

Randall W Myster
Biology Department, Oklahoma State University, Oklahoma City, OK 73107, USA

Stefan Gous
Scion, Forest Biosecurity and Protection, Private Bag 3020, Rotorua, New Zealand

Peter Raal
Department of Conservation, Otago Conservancy, PO Box 5244, Dunedin, New Zealand

Michael S Watt
Scion, PO Box 29237, Fendalton, Christchurch, New Zealand

Michael S Watt
Scion, Fendalton, PO Box 29237, Christchurch, New Zealand

Andrew Meredith and Pete Watt
Indufor Asia-Pacific Ltd, PO Box 105039, Auckland, New Zealand

Aaron Gunn
Blakely Pacific Ltd, PO Box 13–980, Christchurch 8141, New Zealand

Michael S Watt
Scion, PO Box 29237, Fendalton, Christchurch, New Zealand

Thomas Adams
MetService, 30 Salamanca Road, Kelburn, Wellington 6021, New Zealand

Susana Gonzalez Aracil
Interpine, 99 Sala Street, PO Box 1209, Rotorua 3010, New Zealand

Hamish Marshall and Pete Watt
Indufor Asia-Pacific Ltd, PO Box 105039, Auckland, New Zealand